Statistics for Industry and Technology

Goodness-of-Fit Tests and Model Validity

C. Huber-Carol
N. Balakrishnan
M.S. Nikulin
M. Mesbah
Editors

Birkhäuser
Boston • Basel • Berlin

C. Huber-Carol
Laboratoire de Statistique Médicale
Université Rene Descartes—Paris 5
75006 Paris
France

N. Balakrishnan
Department of Mathematics and Statistics
McMaster University
Hamilton, Ontario L8S 4K1
Canada

M. S. Nikulin
Laboratoire Statistique Mathématique
Université Bordeaux 2
33076 Bordeaux Cedex
France
and
Laboratory of Statistical Methods
V. Steklov Mathematical Institute
191011 St. Petersburg
Russia

M. Mesbah
Laboratoire de Statistique Appliquée
Université de Bretagne Sud
56 000 Vannes
France

Library of Congress Cataloging-in-Publication Data

A CIP catalogue record for this book is available from the Library of Congress, Washington D.C., USA.

AMS Subject Classifications: 62-06, 62F03

Printed on acid-free paper.

Birkhäuser

ISBN 0-8176-4209-9 SPIN 10777489
ISBN 3-7643-4209-9

Typeset by the editors in LaTeX.
Printed and bound by Hamilton Printing Company, Rensselaer, NY.
Printed in the United States of America.

9 8 7 6 5 4 3 2 1

Birkhäuser Boston • Basel • Berlin
A member of BertelsmannSpringer Science+Business Media GmbH

Contents

Preface

Commemorating the centennial anniversary of the landmark paper by Karl Pearson on chi-square goodness-of-fit test, an International Conference on *Goodness-of-Fit Tests and Model Validity* was organized at Paris, France, during May 29–31, 2000. This conference successfully attracted numerous statisticians from all over the world, many of them renowned experts in this area of research. The conference thus provided the participants with details on historical developments, elaborate surveys of pertinent topics, information on new research work, and many lively after-lecture discussions. We thank Natacha Heutte, Chantal Guihenneuc, Min Thu Do Hoang, Anouar Benmalek, Jean Marie Tricot, Jean François Petiot, Florence Duguesnoy, Leo Gerville-Reache and Valia Nikoulina for helping us with the organization of the conference. We also thank the French Ministry of Education and Research for their financial support, and the French Group of Biometric Society and the French Statistical Society for their support and cooperation. Thanks are expressed to Habida Mesbah for her patience and her special delivery of forgotten items.

This volume presents a broad spectrum of papers presented at the International Conference. It includes 37 articles in total which, for better presentation as well as convenience of the readers, have been divided into the following eight parts:

Part I – History and Fundamentals
Part II – Chi-Squared Test
Part III – Goodness-of-Fit Tests for Parametric Distributions
Part IV – Regression and Goodness-of-Fit Tests
Part V – Goodness-of-Fit Tests in Survival Analysis and Reliability
Part VI – Graphical Methods and General Goodness-of-Fit Tests
Part VII – Model Validity in Quality of Life
Part VIII – Tests of Hypotheses and Estimation with Applications

The articles in this volume provide a clear testimony to the importance and significance of work relating to the theory, methods and applications of goodness-of-fit tests and model validity. We sincerely hope that the readers will find this volume of interest. It is also our hope that new researchers will gain insight as well as new ideas from this volume, which may possibly encourage them to work in this fertile area of research.

We express our thanks to Lauren Schultz (Birkhäuser, Boston) for taking a keen interest in this project, and to Elizabeth Lowe (of Texniques) for assisting us with the production of this volume. We express our gratitude to all authors for sending in their articles in time and in good form. We express our gratitude to Odile Pons, Ion Grama and some anonymous reviewers for helping us with critically examining merits of the papers during the editorial process. A Mme Curmi, des services financiers de l'université Paris V pour sa compétence et son amabilité. Thanks are also expressed to the different personnel at Université de Paris V, Université de Bretagne Sud, Université de Bordeaux 2 and McMaster University for providing support and help in order to organize the conference smoothly and successfully. Our special thanks go to Debbie Iscoe (Canada) for a fine job in typesetting this entire volume in a camera-ready form.

Paris, France — C. Huber-Carol
Hamilton, Ontario, Canada — N. Balakrishnan
Bordeaux, France; St. Petersburg, Russia — M. S. Nikulin
Vannes, France — M. Mesbah

May 2001

Contributors

Akaira, Masafumi
Institute of Mathematics, University of Tsukuba, Ibaraki 305-8571, Japan
e-mail: *akahira@math.tsukuba.ac.jp*

Arnold, Barry C.
Department of Statistics, University of California, Riverside, California 92521-0138, U.S.A.
e-mail: *barnold@ucrstat2.ucr.edu*

Bagdonavičius, V.
Département de Mathématiques et Sciences Sociales, Université Bordeaux 2, 33076 Bordeaux Cedex, France
e-mail: *Vilijidandas.Bagdonavicius@maf.vu.lt*

Balakrishnan, N.
Department of Mathematics and Statistics, McMaster University, Hamilton, Ontario L8S 4K1, Canada
e-mail: *bala@mcmail.cis.mcmaster.ca*

Baraud, Y.
Ecole Normale Supérieure, Paris, France

Beaver, Robert J.
Department of Statistics, University of California, Riverside, California 92521-0138, U.S.A.
e-mail: *robert.beaver@ucr.edu*

Bosq, Denis
Laboratoire de Probabilités, Université Paris VI, 4, Place Jussieu, 75252 Paris Cedex 05, France
e-mail: *bosq@ccr.jussieu.fr*

Bretagnolle, Jean
Laboratoire de Statistique Appliquée, Université de Paris XI, 91405 Orsay

Cedex 11, France
e-mail: *Jean.Bretagnolle@math.u-psud.fr*

Castillo, Enrique
Department of Applied Mathematics and Sciences, University of Cantabria, E-39005 Santander, Cantabria, Spain
e-mail: *castie@unican.es*

Cox, D. R.
Department of Statistics, Nuffield College, Oxford 0X1 1NF, England, U.K.
e-mail: *david.cox@nuf.ox.ac.uk*

Cuadras, Carles M.
Department of Statistics, University of Barcelona, 08023 Barcelona, Spain
e-mail: *carlesm@porthos.bio.ub.es*

Cuadras, Daniel
University of Barcelona, 08023 Barcelona, Spain

D'Agostino, Ralph B.
Statistics and Consulting Unit, Department of Mathematics and Statistics, Boston University, Boston, Massachusetts 02215, U.S.A.
e-mail: *ralph@bu.edu*

Deheuvels. Paul
L.S.T.A., Université Paris VI, 92340, Bourg-la-Reine, France
e-mail: *pd@ccr.jussieu.fr*

Devarajan, Karthik
Division of Statistics, Northern Illinois University, DeKalb, Illinois 60115, U.S.A.

Diack, Cheikh A. T.
Department of Statistics, University of Warwick, Coventry CV4 7 AL, UK
e-mail: *catdiack@hotmail.com*

Dodge, Yadolah
Groupe de Statistique, University of Neuchâtel, CH-2002 Neuchatel, Switzerland
e-mail:*yadolah.dodge@seco.unine.ch*

Ducharme, Gilles R.
Département des Sciences Mathématiques, Université Montpellier II, 34095 Montpellier Cedex 5, France
e-mail: *ducharme@stat.math.univ-montp2.fr*

Dudley, Richard M.
Department of Mathematics, Massachusetts Institute of Technology, Cambridge, Massachusetts 02215, U.S.A.
e-mail: *rmd@math.mit.edu*

Dufour, Jean-Marie
CIRANO and CRDE, Université de Montréal, Montréal, Quebec H3C 3J7, Canada
e-mail:

Dupuy, Jean-Francois
Department of Applied Statistics, University of South Brittany, 56000 Vannes, France
e-mail: *jean-francois.dupuy@univ-ubs.fr*

Ebrahimi, Nader
Division of Statistics, University of Northern Illinois, DeKalb, Illinois 60115, U.S.A.
e-mail: *nader@math.niu.edu*

Farhat, Abdeljelil
CIRANO, Université de Montréal, Montréal, Quebec H3A 2A5, Canada
e-mail: *farhata@cirano.umontreal.ca*

Frichot, Benoît
Département des Sciences Mathématiques, Université Montpellier II, 34095 Montpellier Cedex 5, France
e-mail: *frichot@math.univ-montp2.fr*

Gerville-Reache, Léo
Département de Mathématiques et Sciences Sociales, Université Bordeaux 2, 33076 Bordeaux Cedex, France
e-mail: *gerville@u-bordeaux2.fr*

Gross, Shulamith T.
Laboratoire de Statistique, Université de Paris V, 75006 Paris, France
e-mail: *gross@citi2.fr*

Gulati, Sneh
Department of Statistics, Florida International University, Miami, Florida 33199, U.S.A.
e-mail: *gulati@fiu.edu*

Hamon, Agnes
Laboratoire SABRES, Université de Bretagne-Sud, 56000 Vannes, France
e-mail: *agnes.hamon@univ-ubs.fr*

Haughton, Dominique M.
Mathematical Sciences, Bentley College, Waltham, Massachusetts 02452-4705, U.S.A.
e-mail: *dhaughton@bentley.edu*

Hoàng, Thu
Laboratoire de Statistique Médicale, Université de Paris V, 75006 Paris, France
e-mail: *hoang@citi2.fr*

Huber, P. J.
P.O. Box 198, CH-7250, Klosters, Switzerland
e-mail: *peter.huber@uni-bayreuth.de*

Huber-Carol, Catherine
Université Paris Vand U472 INSERM, Paris, France
e-mail: *huber@citi2.fr*

Huet, S.
INRA Biométrie, 78352 Jouy-en-Josas Cedex, France

Jurečkovà, Jana
Statistics Department, Charles University, Czech Republic
e-mail: *Jurecko@karlin.mff.cuni.cz*

Kannan, N.
Division of Mathematics and Statistics, The University of Texas at San Antonio, Texas 78249-0664, U.S.A.
e-mail: *kannan@math.utsa.edu*

Keiding, Niels
Department of Biostatistics, University of Copenhagen, 2200 Copenhagen, Denmark
e-mail: *n.keiding@biostat.ku.dk*

Khattree, Ravi
Department of Mathematics and Statistics, Oakland University, Rochester, Missouri 48309-4485, U.S.A.
e-mail: *khattree@oakland.edu*

Laurent, B.
Laboratoire de Statistique, Université de Paris XI, 91405 Orsay Cedex, France
e-mail: *beatrice.laurent@math.u-psud.fr*

Lee, Mei-Ling Ting
Channing Laboratory, Harvard University, Boston, Massachusetts 02115-5804, U.S.A.
e-mail: *stmei@channing.harvard.edu*

Marchetti, Carol E.
Rochester Institute of Technology, Rochester, New York 14623, U.S.A.
e-mail: *cemsma@rit.edu*

Mesbah, Mounir
Department of Applied Statistics, University of South Brittany, 56000 Vannes, France
e-mail: *mounir.mesbah@univ-ubs.fr*

Mohdeb, Zaher
Département de Mathématiques, Université Mentouri Constantine, 25000 Constantine, Algéria
e-mail: *mohdeb@mail.wissal.dz*

Mokkadem, Abdelkader
Department of Mathematics, University of Versailles-Saint-Quentin, 78035 Versailles Cedex, France
e-mail: *mokkadem@math.uvsq.fr*

Mudholkar, G. S.
Department of Statistics, University of Rochester, Rochester, New York 14627-0047, U.S.A.
e-mail: *govind@metro.bst.rochester.edu*

Naik, Dayanand N.
Department of Mathematics and Statistics, Oakland University, Rochester, Missouri 48309-4485, U.S.A.

Nam, Byung-Ho
Statistics and Consulting Unit, Department of Mathematics and Statistics, Boston University, Boston, Massachusetts 02215, U.S.A.
e-mail: *byungho@bu.edu*

Neus, Jordan
Biostatistics, State University of New York at Stony Brook, Stony Brook, New York, U.S.A.
e-mail: *jneus@uhms.sunysb.edu*

Ng, H. K. T.
Department of Mathematics and Statistics, McMaster University, Hamilton, Ontario L8S 4K1, Canada
e-mail: *ngh@math.mcmaster.ca*

Nikulin, M. S.
UFR de Mathématiques, Informatique et Sciences Sociales, Université Bordeaux 2, Bordeaux, France
e-mail: *nikou@mi2s.u-bordeaux2.fr*

Parsons, Van L.
National Center for Health Statistics, Hyattsville, Maryland 20782-2003, U.S.A.
e-mail: *vlp1@cdc.gov*

Pons, Odile
INRA Biométrie, 78352 Jouy-en-Josas Cedex, France
e-mail: *odile.pons@jouy.inra.fr*

Rao, C. R.
Department of Statistics, Pennsylvania State University, University Park, Pennsylvania 16802, U.S.A.
e-mail: *ccr1@psu.edu*

Rayner, G. D.
School of Computing and Mathematics, Deakin University, Geelong, VIC3217 Australia
e-mail: *gdrayner@deakin.edu.au*

Rayner, J. C. W.
School of Mathematics and Applied Statistics, University of Wollongong, Wollongong NSW 2522, Australia
e-mail: *john_rayner@uow.edu.au*

Rousseeuw, P. J.
Department of Mathematics and Computer Science, University of Antwerp, Universiteitsplein 1, B-2610 Antwerp, Belgium
e-mail: *Peter.Rousseeuw@uia.ua.ac.be*

Sarabia, Jose Maria
Economics Department, University of Cantabria, E-39005 Santander, Cantabria, Spain
e-mail: *sarabiaj@unican.es*

Sen, P. K.
Department of Biostatistics, University of North Carolina at Chapel Hill, North Carolina 27599-7400, U.S.A.
e-mail: *pksen@bios.unc.edu*

Seymour, Lynne
Department of Statistics, The University of Georgia, Athens, Georgia 30602-1952, U.S.A.
e-mail: *seymour@stat.uga.edu*

Solev, V.
The Laboratory of Statistical Methods, Steklov Mathematical Institute, St. Petersburg, 19011, Russia
e-mail: *solev@pdmi.ras.ru*

Struyf, Anja
Research Assistant, FWO, 1000, Brussels, Belgium

Whitmore, G. A.
McGill University, Montreal, Quebec H3A 2T5, Canada

Zerbet, Aïcha
Département de Mathématiques et Sciences Sociales, Université Bordeaux 2, 33076 Bordeaux Cedex, France
e-mail: *azerbet@mi2s.u-bordeaux2.fr*

List of Tables

List of Figures

Goodness-of-Fit Tests and Model Validity

PART I

HISTORY AND FUNDAMENTALS

1

Karl Pearson and the Chi-Squared Test

D. R. Cox

Nuffield College, Oxford, England, UK

Abstract: This historical and review paper is in three parts. The first gives some brief details about Karl Pearson. The second describes in outline the 1900 paper which is being celebrated at this conference. The third provides some perspective on the importance, historically and contemporarily, of the chi-squared test.

Keywords and phrases: K. P., history of statistics, chi-squared, goodness-of-fit test, statistical inference

1.1 Karl Pearson 1857–1937: Background to the Chi-Squared Paper

Karl Pearson, K. P. as he is usually referred to, was born of middle-class Yorkshire parents, his father a lawyer and his mother's family connected with shipping. He was at school in London and before going to University had a private tutor in mathematics, Routh, a well-known expert in the theory of elasticity, who introduced K. P. to that subject. He read Mathematics at Cambridge, graduating Third Wrangler in 1879. The two men above him in the Tripos list followed academic careers at Cambridge but there is no evidence that they were research workers of note.

The stereotypic Yorkshireman was and is independent and forthright. While an undergraduate K. P. had a long fight, which he won, with the Authorities of his College securing the abandonment of compulsory attendance at divinity lectures.

It is worth considering the kind of mathematics that K. P. studied at Cambridge. There would, of course, have been no analysis but appreciable calculus, algebra would have meant largely the Theory of Equations and determinants, not matrix algebra, and there would have been substantial emphasis on parts of classical mathematical physics. More importantly the emphasis was strongly on ingenuity and manipulative skill in problem solving rather than on the development of new concepts. There is some evidence that K. P. met, although not necessarily to be taught by, such major figures as Clerk Maxwell, Cayley and Green and more particularly Todhunter. Todhunter had published a History of the Theory of Probability, essentially a long critical essay and review of what had been published on Probability up to that point, and he was engaged in a comparable book on the Theory of Elasticity.

After graduating, K. P. spent an extremely influential year in Germany, studying physics but also philosophy and other aspects of German culture. He was particularly attracted to the 17th century rationalist philosopher Spinoza. During this year he changed the spelling of his name from the English spelling Carl to the Germanic Karl.

After returning to England he qualified as a lawyer and then spent some years, partly supported by a Fellowship from Kings College, Cambridge, in miscellaneous lecturing mostly on such topics as German philosophy and Marxism. He was part of an active world of literary and cultural life in London towards the end of the 19th century. His views seem broadly those of left-wing thought of the time, enlightened in their attitude to women's rights, socialist in political thought, believing in the solution of social problems via rational enquiry and holding views on racial matters that would now widely be regarded as unacceptable. Biographies of major non-scientific figures of the period quite often mention K. P., in passing at least.

He applied for a number of permanent academic jobs and in 1884 was appointed Professor of Engineering and Applied Mathematics at University College London. His primary duty was to teach elementary mathematics to engineers; he is reported as being outstandingly successful in this. He published research papers on the theory of elasticity and collaborated with Todhunter on his History of that field, writing, it is said, much of the second volume.

In 1890 W. F. R. Weldon was appointed Professor of Biology at University College and an intensely active collaboration developed between them lasting until Weldon's early death in 1906. Following the impact on Victorian thought of Charles Darwin and more immediately for K. P. and Weldon of Galton, this was a period of intense interest in genetics and evolutionary biology. Weldon believed that careful collection of observational data on natural variability would provide the key to important issues and K. P. became involved in the analysis of data collected by Weldon (and others) in their extensive field work and in the development of what came to be called the biometric school. Their main technique was the careful study of the shape of univariate and occasionally bivariate frequency distributions and, in discrete cases to the analysis of two-dimensional contingency tables. Recognition that distributions were often far from the normal or Gaussian form led to the development of the flexible system of frequency curves named after Pearson. These were fitted by moments. It was the need for some form of relatively objective way of assessing adequacy of fit that led to the paper [Pearson (1900)] celebrated in this conference.

It was published in *Philosophical Magazine* then as now a respected journal of the physical sciences; indeed it is currently owned by the Physical Society.

Before considering the paper a few comments will be made about K. P.'s work after the chi-squared paper.

1.2 K. P.: After Chi-Squared

In November 1900 Weldon suggested the need for a new journal and in October 1901 the first issue of *Biometrika* appeared. K. P. seems to have been by far the dominant figure in all this, once the initial suggestion had been made. For a period there were joint editors but following Weldon's death K. P. was the sole editor until his death at age 80; indeed he was correcting proof a few weeks before his death and had two papers, one characteristically long and polemical, in the last issue.

In the period up to 1914 K. P. published about 90 papers in *Biometrika* alone, few of them short, and seems to have been the moving force behind many more. Even a cursory glance at these papers reveals K. P.'s astonishing range of interests and his intellectual vigor and originality. Some are methodological papers concerned, for instance, with the analysis of ordinal and nominal data via an underlying Gaussian distribution and with many other topics. By far the majority, however, are substantial pieces of analysis of observational data. The fields range among biology, sociology, criminology, medicine and epidemiology and physical anthropology. The emphasis is always on the data and their interpretation; they are rarely treated merely as exercises in technique.

After the end of the Great War K. P. continued to publish prolifically but his work seems mostly of less current interest; the focus of development of statistics had shifted elsewhere.

1.3 The 1900 Paper

The essence of the paper is as follows. First it is shown by direct transformation of the multiple integrals involved that the distribution of the exponent of a (nonsingular) multivariate normal distribution in d dimensions has what we now call the chi-squared distribution with d degrees of freedom. Evaluation of its tail area is shown to be possible by integration by parts.

Next the covariance matrix of a multinomial distribution with k cells is found and the distribution considered across $k-1$ cells to avoid singularities approximated by a multivariate normal distribution in $k-1$ dimensions.

The exponent of that multivariate normal distribution is then reexpressed in more symmetrical form as the nowadays familiar chi-squared statistic for comparing observed and theoretically known cell probabilities. The informal reasonableness of using this as a test statistic for goodness-of-fit is discussed.

A long verbal discussion then follows recognizing that usually the comparison is with fitted rather than known cell probabilities. It is argued that this replacement will have a relatively small effect. That is the degrees of freedom remain $k-1$. (Indeed even after Greenwood and Yule, and subsequently in generality Fisher, had shown that the degrees of freedom for a 2×2 contingency were one, K. P. insisted that they were three. Note, however, that for many problems of examining distributional form, which were K. P.'s primary motivation, the number of parameters is appreciably less than the number of cells, in which case K. P.'s conclusion is a reasonable first approximation. Another line of explanation of K. P.'s attitude is that he might have regarded the proper chi-squared statistic to be that based on the *theoretical* probabilities which is only estimated by chi-squared from the fitted frequencies. If these are obtained essentially by minimizing a chi-squared measure a bias correction is needed. This leads to the notion that with p adjustable parameters the observed chi-squared plus p should be tested with $k-1$ degrees of freedom and this is not too far from use of $k-p-1$ as the degrees of freedom.)

The paper concludes with a variety of numerical examples.

For very interesting further comments on the paper see Barnard (1991). This is not the place to go into the interchanges between K. P. and other major figures such as Student, Yule and, of course, R. A. Fisher, on which there is an extensive literature.

1.4 Importance of the Chi-Squared Test

For perhaps 70 years following its introduction the chi-squared test was one of the most widely used tools of formal statistical analysis. This may have been not so much because of its application to the originating problem of assessing distributional shape, but rather because the chi-squared test of independence in a contingency table provided the main route for interpretation of qualitative data, especially as they arise in the social sciences. Evidence of departure from independence would then be interpreted descriptively.

More recently the role of the test is less central. There are a number of reasons for this.

First, as compared with K. P.'s time, the primary focus of most studies has shifted from studies of distributional form to studies of dependence. In these studies, issues of distributional shape are of secondary interest in indicating efficient methods of analysis but are not the primary focus. In some contexts, moreover, studies of robustness remove much of the dependence on strong assumptions of distributional form. Even when comparisons with, say, the Poisson distribution are involved, a more relevant and focused question is often whether the variance is equal to the mean or, less commonly, whether minus the log of the probability of zero is equal to the mean. The former is relevant to the estimation of the standard error of a rate and the latter to the analysis of dilution series. Much later Fisher gave the appropriate exact distributional theory and C. R. Rao supplied important complementary results.

Secondly, there has developed a tendency to prefer focused tests rather than what M. S. Bartlett called omnibus tests. This is partly an issue of power but at least as importantly of diagnostic effectiveness. Thus a test for normality based, say on the larger of the standardized third and fourth cumulant, gives a direct indication of the kind of departure from normality involved, whereas an overall chi-squared statistic does not. Even more directly the chi-squared dispersion test for the Poisson distribution is a direct examination of the ratio of the variance to the mean.

Further there has recently developed a preference for procedures based on estimation of interpretable parameters over those yielding primarily a significance test. While chi-squared could be rescaled to estimate a distance measure this would often not be easily interpreted. The preference for log linear models as a route for the interpretation of contingency tables stems partly from this preference for estimation.

There is the following broader issue. Models, as their name implies, are inevitably idealized. Especially with complex biological and social phenomena, it is inconceivable that the systems are precisely described by any mathematical or computer model, especially by the relatively simple models that are commonly used in statistics. Why then should we test goodness-of-fit to a model that we know must be at some level inadequate? One answer with modest amounts of data is that so long as no reasonably significant departure is found the direction of inadequacy is not clearly established by the data. Thus it may be poor strategy either to interpret the departure or to modify the model when, for example, the direction in which we should modify the model is not firmly established. Especially with modest amounts of data it is likely that substantively important departures from the null hypothesis might be present, so that a significant departure from the null hypothesis deserves to be taken seriously. On the other hand, with very large amounts of data the position is often different. Even differences that are quite small in subject-matter terms are likely to be highly statistically significant so that here the issue is often best regarded as considering whether any lack of fit is important in subject-matter terms. Note, however, that very large sets of data often have internal structure which may make assessments of precision based on strong assumptions of independence very misleading.

In many fields the most important aspect of tests of goodness-of-fit lies in checking for various forms of departure from some standard conditions of relative homogeneity. Examples of such tests are those for interaction, for nonlinearity and for heterogeneity when information from different sources is considered for combination.

Despite the generally decreased emphasis on omnibus tests of goodness-of-fit, the notion that such tests are available remains in principle of great importance. It represents the openmindedness of the attitude that our current formulation of a problem may sometimes be shown empirically to be unsatisfactory in ways not clearly formulated *a priori*. The organizers of the present Conference are surely to be congratulated on bringing together this celebration of K. P.'s path-breaking paper.

References

1. Barnard, G. A. (1991). Introduction to Pearson (1900), In *Breakthroughs in Statistics, Vol. 2* (Eds., S. Kotz and N. L. Johnson), pp. 1–10, New York: Springer-Verlag.

2. Pearson, K. (1900). On the criterion that a given system of deviations from the probable in the case of a correlated system of variables is such that it can reasonably be supposed to have risen from random sampling, *Phil. Mag.*, **5**, 157–175.

2

Karl Pearson Chi-Square Test
The Dawn of Statistical Inference

C. R. Rao

Pennsylvania State University, University Park, Pennsylvania

Abstract: Specification or stochastic modeling of data is an important step in statistical analysis of data. Karl Pearson was the first to recognize this problem and introduce a criterion, in a paper published in 1900, to examine whether the observed data support a given specification. He called it chi-square goodness-of-fit test, which motivated research in testing of hypotheses and estimation of unknown parameters and led to the development of statistics as a separate discipline. Efron (1995) says, "Karl Pearson's famous chi-square paper appeared in the spring of 1900, an auspicious beginning to a wonderful century for the field of statistics."

This paper reviews the early work on the chi-square statistic, its derivation from the general theory of asymptotic inference as a score test introduced by Rao (1948), its use in practice and recent contributions to alternative tests. A new test for goodness-of-fit in the continuous case is proposed.

Keywords and phrases: Chi-square test, Jensen difference, likelihood ratio test, quadratic entropy, Rao's score test, Wald test

2.1 Introduction

In an article entitled *Trial by Number*, Hacking says that the goodness-of-fit chi-square test introduced by Karl Pearson (1900), "ushered in a new kind of decision making" and gives it a place among the top 20 discoveries since 1900 considering all branches of science and technology. R. A. Fisher, who was involved in bitter controversies with Pearson, was appreciative of the chi-square test. In his book on *Statistical Methods for Research Workers* (1958, 13th edition, p. 22), Fisher says, "This (chi-square), I believe is the great contribution to statistical methodology which the unsurpassed energy of Professor Pearson's work will be remembered," and devoted one full chapter on numerous ingenious applications of the chi-square test.

Pearson's chi-square is ideally applicable to qualitative data with a finite number, say s, of natural categories and the data are in the form of frequencies of individuals in different categories. The specified hypothesis is of the form

$$\pi_i = \pi_i(\theta), \ i = 1, \dots, s \tag{2.1}$$

where the probability π_i in category i is a given function of a k-vector parameter θ. If p_i is the observed proportion in category i, then a natural test criterion is of the form

$$D(p - \pi(\hat{\theta})) \tag{2.2}$$

for a suitable choice of distance or dissimilarity measure, where $p = (p_1, \dots, p_s)'$, $\pi(\hat{\theta}) = (\pi_1(\hat{\theta}), \dots, \pi_s(\hat{\theta}))'$ and $\hat{\theta}$ is an efficient estimate of θ.

Ideally θ is estimated by

$$\hat{\theta} = \arg\min_{\theta} D(p - \pi(\theta)). \tag{2.3}$$

Various tests of goodness-of-fit proposed in statistical literature differ in the measure (2.2) chosen.

When we have a sample from a continuous distribution with a distribution function $F(x, \theta)$, there are two ways of deriving the goodness-of-fit test. One is to discretize the continuous distribution by choosing class intervals $-\infty = a_0, a_1, \dots, a_{s-1}, a_s = \infty$, and defining

$$\pi_{i+1}(\theta) = F(a_{i+1}, \theta) - F(a_i, \theta), i = 0, \dots, s-1, \tag{2.4}$$

in which case a test of the type (2.2) is applicable. An excellent treatment of such an approach with all the issues involved in the choice of class intervals and the estimation of θ is given in the monograph by Greenwood and Nikulin (1996).

Another is the direct method of estimating θ by an efficient procedure such as the maximum likelihood and constructing a test based on a suitable measure of difference

$$D[F_n(x) - F(x, \hat{\theta})] \tag{2.5}$$

where F_n is the empirical distribution function. A good review of this approach can be found in Durbin (1973).

Section 2.2 of the paper describes three general methods of constructing large sample test criteria of simple and composite hypotheses, viz., likelihood ratio, Wald and Rao's score tests, also referred to as *Holy Trinity*. [See Koenkar (1987, p. 294) and Lehmann (1999, pp. 525–529)]. In Section 2.3, Pearson's chi-square and related tests are shown to be score tests, as observed by A. Bera. The difficulties involved in deriving Wald tests for composite hypotheses are discussed. Alternative tests of goodness-of-fit based on dissimilarity or divergence measures derived from entropy functions are given in Section 2.4. Tests of significance of goodness-of-fit for continuous distributions are reviewed in Section 2.5. A new test is proposed and the possibility of using bootstrap is pointed out.

2.2 Large Sample Criteria: The Holy Trinity

Let $(\mathcal{X}, \beta, P_\theta)$ be a probability space, where P_θ stands for a family of distributions indexed by a k-vector parameter $\theta \in \Theta$. Further let $X = (X_1, \ldots, X_n)'$ be a vector of iid observations from P_θ. We consider the problem of testing simple and composite hypotheses concerning P_θ. When n is large, there are three general methods [Koenkar (1987, p. 294), Lehmann (1999, pp. 525–529)], referred to as the *Holy Trinity*, of constructing large sample test criteria.

2.2.1 Likelihood ratio criterion

Let $L(X, \theta)$ denote the likelihood function based on the sample observations X. Then, to test a simple hypothesis $H_s : \theta = \theta_0$ (a specified value), the likelihood ratio criterion introduced by Neyman and Pearson (1928) is

$$\wedge_s = L(X, \theta_0) \div \sup_{\theta \in \Theta} L(X, \theta). \tag{2.6}$$

To test a composite hypothesis H_c, expressed in the form

$$\theta \in C = \{\theta : f_1(\theta) = 0, \ldots, f_r(\theta) = 0\} \tag{2.7}$$

the likelihood ratio criterion is

$$\wedge_C = \sup_{\theta \in C} L(X, \theta) \div \sup_{\theta \in \Theta} L(X, \theta). \tag{2.8}$$

Large sample properties of the likelihood ratio criterion were studied by Wilks (1938). It was shown that asymptotically

$$-2 \log \wedge_s \sim \chi^2(k) \text{ and } -2 \log \wedge_c \sim \chi^2(r) \tag{2.9}$$

where $\chi^2(b)$ represents chi-square distribution on b degrees of freedom. Note that r is the number of restrictions on θ.

2.2.2 Wald test

Let $\hat{\theta}$ be the maximum likelihood estimate of θ such that the asymptotic normality

$$\sqrt{n}(\hat{\theta}-\theta) \sim N_q(0, I(\theta)^{-1})$$

holds, where $I(\theta)$ is the information matrix for a single observation. Then the Wald (1943) test for $H_s : \theta = \theta_0$ is

$$n(\hat{\theta}-\theta_0)' I(\hat{\theta})(\hat{\theta}-\theta_0) \sim \chi^2(k) \tag{2.10}$$

and for the composite hypothesis H_c defined in (2.7) the Wald test is

$$n[f(\hat{\theta})]'[M(\hat{\theta})I(\hat{\theta})^{-1}M(\hat{\theta})']^{-1}[f(\hat{\theta})] \sim \chi^2(r) \tag{2.11}$$

where

$$f(\cdot) = (f_1(\cdot), \dots, f_r(\cdot))', \text{ an } r\text{-vector},$$
$$M(\theta) = \left(\frac{\partial f_i}{\partial \theta_j}\right), \text{ an } r \times k \text{ matrix.}$$

2.2.3 Rao's score test

The score vector function as defined by Fisher is

$$S(\theta) = \left(\frac{1}{\sqrt{n}}\frac{\partial \log L}{\partial \theta_1}, \dots, \frac{1}{\sqrt{n}}\frac{\partial \log L}{\partial \theta_k}\right). \tag{2.12}$$

The score test of Rao (1948) for the simple hypothesis H_s is

$$[S(\theta_0)]'[I(\theta_0)]^{-1}[S(\theta_0)] \sim \chi^2(k) \tag{2.13}$$

and for the composite hypothesis (2.7) is

$$[S(\tilde{\theta})]'[I(\tilde{\theta})]^{-1}[S(\tilde{\theta})] \sim \chi^2(r) \tag{2.14}$$

where $\tilde{\theta}$ is the m.l. estimate of θ under the restrictions (2.7) of the composite hypothesis.

A variation of the test (2.14) where, instead of the m.l. estimate $\tilde{\theta}$, only $\sqrt{n}$ consistent estimate $\bar{\theta}$ is substituted for θ is called the Neyman-Rao test by Hall and Mathiason (1990). Such a statistic has the same chi-square distribution on r degrees of freedom.

2.3 Specification Tests for a Multinomial Distribution

Let $\pi_1, \ldots, \pi_s, (\sum \pi_i = 1)$ and $O_1, \ldots, O_s, (\sum O_i = n)$ be the cell probabilities and observed frequencies in a sample of size n from an s-cell multinomial distribution. Further let

$$p = (p_1, \ldots, p_s)'; \; p_i = O_i/n, \; \sum_1^s p_i = 1$$

be the estimated probabilities. The variance-covariance matrix of $p = (p_1, \ldots, p_s)'$ is $n^{-1}C$, where

$$C(\pi) = \begin{pmatrix} \pi_1(1-\pi_1) & -\pi_1\pi_2 & \ldots & -\pi_1\pi_s \\ \cdot & \cdot & \ldots & \cdot \\ -\pi_s\pi_1 & -\pi_s\pi_2 & \ldots & \pi_s(1-\pi_s). \end{pmatrix}$$

2.3.1 Test of a simple hypothesis

Let us consider the test of a simple hypothesis

$$H_s : \pi = (\pi_1, \ldots, \pi_s)' = \pi_0 = (\pi_{10}, \ldots, \pi_{s0})'$$

where π_{i0} are specified values.

The likelihood ratio test is

$$(LRT)_S = 2\Sigma O_i \log \frac{O_i}{E_{i0}}; \; E_{i0} = n\pi_{i0}, i = 1, \ldots, s \tag{2.15}$$

which is distributed asymptotically as χ^2 on $(s-1)$ d.f.

The Wald test is

$$W_S = n(p-\pi_0)'[C(p)]^-(p-\pi_0) \sim \chi^2(s-1) \tag{2.16}$$

where $[C(p)]^-$ is a generalized inverse of $C(p)$. An alternative expression for (2.16) is

$$W_s = \sum_1^s \frac{(O_i - n\pi_{i0})^2}{O_i} = \sum_1^s \frac{(O_i - E_{i0})^2}{O_i} \, . \tag{2.17}$$

which is usually called Neyman's modification of Pearson's chi-square defined in (2.21).

For Rao's score test, we compute the scores

$$\frac{\partial \log L}{\partial \pi_i} = \frac{n_i}{\pi_i}, \; i = 1, \ldots, s \tag{2.18}$$

with the variance-covariance matrix

$$c(\pi) = n(\Delta^{-1} - 11'), \; 1' = (1, \ldots, 1) \tag{2.19}$$

where Δ is a diagonal matrix with π_i as the i-th diagonal element. The score statistic is

$$R_s = y'[c(\pi_0)]^- y \tag{2.20}$$

where $y' = (n_1/\pi_{10}, \ldots, n_s/\pi_{s0})$. Observing that $(\Delta(\pi_0) - \pi_0 \pi_0')$ is a g-inverse of $c(\pi_0)$, we find

$$R_s = \left(\sum_1^s \frac{O_i^2}{n\pi_{i0}} \right) - n = \sum_1^s \frac{(O_i - E_{i0})^2}{E_{i0}} \tag{2.21}$$

where $E_{i0} = n\pi_{i0}$, the expected value when $\pi_i = \pi_{i0}, \; i = 1, \ldots, s$. The statistic (2.21) is Pearson's chi-square for testing a simple hypothesis. [Note that in general, the scores have to be computed for independent parameters, $\pi_i, \ldots, \pi_{s-1}$ in the present case, in which case the variance-covariance matrix will be non-singular. The statistic R_s will have the same expression (2.21)].

2.3.2 Tests of a composite hypothesis

Consider the composite hypothesis

$$\pi_i = \pi_i(\theta), \; i = 1, \ldots, s \tag{2.22}$$

where θ is a k-vector unknown parameter. Denote the ml estimate of θ by $\hat{\theta}$.

LRT

The likelihood ratio test of the hypothesis (2.22) is

$$2 \sum O_i \log \frac{O_i}{E_i^*} \sim \chi^2(s - 1 - k) \tag{2.23}$$

where $E_i^* = n\pi_i(\hat{\theta})$, the expected value when $\theta = \hat{\theta}$.

Score test

Rao's score test, obtained by substituting E_i^* for E_{i0} in (2.21), is

$$\sum \frac{(O_i - E_i^*)^2}{E_i^*} \sim \chi^2(s-1-k). \tag{2.24}$$

The results (2.21) and (2.24) show that Pearson's chi-square tests (with the modification made by Fisher for degrees of freedom when θ is estimated) can be obtained as Rao's score tests.

Wald test

The derivation of the Wald test for the composite hypothesis (2.22) is somewhat complicated as it requires the formulation of (2.22) in the form of restrictions

$$g_i(\pi_1, \ldots, \pi_s) = 0, \; i = 1, \ldots, s-1-k$$

on the cell probabilities $\pi_1, \ldots, \pi_s$.

For example, consider the hypothesis that the distribution of male births in families with $(s-1)$ children is binomial, i.e.,

$$\pi_r = \binom{s-1}{r-1} \theta^{r-1}(1-\theta)^{s-r}, \; r = 1, \ldots, s \tag{2.25}$$

where θ is the probability that a child is a male. The equations (2.25) can be written as restrictions

$$\frac{s-1}{1}\frac{\pi_1}{\pi_2} = \frac{s-2}{2}\frac{\pi_2}{\pi_3} = \ldots = \frac{1}{s-1}\frac{\pi_{s-1}}{\pi_s} \tag{2.26}$$

on $\pi_1, \ldots, \pi_s$, which are in the form (2.7) required for applying the Wald test. We use the formula (2.11) to derive the Wald statistic.

It may be noted that there is no unique representation of the restrictions (2.26), and the Wald statistic may depend on the actual form in which the restrictions are expressed. This is one of the drawbacks of the Wald statistic.

2.3.3 Test for goodness-of-fit in a subset of cells

While comparing observed and expected frequencies one may find that the discrepancies are confined to a subset of the cells of the multinomial distribution. A typical case is undercount or overcount in the cell for "zero events" in a binomial or a Poisson distribution. Let us consider the score test for a given specification

$$H_0 : \pi_i = \pi_i(\theta), \; i = 1, \ldots, s \tag{2.27}$$

given the alternative

$$\begin{aligned} H_1 : \pi_i &= \alpha_i \pi_i(\theta)/T, \ i = 1, \dots, r, \\ \pi_i &= \pi_i(\theta)/T, \ i = r+1, \dots, s, \\ T &= \alpha_1 \pi_1(\theta) + \dots + \alpha_r \pi_r(\theta) + \pi_{r+1}(\theta) + \dots + \pi_s(\theta). \end{aligned} \tag{2.28}$$

Under the model H_1, the null hypothesis (2.27) may be stated as

$$\alpha_1 = \alpha_2 = \dots = \alpha_r = 1$$

with $\theta' = (\theta_1, \dots, \theta_k)$ as nuisance parameters. To apply the score test, we need to compute the scores for $\alpha_1, \dots, \alpha_r$ and $\theta_1, \dots, \theta_k$ and the information matrix for the $(r + k)$ parameters at the values $\alpha_1 = \dots = \alpha_r = 1$ and maximum likelihood estimates of θ under H_0.

The scores for α_i and θ_j at estimated values under H_0 are

$$\begin{aligned} a_i &= O_i - n\pi_i(\hat{\theta}), \ i = 1, \dots, r \\ S_j &= 0, \ j = 1, \dots, k. \end{aligned} \tag{2.29}$$

The information matrix is n times

$$\begin{pmatrix} A & B \\ B' & C \end{pmatrix} \tag{2.30}$$

where

$$A = \begin{pmatrix} \pi_1(1-\pi_1) & -\pi_1\pi_2 & \dots & -\pi_1\pi_r \\ \cdot & \cdot & \dots & \cdot \\ -\pi_r\pi_1 & -\pi_r\pi_2 & \dots & \pi_r(1-\pi_r) \end{pmatrix}$$

$$B = \begin{pmatrix} \dfrac{\partial \pi_1}{\partial \theta_1} & \dots & \dfrac{\partial \pi_1}{\partial \theta_k} \\ \cdot & \dots & \cdot \\ \dfrac{\partial \pi_r}{\partial \theta_1} & \dots & \dfrac{\partial \pi_r}{\partial \theta_k} \end{pmatrix}$$

and I is the information matrix for $\theta_1, \dots, \theta_k$. The score statistic for testing H_0 is

$$\chi^2 = n^{-1} a' (A - BI^{-1}B')^{-1} a \tag{2.31}$$

where $a' = (a_1, \dots, a_r)$. The asymptotic distribution of (2.31) is chi-square with r degrees of freedom, if $|A - BI^{-1}B'| \neq 0$. Otherwise it is equal to the rank of $A - BI^{-1}B'$; if the rank is less than r, we use a g-inverse in the definition of (2.31). [Note that $A - BI^{-1}B'$ is the asymptotic variance covariance matrix of a].

2.3.4 Analysis of chi-square

Consider a multinomial distribution in s classes with the following possible specifications for the cell probabilities.

$$H : \pi_i, \dots, \pi_s \quad \text{are arbitrary.}$$
$$H_1 : \pi_i = \pi_i(\theta), \;\; i = 1, \dots, s, \theta \in R^k.$$
$$H_2 : \pi_i = \pi_i(g(\Phi)) = \pi_i(\Phi), \;\; i = 1, \dots, s, \Phi \in R^q, q < k.$$

In practice, there are situations where we first test H_1 against H, and if this holds to test H_2 against H_1, i.e., whether the specification H_1 holds with the restrictions on θ implied by H_2.

Let $O_1, \dots, O_s$ be the observed frequencies in s cells, and $\hat{\theta}$ and $\hat{\Phi}$ be m.l. estimators of θ and Φ under the hypotheses H_1 and H_2 respectively. Then, the chi-square tests for H_1 against H and H_2 against H are

$$\chi_1^2 = \sum \frac{[O_i - n\pi_i(\hat{\theta})]^2}{n\pi_i(\hat{\theta})}, \;\; \text{d.f.} = s - 1 - k \tag{2.32}$$

and

$$\chi_2^2 = \sum \frac{[O_i - n\pi_i(\hat{\Phi})]^2}{n\pi_i(\hat{\Phi})}, \;\; \text{d.f.} = s - 1 - q. \tag{2.33}$$

A test for H_2 given H_1 can be obtained as

$$\chi_{2\cdot 1}^2 = \chi_2^2 - \chi_1^2, \;\; \text{d.f.} = k - q \tag{2.34}$$

which is not a score test, although (2.32 and (2.33) are score tests, but is asymptotically equivalent to the score test for H_2 against H_1. It can be further shown that asymptotically

$$\begin{aligned} \chi_{2\cdot 1}^2 &\doteq \sum \frac{[n\pi_i(\hat{\theta}) - n\pi_i(\hat{\Phi})]^2}{n\pi_i(\hat{\Phi})} \\ &= \sum \frac{(E_{1i} - E_{2i})^2}{E_{2i}} \end{aligned} \tag{2.35}$$

where E_{1i} and E_{2i} are expected values of frequencies in the i-th cell under the hypotheses H_1 and H_2 respectively.

An illustration of such a test for examining the equality of the A and B gene frequencies of the O, A, B, AB blood group system in two communities is given in a paper by Rao (1961).

2.3.5 Some applications of the chi-square test

Chapters 3 and 4 of Fisher's (1958) book contain numerous examples of the use of the chi-square test. In testing for goodness-of-fit, the main problem is the comparison of observed frequencies with those expected under the given specification. As a global test for a given specification, any measure of difference between the observed and expected frequencies can be used. Whether the global test is significant or not at a chosen level of significance, of greater importance is the discovery of patterns in the differences between observed and expected frequencies. (A visual examination of the observed and expected frequencies placed side by side in two columns followed by a confirmatory test of any observed pattern of departure from null hypothesis should be a standard practice in statistical modeling). A brilliant example is provided by the data in Table 11 of Chapter 3 in Fisher's book, where Fisher points out a tendency for families to produce more children of one sex. Another example of interest is the data of Table 4 in the same Chapter of Fisher's book where the observed and expected frequencies agree very closely, yielding a small value of chi-square, indicating that the data are edited.

2.4 Other Tests of Goodness-of-Fit

As observed in earlier sections, any good measure of difference between the vectors of observed and expected frequencies can provide a test of goodness-of-fit. A rich class of tests can be generated by using Csiszár ϕ-divergence between observed (O_i) and expected frequencies (E_i) defined by

$$D_\phi = \sum_1^s E_i \phi\left(\frac{O_i}{E_i}\right) \tag{2.36}$$

for every convex function $\phi : [0, \infty) \to R \cup \{\infty\}$ where $0\phi(0/0) = 0$ and $0\ \phi(p/0) = \lim \phi(u)/u$ as $u \to \infty$. It is shown by Morales, Pardo and Vajda (1995) that

$$\frac{2}{\phi''(1)} D_\phi \sim \chi^2(s-1-k) \tag{2.37}$$

if we choose

$$E_i = n\pi_i(\hat{\theta}),\ \hat{\theta} = \arg\min \sum \pi_i(\theta)\phi\left(\frac{O_i}{n\pi_i(\theta)}\right)$$

where θ is a k-vector parameter.

Read and Cressie (1988) proposed what they call power divergence statistics defined by

$$2nI^{\lambda} = \frac{2}{\lambda(\lambda+1)} \sum O_i \left[\left(\frac{O_i}{E_i} \right)^{\lambda} - 1 \right], -\infty < \lambda < \infty \tag{2.38}$$

where $E_i = n\pi_i(\hat{\theta})$, using a BAN estimator of θ. The statistic (2.38) has the same asymptotic chi-square distribution on $s - 1 - k$ degrees of freedom. This class can be obtained as a special case of (2.36) by choosing

$$\phi(x) = \frac{1}{\lambda(\lambda+1)}(x^{\lambda+1} - x), \ \lambda \neq 0, -1. \tag{2.39}$$

Special choices of λ lead to test criteria such as Pearson's chi-square, Neyman's modified chi-square, log likelihood ratio, modified log likelihood ratio and Freeman-Tukey statistic.

A rich class of diversity measures as test criteria for goodness-of-fit is provided by the Jensen difference defined by Rao (1982)

$$J_H(x, y) = H\left(\frac{x+y}{2} \right) - \frac{1}{2}H(x) - \frac{1}{2}H(y) \tag{2.40}$$

where H is an entropy (i.e., a concave) function of a nonnegative s-vector variable. We can generate an entropy function in many ways. One is to define

$$H(x_1, \dots, x_s) = \sum_{i=1}^{s} \phi(x_i)$$

where $\phi : (0, \infty) \to R$ is a continuous concave function. In such a case the statistic for goodness-of-fit test based on (2.40) is

$$8n \sum_{i=1}^{s} \left[\phi\left(\frac{O_i}{n} + \frac{E_i}{n} \right) - \frac{1}{2}\phi\left(\frac{O_i}{n} \right) - \frac{1}{2}\left(\frac{E_i}{n} \right) \right] \tag{2.41}$$

which is distributed asymptotically as a linear combination of $s-k-1$ chi-square variables on one degree of freedom each, where k is the number of unknown parameters in the specification of the cell probabilities. For a study of the test (2.41) based on Bose-Einstein entropy H defined by Burbea and Rao (1982a, 1982b), the reader is referred to Pardo (1999).

Another possibility is to use the quadratic entropy introduced by Rao (1984)

$$Q(x) = H(x_1, \dots, x_s) = \sum \sum a_{ij} x_i x_j \tag{2.42}$$

where $x' = (x_1, \dots, x_s)$ and the coefficients a_{ij} are chosen such that the matrix

$$(a_{is} + a_{js} - a_{ij} - a_{ss}), \ i, j = 1, \dots, s-1$$

is nonnegative definite. The Jensen difference between

$$O' = (O_1, \ldots, O_k) \text{ and } E' = (E_1, \ldots, E_k)$$

under quadratic entropy (2.42) is

$$D_Q(O, E) = Q\left(\frac{O+E}{2n}\right) - \frac{1}{2}\left(\frac{O}{n}\right) - \frac{1}{2}\left(\frac{E}{n}\right). \tag{2.43}$$

The quantity $\sqrt{D_Q(O, E)}$ is actually a metric on the space of vectors $x' = (x_1, \ldots, x_s)$, with $x_i \geq 0$ and $x_1 + \ldots + x_s = 1$. The goodness-of-fit test based on $D_Q(O, E)$ is

$$n^{-1} \sum \sum a_{ij}(O_i - E_i)(O_j - E_j) \tag{2.44}$$

which is distributed asymptotically as a linear function of chi-square variables on one degree of freedom. The test statistics introduced in this section provide a rich field for further research.

2.5 Specification Tests for Continuous Distributions

The specification tests for the continuous case are somewhat complicated. One method is to replace the continuous distribution by a histogram choosing a set of class intervals, in which case the methods of Section 2.3 are applicable. However, there are no definite rules for the choices of the number and the actual boundaries of class intervals. Some ways of dealing with these problems are described in great detail in the book by Greenwood and Nikulin (1996).

A general criterion for goodness-of-fit in the continuous case is

$$T = \|F_n(x) - F(x, \hat{\theta})\| \tag{2.45}$$

choosing a suitable norm or a discrepancy measure for the difference between $F_n(\cdot)$, the edf (empirical distribution function), and $F(\cdot, \hat{\theta})$, the specified distribution function with an estimated value or a specified value of the parameter θ. Some special cases of (2.45) are the Kolmogorov-Smirnov statistic

$$K_n = \sup_{|x|<\infty} \sqrt{n}|F_n(x) - F(x, \hat{\theta})| \tag{2.46}$$

and Cramér-von Mises statistic

$$W_n = n \int_{-\infty}^{\infty} [F_n(x) - F(x, \hat{\theta}]^2 \Psi(F(x, \hat{\theta}))dF(x, \hat{\theta}) \tag{2.47}$$

where Ψ is some nonnegative function. The asymptotic distributions of (2.46) and (2.47) are very complicated except in the case of a specified value of θ. Some methods of modifying these statistics to simplify their distributions are given by Durbin (1973). The possibility of using bootstrap methodology is discussed in the book by Shao and Tu (1995). Reference may also made to papers by Beran (1986) and Romano (1988). The bootstrap methodology for goodness-of-fit tests in the continuous case is not well developed and some further research is needed.

There are a few other possible tests which are not explicitly mentioned in the statistical literature and which seem to be worth studying.

Let $x' = (x_{(1)}, \dots, x_{(n)})$ be ordered statistics. To test a simple hypothesis that the sample comes from a specified distribution $F(x, \theta_0)$, we compute the theoretical quantiles

$$F(\hat{x}_{(i)}, \theta_0) = \frac{2i-1}{2n}, \quad i = 1, \dots, n \tag{2.48}$$

and compare the vectors x and $\hat{x}' = (\hat{x}_{(1)}, \dots, \hat{x}_{(n)})$. One possible statistic is

$$S_s = \frac{\sqrt{n}\sum(x_{(i)} - \hat{x}_{(i)})^2}{\sum(x_{(i)} - \bar{x})^2 + \sum(\hat{x}_{(i)} - \hat{x})^2}. \tag{2.49}$$

The sampling distribution can be obtained by simulation as $F(x, \theta_0)$ is known.

To test a composite hypothesis that the sample comes from a distribution $F(x, \theta)$, where the function F is specified and the parameter θ is unknown, we can use a statistic S_c which is of the same form as (2.49) with $\hat{x}_{(i)}$ defined by

$$F(\hat{x}_{(i)}, \hat{\theta}) = \frac{2i-1}{2n}, \quad i = 1, \dots, n \tag{2.50}$$

where $\hat{\theta}$ is an efficient estimate of θ. The distribution of S_c may be complicated or may involve the unknown parameter. In such a case, it is worthwhile examining whether the bootstrap method is applicable.

There is some simplification if $F(x, \theta)$ belongs to translation scale family, $F[(x - \mu)/\sigma]$. In such a case, we define $\hat{x}_{(i)}$ as

$$F(\hat{x}_{(i)} | \mu = 0, \sigma = 1) = \frac{2i-1}{2n}, \quad i = 1, \dots, n. \tag{2.51}$$

Then the appropriate statistic for specification is the correlation coefficient r computed from the pairs

$$(x_{(i)}, \hat{x}_{(i)}), \; i = 1, \dots, n.$$

The distribution of r can be obtained by simulation, drawing samples from $F(x | \mu = 0, \sigma = 1)$.

In conclusion, the tests described in this paper may have different power functions depending on the alternatives, but none of them dominates the others. The purpose of data analysis is learning from data, and a global test is only of an exploratory nature. Of greater interest is the study of the pattern of deviations between observed and expected frequencies. No doubt a global test provides some confidence in search for possible alternatives. Any reasonably good global test will do for this purpose. Long live chi-square.

References

1. Beran, R. (1986). Simulated power functions, *Annals of Statistics*, **14**, 151–173.

2. Burbea, J. and Rao, C. R. (1982a). On the convexity of some divergence measures based on entropy functions, *IEEE Transactions Inform. Theory*, **28**, 489–495.

3. Burbea, J. and Rao, C. R. (1982b). On the convexity of higher order Jensen differences based on entropy functions, *IEEE Transactions Inform. Theory*, **28**, 961–963.

4. Csiszár, I. (1963). Eine Informationtheoretishe Ungleichung und ihre Anwendung auf den Beweiss der Ergodizität von Markhoffschen von Ketten, *Publ. Math. Inst. Hung. Acad. Sci., Ser. A8*, 85–108.

5. Durbin, J. (1973). *Distribution Theory for Tests Based on the Sample Distribution Function*, Vol. 9, Philadelphia: SIAM.

6. Efron, B. (1995). The statistical century, *Royal Statistical Society News*, **25**, 1.

7. Fisher, R. A. (1958). *Statistical Methods for Research Workers*, 13th Edition, New York: Hafner.

8. Greenwood, P. E. and Nikulin, M. S. (1996). *A Guide to Chi-squared Testing*, New York: John Wiley & Sons.

9. Hall, W. J. and Mathiason, D. J. (1990). On large sample estimation and testing in parametric models, *International Statistical Review*, **58**, 77–97.

10. Koenkar, R. (1987). A comparison of asymptotic testing methods for l_1-regression, In *Statistical Data Analysis*, (Ed., Y. Dodge), pp. 287–294, North Holland.

11. Lancaster, H. O. (1969). *The Chi-squared Distribution*, New York: John Wiley & Sons.

12. Lehman, E. L. (1999). *Elements of Large Sample Theory*, New York: Springer-Verlag.

13. Morales, D., Pardo, L. and Vajda, I. (1995). Asymptotic divergence of estimates of discrete distributions, *Journal of Statistical Planning and Inference*, **48**, 347–369.

14. Neyman, J. and Pearson, E. S. (1928). On the use and interpretation of certain test criteria, *Biometrika*, **20**, 175–240, 263–294.

15. Pardo, M. C. (1999). On Burbea-Rao divergence based goodness-of-fit tests for multinomial models, *Journal of Multivariate Analysis*, **69**, 65–87.

16. Pearson, K. (1900). On the criterion that a given system of deviations from the probable in the case of a correlated system of variables is such that it can be reasonably supposed to have arisen from random sampling, *Philosophical Magazine*, **50**, 157–172.

17. Rao, C. R. (1948). Large sample tests of statistical hypotheses concerning several parameters with applications to problems of estimation, *Proceedings of the Cambridge Philosophical Society*, **44**, 50–57.

18. Rao, C. R. (1961). A study of large sample test criteria through properties of efficient estimates, *Sankhyā, Series A*, **23**, 25–40.

19. Rao, C. R. (1982). Diversity and dissimilarity coefficients: a unified approach, *Journal of Theoretical Pop. Biology*, **21**, 24–43.

20. Rao, C. R. (1984). Convexity properties of entropy functions and analysis of diversity, In *Inequalities in Statistics and Probability*, IMS Lecture Notes, **5**, 68–77.

21. Read, T. R. C. and Cressie, N. A. C. (1988). *Goodness-of-fit Statistics for Discrete Multivariate Data*, New York: Springer-Verlag.

22. Romano, J. P. (1988). A bootstrap revival of some nonparametric distance tests, *Journal of the American Statistical Association*, **83**, 698–708.

23. Shao, J. and Tu, D. (1995). *The Jackknife and Bootstrap*, New York: Springer-Verlag.

24. Wald, A. (1943). Tests of statistical hypotheses concerning several parameters when the number of observations is large, *Transactions of the American Mathematical Society*, **54**, 426–482.

25. Wilks, S. S. (1938). The large-sample distribution of the likelihood ratio for testing composite hypotheses, *Annals of the Mathematical Statistical Society*, **9**, 60–62.

3

Approximate Models

Peter J. Huber

Klosters, Switzerland

Abstract: The increasing size and complexity of data sets increasingly forces us to deal with less than perfect, but ever more complicated models. I shall discuss general issues of model fitting and of assessing the quality of fit, and the important and often crucial roles of robustness and simulation.

Keywords and phrases: Robustness, simulation, Bayesian modeling, Markov chain, Monte Carlo

3.1 Models

The anniversary of Karl Pearson's paper offers a timely opportunity for a digression and to discuss the role of models in contemporary and future statistics, and the assessment of adequacy of their fit, in a somewhat broader context, stressing necessary changes in philosophy rather than the technical nitty-gritty they involve. The present paper elaborates on what I had tentatively named "postmodern robustness" in Huber (1996, final section).

Karl Pearson's pioneering 1900 paper had been a first step. He had been concerned exclusively with distributional models, and with global tests of goodness-of-fit. He had disregarded problems caused by models containing free parameters: how to estimate such parameters, and how to adjust the count of the number of degrees of freedom. Corresponding improvements then were achieved by Fisher and others. Though, the basic paradigm of distributional models remained in force and still forms the prevalent mental framework for statistical modeling. For example, the classical texts in theoretical statistics, such as Cox and Hinkley (1974) or Lehmann (1986), discuss goodness-of-fit tests only in the context of distributional models. Apart from this, it is curious how current statistical parlance categorizes models into classes — such as "linear models" or "generalized linear models." The reason is of course that each such class permits a specific formal statistical theory and analysis, or, expressing it the other way around: each class constitutes the natural scope of a particular theoretical framework. In my opinion, to be elaborated below, such narrow categorizations can have undesirable consequences.

We now have a decisive advantage over Pearson: whenever there is an arbitrarily complex but supposedly exact model and a test statistic, we can, at least in principle, determine the distribution of that statistic under the null hypothesis to an arbitrary degree of accuracy with the help of simulation. It is this advantage which allows us to concentrate on the conceptual aspects, in particular on the traditionally suppressed problems posed by partial or otherwise inaccurate models.

The word "model" has a bewilderingly wide semantic range, from tiny trains to long-legged girls. Though, it hides a simple common feature: a model is a representation of the essential aspects of some real thing in an idealized, exemplary form, ignoring the inessential ones. The title of this paper is an intentional pleonasm: by definition, a model is not an exact counterpart of the real thing, but a judicious approximation. Mathematical statisticians, being concerned with the pure ideas, sometimes tend to forget this, despite strong admonitions to the contrary, such as by McCullagh and Nelder (1983, p. 6): "all models are wrong." What is considered essential of course depends on the current viewpoint — the same thing may need to be modeled in various different ways. Progress in science usually occurs through thinking in models, they help to separate the essential from the inessential.

A general discussion of the role mathematical models have played in science should help to clarify the issues. They can be qualitative or quantitative, theoretical or empirical, causal or phenomenological, deterministic or stochastic, and very often are a mixture of all. The historical development of the very first non-trivial mathematical models, namely those for planetary motion, illustrates some salient points most nicely, namely the interplay between conceptual/qualitative and phenomenological/quantitative models, the discontinuous jumps from a model class to the next, and the importance of precisely locating the discrepancy rather than merely establishing the existence of discrepancies by a global test of goodness-of-fit.

Early Greek scientists, foremost among them Anaxagoras (ca. 450 BC), had tried to explain the irregular motions of the planets by a "stochastic model," namely by the random action of vortices. This model did not really explain anything, it just offered a convenient excuse for their inability to understand what was going on in the skies. In the 4th century BC, Eudoxos then invented an incredibly clever qualitative model. He managed to explain the puzzling retrograde motion of the planets deterministically in accordance with the philosophical theory that celestial motions ought to be circular and uniform. For each planet, he needed four concentric spheres attached to each other, all rotating uniformly. Quantitatively, the model was not too good. In particular, even after several improvements, it could not possibly explain the varying luminosity of the planets, because in this model their distances from the earth remained constant. About the same time the Babylonians devised empirical models, apparently devoid of any philosophical underpinning; they managed to describe the motions of the moon and the planets phenomenologically through arithmetic schemes involving additive superposition of piecewise linear functions. Around 130 AD, Ptolemy then constructed an even cleverer model than Eudoxos. He retained the politically correct uniform circular motion, but the circles were no longer concentric. With minor adjustments this type of model withstood the tests of observational astronomy for almost 1500 years, until Kepler, with the superior observations of Tycho Brahe at his disposal, found a misfit of merely 8' (a quarter of the apparent diameter of sun or moon, and just barely above observational accuracy) in the motion of a single planet (Mars). He devised a fundamentally new model, replacing the uniform circular by elliptical motions. The laws describing his model later formed the essential basis for Newton's theory of gravitation.

For us, Kepler's step is the most interesting and relevant. First, we note that the geocentric Ptolemaic and the heliocentric Copernican models phenomenologically are equivalent — both belong to the class of epicyclic models, and with properly adjusted parameters the Ptolemaic model renders the phenomena, as seen from the earth, absolutely identically to the Copernican one. But the conceptual step to heliocentricity was a most crucial inducement for Kepler's invention. Second, Kepler's own account shows how difficult it is to overcome modeling prejudices. It required the genius of Kepler to jump over the shadow cast forward by 1500 years of epicyclic modeling. To resist and overcome the temptation to squeeze the data into a preconceived but wrong model class — by piling up a few more uninterpretable parameters — may require comparable independence of mind. I think that here we have a warning tale about the dangers of categorizing models into narrowly specified classes.

3.2 Bayesian Modeling

In the past years, after the paper by Geman and Geman (1984), modeling activity in the statistical literature has concentrated mainly on Bayesian modeling and on Markov Chain Monte Carlo methods; see the survey by Besag *et al.* (1995). There are points of close contact of this activity with the present paper, namely the common concern with complex models and simulation.

Modeling issues provide one of the neatest discriminations between the relative strengths and weaknesses of the Bayesian and frequentist approaches; the two complement each other in a most welcome fashion. The frequentist-

Pearsonian approach has difficulties with the comparative assessment of competing models. P-values, when abused for that purpose, are awkward and unintuitive, to say the least, while the Bayesian approach provides a streamlined mechanism for quantitative comparisons through posterior probabilities.

On the other hand, Bayesian statistics lacks a mechanism for assessing goodness-of-fit in absolute terms. The crux of the matter, alluded to but not elaborated in the excellent discussion of scientific learning and statistical inference by George Box (1980, see p. 383f.), is as follows. Within orthodox Bayesian statistics, we cannot even address the question whether a model Mi under consideration is consonant with the data y. For that, we either must step outside of the Bayes framework and in frequentist-Pearsonian manner perform a goodness-of-fit test of the model against an unspecified alternative (an idea to which Box, being a Bayesian, but not a dogmatic one, obviously does not object), or else apply "tentative overfitting" procedures. The latter are based on the unwarranted presumption that by throwing in a few additional parameters one can obtain a perfectly fitting model. But how and where to insert those additional parameters often is far from obvious, unless one is aided by hindsight and is able to abandon one's tacit prejudices about the class of models. Remember that Kepler rejected the epicyclic Ptolemaic/Copernican models because he could not obtain an adequate fit within that class.

The recent tutorial on Bayesian model averaging by Hoeting *et al.* (1999), through the very tentativeness of their remarks on the choice of the class of models over which to average (what are "models supported by the data"?), further highlights the problems raised by model adequacy. To put the spotlight on the central problem, we consider the special case of a small finite number of discrete models (such that the fit cannot be improved by throwing in a few additional parameters). Before we can meaningfully average, or assign meaningful relative probabilities to the different models by Bayesian methods, we ought to check whether the models collectively make sense. A mere verification by a goodness-of-fit test that the class contains models not rejected by such a test, is not good enough. Averaging over a class of spurious models is not better and safer than treating a single non-rejected model as correct — see the next section — and maybe even worse, because it obscures the underlying problem. Only in very exceptional cases one can check appropriateness of a class of models by a reverse application of a goodness-of-fit test, namely by testing and rejecting the hypothesis that all models under consideration are wrong. For one of the rare cases where such a "badness-of-fit" test works see Huber *et al.* (1982).

In short, while Bayesian modeling offers some food for thought, plus a bag of simulation techniques to borrow from, it cannot help with regard to questions of goodness-of-fit and of model adequacy.

3.3 Mathematical Statistics and Approximate Models

Karl Pearson, as can be seen from a few remarks scattered through his 1900 paper, had been perfectly aware that his distributional models were approximations. For example, he notes that a model fitting well for a particular sample size would fit for smaller, but not necessarily for larger samples.

After Karl Pearson, under the influence of Fisher and others, mathematical statistics developed a particular modus operandi: take a simple, idealized model, create an optimal procedure for this model, and then apply the procedure to the real situation, either ignoring deviations altogether, or invoking a vague continuity principle.[1] By 1960, this approach began to run out of steam: on one hand, one began to run out of manageable simple models, on the other hand, one realized that the continuity principle did not necessarily apply: optimization at an idealized model might lead to procedures that were unstable under minor deviations. The robustness paradigm - explicitly permitting small deviations from the idealized model when optimizing - carried only a few steps further.

Statisticians, as a rule, pay little attention to what happens after a statistical test has been performed. If a goodness-of-fit tests rejects a model, we are left with many alternative actions. Perhaps we do nothing (and continue to live with a model certified to be inaccurate). Perhaps we tinker with the model by adding and adjusting a few more features (and thereby destroy its conceptual simplicity). Or we switch to an entirely different model, maybe one based on a different theory, or maybe, in the absence of such a theory, to a purely phenomenological one.

In the opposite case, if a goodness-of-fit test does not reject a model, statisticians have become accustomed to act as if it were true. Of course this is logically inadmissible, even more so if with McCullagh and Nelder one believes that all models are wrong a priori. The data analytic attitude makes more sense, namely to use such tests merely as warning signs against over-interpretation: renounce attempts at interpreting deviances from the model if a goodness-of-fit test (with a much higher than usual level) fails to reject the model.

Moreover, treating a model that has not been rejected as correct can be misleading and dangerous. Perhaps this is the main lesson we have learned from robustness. A few illustrations follow.

(1) Model-optimality versus fit-optimality. Distributional goodness-of-fit tests typically are based on the value of the distance measured between the em-

[1] There has been curiously little concern in the statistical literature with the optimality of goodness-of-fit tests themselves (i.e. of *2- and F-tests), and it has been relegated to notes, such as by Lehmann (1986, pp. 428–429).

pirical and the model distribution. To fix the idea, let us take the Kolmogorov distance, which makes exposition simpler (but exactly the same arguments apply to a distance based on the chi-2 test statistic):

$$K(\hat{\mu}, \hat{\sigma}) = \sup_x (|F_n(x) - \Phi(x)|)$$

where $\hat{\mu}, \hat{\sigma}$ are either the traditional model-optimal estimates:

(a) $\hat{\mu}, \hat{\sigma} =$ ML estimate for model Φ,

or the fit-optimal estimates:

(b) $\hat{\mu}, \hat{\sigma} = \min_{\mu,\sigma \in R} K(\mu, \sigma)$.

In case (a), the test sometimes correctly rejects the hypothesis of normality for the wrong reason, namely if an outlier inflates the ML estimate for σ, even though the fit, in terms of minimized distance (b), is good. The disturbing fact is that the traditional recommendation, namely to estimate the unknown free parameters by any asymptotically efficient estimate, viz. either maximum likelihood or minimum chi-2, may have very different consequences depending on which estimate one chooses.

(2) Minimax robust estimates. Observational errors in most cases are excellently modeled by the normal distribution, if we make allowance for occasional gross errors (which may be of different origin). If we formalize this in terms of a contamination model, then the normal part represents the essential aspects of the observational process in an idealized form, the contamination part merely models some disturbance factors unrelated to the quantities of interest. But model-optimal estimates for the idealized model, e.g., the sample mean, are unstable under small deviations in the tails of the distribution, while robust estimates, such as judiciously chosen M-estimates, offer stability but lose very little efficiency at the normal model.

Typically, we are not interested in estimating the parameters of the contaminant, or to test it for goodness-of-fit, and mostly not even able, given the available sample size. Note that a least favorable distribution is not intended to model the underlying situation (even though it may approximate the true error distribution better than a normal model), its purpose is to provide a ML estimate that is minimax robust.

(3) Linear fit. Assume that you want to fit a straight line to an approximately linear function that can be observed with errors in a certain interval (a,b). Assume that the goal of the fit is to minimize the integrated mean square deviation between the true, approximately linear function and an idealized straight line. A model-optimal design will put one half of the observations at each of the endpoints of the interval. A fit-optimal design will distribute the observations roughly uniformly over the interval (a,b). The unexpected and

surprising fact is that subliminal deviations of the function from a straight line (i.e. deviations too small to be detected by goodness-of-fit tests) may suffice to make the fit-optimal design superior to the model-optimal design [cf. Huber (1975b)].

In short: for the sake of robustness, we may sometimes prefer a (slightly) wrong to a (possibly) right model.

3.4 Statistical Significance and Relevance

It is a truism that for a sufficiently large sample size any model ultimately will be rejected. But this is merely a symptom of a more serious underlying problem, involving the difference between statistical significance and physical relevance. Since this is a problem of science rather than one of statistics, statisticians unfortunately tend to ignore it. The following example is based on an actual consulting case.

Example: A computer program used by physicists for analyzing certain experiments automatically plotted the empirical spectrum on top of the theoretical spectrum, and in addition it offered a verbal assessment of the goodness-of-fit (based on a chi-2 statistic). Perplexingly, if the fit was visually perfect, it would often be assessed by the program as poor, while a poor visual fit often would be assessed as good. At first, one suspected a programming error in the statistical calculations, but the reason was: the experiments had widely different signal-to-noise ratios, and if the random observational errors were large, the test would not reach statistical significance, and if they were small, the test would reject because of systematic errors (either in the models or in the observations), too small to show up in the plots and probably also too small to be physically relevant.

With increasing data sizes, the paradoxical situation of this example may occur even more often: if a global goodness-of-fit test does not reject, then the observations are too noisy to be useful, and if it rejects, the decision whether or not to accept the model involves a delicate judgment of relevance rather than of statistical significance.

In short, in real-life situations the interpretation of the results of goodness-of-fit tests must rely on judgment of content rather than on P-values. For a traditional mathematical statistician, the implied primacy of judgment over mathematical proof and over statistical significance clearly goes against the grain. John Tukey once said, discussing the future of data analysis (1962, p. 13): 'The most important maxim for data analysts to heed, and one which many statisticians seem to have shunned, is this: "Far better an approximate answer to the right question, which is often vague, than an exact answer to the wrong question, which always can be made precise."' Analogous maxims apply to modeling: a crude answer derived in a messy way from an approximately right model is far better than an exact answer from a wrong but tractable model.

3.5 Composite Models

The problems considered and solved by mathematical statisticians are model problems themselves - they idealize exemplary situations. Unfortunately, these problems and their solutions often are too simple to be directly applicable and then ought to be regarded as mere guidelines. Applied statisticians and data analysts today are confronted by increasingly more massive and more complex real-life data, requiring ever more complex models. Such complex models then must be pieced together from simpler parts.

Composite models facilitate a conceptual separation of the model into important and less important parts, and they also make it easier to locate where the deviations from the model occur. Usually, only some parts of the model are of interest, while other parts are a mere nuisance. Sometimes, in traditional fashion, we may want to test for the presence or absence (perhaps more accurately: lack of importance) of certain features by not including them in the model. But we now have to go beyond. We may be willing to condone a lack of fit in an irrelevant part, provided it does not seriously affect the estimation of the relevant parameters of interest. Once more, this involves questions of judgment rather than of mathematical statistics, and the price to pay is that the traditional global tests of goodness-of-fit lose importance and may become meaningless. Often, despite huge data sizes, preciously few degrees of freedom are relevant for the important parts of the model, while the majority, because of their sheer number, will highlight mere model inadequacies in the less important parts. We may be not willing to spend the effort, or perhaps not even able, to model those irrelevancies. We need local assessments of the relevant parts of the fit, whether informally through plots of various kinds [cf. the remarks by Box (1980)], or formally, through tests.

Parameter estimation in a composite model is tricky. Unfortunately, theoretical results applying to the components rarely carry over to the composition. Usually, some version of backfitting will perform part of the work: assume there are component models A1, ..., An, and you know how to estimate the parameters of Ai separately for each i. Fix preliminary parameter estimates for all models except Ai, improve those of Ai, and cycle repeatedly through all i. This should result in correct values of the parameter estimates. However, backfitting creates devilishly tricky statistical problems with regard to the estimation of standard errors of the estimated parameters, and with counting the number of degrees of freedom for goodness-of-fit tests. The standard errors estimated from the separate models Ai, keeping the others parts of the model fixed, may be serious underestimates. Cross-validation estimates, for example, are invalidated by repeated cycling.

Once I had felt that stochastic modeling, despite its importance, belonged so much to a particular field of application that it was difficult to discuss it in a broad and general framework, and I had therefore excluded it from a discussion of current issues in statistics [Huber (1975a, p. 86)]. I now would modify my former stance. I still believe that an abstract and general discussion will fail because it is practically impossible to establish a common basis of understanding between the partners of such a discussion. On the other hand, a discussion based on, and exemplified by, substantial and specific applications will be fruitful. All my better examples are concerned with the modeling of various applied stochastic processes. This is not an accident: stochastic processes are creating the most involved modeling problems. The following example (on modeling the rotation of the earth) may be the best I have met so far. It shows the intricacies of stochastic models in real situations, in particular how much modeling and data processing sometimes has to be done prior to any statistics, and how different model components must and can be separated despite interactions. The runner-up is in biology rather than in geophysics, and it operates in the time domain rather than in the frequency domain [modeling of circadian cycles, Brown (1988)].

Example: Modeling the length of day [Huber (2000)]. Because of tidal friction, the rotation of the earth slows down: the length of day (LOD) increases by about 2 milliseconds per century. An analysis of medieval and ancient eclipses back to about 700 BC had shown that on top of the systematic slow-down there are very substantial random fluctuations. They must be taken into account when one is extrapolating astronomical calculations to a more distant historical past. In terms of the LOD-process, these poorly determinable fluctuations are compatible with a Brownian motion (or random walk) model, whose increments have a variance of about 0.05 ms2/year.

Being interested in estimating the size of the extrapolation errors, I wondered whether such a millennial Brownian motion component was a long-range effect only, or whether it would be discernible also in the more accurate but much shorter modern series of measurements, and in particular whether those modern measurements might even permit a more accurate estimate of the variance of the increments. I could obtain three such series of different lengths and observational accuracies, listing length-of-day values in intervals of 4 months, 5 days and 1 day, starting in the years 1830, 1962 and 1976 respectively. In the power spectrum of a differenced series of LOD values, a Brownian motion component should manifest itself in the low frequency part of the spectrum as a horizontal tail end. The problem with the modern measurements is that there are many nuisance effects, with periodicities ranging from days to tens of years, and of a size comparable to the putative Brownian motion process.

Aided by hindsight, a reasonably comprehensive description of the components of the LOD-process goes as follows.

(1) Systematic drift of about 2 ms/cy. The "true" rate cannot be estimated very accurately from the data because of the random Brownian motion (2) sitting on top.

(2) Brownian motion (or random walk process). Putative cause: cumulative random changes in the rotational moment of inertia of the earth's mantle, induced by plate tectonics.

(3) Decadal fluctuations, with an amplitude of several milliseconds. See Figure 3.1. The common opinion is that they have to do with damped oscillations on the mantle-core boundary.

(4) Seasonal effects, with an amplitude of about 0.4 ms. Exchange of angular momentum between the atmosphere and the solid earth, caused by seasonal temperature changes and winds. See Figure 3.2. They can be taken out cleanly by fitting a trigonometric polynomial to the LOD-process.

(5) "50-day Madden-Julian oscillation." This is a broad spectral hump near periodicities of 40-50 days, corresponding to damped oscillations with a root-mean-square amplitude of about 0.17 ms. See Figure 3.2. Apparently, it is due to exchange of angular momentum between the atmosphere and the solid earth, mediated by winds and putatively caused by random temperature changes in the atmosphere. The underlying physical mechanism can be modeled quite accurately by an AR(2) process.

(6) Measurement errors. Extremely inhomogeneous, 3–5 ms in the 19th century, dominating the spectrum for periodicities below 8 years in the 4-month series. For most practical purposes they are negligible after 1962, i.e. then they are dominated by the contributions of (5). See Figure 3.1.

(7) Solid earth tides. These are reasonably well understood, deterministic effects. In the later parts of the data series made available to me, namely since 1982, they had been eliminated through preprocessing, but not before (remnants are the peaks in the 10–14 days range in Figure 3.2).

(8) Side effects of preprocessing (suppression of high frequency noise, including measurement errors, through Kalman filtering). They affect the high frequency parts of the spectrum and depress the spectral power for periodicities below 30 days, leading to biased estimates of the parameters of the AR(2) process (5). Their existence was discovered only when analyzing a shorter, very accurate 1-day series ranging from 1984-1996, which showed analogous artifacts, but in the range below 5 days.

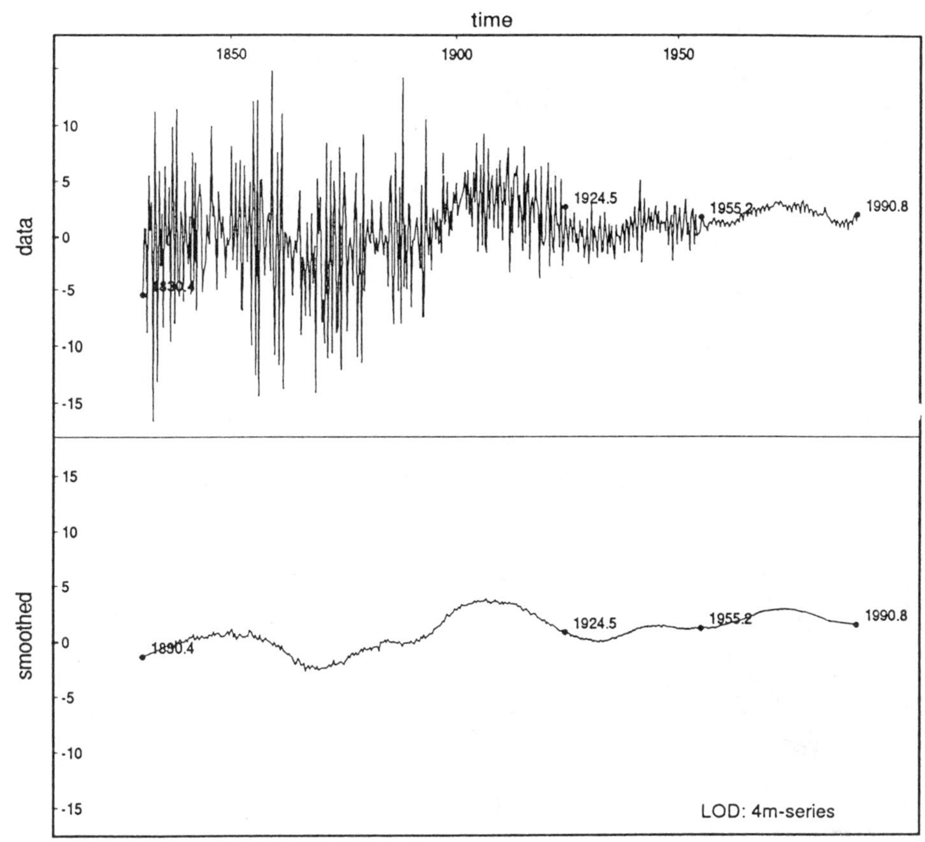

Figure 3.1: The 4-lunation series (covering the years 1830-1990 in 4-month intervals) in the time domain: the actual data in the series, and a smoothed version (obtained by forming moving averages). Note the changing level of the observational noise and the decadal waves

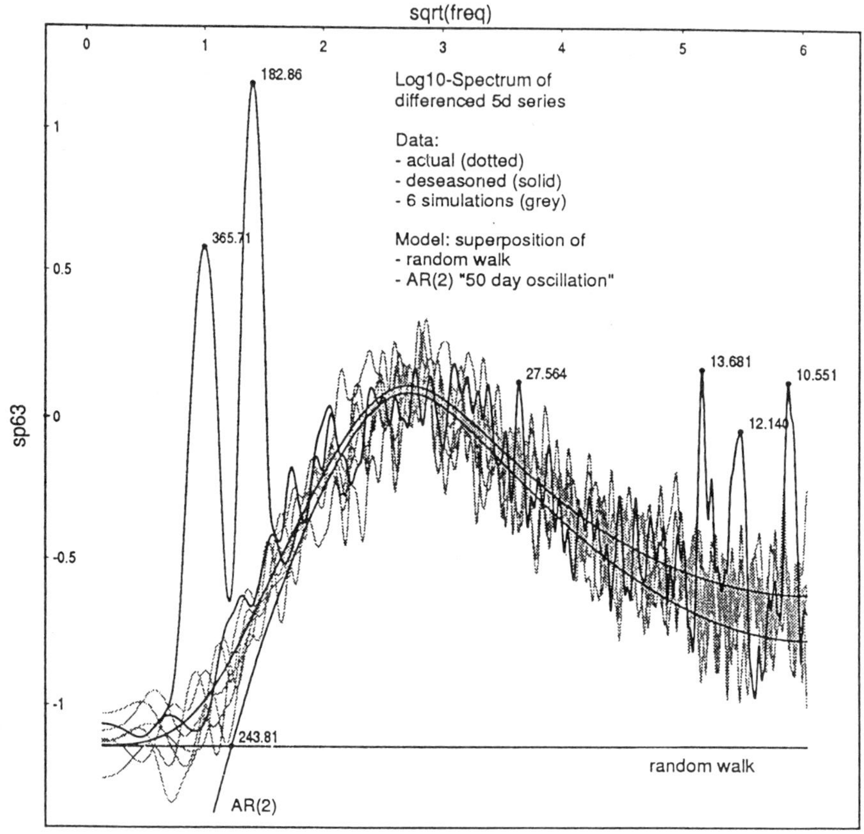

Figure 3.2: Log10-spectrum of the differenced 5-day series (covering the years 1962-1995 in 5-day intervals). The cross-over between the random walk process and the AR(2) model occurs near 8 months (243.81 days). On purpose, only the two most prominent components (2) and (5) of the model are used

There is a delicate interplay between the components (2) and (3). We note that random changes in the rotational moment of inertia of the earth's mantle, as postulated in (2), by preservation of angular momentum cause wiggles in the rotation rate of the mantle. These wiggles excite damped oscillations on the mantle-core boundary, with a resonance in the decadal range. High-frequency components only wiggle the mantle, but in the low-frequency range, mantle and core move together as one solid body. Even though the exact coupling mechanisms are not known, the net effect is that the spectrum of the differenced LOD-series will be flat both below and above the resonance frequency, with a 19% smaller value in the low frequency range (the size of the drop is determined by the known ratio between the moments of inertia of mantle and core), with a hump of poorly determined shape in between.

The feature of interest in the spectrum of Figure 3.2 is the putative Brownian motion (or random walk) component, which should manifest itself in a flat low-frequency spectrum. We would like to check whether a Brownian motion model fits, and we would like to estimate the size of its contribution. In the 5-day series, the AR(2)-contribution corresponding to the 50-day Madden-Julian oscillation dominates the spectrum of the deseasoned series for periodicities shorter than about 8 months. Information about the putative Brownian motion part can be gleaned only from the tiny low-frequency tail end of the spectrum (which has been stretched out in Figure 3.2 for better visibility by plotting the power spectra against the square root of the frequency rather than against the frequency itself). It is obviously non-trivial to separate the Brownian motion contribution from the AR(2) component. When estimating the AR(2) parameters we must rely on the middle frequency range (where the 50-day oscillation is dominant), and we must remember the tricks of the robustness trade and make sure for example that the irrelevant peaks in the 10–14 day range do not bias our parameter estimates. Then, in the low frequency range, where the Brownian motion is dominant, the expected contribution of the AR(2) model must be subtracted as a correction from the total spectrum, in order to estimate the contribution of the Brownian motion. As an added complication, the very low end of the data spectrum may already be somewhat inflated by the contributions of decadal fluctuations (they were not modeled in the simulations depicted in Figure 3.2). In order to reduce processing artifacts caused by smoothing, the actual parameter estimation was not based on the spectra shown in Figure 3.2, but on the periodogram values themselves. This yielded an estimate of the variance of the increments of 0.072 ms2/year, valid above the resonance frequency. For the millennial range (below the resonance frequency), this translates into 0.058 ms2/year, with a 95% confidence interval (0.040, 0.089). - By the way, for somebody reared on Box-Jenkins time-series analysis, it is quite an educational experience if he has to devise his own ARMA estimates based on a periodogram segment!

In the 4-lunation series the high level of measurement noise creates complications: it dominates the spectrum for periodicities shorter than 8-10 years. This leaves deplorably few periodogram ordinates for estimating the Brownian motion contribution in the spectrum. They straddle the decadal resonance hump (3); after subtracting a somewhat crudely estimated contribution from measurement errors leaking into that range, the average spectral power there is 0.13 ms2/year. While this value is boosted by the decadal hump, it still is just barely significantly larger than the value 0.072 ms2/year estimated from the 5-day series. If we model (3) by a damped harmonic oscillator excited by that Brownian motion and estimate the coupling parameters from the data (in statistical parlance this amounts to modeling the mantle-core oscillations by an AR(2) process), we can get an essentially perfect fit of the spectrum in that range.

Among other things, this example illustrates that for some parts of the model (usually the less interesting ones) we may have an abundance of degrees of freedom, and a scarcity for the interesting parts.

3.6 The Role of Simulation

Following Karl Pearson's classical example, goodness-of-fit usually is assessed in terms of weighted sums of squared deviations of the observations from their expected values; their distribution then is approximated by a chi-2 distribution with a suitable number of degrees of freedom. As already Pearson had found, the distribution theory of this test statistic is not exactly trivial, and with complex, composite models, this kind of comparison is no longer feasible for multiple reasons. The fact that iterative backfitting defies distribution theory (except in trivial cases) is only one of them. Another serious problem is due to artifacts caused by preprocessing and processing of the data. Remember that a statistician rarely sees the raw data themselves - most large data collections in the sciences are being heavily preprocessed already in the collection stage, and the scientists not only tend to forget to mention it, but sometimes they also forget exactly what they had done.

A possible, and often the only escape from this quandary is simulation: compare the data with simulations of the model. (Note that in the goodness-of-fit context resampling methods are inappropriate, quite apart from the fact that they do not work for highly structured data.) This way, one can derive a test from any arbitrary statistic: calculate the statistic and adjust nuisance parameters in exactly the same way for the data and for a large number of simulations of the model. Reject if the value of the statistic derived from the data is sufficiently far out in the tails of the set of values derived from the simulations. In particular, the spread of the simulated values of a point estimate permits to supplement the latter with an estimated standard error.

Of course, just like the classical approaches, also the simulation methods only measure stochastic variabilities intrinsic to a model assumed to be correct: they implicitly assume that the estimated parameter values are so close to the true ones that the latter can be replaced by the former without committing a serious error when assessing the variability of an estimate or test statistic. Admittedly, in practice there may be problems; for example, parameter estimation may fail to converge for a small percentage of the samples, and this may selectively affect the tails.

Under mild monotonicity assumptions, but at a relatively high computing cost, it is possible to supplement point estimates with somewhat more reliable confidence interval estimates than by the method just described. For example, in order to find a lower confidence bound, one takes the model and replaces the estimated value θ of the model parameter of interest (in the example of the preceding section: the variance of the increments of the Brownian motion) by a suitably chosen smaller value θ_0. Then one uses the thus modified model for simulating 1000 data sets, and derives new parameter estimates from each of these sets. If, say, 950 of the newly estimated values of θ then are smaller, and 50 larger than the original estimate of θ, then θ_0 constitutes an approximate one-sided 95% lower confidence bound. The determination of a suitable θ_0 is expensive because it requires a considerable amount of trial and error.

But thanks to simulation, a judgmental assessment of goodness-of-fit need not even be based on a test statistic (whose selection always is delicate). The principle is simple and inspired by the line-up methods used by the police: if the actual data hides well among half a dozen or a dozen simulations of the model, the model is judged acceptable; if the actual data sticks out like a sore thumb, the model is no good. Figure 3.2 illustrates the approach by showing the spectrum estimates resulting from 6 simulations of the model, indicating a good fit in the interesting low frequency range. But the example also illustrates a general problem of any global approaches to goodness-of-fit, namely the peaks in the 10–14 day range, which make the actual data set stick out like the sore thumb mentioned above. In this case the origin of the discrepancy is understood, and it is irrelevant because it lies in an uninteresting frequency range. A more sophisticated version of the line-up method, permitting approximate significance tests, is to create a pool composed of the actual data set and, say, 99 simulated sets. Somebody not knowing which is which has 5 attempts to pick the actual set out of the pool, using any tools of his or her choice. If the actual data set is not among the selected five, the model is deemed to be adequate.

3.7 Summary Conclusions

If we want to be able to deal with increasingly larger and more complex data sets, we need to go beyond the current, overly narrow ingrained modeling concepts of statistics. The models will become more complex, too, but a clean statistical theory with mathematically rigorous results is possible only for clean and simple models. We will have to pay a price, but we will also gain something in the process. The questions will shift more and more from a mere yes-or-no global check whether the model is adequate (I prefer the term "model adequacy" to "model validity" - a model adequately rendering the observations need not be valid in any intrinsic sense), to a detailed assessment of the quality of the fit, to questions of interpreting the fit, and in particular to the need to locate and interpret deviations from a model which is known to be imprecise, and to separate essential deviations from irrelevant ones.

References

1. Besag, J., Green, P., Higdon, D., and Mengersen, K. (1995). Bayesian computations and stochastic systems, *Statistical Science*, **10**, 1–66.

2. Box, G. E. P. (1980). Sampling and Bayes' inference in scientific modelling and robustness, *Journal of the Royal Statistical Society, Series A*, **143**, 383–430.

3. Brown, E. N. (1988). Identification and estimation of differential equation models for circadian data, *Ph.D. Thesis*, Department of Statistics, Harvard University.

4. Cox, D. R. and Hinkley, D. V. (1974). *Theoretical Statistics*, London: Chapman and Hall.

5. Geman, S. and Geman, D. (1984). Stochastic relaxation, Gibbs distributions, and the Bayesian restoration of images, *IEEE Transactions on Pattern Analysis and Machine Intelligence*, **6**, 721–741.

6. Hoeting, J. A., Madigan, D., Raftery, A. E., and Volinsky, C. T. (1999). Bayesian model averaging: A tutorial, *Statistical Science*, **14**, 382–417.

7. Huber, P. J. (1975a). Applications vs. abstraction: the selling out of mathematical statistics? In *Proceedings of the Conference on Directions for Mathematical Statistics* (Ed., S. G. Ghurye), Special Supplement to *Advances in Applied Probability*.

8. Huber, P. J. (1975b). Robustness and designs, In *A Survey of Statistical Design and Linear Models* (Ed., J. N. Srivastava), pp. 287–301, North-Holland Publishing Company.

9. Huber, P. J. (1996). *Robust Statistical Procedures*, Second Edition, Philadelphia, PA.: SIAM.

10. Huber, P. J. (2000). Modeling the length of day and extrapolating the rotation of the earth, In *Astronomical Amusements, Papers in Honor of Jean Meeus* (Eds., F. Bonoli, S. De Meis, and A. Panaino), Milano: IsIAO.

11. Huber, P. J., Sachs, A., Stol, M., Whiting, R. M., Leichty, E., Walker, C. B. F., and vanDriel, G. (1982). *Astronomical Dating of Babylon I and Ur III. Occasional Papers on the Near East*, Vol. 1, Issue 4, Malibu: Undena Publications.

12. Lehmann, E. L. (1986). *Testing Statistical Hypotheses.* 2nd edition. New York: John Wiley & Sons.

13. McCullagh, P. and Nelder, J. A. (1983). *Generalized Linear Models*, London: Chapman and Hall.

14. Tukey, J. W. (1962). The future of data analysis, *Annals of Mathematical Statistics*, **33**, 1–67.

PART II

Chi-Squared Test

4

Partitioning the Pearson-Fisher Chi-Squared Goodness-of-Fit Statistic

G. D. Rayner

Deakin University, Geelong, Australia

Abstract: This paper presents an overview of Rayner and Best's (1989) categorised Neyman smooth goodness-of-fit score tests, along with an explanation of recent work into how these tests can be used to construct components of the Pearson-Fisher chi-squared test statistic in the presence of unknown nuisance parameters. A short simulation study examining the size of these component test statistics is also presented.

Keywords and phrases: Pearson-Fisher, chi-squared statistic decomposition, Neyman smooth test, categorized composite null hypothesis

4.1 Introduction

Pearson's (1900) chi-squared test statistic

$$X_P^2 = \sum (observed - expected)^2 / expected$$

was the first, is the most well known, and is probably the most frequently used test for goodness-of-fit. This test is essentially an *omnibus* test in that it is sensitive to a wide variety of different ways in which the data can be different to the hypothesized distribution. For example, the chi-squared test is able to detect data that differs from the hypothesized distribution in terms of any of location, scale, shape, etc. It is interesting to try and consider the *component* test statistics, sensitive only to more specific departures, that might combine to produce the chi-squared test statistic.

For a sample space broken into m classes, let N_j $(j = 1, \ldots, m)$ be the number of observations from the sample (of size $n = \sum_j N_j$) that fall into the

j-th class. If p_j is the probability of an observation falling into the j-th class under the *completely specified* hypothesized distribution, then the Pearson chi-squared test statistic

$$X_P^2 = \sum_{j=1}^{m} (N_j - np_j)^2/(np_j) \tag{4.1}$$

is asymptotically χ^2_{m-1} distributed. A value of X_P^2 larger than the appropriate critical value indicates that the observations are not distributed according to the null hypothesis.

When the hypothesized distribution is *not* completely specified (that is, q unknown parameters are present), then obtaining $\hat{p}_j$ (the estimated probability of an observation falling into the j-th class) requires somehow finding these q unknown nuisance parameters. Different methods of estimating the nuisance parameters result in different tests [see Rayner and Best (1989, chapter 2)]. For example, if the q nuisance parameters are estimated using MLE's based on the grouped data, then we obtain the Pearson-Fisher chi-squared test statistic

$$X_{PF}^2 = \sum_{j=1}^{m} (N_j - n\hat{p}_j)^2/(n\hat{p}_j). \tag{4.2}$$

This statistic is asymptotically χ^2_{m-q-1} distributed.

This paper begins in Section 4.2 with an introduction to Rayner and Best's (1989) smooth goodness-of-fit tests for categorized data, both when there are no nuisance parameters (the simple case), and when nuisance parameters are present (the composite case). Rayner and Best's (1989) approach to partitioning the Pearson-Fisher chi-squared statistic X_{PF}^2 is explained, along with its major drawback: they supply only *conditions* for obtaining the component test statistics, rather than a constructive method for doing so.

Section 4.3 outlines my recently presented method (Rayner, 2000a) for constructing an interpretable decomposition of X_{PF}^2 from the restrictions given by Rayner and Best (1989). Finally, size studies of these component test statistics obtained under several different categorisations are presented and discussed in Sections 4.4 and 4.5.

4.2 Neyman Smooth Goodness-of-Fit Tests

Rayner and Best (1989) base their smooth goodness-of-fit tests on an idea of Neyman's (1937), where the null hypothesis density is embedded in a k-parameter *Neyman smooth alternative*, such that when the vector of these k parameters $\theta = 0$, this alternative is the same as the hypothesized distribution. The score test for $\theta = 0$ versus $\theta \neq 0$, with all its desirable asymptotic properties, is then used as a goodness-of-fit test.

4.2.1 Smooth goodness-of-fit tests for categorized data

In the simple categorized case, where no unknown (nuisance) parameters are present, Rayner and Best's (1989) Neyman smooth tests of goodness-of-fit are obtained by deriving the score test statistic for the category probabilities $p = (p_1, \dots, p_m)^T$. Let

$$\pi_j = C(\theta) \exp\left\{ \sum_{i=1}^{k} \theta_i h_{i,j} \right\} p_j, \quad j = 1, \dots, m$$

where $\theta = (\theta_1, \dots, \theta_k)^T$ are $k \leq m-1$ real parameters and $C(\theta)$ is a normalizing constant (such that $\sum_j \pi_j = 1$), and for each i, $h_{i,j}$ are the values taken by a random variable H_i with $P(H_i = h_{i,j}) = \pi_j$. Testing

$$\begin{aligned} H_0 &: \theta = 0, \text{ versus} \\ H_1 &: \theta \neq 0 \end{aligned}$$

decides between the null hypothesis probabilities p and the *k-dimensional Neyman smooth alternative* probabilities $\pi = (\pi_1, \dots, \pi_m)$.

Put $N = (N_1, \dots, N_m)^T$ as the vector of observed counts in each category, $n = \sum_j N_j$ the sample size, $D = \text{diag}(p_1, \dots, p_m)$, and the $k \times m$ matrix H as having entries $h_{i,j}$. Rayner and Best (1989) calculate the score statistic for this situation as

$$S_k = (N - np)^T H^T \Sigma^{-1} H(N - np)/n$$

where Σ is the covariance matrix of the random variables H_i under the null hypothesis.

For $k = m - 1$, requiring that H satisfies $HDH^T = I_{m-1}$ and $Hp = 0$ gives $\Sigma = I_{m-1}$, where I_k is the $k \times k$ identity matrix. This means that the score test statistic S_k can be written as

$$\begin{aligned} S_{m-1} &= (N - np)^T H^T H(N - np)/n \\ &= (N - np)^T (D^{-1} - D^{-1} pp^T D^{-1})(N - np)/n \\ &= \sum_{i=1}^{m} (N_i - np_i)^2/(np_i) = X_P^2, \end{aligned}$$

the Pearson chi-squared statistic (see equation (4.1)). Defining

$$V = H(N - np)/\sqrt{n} = HN/\sqrt{n}$$

points the way towards partitioning $X_P^2 = V^T V$ into $m - 1$ asymptotically independent standard normal component test statistics

$$V_r = \sum_{j=1}^{m} h_{r,j} N_j/\sqrt{n}, \quad r = 1, \dots, m - 1.$$

The interpretation of these component test statistics depends entirely on H. In the absence of a compelling reason otherwise, Rayner and Best (1989) recommend selecting $h_{r,j}$ to be the r-th degree orthogonal polynomial evaluated at j, or $h_{r,j} = h_r(j)$. Here, the restrictions on H become

$$\sum_{j=1}^{m} h_r(j)p_j = \left\{ \begin{array}{ll} 1, & r=0 \\ 0, & r \neq 0 \end{array} \right\} \text{ and } \sum_{j=1}^{m} h_r(j)h_s(j)p_j = \left\{ \begin{array}{ll} 1, & r=s \\ 0, & r \neq s \end{array} \right\}, \quad (4.3)$$

which define these orthogonal polynomials. Rayner and Best (1989) suggest that this selection allows an *r-th order moment departure* interpretation for the r-th component test statistic V_r. That is, a significantly large value for V_r indicates that the data departs from the hypothesized distribution in terms of moments of order less than or equal to r.

Rayner and Best (1989) clearly describe how these component statistics should be used: either (1), in an EDA fashion to examine *how* the data differ from the hypothesized distribution; or (2), when *testing* for the hypothesized distribution, only the first few components along with a residual statistic (say, V_1, V_2, V_3, V_4 and $X_P^2 - V_1^2 - V_2^2 - V_3^2 - V_4^2$) should be used. It is important to avoid *post mortem* testing using what are *discovered* to be the most significant components.

4.2.2 Partitioning the Pearson-Fisher chi-squared statistic

In the composite case, q unknown (nuisance) parameters $\beta = (\beta_1, \ldots, \beta_q)$ are present, so the category probabilities must be expressed in terms of them, that is, $p = p(\beta)$. As before, Rayner and Best's (1989) Neyman smooth tests of goodness-of-fit are obtained by deriving the score test statistic for

$$\begin{aligned} H_0 &: \theta = 0, \text{ versus} \\ H_1 &: \theta \neq 0 \end{aligned}$$

using *k-dimensional Neyman smooth alternative* probabilities $\pi = (\pi_1, \ldots, \pi_m)$ which must now be expressed as

$$\pi_j(\beta) = C(\theta, \beta) \exp\left\{ \sum_{i=1}^{k} \theta_i h_{i,j}(\beta) \right\} p_j(\beta), \quad j = 1, \ldots, m.$$

For $\hat{\beta}$ the MLE's of the nuisance parameters β, then $\hat{p} = p(\hat{\beta})$ and $\hat{H} = H(\hat{\beta})$. Now the score statistic is

$$\hat{S}_k = (N - n\hat{p})^T \hat{H}^T \hat{\Sigma}^{-1} \hat{H} (N - n\hat{p})/n$$

where $\hat{\Sigma} = \Sigma(\hat{\beta})$.

Define the $q \times m$ matrix $\hat{W}$ so that $\hat{W}_{u,j} = \partial p_j / \partial \beta_u$ (for $u = 1, \ldots, q$ and $j = 1, \ldots, m$) evaluated at $\hat{p}$. For $k = m - q - 1$, requiring that $\hat{H}$ satisfies

$$\hat{H}\hat{p} = 0, \ \hat{H}\hat{W}^T = 0 \text{ and } \hat{H}\hat{D}\hat{H}^T = I_{m-q-1} \tag{4.4}$$

means that the score test statistic S_k can be written as

$$\begin{aligned} \hat{S}_{m-q-1} &= (N - n\hat{p})^T \hat{H}^T \hat{H} (N - n\hat{p})/n \\ &= (N - n\hat{p})^T \hat{D}^{-1} (N - n\hat{p})/n = X_{PF}^2. \end{aligned}$$

See Rayner and Best (1989) for details.

This is X_{PF}^2, the Pearson-Fisher chi-squared statistic (see equation (4.2)), obtained by substituting MLE's for unknown parameters in the usual Pearson chi-squared statistic. The $m - q - 1$ components of X_{PF}^2 are therefore given by

$$\hat{V} = \hat{H}(N - n\hat{p})/\sqrt{n} = \hat{H}N/\sqrt{n}$$

since $X_{PF}^2 = \hat{V}^T\hat{V}$. However, Rayner and Best (1989) do not describe how to obtain $\hat{H}$ from the restrictions in equation (4.4).

In Rayner and McAlevey (1990) and Rayner and Best (1990) some examples are provided that use this construction. Even here however, only a set of restrictions on the test statistic are provided, and although the component test statistics have evidently been calculated in these examples, the method used to do so is not discussed in sufficient detail that it can be duplicated.

4.3 Constructing the Pearson-Fisher Decomposition

There are many possible decompositions of the form outlined in Section 4.2.2. Recently [Rayner (2000a)] I demonstrated how each possible decomposition corresponds precisely with the orthonormal scheme used. This allows $m - q - 1$ component statistics $\hat{V}_{q+1}, \ldots, \hat{V}_{m-1}$ to be constructed, where $\hat{V}_r$ is a linear combination of the r-th order orthogonal polynomials defined in equation (4.3). The particular linear combination is chosen to ensure compatibility with the simple case (where no nuisance parameters are present). See Rayner (2000a) for details.

Define

$$F = \hat{D}^{-1} - \hat{D}^{-1}\hat{p}\hat{p}^T\hat{D}^{-1} - \hat{D}^{-1}\hat{W}^T \left(\hat{W}\hat{D}^{-1}\hat{W}^T\right)^{-1} \hat{W}\hat{D}^{-1}. \tag{4.5}$$

Obtain the $m - q - 1$ non-zero eigenvalues $\lambda_1, \ldots, \lambda_{m-q-1}$ and normalized corresponding eigenvectors $f_1, \ldots, f_{m-q-1}$ of F (arranged in non-decreasing eigenvalue order). Define the $m \times m$ matrix $\Lambda = \text{diag}(\lambda_1, \ldots, \lambda_{m-q-1}, 0, \ldots, 0)$. Also let $U_1 = (f_1, \ldots, f_{m-q-1})$ and $U = (U_1, U_2)$ where U_2 is an arbitrary $m \times (q+1)$ matrix of normalized column vectors chosen to be orthogonal to $f_1, \ldots, f_{m-q-1}$. With U and Λ defined in this way, $FU = U\Lambda$ and $UU^T = U^TU = I_m$ so that

$$U^TFU = \Lambda \text{ and } F = U\Lambda U^T. \tag{4.6}$$

For H_0 an $(m-q-1) \times m$ matrix, choose its rows to be values of $m-q-1$ selected orthogonal polynomials $h_r(j)$ $(0 < r < m)$ evaluated at $j = 0, \ldots, m-1$ and defined by equation (4.3) above (for calculation details see Emerson, 1968). Then define $F_0 = H_0^T H_0$ and obtain U_0 and Λ_0 from this F_0 in the same way as U and Λ are obtained from F (see equation (4.6)) to give $F_0 = U_0\Lambda_0 U_0^T$. Then let

$$M = U_0 \Lambda_0^{*1/2} \Lambda^{1/2} U^T$$

where Λ_0^* is the inverse of the first $m-q-1$ rows and columns of Λ_0 embedded in the first $m-q-1$ rows and columns of an $m \times m$ matrix of zeros. Using $\hat{H} = H_0 M$ and

$$\hat{V} = \hat{H}(N - n\hat{p})/\sqrt{n} = \hat{H}N/\sqrt{n} = H_0 MN/\sqrt{n} \tag{4.7}$$

provides the desired $m-q-1$ component test statistics of X_{PF}^2 [Rayner (2000a)]. A program using the free statistical package R [Ihaka and Gentleman (1996)] to obtain the component test statistics and p-values when testing for the normal distribution is available from the author's website [Rayner (2000b)].

4.4 Simulation Study

A small size study was carried out to assess nominally 0.5%, 1%, 5%, and 10% level tests for normality based on samples of size $n = 10$, 20, 30 and 50. For each sample size, $R = 10,000$ simulated samples were taken from the standard normal distribution. The tests examined were the component tests $\hat{V}_3, \ldots, \hat{V}_6$ under three different categorisation schemes:

(i) First, the *uncategorised* data was used, and the composite uncategorised tests for normality of Rayner and Best (1989, chapter 6) were calculated to obtain $\hat{V}_{3,u}, \ldots, \hat{V}_{6,u}$;

(ii) Then the data were *moderately categorised* into the $m_1 = 10$ classes $(-\infty, -3]$, $(-3, -2]$, $(-2, -1.5]$, $(-1.5, -0.5]$, $(-0.5, 0]$, $(0, 0.5]$, $(0.5, 1.5]$, $(1.5, 2]$, $(2, 3]$, $(3, \infty)$ to obtain $\hat{V}_{3,c_1}, \ldots, \hat{V}_{6,c_1}$ using the method in Section 4.3;

(iii) Finally, the data were *coarsely categorised* into $m_2 = 6$ classes $(-\infty, -2]$, $(-2, -1]$, $(-1, 0]$, $(0, 1]$, $(1, 2]$, $(2, \infty)$ to obtain $\hat{V}_{3,c_2}, \dots, \hat{V}_{5,c_2}$, also using the method in Section 4.3.

While implementing my method of Section 4.3, there were infrequent problems when some of the smaller sized samples were categorised into $m_2 = 6$ classes in such a way that the $\hat{W}\hat{D}^{-1}\hat{W}^T$ term in equation (4.5) was singular (1.99% of samples when $n = 10$ and 0.03% of samples when $n = 20$). This problem seldom occurred for the more moderate categorisation into $m_1 = 10$ classes, though for $n = 10$, 0.06% of all samples caused problems under both categorisations. Samples caused problems with the moderate categorisation only when they also caused problems with the coarser categorisation.

For these problem samples, only two adjacent central categories had any observations. This meant that since $\hat{\mu}$ and $\hat{\sigma}$ were being fit to the sample, the estimated scale parameter $\hat{\sigma}$ could be made sufficiently small so as to all but completely match the observed and expected counts, giving an extremely small chi-squared statistic X^2_{PF} (the largest X^2_{PF} observed for these samples was $1.21 * 10^{-13}$). However, as $\hat{\sigma} \to 0$ the $\hat{W}\hat{D}^{-1}\hat{W}^T$ term in equation (4.5) (and thus the estimated covariance matrix $\hat{\Sigma}$) becomes singular, preventing the decomposition of X^2_{PF}. Due to numerical limitations, this started to occur when $\hat{\sigma} \approx 0.1$. In practice, the estimated covariance matrix would be singular for this reason only if an unrealistic categorisation regime was used.

4.5 Results and Discussion

Figure 4.1 compares the sampling distribution of the component statistics $\hat{V}_{r,u}$, $\hat{V}_{r,c_1}$, and $\hat{V}_{r,c_2}$ $(r = 3, \dots, 6)$ for samples of size $n = 20$ taken from the standard normal distribution. The asymptotic normality of the $\hat{V}_r \sim N(0, 1)$ statistics seems to be manifesting itself already. Table 4.1 shows simulated component test sizes for various different categorisations, sample sizes and nominal significance levels using the asymptotic χ^2_1 critical values. Table 4.2 gives simulated critical values for these tests.

Carolan and Rayner (2000) observe that when performing *composite* tests for normality (where the mean and standard deviation must be estimated) the *uncategorised* smooth test statistics converge only slowly to their asymptotic χ^2_1 distribution. Tables 4.1 and 4.2 confirm that this is also the case for my *composite categorised* smooth test statistic components. The component test sizes are almost always too small and the asymptotic distribution overestimates the critical value almost every time. In practice using this asymptotic critical value would result in a p-value that is too small so normality would be rejected too often.

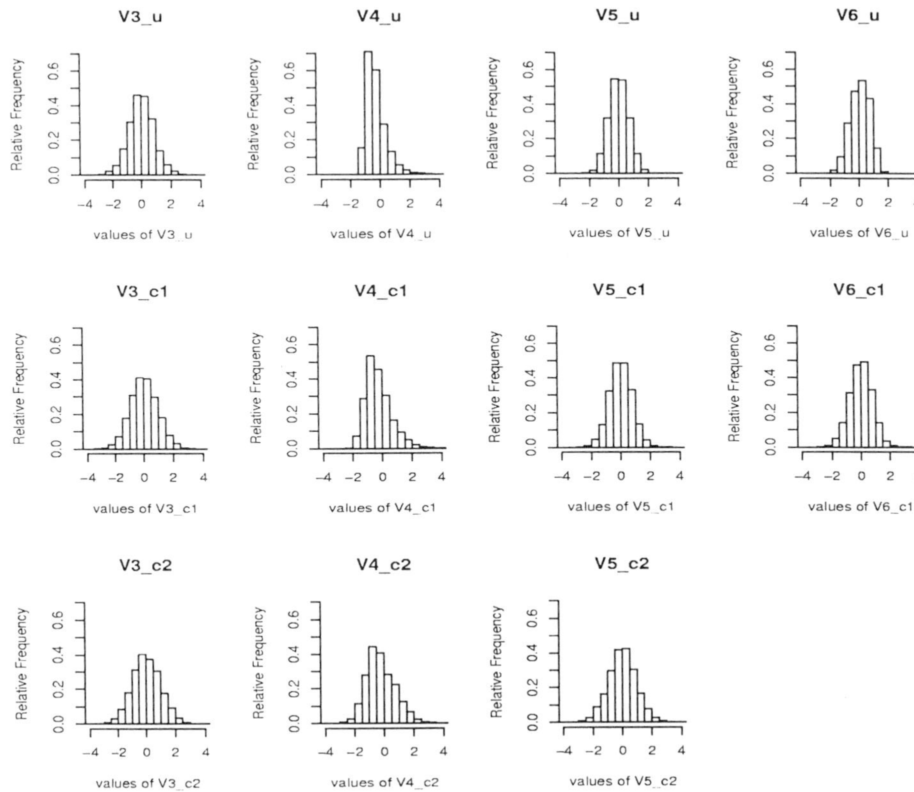

Figure 4.1: Sampling distribution of the $\hat{V}_3$, $\hat{V}_4$, $\hat{V}_5$, $\hat{V}_6$ statistics obtained from $R = 10,000$ samples of size $n = 20$ taken from the standard normal distribution. The top row is for the uncategorised data (u) using Rayner and Best's method (1989, Chapter 6), and the other rows use my categorised method (Section 4.3) with $m_1 = 10$ categories (middle, c_1) and $m_2 = 6$ categories (bottom, c_2)

Interestingly, this is in contrast to the fairly fast convergence of the distribution of $\hat{V}_{r,c}$ that has been observed for *simple* (when there are no nuisance parameters to estimate) *categorised* data. See, for example, the discussion of size in Rayner and Rayner (2000). Their discussion is concerned with testing for uniformity, but in the absence of nuisance parameters the distribution function transformation can be used to state any goodness-of-fit problem in terms of testing for uniformity.

Table 4.1: Simulated percentage test sizes using (asymptotic) χ_1^2 critical values for the component tests $\hat{V}_3^2$, $\hat{V}_4^2$, $\hat{V}_5^2$, and $\hat{V}_6^2$ under different categorisations of the data: the uncategorised method (u) [Rayner and Best (1989, Chapter 6)], and my method under two different categorisation schemes (c_1 and c_2)

Tests	Nominal size (significance level α) 0.5%				1%				5%				10%			
n	10	20	30	50	10	20	30	50	10	20	30	50	10	20	30	50
u - uncategorised																
$\hat{V}_{3,u}^2$	0.01%	0.22%	0.48%	0.55%	0.05%	0.47%	0.89%	0.99%	1.18%	3.05%	3.87%	4.33%	3.52%	6.29%	7.40%	8.48%
$\hat{V}_{4,u}^2$	0.00%	0.34%	0.60%	0.76%	0.00%	0.53%	0.84%	1.13%	0.08%	1.38%	1.90%	2.65%	0.30%	2.13%	3.02%	4.12%
$\hat{V}_{5,u}^2$	0.00%	0.02%	0.13%	0.49%	0.00%	0.02%	0.24%	0.61%	0.10%	0.35%	0.87%	1.43%	0.97%	1.35%	1.99%	2.82%
$\hat{V}_{6,u}^2$	0.00%	0.01%	0.02%	0.15%	0.00%	0.01%	0.04%	0.18%	0.01%	0.19%	0.29%	0.65%	0.55%	0.90%	1.02%	1.63%
c_1 - categorised, $m_1 = 10$																
$\hat{V}_{3,c_1}^2$	0.12%	0.25%	0.36%	0.41%	0.29%	0.60%	0.69%	0.80%	2.43%	3.72%	4.20%	4.59%	5.76%	7.76%	9.19%	9.69%
$\hat{V}_{4,c_1}^2$	0.07%	0.50%	0.56%	0.53%	0.14%	0.71%	0.85%	0.89%	0.76%	2.18%	3.00%	3.81%	2.05%	5.37%	6.53%	8.13%
$\hat{V}_{5,c_1}^2$	0.02%	0.17%	0.58%	0.89%	0.02%	0.34%	0.89%	1.24%	0.60%	1.53%	2.85%	3.53%	2.88%	3.72%	5.13%	6.35%
$\hat{V}_{6,c_1}^2$	0.04%	0.09%	0.50%	0.89%	0.11%	0.19%	0.74%	1.32%	1.15%	1.30%	2.14%	3.36%	2.84%	3.57%	4.52%	5.96%
c_2 - categorised, $m_2 = 6$																
$\hat{V}_{3,c_2}^2$	0.13%	0.23%	0.46%	0.27%	0.45%	0.58%	0.81%	0.79%	3.35%	4.27%	5.09%	4.77%	8.19%	9.54%	10.60%	10.32%
$\hat{V}_{4,c_2}^2$	0.08%	0.34%	0.31%	0.32%	0.43%	0.71%	0.69%	0.79%	2.90%	3.82%	4.05%	4.74%	4.86%	7.97%	9.48%	9.99%
$\hat{V}_{5,c_2}^2$	0.13%	0.22%	0.37%	0.65%	0.37%	0.55%	0.87%	1.00%	2.80%	3.71%	4.23%	5.05%	6.50%	8.38%	9.00%	10.42%

Table 4.2: Simulated and asymptotic (χ_1^2) critical values for $\hat{V}_3^2$, $\hat{V}_4^2$, $\hat{V}_5^2$, and $\hat{V}_6^2$ under different categorisations of the data: the uncategorised method (u) [Rayner and Best (1989, Chapter 6)], and my method under two different categorisation schemes (c_1 and c_2)

Tests	Nominal size (significance level α) 0.5%				1%				5%				10%			
n	10	20	30	50	10	20	30	50	10	20	30	50	10	20	30	50
χ_1^2	7.879	7.879	7.879	7.879	6.635	6.635	6.635	6.635	3.841	3.841	3.841	3.841	2.706	2.706	2.706	2.706
u - uncategorised																
$\hat{V}_{3,u}^2$	4.873	6.576	7.731	8.033	4.012	5.542	6.275	6.539	2.312	3.044	3.333	3.603	1.543	2.035	2.229	2.462
$\hat{V}_{4,u}^2$	2.320	6.763	8.963	9.982	1.660	4.556	5.942	7.021	0.956	1.491	1.894	2.411	0.792	1.138	1.367	1.651
$\hat{V}_{5,u}^2$	3.185	3.534	4.542	7.783	2.693	2.940	3.615	4.719	1.739	1.735	1.866	1.974	1.325	1.227	1.286	1.321
$\hat{V}_{6,u}^2$	2.751	3.134	3.337	4.011	2.379	2.641	2.747	3.255	1.573	1.610	1.699	1.777	1.185	1.196	1.210	1.228
c_1 - categorised, $m_1 = 10$																
$\hat{V}_{3,c_1}^2$	5.814	6.849	7.144	7.385	4.965	5.795	6.131	6.085	2.844	3.368	3.569	3.724	1.868	2.380	2.588	2.652
$\hat{V}_{4,c_1}^2$	4.350	7.778	8.060	7.929	3.434	5.365	6.120	6.390	2.022	2.799	3.048	3.404	1.523	2.037	2.164	2.415
$\hat{V}_{5,c_1}^2$	4.060	5.821	8.124	9.469	3.659	4.412	6.289	7.295	2.189	2.385	2.741	3.138	1.616	1.619	1.810	1.978
$\hat{V}_{6,c_1}^2$	4.729	4.898	7.774	11.673	3.942	4.069	5.825	7.646	2.083	2.334	2.545	2.985	1.420	1.612	1.720	1.890
c_2 - categorised, $m_2 = 6$																
$\hat{V}_{3,c_2}^2$	6.404	6.757	7.737	7.319	5.480	6.022	6.152	6.205	3.177	3.561	3.859	3.766	2.387	2.588	2.773	2.753
$\hat{V}_{4,c_2}^2$	5.611	6.855	7.284	7.062	4.649	6.219	5.979	6.330	2.655	3.537	3.611	3.703	2.281	2.407	2.661	2.703
$\hat{V}_{5,c_2}^2$	5.958	6.781	7.201	8.118	5.003	5.574	6.379	6.634	3.084	3.481	3.649	3.856	2.059	2.426	2.544	2.759

References

1. Carolan, A. C. and Rayner, J. C. W. (2000). A note on the asymptotic behaviour of smooth tests of goodness-of-fit, *Manuscript in preparation.*

2. Emerson, P. L. (1968). Numerical construction of orthogonal polynomials from a general recurrence formula, *Biometrics*, **24**, 695–701.

3. Ihaka, R. and Gentleman, R. (1996). R: a language for data analysis and graphics, *Journal of Computational and Graphical Statistics*, **5**, 299–314.

4. Neyman, J. (1937). "Smooth" test for goodness of fit, *Skand. Aktuarietidskr.*, **20**, 150–199.

5. Pearson, K. (1900). On the criterion that a given system of deviations from the probable in the case of a correlated system of variables is such that it can reasonable be supposed to have arisen from random sampling, *Philosophical Magazine*, 5th ser., **50**, 157–175.

6. Rayner, G. D. (2000a). Components of the Pearson-Fisher chi-squared statistic, *Statistics & Probability Letters* (submitted).

7. Rayner, G. D. (2000b). Pearson-Fisher chi-squared decomposition programs and examples [Web documents], <http://www3.cm.deakin.edu.au/~gdrayner/mypage/research/findH/>, [Accessed: 30/11/2000].

8. Rayner, J. C. W. and Best, D. J. (1989). *Smooth Tests of Goodness of Fit*, New York: Oxford University Press.

9. Rayner, J. C. W. and Best, D. J. (1990). Smooth tests of goodness of fit: an overview, *International Statistical Review*, **58**, 9–17.

10. Rayner, J. C. W. and McAlevey, L. G. (1990). Smooth goodness of fit tests for categorised composite null hypotheses, *Statistics & Probability Letters*, **9**, 423–429.

11. Rayner, G. D. and Rayner, J. C. W. (2000). Class construction in Neyman smooth categorised testing for uniformity, *Communications in Statistics—Simulation and Computation* (submitted).

5

Statistical Tests for Normal Family in Presence of Outlying Observations

Aïcha Zerbet

Université Bordeaux 2, Bordeaux, France

Abstract: A package of programs in the Fortran software is available for statistical analysis of normal data in presence of outlying observations. At first the Bol'shev test, based on the Chauvenet rule, is applied for detecting all outlying observations in a sample. After the chi-squared type test, based on the statistic of Nikulin-Rao-Robson-Moore with the Neyman-Pearson classes for grouping of data, is applied for testing of normality. We include a practical application of our software to treat the data of Milliken and the data of Daniel. The power of the test for testing normality against the family of logistic distributions, formed on the Neyman-Pearson classes, is also studied.

Keywords and phrases: Bol'shev test, chi-squared testing, Chauvenet rule, logistic distribution, maximum likelihood estimator, Neyman-Pearson classes, Nikulin-Rao-Robson-Moore statistic, normal distribution, outliers

5.1 The Chi-Squared Test of Normality in the Univariate Case

Consider the problem of testing the hypothesis H_0 according to which the distribution function of the independent identically distributed random variables $X_1, ..., X_n$ is $\Phi(\frac{x-\mu}{\sigma})$, $|\mu| < \infty$, $\sigma > 0$, where $\Phi(x)$ is the distribution function of the standard normal law.

Let $\mathbf{p} = (p_1, ..., p_r)$ be the vector of positive probability, $p_1 + ... + p_r = 1$, and define the x_i by

$$x_0 = -\infty, \quad x_r = +\infty, \quad x_j = \Phi^{-1}(p_1 + ... + p_j), \quad j = 1, ..., r-1\,;\ \varphi(x) = \Phi'(x).$$

The maximum-likelihood estimator of $\theta = (\mu, \sigma)$ is $\hat{\theta}_n = (\bar{X}_n, s_n^2)$, where

$$\bar{X}_n = \frac{1}{n}\sum_{i=1}^{n} X_i, \quad s_n^2 = \frac{1}{n}\sum_{i=1}^{n}(X_i - \bar{X}_n)^2.$$

Let

$$Y_i = \frac{X_i - \bar{X}_n}{s_n}, \quad i = 1, ..., n.$$

We note that under H_0 the statistic Y_i follows the so-called Thompson distribution with $n-2$ degrees of freedom:

$$\mathbf{P}\{Y_i \leq y\} = T_{n-2}(y) \;=\; \frac{1}{\sqrt{\pi(n-1)}}\frac{\Gamma(\frac{n-1}{2})}{\Gamma(\frac{n-2}{2})}\int_{-\sqrt{n-1}}^{y}\left(1 - \frac{y^2}{n-1}\right)^{\frac{n-4}{2}}, \quad |y| < \sqrt{n-1},$$

which does not depend on μ and σ^2 ; $T_{n-2}\left(\frac{x-\bar{X}_n}{s_n}\right)$ is the MVUE for $\Phi\left(\frac{x-\mu}{\sigma}\right)$. We consider $\nu^\star = (\nu_1^\star, ..., \nu_r^\star)$ the frequency vector obtained by grouping $Y_1, Y_2, \ldots, Y_n$ over the intervals $(x_0, x_1], (x_1, x_2], ..., (x_{r-1}, x_r)$. For testing H_0 we consider, following Drost (1988) and Zhang (1999), the statistic of Nikulin-Rao-Robson-Moore given by

$$Y_n^2 = X^2 + \frac{\lambda_1\beta^2(\nu^\star) - 2\lambda_3\alpha(\nu^\star)\beta(\nu^\star) + \lambda_2\alpha^2(\nu^\star)}{n[\lambda_1\lambda_2 - \lambda_3^2]},$$

where

$$X^2 = \sum_{i=1}^{r}\frac{(\nu_i^\star - np_i)^2}{np_i}, \; \alpha(\nu^\star) = \sum_{j=1}^{r}\frac{\nu_j^\star(\varphi(x_j) - \varphi(x_{j-1}))}{p_j}$$

$$\beta(\nu^\star) = \sum_{j=1}^{r}\frac{\nu_j^\star(-x_j\varphi(x_j) + x_{j-1}\varphi(x_{j-1}))}{p_j},$$

$$\lambda_1 = 1 - \sum_{j=1}^{r}\frac{(\varphi(x_j) - \varphi(x_{j-1}))^2}{p_j}, \; \lambda_2 = 2 - \sum_{j=1}^{r}\frac{(-x_j\varphi(x_j) + x_{j-1}\varphi(x_{j-1}))^2}{p_j},$$

$$\lambda_3 = \sum_{j=1}^{r}\frac{(\varphi(x_j) - \varphi(x_{j-1}))(-x_j\varphi(x_j) + x_{j-1}\varphi(x_{j-1}))}{p_j}.$$

Theorem 5.1.1 . *Under H_0, the statistic Y_n^2 has in the limit, when $n \longrightarrow \infty$, the chi-squared distribution with $r-1$ degrees of freedom.*

Remark 5.1.1 We recommend choosing r such that

$$r \leq \min(\frac{1}{\alpha}, \log n),$$

where α is the significance level $(0 < \alpha < 0.5)$. Note that Sturges' empirical rule suggests

$$r = 1 + \log n.$$

With this choice of r, the expected number of observations in each class is not small. If there is no alternative for H_0, then it is reasonable to choose $p_i = 1/r$. For more details, see Drost (1988), and Greenwood and Nikulin (1996).

5.1.1 Example: Analysis of the data of Milliken

We consider the data of Milliken [Linnik (1962)] from his famous experiment on the evaluation of the charge of an electron :

4.781	4.795	4.769	4.792	4.770	4.775	4.772	4.791	4.782	4.767	4.764	4.776
4.771	4.789	4.772	4.789	4.764	4.774	4.778	4.791	4.777	4.765	4.785	4.805
4.768	4.801	4.785	4.783	4.808	4.771	4.809	4.790	4.779	4.788	4.772	4.791
4.788	4.783	4.740	4.775	4.761	4.792	4.758	4.764	4.810	4.799	4.799	4.797
4.790	4.747	4.769	4.806	4.779	4.785	4.790	4.777	4.749	4.781		

We suppose, our hypothesis H_0, that the data are the realizations of a normal $N(\mu, \sigma^2)$ sample of size 58. On the basis of this data we have $\bar{X}_n = 4.7808$ and $s_n^2 = 22980 * 10^{-8}$. To test H_0, we construct a chi-squared test based on the statistic Y_n^2 with equiprobable classes. This hypothesis should be rejected if $Y_n^2 > x_\alpha$, where x_α is the α-upper quantile of the chi-squared with $(r - 1)$ degrees of freedom. If we choose, for example, $r = 3$ and $\alpha = 0.1$ then the results of computations are as follows :

x_1	x_2	$\nu_1^\star$	$\nu_2^\star$	$\nu_3^\star$	λ_1	λ_2	λ_3	X^2	Y_{58}^2
-0.4307	0.43073	20	16	22	0.2068	1.5585	0	0.9655	1.6066

So, in this case we have $x_\alpha = 4.6052$. Since the observed value of Y_{58}^2 is less than x_α then we accept the hypothesis of normality H_0.

5.2 Bol'shev Test for Outliers

We present here the test of Bol'shev [see, Bol'shev and Ubaidullaeva (1974)], based on the rule of Chauvenet (1863) for the detection of outlying observations in a set of the experimental data. We underline also here the importance of the test of Bol'shev which does not suppose that one knows the exact number of the outliers, but only that it does not exceed a maximum number s, contrary to the other tests, like those of Chauvenet (1863), Pearson and Chandrasekar (1936), Grubbs (1950) and Wilks (1963) which suppose in advance that one knows this number exactly.

5.2.1 Stages of application of the test of Bol'shev

Being given a normal $N(\mu,\sigma)$ sample $\mathbf{X} = (X_1, X_2.., X_n)$, first of all, we construct the statistics $V_1, \ldots, V_n$ which are uniform on the interval $[0, n]$, where

$$V_i = \begin{cases} n[1 - T_{n-2}(Y_i)], & \text{in the unilateral case,} \\ n[1 - T_{n-2}(|Y_i|)], & \text{in the bilateral case,} \end{cases} \qquad \text{for } i = 1, ..., n.$$

After using the statistics $V_1, \ldots, V_n$, we construct the vector of the order statistics

$$V^{(.)} = (V_{(1)}, \ldots, V_{(n)}), \quad V_{(1)} \leq \cdots \leq V_{(n)}.$$

Then, supposing α is fixed $(0 \leq \alpha \leq 0.5)$, we compute $\dfrac{V_{j(i)}}{j(i)}$ for all $X_i, i = 1, \ldots, n$, where $j(i)$ is the number of $V_{(j)}$ corresponding to X_i. If

$$\frac{V_{j(i)}}{j(i)} \leq \frac{\alpha}{\lambda},$$

then we declare that X_i is an outlier observation ($\lambda = 1$ in the unilateral case; $\lambda =$ 2 in the bilateral case).

5.2.2 Example 2: Analysis of the data of Daniel (1959)

Daniel records the results of an experiment factorille form 2^5:5 factors, each one on two levels, where the 31 contrasts, in ascending order of the absolute value, are

0	0.0281	-0.0561	-0.0842	-0.0982	0.1263	0.1684	0.1964	0.2245	-0.2526
0.2947	-0.3087	0.3929	0.4069	0.4209	0.4350	0.4630	-0.4771	0.5472	0.6595
0.7437	-0.7437	-0.7577	-0.8138	-0.8138	-0.8980	1.080	-1.305	2.147	-2.666
-3.143									

By applying the test of chi-squared for testing normality to these data ($\alpha = 0.1$), we obtain for $r = 4$ the following results :

x_1	x_2	x_3	$\nu_1^\star$	$\nu_2^\star$	$\nu_3^\star$	$\nu_4^\star$	λ_1	λ_2	λ_3	X^2	Y_{31}^2
-0.6745	0.0	0.6745	6.0	5.0	15.0	5.0	0.1394	1.2650	0.0	9.1290	11.5502

Since the value of the statistic Y_n^2 is higher than the quantile $x_\alpha = 6.2514$ of the law of chi-squared with 3 degrees of freedom corresponding to the level of significance $\alpha = 0.1$, we must reject the null hypothesis.

Carrying out the test of Bol'shev to detect the outliers on the same level of significance $\alpha = 0.1$, we conclude that the observation $X_{31} = -3.143$ is an outlier. We apply again the chi-squared test on the remainder of the data after elimination of outlier X_{31}. At the end we obtain the following results:

x_1	x_2	x_3	$\nu_1^\star$	$\nu_2^\star$	$\nu_3^\star$	$\nu_4^\star$	λ_1	λ_2	λ_3	X^2	Y_{30}^2
-0.6745	0.0	0.6745	7.0	6.0	12.0	5.0	0.1394	1.2650	0.0	3.8667	4.6484

This time, we must accept the null hypothesis as long as $Y_{30}^2 < 6.2514$, noting that earlier the rejection of the hypothesis H_0 was due to the presence of one outlier.

5.3 Power of the Chi-Squared Test

In this section, we wish to test the hypothesis H_0, against H_n, according to which the density function of X_1 is the mixture between the normal and logistic density

$$\frac{1}{\sigma}f_n(\frac{x-\mu}{\sigma}) = (1-\frac{1}{\sqrt{n}})\frac{1}{\sigma}\varphi(\frac{x-\mu}{\sigma}) + \frac{1}{\sigma\sqrt{n}}g(\frac{x-\mu}{\sigma}),$$

where g is the logistic density function and

$$g(x) = \frac{\pi\exp\{-\frac{\pi}{\sqrt{3}}x\}}{\sqrt{3}\left[1+\exp\{-\frac{\pi}{\sqrt{3}}x\}\right]^2}.$$

Since we test H_0 against H_n, it is better [see, for example, Aguirre and Nikulin (1994)] to consider the Neyman-Pearson classes I_1 and I_2 for grouping data, where

$$I_1 = \left\{x : \varphi(\frac{x-\mu}{\sigma}) \geq f_n(\frac{x-\mu}{\sigma})\right\} \quad \text{and} \quad I_2 = \left\{x : \varphi(\frac{x-\mu}{\sigma}) < f_n(\frac{x-\mu}{\sigma})\right\}$$

for which we need to solve the equation

$$\varphi(\frac{x-\mu}{\sigma}) = f_n(\frac{x-\mu}{\sigma}), \text{ which is equivalent to } \varphi(\frac{x-\mu}{\sigma}) = g(\frac{x-\mu}{\sigma}).$$

Let $a_i = a_i(\hat{\theta}_n)$, $(i = 1, ..., m)$ be the roots of the last equation and let

$$c_j(\hat{\theta}_n) = \int_{a_{j-1}(\hat{\theta})}^{a_j(\hat{\theta})} \frac{1}{s_n}\left(g(\frac{x-\bar{X}_n}{s_n}) - \varphi(\frac{x-\bar{X}_n}{s_n})\right)dx,$$

$$\int_{a_{j-1}(\hat{\theta})}^{a_j(\hat{\theta})} \frac{1}{s_n}f_n(\frac{x-\bar{X}_n}{s_n}) = p_j + \frac{1}{\sqrt{n}}c_j(\hat{\theta}_n), \ (j = 1, 2, ..., m).$$

For example, on the Figure 5.1 we have

$$I_1 = [a_1, a_2)\bigcup[a_3, a_4) \quad \text{and} \quad I_2 = \mathbf{R}^1\ I_1 = (-\infty, a_1)\bigcup[a_2, a_3)\bigcup[a_4, = \infty).$$

The limit distribution of the test-statistic is given by the following

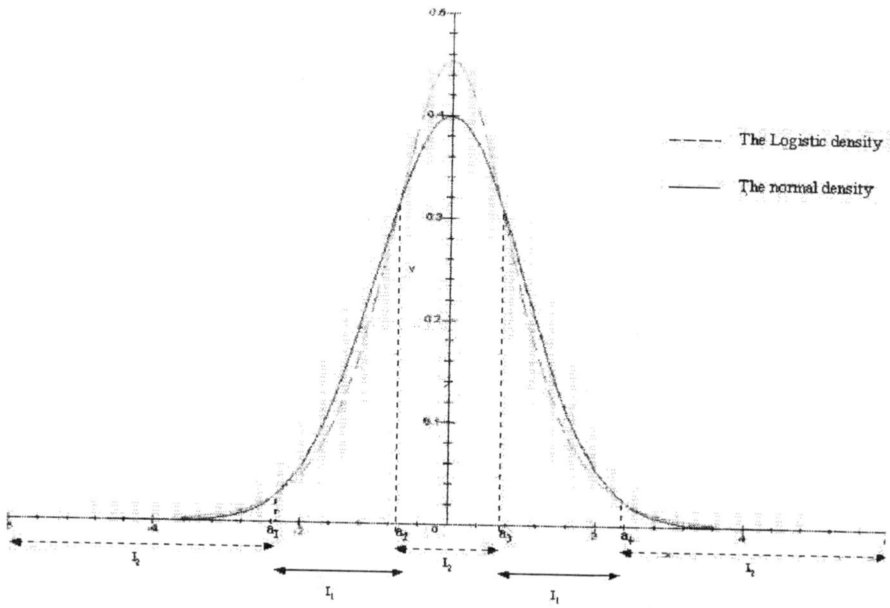

Figure 5.1: Neyman-Pearson classes

Theorem 5.3.1

$$\lim_{n\longrightarrow+\infty} \mathbf{P}\{Y_n^2(\hat{\theta})) \geq x|H_n\} = \mathbf{P}\{\chi_1^2(\lambda) \geq x\}$$

where

$$\lambda = \lambda(\theta) = \sum_{j=1}^{m} \frac{c_j^2(\theta)}{p_j} + d^T(\theta)G^{-1}(\theta)d(\theta), \;\; d_j(\theta) = \sum_{i=1}^{m} \frac{w_{ji}(\theta)c_i(\theta)}{p_i}, \; j = 1, 2,$$

the elements g_{kl} *of* G *are*

$$g_{kl} = I_{kl} - \sum_{j=1}^{m} \frac{w_{kj}w_{lj}}{p_j},$$

and

$$w_{1j} = \int_{a_{j-1}(\theta)}^{a_j(\theta)} \frac{\partial f_n(\frac{x-\mu}{\sigma})}{\partial\mu}, \quad w_{2j} = \int_{a_{j-1}(\theta)}^{a_j(\theta)} \frac{\partial f_n(\frac{x-\mu}{\sigma})}{\partial\sigma},$$

and I_{kl} *are the elements of the Fisher information matrix.*

The asymptotic distribution of the statistic Y_n^2 is the noncentral chi-squared distribution which can be approximated for small λ by a central chi-squared distribution using Patnaik's method.

Acknowledgements. This research was supported by the Conseil Regional d'Aquitane, Grant 20000204009, and by the Comite Mixte Inter Universitaire Franco-Marocain (C.M.I.F.M.), N^0 00/211/MA.

References

1. Aguirre, N. and Nikulin, M. S. (1994). Goodness-of-fit test for the family of logistic distributions, *Questiio*, **18**, 317–335.

2. Bol'shev, L. N. and Ubaidullaeva, M. (1974). Chauvenet's test in the classical theory of errors, *Theory of Probability and its Applications*, **19**, 683–692.

3. Chauvenet, W. (1863). *A Manual of Spherical and Practical Astronomy*, II, Philadelphia.

4. Daniel, C. (1959). Use of half-normal plots in interpreting factorial two-level experiments, *Technometrics*, **1**, 311-341.

5. Drost, F. C. (1988). Asymptotics for generalised chi-square goodness-of-fit tests, *In Amsterdam: Centre for Mathematics and Computer Sciences,* CWI tracts, **48**.

6. Greenwood, P. and Nikulin, M. (1996). *A Guide to Chi-squared Testing*, New York: John Wiley & Sons.

7. Grubbs, F. E. (1950). Sample criteria for testing outlying observations, *Annals of Mathematical Statistics*, **21**, 27–58.

8. Linnik, Yu. V. (1962). *The Method of Least Squares and the Principles of the Mathematical-Statistical Theory of Processing of Observations*, Second revised and augmented edition. Moscow: Gosudarstv. Izdat. Fiz. -Mat. Lit.

9. Moore, D. S. and Spruill, M. C. (1975). Unified large-sample theory of general chi-squared statistics for tests of fit, *Annals of Statistics*, **3**, 599–616.

10. Pearson, E. S. and Chandrasekar, C. (1936). The efficiency of statistical tools and a criterion for rejection of outlying observation, *Biometrika*, **28**, 308–320.

11. Rao, K. C. and Robson, D. S. (1974). A chi-squared statistic for goodness-of-fit tests within the exponential family, *Communications in Statistics*, **3**, 1139–1153.

12. Zhang, B. (1999). A chi-squared goodness-of-fit test for logistic regression models based on case-control data, *Biometrika*, **86**, 531–539.

13. Wilks, S. S. (1963). Multivariate statistical outliers, *Sankhyā, Series A*, **25**, 407–426.

6

Chi-Squared Test for the Law of Annual Death Rates: Case with Censure for Life Insurance Files

Léo Gerville-Reache

Université Bordeaux 2, Bordeaux, France

Abstract: The object of this article is to set up a chi-squared test in the case of a law of multidimensional binomial type when the observations are censured. This problem arises, for example, in the construction of a goodness-of-fit test for the law of annual death rates used in life insurance to calculate various premiums. For the non-censured case with a complete example for the Makeham law, see Gerville-Réache and Nikulin (2000).

Keywords and phrases: Chi-squared test, censure, composite hypothesis, conditional probabilities

6.1 Introduction

Consider an individual of age x (in years) at time 0 taken as the origin. Denote by T_x his residual lifetime counted from this origin which is a random variable. One characterizes the law of probability of T_x by the probability of death :

$$_tq_x = \mathbf{P}\left\{T_x \leq t\,|T_x > 0\right\}, \qquad t > 0, \quad x > 0. \tag{6.1}$$

Then, we define the annual death rate

$$q_x = \mathbf{P}\left\{T_x \leq 1\,|T_x > 0\right\}, \qquad x > 0. \tag{6.2}$$

Also we introduce the instantaneous death rate μ_x as

$$q_x = 1 - \exp\left(-\int_x^{x+1} \mu_y \, dy\right), \qquad x > 0.$$

Makeham proposed the following formula:

$$\mu_x = \alpha + \beta \cdot c^x, \qquad \alpha > 0, \quad \beta > 0, \quad c > 1. \tag{6.3}$$

So, we obtain easily the annual death rate for Makeham law as

$$q_x = 1 - \exp\left(-\alpha - \beta\frac{(c-1)}{\log(c)}c^x\right). \qquad \alpha > 0, \quad \beta > 0, \quad c > 1. \tag{6.4}$$

6.2 Chi-Squared Goodness-of-Fit Test

6.2.1 Statistics with censure

One observes over a period of a year a population of N people that are supposed to be independent compared to the property "death". One gathers the people of the same age and one obtains at the beginning of the observation ω groups of people : the group of age G_x contains l_x people of age x ($x = 0, ..., \omega - 1$). It is supposed that one does not observe people of age higher than ω. Thus, we obtain ω independent groups. In the group G_x, each individual is supposed to have the same probability q_x of dying during the year.

Let t_{ix} be the relative date of input in the study of the individual ix of age x and s_{ix} its date of output (death or disappearance of files) with $0 \leq t_{ix} < s_{ix} \leq 1$. Let I_x be the set of indices of individuals of age x whose deaths were observed during the time studied. This is the principle of censure.

At the end of the observation, one counts for each group G_x, the number of deaths D_x^*, using which we define the annual death rate observed in age x [see Gerber (1990, pp. 109–114)] as

$$Q_x^* = \frac{D_x^*}{\sum_{ix=1}^{lx}(s_{ix} - t_{ix}) + \sum_{\mathrm{I}_x}(1 - s_{ix})}. \tag{6.5}$$

We notice that if $s_{ix} = 1$ ($\forall\ ix$ not belonging to I_x) and $t_{ix} = 0$ ($\forall\ ix$), then $Q_x^* = D_x^*/l_x$ (this is the case without censure). It is supposed that censure is independent of death and age.

Let

$$\overline{s_x} = \frac{1}{l_x}\sum_{ix=1}^{lx} s_{ix}\ ,\ \overline{t_x} = \frac{1}{l_x}\sum_{ix=1}^{lx} t_{ix} \text{ and } \overline{s_{\mathrm{I}_x}} = \frac{1}{D_x^*}\sum_{\mathrm{I}_x} s_{ix};$$

we obtain then

$$Q_x^* = \frac{D_x^*}{l_x(\overline{s_x} - \overline{t_x}) + D_x^*(1 - \overline{s_{\mathrm{I}_x}})}.$$

As D_x^* is small compared to l_x and for a reasonable rate of death, $D_x^*(1-\overline{s_{\mathrm{I}_x}})$ is negligible compared to $l_x^* = \lceil l_x(\overline{s_x} - \overline{t_x})\rceil$, we note that

$$Q_x^* \approx \frac{D_x^*}{l_x(\overline{s_x} - \overline{t_x})} = \frac{D_x^*}{l_x^*}.$$

Let T_{ix} be the residual lifetime of the individual ix, then

$$q_x = \mathbf{P}\left\{T_{ix} \le 1 \,|T_{ix} > 0\right\}. \tag{6.6}$$

It is reasonable then to consider the following approximation:

$$q^*_{ix} = \mathbf{P}\left\{T_{ix} \le s_{ix} \,|T_{ix} > t_{ix}\right\} \approx (s_{ix} - t_{ix})q_x. \tag{6.7}$$

We obtain by taking an average that D^*_x follows approximatively a binomial law with parameters

$$(l_x; (\overline{s_x} - \overline{t_x})q_x).$$

For rather large D^*_x, we obtain that D^*_x follows asymptotically a normal law with parameters $(l^*_x q_x; l^*_x q_x(1 - q_x))$. Because $1 - q_x$ is very close to $1 - q_x(\overline{s_x} - \overline{t_x})$ for a reasonable censure, we deduce from it that Q^*_x is thus an efficient and asymptotically normally distributed estimator of q_x .

We obtain naturally that

$$X^2_\omega = \sum_{x=0}^{\omega-1} \frac{(D^*_x - l^*_x q_x)^2}{l^*_x q_x(1 - q_x))} \tag{6.8}$$

Follows asymptotically a chi-squared law with ω degrees of freedom.

6.2.2 Goodness-of-fit test for a composite hypothesis

One wants to test the composite hypothesis H_0 according to which q_x comes from a parametric family of functions

$$q_x = q_x(\boldsymbol{\theta}), \text{ where } \boldsymbol{\theta} = (\theta_1, \theta_2, ..., \theta_s)^t \in \Theta \subseteq R^s,$$

with $s < \omega$. In this case, we must initially build an estimator $\hat{\theta}$ of θ, and then to establish the asymptotic distribution of

$$X^2_\omega(\hat{\boldsymbol{\theta}}) = \sum_{x=0}^{\omega-1} \frac{(D^*_x - l^*_x q_x(\hat{\boldsymbol{\theta}}))^2}{l^*_x q_x(\hat{\boldsymbol{\theta}})(1 - q_x(\hat{\boldsymbol{\theta}}))}. \tag{6.9}$$

It should be noted here that the statistic $X^2_\omega(\hat{\boldsymbol{\theta}}\ )$ is different from the traditional statistic of Pearson insofar as we define a goodness-of-fit test on conditional probabilities. We saw previously that D^*_x are independent random variables which follow binomial laws with parameters l_x and $(\overline{s_x} - \overline{t_x})q_x$.

The likelihood function of $(D^*_0, D^*_1, ..., D^*_{\omega-1})^t$ is

$$L(\theta) = \prod_{x=0}^{\omega-1} C_{l_x}^{D^*_x} \left[(\overline{s_x} - \overline{t_x})q_x(\boldsymbol{\theta})\right]^{D^*_x} \left[(1 - (\overline{s_x} - \overline{t_x})q_x(\boldsymbol{\theta}))\right]^{l_x - D^*_x}. \tag{6.10}$$

One takes the estimator which maximizes the likelihood function

$$\widehat{\theta} = \arg\max_{\Theta} L(\theta). \tag{6.11}$$

Theorem 6.2.1 *Suppose that $q_x(\boldsymbol{\theta})$, a function of $\boldsymbol{\theta} = (\theta_1, \theta_2, ..., \theta_s)^t \in \Theta \subseteq R^s$, admits partial continuous derivatives. Also, let us suppose that*

$$\lim_{N^*=(\overline{s_x}-\overline{t_x})^2 N\to\infty} \frac{\max_x(l_x^*)}{\min_x(l_x^*)} \leq c \text{ (a strictly positive constant.)}$$

Let $\alpha_x = l_x^*/N^* (\approx l_x/N)$ and

$$\mathbf{F} = \left\| \sum_{x=0}^{\omega-1} \frac{\alpha_x}{q_x(\boldsymbol{\theta})(1-q_x(\boldsymbol{\theta}))} \frac{\partial q_x(\boldsymbol{\theta})}{\partial \theta_i} \frac{\partial q_x(\boldsymbol{\theta})}{\partial \theta_j} \right\|_{s\times s}.$$

Then, for $N^* \to \infty$, $(\overline{s_x} - \overline{t_x})\sqrt{N}(\hat{\boldsymbol{\theta}} - \boldsymbol{\theta})$ asymptotically follows a normal law $\mathcal{N}_s(\mathbf{0}_s, \mathbf{F}^{-1})$, and $X_\omega^2(\hat{\boldsymbol{\theta}})$ asymptotically follows a chi-squared law with $\omega - s$ degrees of freedom.

6.3 Demonstration

Let $q_x^*(\boldsymbol{\theta}) = (\overline{s_x} - \overline{t_x}) q_x(\boldsymbol{\theta})$. We have seen that the likelihood function of $(D_0^*, D_1^*, ..., D_{\omega-1}^*)^t$ is

$$L(\theta) = \prod_{x=0}^{\omega-1} C_{l_x^*}^{D_x^*} \left[q_x^*(\boldsymbol{\theta})\right]^{D_x^*} \left[1 - q_x^*(\boldsymbol{\theta})\right]^{l_x^* - D_x^*} \tag{6.12}$$

which yields

$$\ln\left[L(\theta)\right] = \sum_{x=0}^{\omega-1} \left[\ln\left[C_{l_x}^{D_x^*}\right] + D_x^* \cdot \ln\left[q_x^*(\theta)\right] + (l_x - D_x^*) \cdot \ln\left[(1 - q_x^*(\ \theta))\right]\right].$$

Under the assumption that $q_x(\theta)$, a function of $\theta = (\theta_1, \theta_{,2}, ..., \theta_s)^t \in \Theta \subseteq R^s$, admits partial continuous derivatives, a necessary condition for $\widehat{\theta}$ to be the maximum likelihood estimator of θ is

$$\frac{\partial}{\partial\theta} \ln\left[L(\theta)\right] = \mathbf{0}_s$$

$$\Leftrightarrow \quad \frac{\partial}{\partial\theta_i} \ln\left[L(\ \theta)\right] = 0, \ \forall i = 1, ..., s$$

$$\Leftrightarrow \quad \sum_{x=0}^{\omega-1} \frac{D_x^*}{q_x^*(\theta)} \cdot \frac{\partial}{\partial\theta_i} q_x^*(\theta) - \frac{(l_x - D_x^*)}{(1 - q_x^*(\theta))} \cdot \frac{\partial}{\partial\theta_i} q_x^*(\theta) = 0, \ \forall i = 1, ..., s$$

$$\Leftrightarrow \quad \sum_{x=0}^{\omega-1} \left[\frac{D_x^*}{q_x(\theta)} - \frac{(l_x - D_x^*)}{(1/(\overline{s_x} - \overline{t_x}) - q_x(\theta))}\right] \cdot \frac{\partial}{\partial\theta_i} q_x(\theta) = 0, \ \forall i = 1, ..., s$$

$$\Leftrightarrow \quad \sum_{x=0}^{\omega-1} \frac{D_x^*/(\overline{s_x} - \overline{t_x}) - l_x q_x(\theta)}{q_x(\theta)(1/(\overline{s_x} - \overline{t_x}) - q_x(\theta))} \cdot \frac{\partial}{\partial\theta_i} q_x(\theta) = 0, \ \forall i = 1, ..., s.$$

If D_x is the number of deaths which would have been observed without censure, then we have the approximate equality $D_x \approx D_x^*/(\overline{s_x} - \overline{t_x})$. Moreover, $q_x(\theta)$ is small as compared to 1. So, with a reasonable censure, we have

$$\frac{1}{(\overline{s_x} - \overline{t_x})} - q_x(\theta) \approx \frac{1}{(\overline{s_x} - \overline{t_x})}(1 - q_x(\theta)).$$

Hence,

$$\frac{\partial}{\partial \theta} \ln [L(\theta)] = \mathbf{0}_s$$

$$\Leftrightarrow \quad (\overline{s_x} - \overline{t_x}) \sum_{x=0}^{\omega-1} \frac{D_x - l_x q_x(\theta)}{q_x(\theta)(1 - q_x(\theta))} \cdot \frac{\partial}{\partial \theta_i} q_x(\theta) = 0, \ \forall i = 1, ..., s$$

$$\Leftrightarrow \quad \sum_{x=0}^{\omega-1} \frac{D_x - l_x q_x(\theta)}{q_x(\theta)(1 - q_x(\theta))} \cdot \frac{\partial}{\partial \theta_i} q_x(\theta) = 0, \ \forall i = 1, ..., s.$$

This is the equation which checks whether the estimator is a maximum likelihood estimator in the case when there is no censure as shown by in Gerville-Réache and Nikulin (2000). Therefore, we can follow the derivation of the asymptotic law of the statistic $X_\omega^2(\hat{\boldsymbol{\theta}})$ of the non-censured case simply by replacing N by N^*, l_x by l_x^*, D_x by D_x^*, and Q_x by Q_x^* in the result of Gerville-Réache and Nikulin (2000).

References

1. Chernoff, H. and Lehmann, E. L. (1954). The use of maximum likelihood estimates in chi-square test of goodness-of-fit, *Annals of Mathematical Statistics*, **25**, 579–586.

2. Courgeau, D. and Lelièvre, E. (1989). *Analyse démographique des biographies*, Edition de l'INED, Paris.

3. Gerber, H. U. (1990). *Life Insurance Mathematics*, New York: Springer-Verlag.

4. Gerville-Réache, L. and Nikulin, M. S. (2000). *Analyse statistique du modèle démogra–phique de Makeham*, Revue Roumaine de Mathématiques Pures et Appliquées (to appear).

5. Greenwood, P. E. and Nikulin M. S. (1996). *A Guide to Chi-Square Testing*, New York: John Wiley & Sons.

6. Petauton, P. (1991). *Théorie et pratique de l'assurance vie*, Paris: Dunod.

7. Voinov, V. G. and Nikulin, M. S. (1993). *Unbiased Estimators and Their Applications—1: Univariate Case*, Dordrecht: Kluwer.

PART III

Goodness-of-Fit Tests for Parametric Distributions

7

Shapiro-Wilk Type Goodness-of-Fit Tests for Normality: Asymptotics Revisited

Pranab Kumar Sen

University of North Carolina at Chapel Hill, North Carolina

Abstract: Regression type goodness-of-fit tests for normality based on L-statistics, proposed by Shapiro and Wilk (1965), are known to possess good power properties. However, due to some distributional problems (particularly for large sample sizes), various modifications have been considered in the literature. The intricacies of these asymptotics are presented here in a general setup, and in the light of that some theoretical explanations are provided for the asymptotic power of related tests.

Keywords and phrases: BLUE, degenerate U-statistics, FOE, L-statistics, regression GOF tests, SOADR

7.1 Introduction

An omnibus goodness-of-fit (GOF) test for normality, with nuisance μ, σ, location and scale parameters, is due to Shapiro and Wilk (1965). Their ingenious test is based on the regression of the observed sample order statistics on the expected values of order statistics in a sample of the same size from the standard normal distribution. Based on extensive numerical studies it has been revealed that their test has good power properties against a broad class of alternatives. However, the actual distribution of their test statistic, even under the null hypothesis, is quite involved; tables have been provided up to sample size 50, and beyond that suitable approximations have been incorporated to approximate them well (Shapiro 1998). In this context, some asymptotic distributional problems have also been discussed by De Wet and Venter (1973), though that provides very little simplicity in this respect. It has been thoroughly discussed in Shapiro (1998) that generally such asymptotic approximations entail some loss of power. One of the objectives of the current study is to focus on some asymptotics that would provide good explanation for this shortcoming.

Basically, the asymptotic distribution of the Shapiro-Wilk type of tests is governed by a second-order asymptotic distributional representation (SOADR) property that has been systematically presented in Jurečková and Sen (1996, Ch. 4). Borrowing strength from such results, Jurečková and Sen (2000) considered a general class of GOF-tests for a class of underlying distributions (including the normal one as a notable case), and proposed alternative tests based on a pair of location estimators that are first-order equivalent (FOE). In their set-up too, the SOADR results play a vital role. We refer to Jurečková, Picek, and Sen (2001) for some numerical studies relating to such robust GOF tests.

In view of the fact that such GOF tests are for composite hypotheses and the alternatives are not necessarily contiguous, there may not be an omnibus test having better power property for the entire class of alternatives. In the same vein, the usual (asymptotic) optimality properties of likelihood ratio type tests may not be tenable here. As such, we find it more convincing to stress the simplicity of the asymptotic null hypothesis distribution and other robustness properties. In this context as well, there is a basic role of the SOADR results most of which are known by this time. Along with the preliminary notion, the Shapiro-Wilk (1965) type of test statistics are presented in Section 7.2. SOADR results are presented in Section 7.3. In the light of these results, in Section 7.4, the contemplated asymptotics are discussed. The last section is devoted to some concluding remarks.

7.2 Preliminary Notion

Let $X_1, \ldots, X_n$ be n independent observations from a distribution with a probability density function (pdf) $f(x) = \sigma^{-1} f_o((x-\mu)/\sigma)$ where f_o is a pdf free from the nuisance location and scale parameters μ, and σ. We denote the standard normal pdf by $\phi(x)$ and the corresponding distribution function by $\Phi(x)$. In GOF testing for normality, we set

$$H_0 : f_o = \phi \text{ against } H_1 : f_o \neq \phi, \tag{7.1}$$

treating μ, σ as nuisance parameters. We arrange the X_i in an ascending order, and denote the order statistics by $X_{n:1} < \cdots < X_{n:n}$. Also, in a sample of size n from the standard normal distribution, we denote the order statistics by $Z_{n:i}$, $i = 1, \ldots, n$. Let then

$$\begin{aligned} m_{ni} &= E\{Z_{n:i}\}, \quad i = 1, \ldots, n; \\ v_{nij} &= Cov\{Z_{n:i}, Z_{n:j}\}, \quad i, j = 1, \ldots, n. \end{aligned} \tag{7.2}$$

Also let $\mathbf{m}_n = (m_{n1}, \ldots, m_{nn})'$ and $\mathbf{V}_n = ((v_{nij}))$ be the vector of expected order statistics and their covariance matrix respectively. Then the best linear unbiased estimator (BLUE) of σ is given by

$$\tilde{\sigma}_n = \sum_{i=1}^{n} a_{ni} X_{n:i}, \tag{7.3}$$

where $\mathbf{a}_n = (a_{n1}, \ldots, a_{nn})'$ is given by

$$\mathbf{a}_n = \{\mathbf{m}_n' \mathbf{V}_n^{-1} \mathbf{m}_n\}^{-1} (\mathbf{V}_n^{-1} \mathbf{m}_n). \tag{7.4}$$

Note that by virtue of the symmetry of Φ, we have $\mathbf{a}_n' \mathbf{1}_n = 0$, but $\mathbf{a}_n' \mathbf{a}_n$ may not be strictly equal to one. For that reason, Shapiro and Wilk (1965) considered a modified estimator wherein they let

$$\mathbf{a}_n^o = \{\mathbf{m}_n' \mathbf{V}_n^{-1} \mathbf{V}_n^{-1} \mathbf{m}_n\}^{-1/2} (\mathbf{V}_n^{-1} \mathbf{m}_n) \tag{7.5}$$

so that $\mathbf{a}_n^{o'} \mathbf{a}_n^o = 1$. We also note that the BLUE of μ is $\bar{X}_n = n^{-1} \sum_{i=1}^{n} X_i$, which is also the maximum likelihood estimator (MLE) of μ. The MLE of σ is S_n^o where

$$S_n^{o2} = n^{-1} \sum_{i=1}^{n} (X_i - \bar{X}_n)^2. \tag{7.6}$$

The Shapiro-Wilk (SW) test statistic is then expressed as

$$W_n = [\sum_{i=1}^{n} a_{ni}^o X_{n:i}]^2 / [\sum_{i=1}^{n} (X_{n:i} - \bar{X}_n)^2]. \tag{7.7}$$

Note that W_n is bounded from above by 1 and non-normality is indicated by a shift of the test statistic to lower values. For this reason, and from the SOADR results to be presented later, we shall take the SW test statistic in the equivalent form

$$W_n^* = (n-1)\{1 - W_n\}, \tag{7.8}$$

rejecting the null hypothesis for higher values of W_n^*. Later on, we shall consider a slightly different form of W_n^* that appears to have a simpler asymptotic distribution. But the present form serves better the discussion to follow. For later convenience, we set

$$X_i = \mu + \sigma e_i, \ i = 1, \ldots, n, \tag{7.9}$$

where the e_i are independent and identically distributed random variables with zero mean and unit variance; under H_0, the e_i have the standard normal distribution. Further, let $e_{n:i}$, $i = 1, \ldots, n$ be the associated order statistics, so that $X_{n:i} = \mu + \sigma e_{n:i}$, $i = 1, \ldots, n$. We also use the notation

$$\mathbf{e}_n^* = (e_{n:1}, \ldots, e_{n:n})'. \tag{7.10}$$

Note that under the null hypothesis $e^*_{n:i}$ has the expectation m_{ni}, so that rewriting W^*_n as

$$W^*_n = (n-1)\frac{\mathbf{e}^{*'}_n[\mathbf{I}_n - n^{-1}\mathbf{1}_n\mathbf{1}'_n - \mathbf{a}^o_n\mathbf{a}^{o'}_n]\mathbf{e}^*_n}{\mathbf{e}^{*'}_n[\mathbf{I}_n - n^{-1}\mathbf{1}_n\mathbf{1}'_n]\mathbf{e}^*_n}, \tag{7.11}$$

we claim that its distribution is free from the nuisance parameters, and under the null hypothesis W^*_n would be stochastically smaller (still nonnegative) [as may be verified by invoking the idempotent nature of the two matrices in the numerator and denominator of (7.11)]. Under alternatives, the expectation vector would differ from $\mathbf{m}_n$, and as a result, the expectation of the numerator quadratic form would be away from zero, so that W^*_n would be $O_p(n)$. This provides the rationality of the SW-test. In the same way, for a suitable approximation to $\mathbf{a}^o_n$, say denoted by $\mathbf{b}_n$, the corresponding analogous form of the modified SW-test statistic can be expressed as

$$W^{**}_n = (n-1)\frac{\mathbf{e}^{*'}_n[\mathbf{I}_n - n^{-1}\mathbf{1}_n\mathbf{1}'_n - \mathbf{b}_n\mathbf{b}'_n]\mathbf{e}^*_n}{\mathbf{e}^{*'}_n[\mathbf{I}_n - n^{-1}\mathbf{1}_n\mathbf{1}'_n]\mathbf{e}^*_n}, \tag{7.12}$$

where because of the symmetry of F, we must have $\mathbf{b}'_n\mathbf{1}_n = 0$, and by choice $\mathbf{b}'_n\mathbf{b}_n = 1$. In this context, we focus on the modified SW-test statistic considered by Shapiro and Francia (1972) where $\mathbf{b}_n$ is chosen as $(\mathbf{m}'_n\mathbf{m}_n)^{-1/2}\mathbf{m}_n$, and for which other simple and good approximations are available. Shapiro (1998) commented on some loss of power due to the use of such modified test statistics, and through some modifications, we will suggest some improvements for large n. A crucial factor in this context is the functional dependence of $\mathbf{b}_n$ on $\mathbf{m}_n$ through a linear transformation (as is the case with the Shapiro-Francia (1972) modification). There are some delicate asymptotic issues that involve second-order asymptotics and will be discussed later. Thus the crux of the problem is to study the distribution theory of the ratio of such quadratic forms in order statistics. We refer to Shapiro and Wilk (1965) and Shapiro (1998) for an extensive discussion of the small and moderate sample properties of their GOF test statistics and its relationship with various approximations that have been considered during the past three decades by a host of researchers. We intend to concentrate on large sample properties and on ramifications of the Shapiro-Wilk type of tests. Actually, based on some SOADR results to be presented in the next two sections, we advocate a slightly different reformulation of the SW-test statistic which has a simpler asymptotic null distribution, and this also explains how similar reformulations can be made of the modified forms of SW-test statistics, such as W^{**}_n.

7.3 SOADR Results for BLUE and LSE

We reiterate the SOADR properties of the BLUE and the LSE of the scale parameter, mostly along the lines of Jurečková and Sen (1996, 2000), and incorporate these findings in a unified treatise of the general asymptotics of the SW-type GOF tests. We present some SOADR results here, avoiding derivation by cross-reference to existing literature.

First consider the MLE (under normality) of σ^2, for which an unbiased version is

$$S_n^2 = (n-1)^{-1}\sum_{i=1}^{n}(X_i - \bar{X}_n)^2. \tag{7.13}$$

Note that if we consider a kernel $g(x,y) = \frac{1}{2}(x-y)^2$, then $\sigma^2 = E_F\{g(X_1, X_2)\}$ and

$$S_n^2 = \binom{n}{2}^{-1} \sum_{\{1\leq i<j\leq n\}} g(X_i, X_j) \tag{7.14}$$

is a U-statistic [Hoeffding (1948)] of degree 2. As such, if we use the classical Hoeffding-decomposition of U-statistics, we obtain the following representation:

$$S_n^2 - \sigma^2 = U_n^{(1)} + U_n^{(2)}, \tag{7.15}$$

where letting $g_1(x) = E[g(X,Y)|X=x)] = \frac{1}{2}[(x-\mu)^2 + \sigma^2]$,

$$U_n^{(1)} = n^{-1}\sum_{i=1}^{n} 2[g_1(X_i) - Eg_1(X_i)] = n^{-1}\sum_{i=1}^{n}[(X_i-\mu)^2 - \sigma^2)] \tag{7.16}$$

and writing $g_2(x,y) = g(x,y) - g_1(x) - g_1(y) + \sigma^2$,

$$\begin{aligned} U_n^{(2)} &= \binom{n}{2}^{-1} \sum_{\{1\leq i<j\leq n\}} g_2(X_i, X_j) \\ &= \frac{1}{n}(S_n^2 - \sigma^2) - [(\bar{X}_n - \mu)^2 - \frac{1}{n}\sigma^2]. \end{aligned} \tag{7.17}$$

Note that $U_n^{(1)}$ is an average of n centered i.i.d. random variables, so that the classical central limit theorem applies. In particular, when the d.f. F is normal, we have $\sqrt{n}U_n^{(1)}$ asymptotically normal with zero mean and variance $2\sigma^4$. Moreover, in this special case, we therefore claim that

$$(n-1)U_n^{(2)}/\sigma^2 = 1 - Z_1^2 + O_p(n^{-1/2}), \tag{7.18}$$

where $Z_1^2 = n(\bar{X}_n - \mu)^2/\sigma^2$ has a chi square distribution with 1 degree of freedom (DF), independently of S_n^2. As a result, in the normal case, we obtain the following SOADR result for S_n^2.

$$S_n^2/\sigma^2 = 1 + U_n^{(1)}/\sigma^2 + \frac{1}{n-1} - \frac{1}{n-1}Z_1^2 + O_p(n^{-3/2}). \tag{7.19}$$

It is of course possible to exploit the exact distribution of $(n-1)S_n^2/\sigma^2$ under normality (which is chi square with $n-1$ DF to obtain a parallel SOADR result (under normality), but the above representation would be more useful for alternatives as well as for the study of limiting distribution of the SW-GOF test statistic in a unified way.

Let us next consider the case of the BLUE of σ. Recall that whereas S_n^2 is an unbiased estimator of σ^2, the BLUE is for σ, and hence, its unbiasedness and SOADR properties could be affected in estimating σ^2 instead of σ. For that reason, we consider first the SOADR for the BLUE of σ given by (7.3)–(7.4) and modify that for the Shapiro-Wilk estimator of σ^2 with scores given by (7.5).

Note that $\tilde{\sigma}_n$ in (7.3) is an unbiased estimator of σ, and can be expressed as

$$\tilde{\sigma}_n = \int_{\mathbf{R}} x\psi_n(F_n(x))dF_n(x), \tag{7.20}$$

where $F_n(x)$ is the empirical d.f of the X_i, and the score function $\phi_n(u)$ converges to a smooth score function $\psi(u) = \Phi^{-1}(u)$, the quantile function of the standard normal d.f., for all $u \in (0,1)$. In fact, $\psi_n(u)$ can be approximated well by $\Phi^{-1}(nu/(n+1))$, $u \in (0,1)$; we refer to Jung (1955) for an excellent motivation and useful illustration. Although, in Jung's case there was a smooth score function while we have here the scores generated by the expected order statistics values and their covariance matrix, smoothness can be imported up to the second order terms by using the Hájek (1968) projection results and representations on the projected terms. Thus, side by side, we let $\xi(F) = \int_0^1 F^{-1}(u)\psi(u)du$ and note that when F is normal (with a scale parameter σ), $\xi(F) = \sigma$. Then, proceeding as in Section 7.4 of Sen (1981) [viz., (7.4.30) to (7.4.37)], we obtain that

$$\tilde{\sigma}_n - \xi(F) = Z_{n1} + Z_{n2}, \tag{7.21}$$

where $Z_{n1} = n^{-1}\sum_{i=1}^n \psi_1(X_i)$ with

$$\psi_1(x) = -\int_{-\infty}^{\infty} [I(y \geq x) - F(y)]\phi(F(y))dy, \ x \in \mathbf{R}, \tag{7.22}$$

so that Z_{n1} is an average of i.i.d.r.v.'s with zero mean and a finite positive variance, say γ^2. The component Z_{n2} involves second-order terms; by virtue of unbiasedness of $\tilde{\sigma}_n$ when F is normal, we have $E(Z_{n2}) = 0$, for F normal, and proceeding as in Theorem 4.3.1 of Jurečková and Sen (1996), it can be shown that $Z_{n2} = O_p(n^{-1})$. Actually, their Theorem 4.5.2 (p. 155) gives a SOADR result for $\tilde{\sigma}_n$. In passing we may remark that when F is itself normal, we have $Z_{n1} = (1/2)U_n^{(1)}$, where the latter is defined by (7.16).

Let us now look into (7.5), and define c_n by letting

$$c_n = E(\mathbf{a}_n^{o'}\mathbf{e}_n^*)^2 = E(\mathbf{a}_n^{o'}[\mathbf{I}_n - \frac{1}{n}\mathbf{1}_n\mathbf{1}_n^{'}]\mathbf{e}_n^*)^2, \tag{7.23}$$

where $\mathbf{e}_n^*$ is defined by (7.9) and (7.10). Using the Cauchy-Schwarz inequality, we have then

$$c_n \le \mathbf{a}_n^{o'}\mathbf{a}_n^o E(\mathbf{e}_n^{*'}[\mathbf{I}_n - \frac{1}{n}\mathbf{1}_n\mathbf{1}_n^{'}]\mathbf{e}_n^*) = (n-1). \tag{7.24}$$

On the other hand, by definition,

$$c_n = \frac{(\mathbf{m}_n^{'}\mathbf{V}_n^{-1}\mathbf{m}_n)^2}{\mathbf{m}_n^{'}\mathbf{V}_n^{-2}\mathbf{m}_n} + \frac{\mathbf{m}_n^{'}\mathbf{V}_n^{-1}\mathbf{m}_n}{\mathbf{m}_n^{'}\mathbf{V}_n^{-2}\mathbf{m}_n}, \tag{7.25}$$

so that the right hand side of (7.25) can not be greater than $(n-1)$. It is to be noted further that by definition

$$\mathbf{m}_n^{'}\mathbf{m}_n = n - \sum_{i=1}^{n} v_{nii} = n - \text{trace}(\mathbf{V}_n), \tag{7.26}$$

so that

$$c_n = \mathbf{m}_n^{'}\mathbf{m}_n\rho_n^2 + \frac{\mathbf{m}_n^{'}\mathbf{V}_n^{-1}\mathbf{m}_n}{\mathbf{m}_n^{'}\mathbf{V}_n^{-2}\mathbf{m}_n}, \tag{7.27}$$

where $\rho_n^2 = \{(\mathbf{m}_n^{'}\mathbf{V}_n^{-1}\mathbf{m}_n)^2\}/\{(\mathbf{m}_n^{'}\mathbf{m}_n)(\mathbf{m}_n^{'}\mathbf{V}_n^{-2}\mathbf{m}_n)\}$ is bounded from above by 1. Note further that

$$\frac{\mathbf{m}_n^{'}\mathbf{V}_n^{-1}\mathbf{m}_n}{\mathbf{m}_n^{'}\mathbf{V}_n^{-2}\mathbf{m}_n} = \rho_n^2 \frac{\mathbf{m}_n^{'}\mathbf{m}_n}{\mathbf{m}_n^{'}\mathbf{V}_n^{-1}\mathbf{m}_n}, \tag{7.28}$$

and as a result,

$$c_n = \rho_n^2\mathbf{m}_n^{'}\mathbf{m}_n\{1 + (\mathbf{m}_n^{'}\mathbf{V}_n^{-1}\mathbf{m}_n)^{-1}\}. \tag{7.29}$$

This displays the role of the score function $\mathbf{a}_n$ in relation to the constant c_n. This result will be useful in the sequel.

For some $\mathbf{b}_n$ other than $\mathbf{a}_n^o$, if we define c_n^* as in (7.23), we would have (7.24) intact, though (7.25) and (7.27) would be somewhat different. Specifically, parallel to (7.25), we would have

$$c_n^* = (\mathbf{b}_n^{'}\mathbf{m}_n)^2 + \mathbf{b}_n^{'}\mathbf{V}_n\mathbf{b}_n, \tag{7.30}$$

where $\mathbf{b}_n^{'}\mathbf{b}_n = 1$. In particular, if following Shapiro and Francia (1972), we choose $\mathbf{b}_n = (\mathbf{m}_n^{'}\mathbf{m}_n)^{-1/2}\mathbf{m}_n$, (3.18) reduces to

$$c_n^* = \mathbf{m}_n^{'}\mathbf{m}_n + \frac{\mathbf{m}_n^{'}\mathbf{V}_n\mathbf{m}_n}{\mathbf{m}_n^{'}\mathbf{m}_n}. \tag{7.31}$$

Note that by the Cauchy-Schwarz inequality,

$$(\mathbf{m}_n'\mathbf{V}_n\mathbf{m}_n)(\mathbf{m}_n'\mathbf{V}_n^{-1}\mathbf{m}_n) \geq (\mathbf{m}_n'\mathbf{m}_n)^2, \tag{7.32}$$

and hence, we obtain that

$$c_n^* - c_n = \mathbf{m}_n'\mathbf{m}_n\{(1-\rho_n^2) + \frac{\mathbf{m}_n'\mathbf{V}_n\mathbf{m}_n}{(\mathbf{m}_n'\mathbf{m}_n)^2} - \frac{\rho_n^2}{\mathbf{m}_n'\mathbf{V}_n^{-1}\mathbf{m}_n}\} \geq 0. \tag{7.33}$$

Consequently, $c_n^* \geq c_n$, and this has a bearing on the asymptotic behaviour of the Shapiro-Wilk type test. However, as we shall see later on, the basic difference comes from the SOADR results for the associated linear estimates of σ that are FOE but not necessarily BLUE. In this context, we need the following results that are presented first.

Consider the case of a linear estimate of σ for which $\mathbf{b}_n = \mathbf{B}_n\mathbf{m}_n$ for some symmetric p.d. matrix $\mathbf{B}_n$ such that $\mathbf{b}_n'\mathbf{1}_n = 0$ and $\mathbf{b}_n'\mathbf{b}_n = 1$. Note that by virtue of the expression of the variance of the BLUE estimator [Jung (1955)], we have $\mathbf{m}_n'\mathbf{V}_n^{-1}\mathbf{m}_n = 2n$, while by (7.26), we have $\mathbf{m}_n'\mathbf{m}_n = n - \text{trace}(\mathbf{V}_n)$. Therefore, we may write

$$\mathbf{m}_n'\mathbf{V}_n^{-1}\mathbf{m}_n = \gamma_n\mathbf{m}_n'\mathbf{m}_n;\ \gamma_n = \frac{2n}{n - \text{trace}(\mathbf{V}_n)}\ (> 2). \tag{7.34}$$

We rewrite this equation as follows:

$$\mathbf{m}_n'[\mathbf{V}_n^{-1} - \gamma_n\mathbf{I}_n]\mathbf{m}_n = 0. \tag{7.35}$$

Or, in other words, γ_n is an eigenvalue of $\mathbf{V}_n^{-1}$ with respect to the eigenvector $\mathbf{m}_n$. As such, we have

$$\mathbf{V}_n^{-1}\mathbf{m}_n = \gamma_n\mathbf{m}_n, \tag{7.36}$$

and repeated use of this equation leads us to the identity

$$\rho_n^2 = 1 \text{ so that } \ c_n^* = c_n = n - \text{trace}(\mathbf{V}_n), \tag{7.37}$$

where ρ_n^2 is defined after (7.27) and c_n^* refers to the specific case of the Shapiro-Francia (1972) modification. Let us consider now the general form of c_n^* when $\mathbf{b}_n = \mathbf{B}_n|bfm_n$ where $\mathbf{B}_n$ is a symmetric positive definite matrix such that $\mathbf{1}_n'\mathbf{B}_n\mathbf{m}_n = 0$ and $\mathbf{m}_n'\mathbf{B}_n\mathbf{B}_n\mathbf{m}_n = 1$. Let then $\mathbf{m}_n'\mathbf{B}_n\mathbf{m}_n = \delta_n\mathbf{m}_n'\mathbf{m}_n$, so that by similar argument we claim that δ_n is an eigenvalue of $\mathbf{B}_n$ with respect to the eigenvector $\mathbf{m}_n$. Then, it follows from the above that for the entire class of such $\mathbf{B}_n$, the corresponding c_n^* will be equal to c_n in (7.25), and by (7.29), we write this as

$$c_n = (2n+1)[n - \text{trace}(\mathbf{V}_n)]/2n. \tag{7.38}$$

This characterization is essentially based on the properties of the eigenvalues and eigenvectors. On the other hand, for some other $\mathbf{b}_n$ which can not be expressed exactly as $\mathbf{B}_n\mathbf{m}_n$, we could write $\mathbf{b}_n = \mathbf{B}_n\mathbf{m}_n^*$ for some scores vector $\mathbf{m}_n^*$ that replaces $\mathbf{m}_n$. Although, for large sample sizes, $\mathbf{m}_n$ and $\mathbf{m}_n^*$ could be very close to each other (as is the case with some other modifications of the SW-test proposed in the literature [Shapiro (1998)]), we would have an eigenvector different from $\mathbf{m}_n$, and as a result, the corresponding ρ_n^2 may not be strictly equal to one, and this in turn may also cause perturbation in the associated c_n^*. This point will be made clear in the next section.

7.4 Asymptotics for W_n^*

By virtue of (7.11), for the desired asymptotics, we may assume without any loss of generality that $\mu = 0$, $\sigma = 1$. Thus, effectively, we work with the reduced order statistics $e_{n:i}$; then S_n^2 can be defined as $(n-1)^{-1}\mathbf{e}_n^{*'}[\mathbf{I}_n - n^{-1}\mathbf{1}_n'\mathbf{1}_n]\mathbf{e}_n^*$, and we have the following SOADR result:

$$\begin{aligned} S_n^2 &= 1 + U_n^{(1)} + U_n^{(2)}, \\ U_n^{(2)} &= \frac{1}{n-1}\{1 - Z_1^2\} + O_p(n^{-3/2}), \end{aligned} \tag{7.39}$$

where $U_n^{(1)}$ is defined by (7.16) (but based on the $e_{n:i}$), and $Z_1^2 = n\bar{e}_n^2$ has the chi square distribution with 1 DF, independently of $U_n^{(1)}$. Note that S_n^2, Z_1 are jointly sufficient statistics for the normal F, so that $U_n^{(2)}$ is also a function of the sufficient statistics.

We rewrite $\tilde{\sigma}_n$, defined by (7.3) and (7.4) as $\tilde{\sigma}_n = (\mathbf{m}_n'\mathbf{V}_n^{-1}\mathbf{e}_n^*)/(\mathbf{m}_n'\mathbf{V}_n^{-1}\mathbf{m}_n)$ and note that here $\tilde{\sigma}_n$ has expectation 1. Let then

$$\begin{aligned} Q_{ni} &= E\{\tilde{\sigma}_n|e_i\} = \sum_{j=1}^{n} a_{nj}E[e_{n:j}|e_i] \\ &= \sum_{j=1}^{n} a_{nj}q_{nj}(e_i) = q_n(e_i),\ i = 1, \ldots, n, \end{aligned} \tag{7.40}$$

so that by the Hájek (1968) projection result,

$$\begin{aligned} \tilde{\sigma}_n - 1 &= \sum_{i=1}^{n}[q_n(e_i) - 1] + R_n; \\ E(\tilde{\sigma}_n - 1 - \sum_{i=1}^{n}[q_n(e_i) - 1])^2 &= E(\tilde{\sigma}_n - 1)^2 - nE(q_n(e_1) - 1)^2, \\ E(R_n^2) &= O(n^{-2}). \end{aligned} \tag{7.41}$$

Further, comparing (7.41) with (7.21) and (7.22), we conclude that

$$\sum_{i=1}^{n}[q_n(e_i) - 1] = Z_{n1}, \tag{7.42}$$

where Z_{n1} is defined as in (7.22), but now based on the e_i instead of the X_i. We write

$$\tilde{\sigma}_n - 1 = Z_{n1} + R_n, \text{ and } 2Z_{n1} = U_n^{(1)}. \tag{7.43}$$

Therefore, defining $\mathbf{a}_n^o$ as in (7.5), we have

$$\begin{aligned}(\mathbf{a}_n^{o'}\mathbf{e}_n^*)^2 &= \frac{(\mathbf{m}_n'\mathbf{V}_n^{-1}\mathbf{m}_n)^2}{\mathbf{m}_n'\mathbf{V}_n^{-2}\mathbf{m}_n}\{1 + Z_{n1} + R_n\}^2 \\ &= (\mathbf{m}_n'\mathbf{m}_n)\{1 + 2Z_{n1} + 2R_n + Z_{n1}^2 + + O_p(n^{-3/2})\}, \end{aligned} \tag{7.44}$$

where $\mathbf{m}_n'\mathbf{m}_n = n - \text{trace}(\mathbf{V}_n)$. Based on these SOADR results, we have by some simple steps

$$\begin{aligned} W_n^* &= (n-1) - (n - \text{trace}(\mathbf{V}_n))\{1 + 2Z_{n1} + 2R_n + Z_{n1}^2 - U_n^{(1)} - U_n^{(2)} \\ &\quad + U_n^{(1)2} - 2Z_{n1}U_n^{(1)} + O_p(n^{-3/2})\} \\ &= (\text{trace}(\mathbf{V}_n) - 1)\} \\ &\quad + (n - \text{trace}(\mathbf{V}_n))\{(U_n^{(1)} - 2Z_{n1}) - (U_n^{(1)} - Z_{n1})^2 \\ &\quad + (U_n^{(2)} - 2R_n)\} + O_p(n^{-1/2}) \\ &= (\text{trace}(\mathbf{V}_n) - 1)\} \\ &\quad + (n - \text{trace}(\mathbf{V}_n))\{\frac{1}{(n-1)}[1 - Z_1^2] - \frac{1}{4}U_n^{(1)2} - 2R_n\} + o_p(1), \end{aligned} \tag{7.45}$$

where in the last step, we make use of (7.43) and (7.39). It may be noted that $\frac{1}{2}nU_n^{(1)2}$ has asymptotically a chi square distribution with 1 DF, independently of Z_1^2, and hence, the distribution of W_n^*, under the null hypothesis of normality, depends on (i) the nonstochastic trace$(\mathbf{V}_n)$ (which is ≥ 1), and (ii) the stochastic Z_1^2, $nU_n^{(1)2}$ as well as the residual R_n (which is $O_p(n^{-1})$). On using the basic results of Hoeffding (1953) on expected order statistics, we claim that as $n \to \infty$,

$$\frac{1}{n}(\mathbf{m}_n'\mathbf{m}_n) \to 1, \text{ so that } \frac{1}{n}\text{trace}(\mathbf{V}_n) \to 0. \tag{7.46}$$

With this simplification, we note that

$$\begin{aligned} W_n^* &= \{\text{trace}(\mathbf{V}_n) - 1\} - W_{no}^*; \\ W_{no}^* &= \mathbf{m}_n'\mathbf{m}_n\{\tilde{\sigma}_n^2 - S_n^2\}/S_n^2. \end{aligned} \tag{7.47}$$

We shall see that W_{no}^* has a simpler asymptotic distribution. To see this, we write

$$\begin{aligned} W_{no}^* &= \frac{\mathbf{m}_n'\mathbf{m}_n}{\sigma^2}\{\tilde{\sigma}_n^2 - S_n^2\}\frac{\sigma^2}{S_n^2} \\ &= W_n^{o*} \cdot (S_n^2/\sigma^2)^{-1} \text{ say,} \end{aligned} \tag{7.48}$$

where the last factor on the right hand side is, by (7.39), $1+O_p(n^{-1/2})$. Consequently, by the Slutzky theorem, W^*_{no} and W^{o*}_n both have the same asymptotic distribution, if they have any at all. To study this, we incorporate (7.39) and (7.43), and conclude that W^{o*}_n has the asymptotic representation

$$W^{o*}_n = \mathbf{m}_n{}'\mathbf{m}_n\{2R_n + Z^2_{n1} - U^{(2)}_n\} + o_p(1). \tag{7.49}$$

At this stage, noting that for normal F, $E(R_n) = 0$, we make use of Theorem 4.5.2 of Jurečková and Sen (1996) (after verifying that for the normal distribution, the associated scores $\mathbf{a}_n$ satisfy the needed regularity conditions), and obtain a SOADR for R_n. We may write

$$2nR_n = n\int\int \tau^{(2)}(F;x,y)d[F_n(x) - F(x)]d[F_n(y) - F(y)] + o_p(1), \tag{7.50}$$

where $\tau^{(2)}(F,..)$ stands for a functional (second) derivative of the L-functional that appears in (7.20). Thus, keeping in mind the weak convergence of $\sqrt{n}(F_n - F)$ to a Brownian Bridge, we have a typical quadratic variation of a Wiener process, for which a Cramér-von Mises type asymptotic distribution, as has been suggested in De Wet and Venter (1973) and others hold. Combining all these asymptotics, we conclude that as n increases, for a normal F,

$$W^*_{no} \to_{\mathcal{D}} \sum_{k\geq 1} \lambda_k(Z^2_k - 1), \tag{7.51}$$

where the λ_k are nonnegative Fourier coefficients and the Z_k are independent standard normal deviates. Here $\lambda_1 = 1$, $\lambda_2 = 1/2$. This form is in close proximity with the form suggested by De Wet and Venter (1973) for some allied forms of W_n. We provide here a clear representation involving appropriate SOADR results that have been studied in the literature, mostly, in the past decade.

There are certain distinct advantages in writing W^*_{no} in terms of such a degenerate U-statistics. It not only provides access to the study of the asymptotic properties of the SW-test statistic, but also allows us to make use of suitable resampling plans to generate asymptotic critical levels of the SW-test statistic - a much needed task to make the SW-test applicable in large sample sizes. In this context, we may refer to Hušková and Janssen (1993) where the validity of bootstrapping for degenerate U-statistics has been critically examined, and we may adopt with some advantage their methodology to generate the asymptotic null distribution of W^*_{no}. The asymptotic distribution for W^*_n can be readily obtained from that of W^*_{no}.

Let us now discuss the case of the Shapiro-Francia (1972) modification of the SW-test statistic, considered in (7.29). If we define their estimate of σ by

$$\tilde{\sigma}^*_n = \frac{(\mathbf{m}'_n\mathbf{e}^*_n)}{(\mathbf{m}'_n\mathbf{m}_n)} \tag{7.52}$$

and note that it is unbiased but not the BLUE of σ (while $\tilde{\sigma}_n$ is BLUE), exploiting the BLUE characterization, and noting that $\tilde{\sigma}_n^*$ is asymptotically first-order efficient (FOE), we can write

$$\tilde{\sigma}_n^* = 1 + Z_{n1} + Z_{n2} + Z_{n3} \tag{7.53}$$

where Z_{n1}, Z_{n2} are defined as in (7.20) and Z_{n3} is orthogonal to Z_{n2} (and is also $O_p(n^{-1})$). Also the SOADR result applies to $\tilde{\sigma}_n^*$ as well. As such, if we proceed as in (7.19) through (7.50), we obtain the following representation for W_{no}^{**}.

$$W_{no}^{**} \to_{\mathcal{D}} \sum_{k \geq 1} \lambda_k^* (Z_k^2 - 1), \tag{7.54}$$

where $\lambda_1^* = \lambda_1 = 1, \lambda_2^* = \lambda_2 = 1/2$, and by the convolution property, $\lambda_k^* \geq \lambda_k$, $\forall k \geq 3$. Comparing (7.50) and (7.54), we may gather that when F is normal (i.e., under H_0),

$$W_n^* \text{ is stochastically more dispersed than } W_n^{**}. \tag{7.55}$$

This explains why in considering a modified form of the SW-test statistic, there is a need to adjust for the critical values, and doing that might make those modifications more competitive with the SW-test itself. Of course, speaking for the power properties, even for large sample sizes, we need to pay adequate attention to the intricate asymptotics, and these are considered in the next section.

The asymptotics for the Shapiro-Francia test in general go over to other cases where $\mathbf{b}_n$ are related to $\mathbf{m}_n$ by suitable matrix multiplication, as has been discussed after (7.37). However, we need to assume that they are FOE and admit SOADR. If we consider other $\mathbf{b}_n$ that are related to various approximations to $\mathbf{m}_n$ and $\mathbf{V}_n$, as has been discussed in Shapiro (1998, pp. 481–482), the eigenvalues will be different, though quite close to the ones discussed earlier. Further, because of the fact that in such a case, ρ_n^2 defined in an analogous way, will be typically less than one (though quite close to 1), while $n^{-1}(\mathbf{m}_n'\mathbf{m}_n) \to 1$, we could see that there will be additional variation due to ρ_n^2 being less than one, and more so due to other terms that appear in the second-order expansion. As such, the SW-approximation may not generally apply here very satisfactorily. Although their critical values can be estimated by similar resampling methods, because of more variable second-order terms, their distribution will be more dispersed, and as a result, there could be some loss of power.

7.5 Asymptotics Under Alternatives

The basic strength of the SW-test relates to its suitability for a very broad class of alternatives (to the null hypothesis that F is normal), and in this context, the location and scale parameters are treated as nuisance parameters. As such, this class of alternatives does not include the conventional location-scale family. In addition, such alternatives need not be of the contiguous type in the sense that they may not be brought closer to the null hypothesis by considering a suitable sequence of local parametric functions. For example, in testing for F normal (with nuisance mean and variance), we may be interested in alternatives that F is logistic, or Laplace or some other d.f., admitting nuisance location and scale parameters. In order to exploit fully the properties of BLUE, a minimal requirement for this class of alternatives is that the associated d.f.'s admit finite second order moments so that the expected order statistics and the associated dispersion matrix are defined properly. Also, the appropriateness of the SW-type of tests when the null hypothesis relates to F being some other nonnormal d.f. needs to be assessed properly. In that setup, though it might be possible to obtain BLUE of the scale parameter, the sample variance may not be generally the efficient estimator (MLE) of the (squared) scale parameter. In that way, a stipulated BLUE estimator (specifically for the assumed F) would be FOE but the conventional sample standard deviation may not be FOE. As a result, the asymptotic distribution (and even the finite sample ones, tabulated extensively for the normal case) could be quite different. Faced with this situation, Jurečková and Sen (2000) considered an alternative approach based on a pair of robust estimators of location parameter which are FOE under the assumed F (under H_0), and developed GOF tests that are usable for a broad class of F including the normal distribution as an important member. The numerical work presented in Jurečková, Picek, and Sen (2001) casts light on the competitive nature of such robust tests (though for testing for normality, as expected, the SW-test performs slightly better).

Let us concentrate on the nature of asymptotics that crops up in the study of the SW-type GOF-tests for normality. Let us denote the distribution under the alternative by $G(x)$ and its pdf by $g(x)$; we assume that G admits finite second moment so that the order statistics have finite expectations and variance covariance terms. Also, without any loss of generality, we let the location parameter be 0 and the scale parameter equal to 1. Compared to (7.2), we now introduce the order statistics in a sample of size n from this standardized G by $Z^*_{n:i}$, $i = 1, \ldots, n$. Let then $\boldsymbol{\gamma}_n = (\gamma_{n1}, \ldots, \gamma_{nn})'$ where

$$\gamma_{ni} = E\{Z^*_{n:i}\}, \ i = 1, \ldots, n. \tag{7.56}$$

Let us recall the notation introduced in (7.3) through (7.7). Note that under the alternative hypothesis, $\tilde{\sigma}_n$ is an unbiased estimator of

$$\delta_n = \frac{(\mathbf{m}_n' \mathbf{V}_n^{-1} \gamma_n)}{(\mathbf{m}_n' \mathbf{V}_n^{-1} \mathbf{m}_n)}. \tag{7.57}$$

On the other hand, S_n^2, defined by (7.13), is an unbiased and consistent estimator of the second moment (say ν_2) of the standardized d.f. G (whose scale parameter is taken as 1); ν_2 may not necessarily be equal to 1; take for example, G logistic with unit scale parameter. Further, using the results of Hoeffding (1953) on expected order statistics, we claim that

$$\frac{1}{n}(\gamma_n' \gamma_n) = \nu_2 - o(1), \text{ as } n \to \infty. \tag{7.58}$$

With these results in hand, let us define

$$\Delta_n = \frac{(\mathbf{m}_n' \mathbf{V}_n^{-1} \gamma_n)^2}{(\mathbf{m}_n' \mathbf{V}_n^{-2} \mathbf{m}_n)(\gamma_n' \gamma_n)}. \tag{7.59}$$

Using the Cauchy-Schwarz inequality, it is easy to show that

$$0 \le \Delta_n \le 1, \ \forall n, \text{ and } G, \tag{7.60}$$

while using the Hoeffding (1953) results, along with the fact [Jung (1955)] that $\mathbf{V}_n^{-1}\mathbf{m}_n$ can be well approximated by $\mathbf{m}_n$, we conclude that as n increases,

$$\Delta_n \to \frac{(\int_0^1 F_o^{-1}(u) G^{-1}(u) du)^2}{\nu_2} = \Delta, \text{ say}, \tag{7.61}$$

where F_o is the standard normal d.f., so that its second moment is equal to 1, and by definition, $\nu_2 = \int_0^1 (G^{-1}(u))^2 du$. As a result, it follows by some standard steps that under the alternative that G is the true d.f.,

$$W_n \to_P \Delta, \text{ as } n \to \infty. \tag{7.62}$$

On the other hand, from the results in Section 7.4, we conclude that under the null hypothesis of normality,

$$W_n = 1 - (n-1)^{-1} W_n^* \to_P 1, \text{ as } n \to \infty. \tag{7.63}$$

As a result we conclude that the SW-test for normality is consistent for the entire class of alternatives for which Δ is strictly less than one. This is the case when the two quantile functions $\Phi^{-1}(p)$ and $G^{-1}(p)$ (for $p \in (0,1)$) do not coincide for all values of p. In that way, the domain of consistency of the SW-test includes all nonnormal distributions admitting finite second order moments (so that the normal BLUE of σ converges to a limit other than the scale parameter of such a distribution). This is certainly a very strong result in the sense that it includes separable families of alternatives in a very natural way, and it includes mixture models also in the same vein. For example, against F normal, we might be interested in the set of alternatives that it is contaminated d.f., namely,

$$F(x) = (1-\eta)\Phi(x) + \eta H(x), \tag{7.64}$$

where $\eta > 0$ is small, and $H(x)$ has a heavier tail; it could be a normal d.f. with a larger variance or even some other one like the Laplace that has a heavier tail than a normal one. It is also possible to treat η as a sequence converging to 0, and in that way local contamination models are also contemplated in this setup. However, the consistency is a minimum requirement for any GOF test, and it should not be overemphasized. There may not be a unique GOF test for normality with power-optimality against such a broad class of alternatives. For this reason, Jurečková and Sen (2000) discussed such asymptotic power pictures for other tests.

As regards the consistency property of modified SW-type tests (as discussed in earlier sections), the picture is the same. It is only with respect to power properties there could be some difference. In view of (7.55), there is a need to calibrate the critical levels of such modified test statistics (such as W_n^{**}) as otherwise for local alternatives there might not be a perceptible difference particularly when the sample size is large. However, if we consider a fixed alternative (that is more appropriate in the present context), then the rate at which the power function goes to one [in the Bahadur (1960) sense] might be different. The basic difficulty for such a study stems from the fact that due to their complicated null hypothesis distributions the exact Bahadur-slopes for such statistics are not that simple to formulate, while the approximate Bahadur-slope comparisons are known to be deficient in certain respects. As such, the empirical evidence acquired from extensive numerical studies made so far [viz., Shapiro (1998)] should be used as a stepping stone for further comparative studies.

Acknowledgements. This work was supported by the US-Czech Collaborative Research Grant NSF INT-96000518. Thanks are due to the organizers of the GOF 2000 Conference in Paris, France for the invitation and support for the presentation of the manuscript in the meeting.

References

1. Bahadur, R. R. (1960). Stochastic comparison of tests. *Annals of Mathematical Statistics*, **31**, 276–295.

2. De Wet, T. and Venter, J. H. (1973). Asymptotic distributions of quadratic forms with application to test of fit, *Annals of Statistics*, **1**, 380–387.

3. Hájék, J. (1968). Asymptotic normality of simple linear rank statistics under alternatives, *Annals of Mathematical Statistics*, **39**, 325–346.

4. Hoeffding, W. (1948). On a class of statistics with asymptotically normal distribution, *Annals of Mathematical Statistics*, **19**, 293–325.

5. Hoeffding, W. (1953). On the distribution of the expected values of the order statistics, *Annals of Mathematical Statistics*, **24**, 93–100.

6. Hušková, M. and Janssen, P. (1993). Consistency of the generalized bootstrap for degenerate U-statistics, *Annals of Statistics*, **21**, 1811–1823.

7. Jung, J. (1955). On linear estimates defined by a continuous weight function, *Ark. Mat.* Band 3, no. 15, 199–209.

8. Jurečková, J., Picek, J., and Sen, P. K. (2001). A goodness-of-fit test with nuisance parameters: numerical performance, *Journal of Statistical Planning and Inference*, (to appear).

9. Jurečková, J. and Sen, P. K. (1996). *Robust Statistical Procedures: Asymptotics and Interrelations.* New York: John Wiley & Sons.

10. Jureķová, J. and Sen, P. K. (2000). Goodness-of-fit tests and second-order asymptotic relations, *Journal of Statistical Planning and Inference*, **91**, in press.

11. Sen, P. K. (1981). *Sequential Nonparametrics: Invariance Principles and Statistical Inference*, New York: John Wiley & Sons.

12. Shapiro, S. S. (1998). Distribution assessment, In *Handbook of Statistics, Vol. 17: Order Statistics: Applications* (Eds., N. Balakrishnan and C. R. Rao), pp. 475–494, Amsterdam: Elsevier.

13. Shapiro, S. S. and Francia, R. S. (1972). Approximate analysis of variance test for normality, *Journal of the American Statistical Association*, **67**, 215-225.

14. Shapiro, S. S. and Wilk, M. B. (1965). An analysis of variance test for normality (complete samples), *Biometrika*, **52**, 591–611.

8

A Test of Exponentiality Based on Spacings for Progressively Type-II Censored Data

N. Balakrishnan, H. K. T. Ng, and N. Kannan

McMaster University, Hamilton, Ontario, Canada
McMaster University, Hamilton, Ontario, Canada
The University of Texas at San Antonio, San Antonio, Texas

Abstract: There have been numerous tests proposed in the literature to determine whether or not an exponential model is appropriate for a given data set. These procedures range from graphical techniques, to tests that exploit characterization results for the exponential distribution. In this article, we propose a goodness-of-fit test for the exponential distribution based on general progressively Type-II censored data. This test based on spacings generalizes a test proposed by Tiku (1980). We derive the exact and asymptotic null distribution of the test statistic. The results of a simulation study of the power under several different alternatives like the Weibull, Lomax, Lognormal and Gamma distributions are presented. We also discuss an approximation to the power based on normality and compare the results with those obtained by simulation. A wide range of sample sizes and progressive censoring schemes have been considered for the empirical study. We also compare the performance of this procedure with two standard tests for exponentiality, viz. the Cramer-von Mises and the Shapiro-Wilk test. The results are illustrated on some real data for the one- and two-parameter exponential models. Finally, some extensions to the multi-sample case are suggested.

Keywords and phrases: Exponential distribution, goodness-of-fit, lifetime

8.1 Introduction

The exponential distribution is one of the most widely used life-time models in the areas of life testing and reliability. The volume by Balakrishnan and Basu (1995) [see also Johnson, Kotz, and Balakrishnan (1994, Chapter 19)] provides an extensive review of the genesis of the distribution and its properties, including several characterization results. Because of its wide applicability and its relations to other distributions like the gamma and Weibull, there have been numerous tests proposed in the literature to determine whether or not an exponential model is indeed appropriate for a given sample.

The history of goodness-of-fit tests originated with the seminal paper by Karl Pearson in 1900 on the chi-squared test. Tests based on the empirical distribution function (EDF) like the Kolmogorov-Smirnov, Cramer-von Mises and their variants are applicable for testing the hypothesis that the random sample comes from some arbitrary distribution. The properties of these "omnibus" tests under various scenarios have been investigated by several authors; see D'Agostino and Stephens (1986) for a detailed bibliography. These tests are intuitive, and easily modified in the event of censored data.

However, if the investigator is interested in testing whether a particular model like the normal or exponential is appropriate, it may be appropriate to use the properties of the underlying distribution to derive a more specific (hopefully, more powerful) test. For the exponential distribution, one can exploit the fact that the hazard function is constant or that the logarithm of the survival function is linear. Shapiro (1995) and Stephens (1986) provide a fairly extensive review of the literature on tests for the exponential distribution. Spinelli and Stephens (1987) discuss tests for the two-parameter exponential distribution when the parameters are unknown.

In this article, we propose a test for exponentiality based on spacings under Progressive Type-II censoring. In Section 8.2, we will briefly describe the idea of Progressive Type-II censoring and some basic results. In Section 8.3, we will propose a test statistic for exponentiality based on spacings. We will derive the exact and asymptotic null distributions of the test statistic. In Section 8.4, we present results of a simulation study to investigate the power of this test under several different alternatives. We also discuss an approximation to the power and compare the approximate values with those obtained by simulations. In Section 8.5, we examine two standard tests (Cramer-von Mises A^2 and the Shapiro-Wilk W_E) for exponentiality discussed extensively in the literature, and compare the power performance of all three procedures. Section 8.6 considers tests for the two-parameter exponential distribution. We illustrate the test procedures proposed here using some numerical examples in Section 8.7. Section 8.8 discusses the multi-sample extension of this procedure. Finally, we conclude with some comments and suggestions for further research in Section 8.9.

8.2 Progressive Censoring

Conventional Type-I and Type-II censoring schemes do not allow for removal of units at points other than the terminal point of the experiment. We consider a more general censoring scheme called *Progressive Type-II Right censoring* as follows: Consider an experiment in which n units are placed on a life test. At the time of the first failure, R_1 units are randomly removed from the remaining $n-1$ surviving units. At the second failure, R_2 units from the remaining $n-2-R_1$ units are randomly removed. The test continues until the mth failure. At this time, all remaining $R_m = n-m-R_1-R_2-\ldots-R_{m-1}$ units are removed. The $R_i^{\prime}s$ are fixed prior to the study. If $R_1 = R_2 = \ldots = R_m = 0$, we have $n = m$ which corresponds to the complete sample. If $R_1 = R_2 = \ldots = R_{m-1} = 0$, then $R_m = n-m$ which corresponds to the conventional Type-II right censoring scheme. The idea of progressive censoring is due to Cohen (1963, 1966).

We may introduce a further generalization: Suppose the failure times of the first r units are not observed. At the $(r+1)^{th}$ failure, R_{r+1} units are randomly removed. At successive failures, we remove units randomly as before. This is called *General Progressive Type-II censoring.* If $r = 0$, this is the scheme outlined above. All the procedures outlined in this paper may be easily modified to deal with this general case. However, for simplicity we will consider only Progressively Type-II right censored data.

Let $X_{i:m:n}^{(R_1,\ldots,R_m)}$ denote the ith failure time. These failure times are referred to as progressively Type-II right censored order statistics. The joint probability density function of these order statistics may be written using probability arguments. However, the marginal distributions do not have the same simple form as in the case of the usual order statistics. For an exhaustive list of references and further details on progressive censoring, the reader may refer to the recent book by Balakrishnan and Aggarwala (2000).

The standard goodness-of-fit tests can be easily modified to deal with Type-I and Type-II censored data. However, the loss of information, especially in the tails of the distribution results in a significant loss of discriminatory power in many cases. In the next section, we propose a test for exponentiality using spacings under progressive censoring and examine its properties.

8.3 Test for Exponentiality

Let us assume that the failure times have an exponential distribution with probability density function (p.d.f.)

$$f(x;\sigma) = \frac{1}{\sigma}\exp\left(-\frac{x}{\sigma}\right), \qquad x > 0, \tag{8.1}$$

and with cumulative distribution function (c.d.f.)

$$F(x;\sigma) = 1 - \exp\left(-\frac{x}{\sigma}\right), \qquad x > 0, \tag{8.2}$$

where $\sigma > 0$ is an unknown scale parameter.

Let $X_{1:m:n}^{(R_1,\ldots,R_m)}, X_{2:m:n}^{(R_1,\ldots,R_m)}, \ldots, X_{m:m:n}^{(R_1,\ldots,R_m)}$ denote a progressively Type-II right censored sample. We would like to test whether such a sample comes from an exponential distribution with p.d.f. (8.1) with σ being unknown. In other words, we want to test the hypotheses

$$\begin{aligned} & H_0 : X \stackrel{d}{=} Exp(\sigma) \\ \text{against} \quad & H_1 : X \stackrel{d}{\neq} Exp(\sigma). \end{aligned} \tag{8.3}$$

For convenience, we will suppress the censoring scheme in the notation of the $X'_{i:m:n}$s.

Define the normalized spacings $S_1, S_2, \ldots, S_m$ as

$$\begin{aligned} S_1 &= nX_{1:m:n}^{(R_1,\ldots,R_m)}, \\ S_2 &= (n - R_1 - 1)(X_{2:m:n}^{(R_1,\ldots,R_m)} - X_{1:m:n}^{(R_1,\ldots,R_m)}), \\ S_3 &= (n - R_1 - R_2 - 2)(X_{3:m:n}^{(R_1,\ldots,R_m)} - X_{2:m:n}^{(R_1,\ldots,R_m)}), \\ \ldots & \quad \ldots \\ S_m &= (n - R_1 - \ldots - R_{m-1} - m + 1)(X_{m:m:n}^{(R_1,\ldots,R_m)} - X_{m-1:m:n}^{(R_1,\ldots,R_m)}). \end{aligned} \tag{8.4}$$

If the underlying distribution is exponential, $S_1, S_2, \ldots, S_m$ defined in (8.4) are all independent and identically distributed as exponential with scale parameter σ; see Balakrishnan and Aggarwala (2000) for details.

Consider the test statistic given by

$$T = \frac{\sum\limits_{i=1}^{m-1}(m-i)S_i}{(m-1)\sum\limits_{i=1}^{m} S_i}. \tag{8.5}$$

The numerator of the test statistic is a linear combination of the spacings with decreasing weights, and the denominator is the sum of the spacings. The test statistic is clearly scale invariant, with small and large values of T leading to the rejection of H_0. The statistic T was suggested by Tiku (1980) for complete and doubly Type-II censored samples. Balakrishnan (1983) studied the power of the test against a variety of alternatives, and showed that the test (for complete samples) performs well compared to standard tests in the literature.

8.3.1 Null distribution of T

To derive the null distribution of the test statistic T, we first write T in the following form:

$$\begin{aligned} T &= \frac{1}{m-1}\left[\frac{S_1}{\sum_{i=1}^m S_i} + \frac{S_1+S_2}{\sum_{i=1}^m S_i} + \frac{S_1+S_2+S_3}{\sum_{i=1}^m S_i} + \ldots + \frac{S_1+\ldots+S_{m-1}}{\sum_{i=1}^m S_i}\right] \\ &= \frac{1}{m-1}\sum_{j=1}^{m-1} Z_j, \end{aligned} \tag{8.6}$$

where

$$Z_j = \frac{\sum_{i=1}^j S_i}{\sum_{i=1}^m S_i}, \quad j = 1, 2, \ldots, m-1.$$

Since S_1, S_2, ..., S_m are all independent and identically distributed as exponential with scale parameter σ, the joint p.d.f. of S_1, S_2, ..., S_m is given by

$$f_{S_1,\ldots,S_m}(s_1,\ldots,s_m) = \frac{1}{\sigma^m}\exp\left(-\frac{1}{\sigma}\sum_{i=1}^m s_i\right), \quad s_i > 0, i = 1, 2, \ldots, m.$$

Consider the transformation

$$\begin{aligned} Z_j &= \frac{\sum_{i=1}^j S_i}{\sum_{i=1}^m S_i}, \quad j = 1, 2, \ldots, m-1, \\ Z_m &= \sum_{i=1}^m S_i. \end{aligned}$$

We then have

$$S_1 = Z_1 Z_m$$

$$\begin{aligned} S_2 &= Z_2 Z_m - Z_1 Z_m \\ S_3 &= Z_3 Z_m - Z_2 Z_m \\ \cdots & \quad \cdots \\ S_{m-1} &= Z_{m-1} Z_m - Z_{m-2} Z_m \\ S_m &= Z_m - Z_{m-1} Z_m. \end{aligned}$$

The Jacobian of this transformation is

$$|J| = \det \begin{bmatrix} z_m & 0 & 0 & 0 & \cdots & \cdots & 0 & z_1 \\ -z_m & z_m & 0 & 0 & \cdots & \cdots & 0 & z_2 - z_1 \\ 0 & -z_m & z_m & 0 & \cdots & \cdots & 0 & z_3 - z_2 \\ \vdots & & \ddots & \ddots & \ddots & & \vdots & \vdots \\ \vdots & & & \ddots & \ddots & \ddots & \vdots & \vdots \\ \vdots & & & & \ddots & \ddots & 0 & \vdots \\ 0 & \cdots & 0 & \cdots & 0 & -z_m & z_m & z_{m-1} - z_{m-2} \\ 0 & \cdots & 0 & \cdots & \cdots & 0 & -z_m & 1 - z_{m-1} \end{bmatrix}$$

which can be shown to equal z_m^{m-1}.

Therefore, the joint density of Z_1, Z_2, ..., Z_m is given by

$$\begin{aligned} & f_{Z_1,\ldots,Z_m}(z_1, \ldots, z_m) \\ & = \frac{1}{\sigma^m} e^{-z_m/\sigma} z_m^{m-1}, \qquad 0 < z_1 < z_2 < \ldots < z_{m-1} < 1, z_m > 0, \end{aligned}$$

which yields the joint density of Z_1, Z_2, ..., Z_{m-1} to be

$$\begin{aligned} f_{Z_1,\ldots,Z_{m-1}}(z_1, \ldots, z_{m-1}) &= \int_0^\infty \frac{1}{\sigma^m} e^{-z_m/\sigma} z_m^{m-1}\, dz_m \\ &= (m-1)!, \qquad 0 < z_1 < z_2 < \ldots < z_{m-1} < 1. \end{aligned}$$

The joint distribution of Z_1, Z_2, ..., Z_{m-1} is thus the same as the joint distribution of the $(m-1)$ order statistics (say, $U_{(1)}, \ldots, U_{(m-1)}$) obtained from a random sample of size $(m-1)$ from the Uniform (0,1) distribution (say, $U_1, \ldots, U_{m-1}$). Hence, we immediately have

$$(m-1)T \stackrel{d}{=} \sum_{i=1}^{m-1} Z_i \stackrel{d}{=} \sum_{i=1}^{m-1} U_{(i)} \stackrel{d}{=} \sum_{i=1}^{m-1} U_i.$$

This implies that the null distribution of the test statistic T is exactly the same as the average of $m-1$ i.i.d. Uniform(0,1) random variables. Therefore, the null distribution of T tends to normality very rapidly as m increases. It is readily verified that the mean of the limiting distribution is $E(T) = \frac{1}{2}$ and variance $Var(T) = \frac{1}{12(m-1)}$.

Remark 8.3.1 The above expressions of $E(T) = \frac{1}{2}$ and $\text{Var}(T) = \frac{1}{12(m-1)}$ can also be derived by taking expectations on both sides of

$$T(m-1)\sum_{i=1}^{m} S_i = \sum_{i=1}^{m-1}(m-i)S_i$$

and using Basu's theorem with the facts that $2\sum_{i=1}^{m} S_i/\sigma$ is distributed as χ^2_{2m}, $2S_i/\sigma$ is distributed as χ^2_2, and that the ancillary statistic T is independent of the complete sufficient statistic $\sum_{i=1}^{m} S_i$.

8.4 Power Function Approximation and Simulation Results

8.4.1 Approximation of power function

The power function of the test is given by

$$\Pr\left(T \geq \frac{1}{2} + z_{\alpha/2}\sqrt{\frac{1}{12(m-1)}}\,\middle|\, H_1\right) + \Pr\left(T \leq \frac{1}{2} - z_{\alpha/2}\sqrt{\frac{1}{12(m-1)}}\,\middle|\, H_1\right).$$

To compute the power under different alternatives, we need to compute probabilities of the form $\Pr(T \geq c)$, for c being some constant. Since $(m-1)\sum\limits_{i=1}^{m} S_i$ is a positive quantity, we may write

$$\begin{aligned}\Pr(T \geq c) &= \Pr\left[\frac{\sum\limits_{i=1}^{m-1}(m-i)S_i}{(m-1)\sum\limits_{i=1}^{m} S_i} \geq c\right] \\ &= \Pr(L \geq 0),\end{aligned}$$

where

$$L = \sum_{i=1}^{m-1}(m-i)S_i - c(m-1)\sum_{i=1}^{m} S_i.$$

From (8.4), L may be written as a linear combination of the progressively Type-II right censored order statistics as

$$\begin{aligned}L &= \sum_{i=1}^{m-1}(m-i)S_i - c(m-1)\sum_{i=1}^{m} S_i \\ &= \sum_{i=1}^{m} a_i X_{i:m:n}^{(R_1,\ldots,R_m)},\end{aligned}$$

where

$$\begin{aligned}
a_i &= [(m-i)-c(m-1)](R_i+1)+(n-i-R_1-\ldots-R_i), \\
&\qquad\qquad\qquad\qquad\qquad i=1,\ldots,m-1, \\
a_m &= -c(m-1)(R_m+1).
\end{aligned}$$

For large values of m, we may approximate the probability by

$$\Pr(L \geq 0) \approx \Pr\left[Z \geq -\frac{\theta}{\tau}\right], \tag{8.7}$$

where Z is a standard normal random variable, and

$$\begin{aligned}
\theta &= E(L) = \sum_{i=1}^{m} a_i E\left[X_{i:m:n}^{(R_1,\ldots,R_m)}\right], \\
\tau^2 &= Var(L) = \sum_{i=1}^{m} a_i^2 Var\left[X_{i:m:n}^{(R_1,\ldots,R_m)}\right] \\
&\quad +2\sum_{i=1}^{m-1}\sum_{j=i+1}^{m} a_i a_j Cov\left[X_{i:m:n}^{(R_1,\ldots,R_m)}, X_{j:m:n}^{(R_1,\ldots,R_m)}\right].
\end{aligned}$$

The single and product moments of progressively Type-II right censored order statistics occurring in the above expression may be obtained by first-order approximations; see Balakrishnan and Rao (1997). The idea is to use the probability integral transformation

$$X_{i:m:n}^{(R_1,\ldots,R_m)} \stackrel{d}{=} F^{-1}(U_{i:m:n}^{(R_1,\ldots,R_m)}), \tag{8.8}$$

where $U_{i:m:n}^{(R_1,\ldots,R_m)}$ is the ith progressively Type-II right censored order statistic from the uniform $U(0,1)$ distribution, and F^{-1} is the inverse cdf of the underlying distribution.

The mean, variance, and covariance for progressively Type-II order statistics from the Uniform $U(0,1)$ distribution are given by [see Balakrishnan and Aggarwala (2000)]

$$E(U_{i:m:n}) = \alpha_{i:m:n} = 1-b_i, \qquad i=1,\ldots,m, \tag{8.9}$$

$$Var(U_{i:m:n}) = a_i b_i, \qquad i=1,\ldots,m, \tag{8.10}$$

$$Cov(U_{i:m:n}, U_{j:m:n}) = a_i b_j, \qquad 1 \leq i \leq j \leq m, \tag{8.11}$$

where

$$\begin{aligned}
b_i &= \prod_{k=1}^{i} \frac{R_k+R_{k+1}+\ldots+R_m+m-k+1}{R_k+R_{k+1}+\ldots+R_m+m-k+2}, \\
a_i &= \prod_{k=1}^{i} \frac{R_k+R_{k+1}+\ldots+R_m+m-k+2}{R_k+R_{k+1}+\ldots+R_m+m-k+3} \\
&\quad - \prod_{k=1}^{i} \frac{R_k+R_{k+1}+\ldots+R_m+m-k+1}{R_k+R_{k+1}+\ldots+R_m+m-k+2}.
\end{aligned}$$

Expanding $F^{-1}(U_{i:m:n})$ in a Taylor series (keeping only the first term), we have

$$\begin{aligned} E(X_{i:m:n}) &\approx F^{-1}(\alpha_{i:m:n}), & (8.12)\\ Var(X_{i:m:n}) &\approx \left\{F^{-1(1)}(\alpha_{i:m:n})\right\}^2 Var(U_{i:m:n}), & (8.13)\\ Cov(X_{i:m:n}, X_{j:m:n}) &\approx \left\{F^{-1(1)}(\alpha_{i:m:n})\right\}\left\{F^{-1(1)}(\alpha_{j:m:n})\right\} & \\ &\quad \times Cov(U_{i:m:n}, U_{j:m:n}), & (8.14) \end{aligned}$$

where $F^{-1(1)}(u) = \frac{dF^{-1}(u)}{du} = \frac{1}{f(F^{-1}(u))}$. Balakrishnan and Rao (1997) used these results to derive expressions for the approximate best linear unbiased estimators for an arbitrary location-scale family of distributions.

We would like to point out that even though limiting results for linear combinations of regular order statistics are available [see, for example, David (1981)], such results under progressive censoring have not been studied yet. It is unclear whether the results in the regular case can be easily extended to progressive censoring.

Instead of rewriting the test statistic as a linear combination of the progressively censored order statistics, we may directly approximate the power by considering the test statistic T. We may write

$$T = \frac{\sum_{i=1}^{m-1}(m-i)S_i}{(m-1)\sum_{i=1}^{m} S_i} = \frac{W_1}{W_2}. \tag{8.15}$$

We then have,

$$E(T) = E\left[\frac{W_1}{W_2}\right] \approx \frac{E(W_1)}{E(W_2)}, \tag{8.16}$$

$$Var(T) \approx \left[\frac{E(W_1)}{E(W_2)}\right]^2 \left[\frac{Var(W_1)}{E^2(W_1)} + \frac{Var(W_2)}{E^2(W_2)} - \frac{2Cov(W_1, W_2)}{E(W_1)E(W_2)}\right]. \tag{8.17}$$

See Kendall and Stuart (1969) for details. We may then approximate the distribution of T by a normal distribution with mean and variance given by the above expressions.

8.4.2 Monte Carlo power comparison

In order to assess the power properties of the test statistic T, a Monte Carlo simulation study was conducted to determine the power under different alternatives. The following lifetime distributions were used as alternatives to the exponential distribution:

1. Weibull distribution with shape parameter 0.5, 2.0;
2. Lomax distribution with shape parameter 0.5, 2.0;
3. Lognormal distribution with shape parameter 0.5, 1.0;
4. Gamma distribution with shape parameter 0.75, 2.0.

For a detailed discussion on various properties of these distributions, one may refer to Johnson, Kotz, and Balakrishnan (1994). For different choices of sample sizes and progressive censoring schemes, we generated 100,000 sets of data in order to obtain the estimated power values. These values are tabulated in Tables 8.2–8.5 for $n = 20$ $(m = 8, 12, 16), 40$ $(m = 10, 20, 30)$ and 60 $(m = 20, 40, 50)$ with three different progressive censoring schemes in each case. For convenience, Table 8.9 lists the different censoring schemes (c.s.) used in the simulation study.

The power values presented in Tables 8.2–8.5 clearly show that the test proposed performs very well for all the alternatives considered. The power increases with m for a fixed n, and also increases as n increases. We also calculated the power values of T from the normal approximation using the two methods and found them to be close to the simulated power values for large values of m. The approximations are also presented in Tables 8.2–8.5 for comparison. It is important to note from these tables that the approximation in (8.7) does not work well for small values of m, even when the value of n is large.

To demonstrate the accuracy of the Monte Carlo simulations, we also tabulate in Table 8.6 the null probabilities for the exponential distribution at levels 2.5(2.5)50%. Since the critical values are independent of n and the progressive censoring schemes, we only present the value for different values of m. We can see that the simulated probabilities under the null distribution are very close to the pre-fixed levels, which suggests that the Monte Carlo method provides a very good approximation. The results in Table 8.6 also provide ample evidence to the accuracy of the normal approximation to the null distribution of the test statistic. If we have to report the p-value of the test, we are then justified in computing tail probabilities using the normal approximation.

8.5 Modified EDF and Shapiro-Wilk Statistics

As we have mentioned in the introduction, there have been several goodness-of-fit tests for exponentiality proposed in the literature. Spinelli and Stephens (1987) compared the performance of several test procedures based on the EDF as well as those based on regression methods. They concluded that, in particular, two statistics, viz. Cramer-von Mises A^2 and the Shapiro-Wilk W_E, had overall better power performance. In this section, we will modify the two statistics in the case of progressively Type-II censored data, and compare their performance with the test based on spacings proposed in Section 8.3.

Testing the null hypothesis that the sample comes from an exponential distribution is equivalent to testing the hypothesis that the spacings $S_1, S_2, \ldots, S_m$ are distributed as scaled exponential. We can then apply the procedures in Spinelli and Stephens (1987) to the S_i's as follows: Let $S_{(1)}, S_{(2)}, \ldots, S_{(m)}$ be the ordered spacings. Let $\hat{\sigma} = \bar{S} = \frac{1}{m}\sum_{i=1}^{m} S_i$ denote the estimator of σ based on generalized least squares. Define

$$w_i = S_{(i)}/\hat{\sigma} \qquad \text{and} \qquad z_i = 1 - \exp(-w_i), \qquad i = 1, \ldots, m.$$

The test statistic A^2 is then defined as

$$A^2 = -\frac{1}{m}\sum_{i=1}^{m}(2i-1)\{\ln z_i + \ln(1-z_i)\} - m. \tag{8.18}$$

Large values of A^2 lead to rejection of the null hypothesis that the sample comes from an exponential distribution.

An alternative test was introduced by Shapiro and Wilk (1972) that compares the generalized least squares estimator of σ with the estimator obtained from the sample variance. The resulting test statistics W_E is defined as

$$W_E = \bar{S}^2 / \sum_{i=1}^{m} S_i^2. \tag{8.19}$$

This test is a two-tailed test.

A Monte Carlo simulation study was conducted to compare the three procedures. We simulated the 5 and 10 percentage points for the test statistics A^2 and W_E, and used the values to compute the power for different alternatives. The results presented in Tables 8.2–8.5 show that for all the alternatives considered, the test based on spacings performs significantly better than either of the other two procedures.

8.6 Two-Parameter Exponential Case

We may also consider a test for the two-parameter exponential distribution (location-scale model) with p.d.f.

$$f(x;\mu,\sigma) = \frac{1}{\sigma}\exp\left[-\left(\frac{x-\mu}{\sigma}\right)\right], \qquad x > \mu, \tag{8.20}$$

where the scale $\sigma > 0$, and the location μ are unknown parameters. In this case, the progressively Type-II right censored spacings S_1^*, S_2^*, ..., S_m^* are defined as

$$\begin{aligned} S_1^* &= n(X_{1:m:n}^{(R_1,\ldots,R_m)} - \mu), \\ S_i^* &= S_i, \qquad i = 2, 3, \ldots, m \end{aligned} \tag{8.21}$$

where S_i's are as defined earlier in (8.4). Once again, S_1^*, S_2^*, ..., S_m^* are all independent and identically distributed as exponential with scale parameter σ. Since the first spacing S_1^* involves the unknown parameter μ, the test statistic T proposed earlier in (8.5) may be modified as

$$T^* = \frac{\sum_{i=2}^{m-1} (m-i) S_i^*}{(m-2) \sum_{i=2}^{m} S_i^*}. \tag{8.22}$$

Following the same procedure outlined in Section 8.3, the null distribution of the test statistic T^* can be derived. The distribution of T^* is the same as the distribution of the average of $(m-2)$ i.i.d. Uniform(0,1) random variables. Hence, the asymptotic null distribution of T^* is normal with mean $E(T^*) = \frac{1}{2}$ and variance $Var(T^*) = \frac{1}{12(m-2)}$. Furthermore, the power approximation procedure discussed in Section 8.4 can also be adapted to this two-parameter exponential case.

8.7 Illustrative Examples

8.7.1 Example 1: One-parameter exponential case

In this section, we present two examples to illustrate the use of the statistics T and T^* in testing for the validity of the one- and two-parameter exponential distributions for an observed progressively Type-II right censored sample.

We consider the following progressively Type-II right censored data giving the times to breakdown of an insulating fluid tested at 34 kilovolts. This data is taken from Table 6.1 of Nelson (1982), and has been considered earlier by Viveros and Balakrishnan (1994). The observations in the original time scale, the progressive censoring pattern and the spacings computed from Eq. (8.4) are as follows:

Progressively censored sample presented by Viveros and Balakrishnan (1994)

i	1	2	3	4	5	6	7	8
$x_{i:m:n}$	0.18999	0.77997	0.95993	1.30996	2.77986	4.84962	6.49999	7.35000
R_i	0	0	3	0	3	0	0	5
S_i	3.60975	10.61969	3.05924	4.55051	17.63873	16.55808	11.55257	5.10007

Nelson (1982) and Viveros and Balakrishnan (1994) considered a Weibull model for these data and constructed confidence intervals for the Weibull shape and scale parameters based on the complete sample and the progressively censored sample, respectively. In both cases, the confidence interval for the shape parameter contained the value of 1 (the shape parameter value for the exponential case) leading to the conclusion that the data are consistent with an exponential distribution.

In this example, we have $n = 19$, $m = 8$. The test statistic is computed as

$$T = \frac{\sum_{i=1}^{m-1}(m-i)S_i}{(m-1)\sum_{i=1}^{m} S_i} = \frac{220.06957}{508.82052} = 0.43251,$$

and the p-value is

$$2\Phi\left(\frac{0.43251 - 0.5}{\sqrt{1/84}}\right) = 2 \times 0.26810 = 0.53620.$$

Based on this p-value, we fail to reject the null hypothesis that the random sample is from an exponential distribution. This is consistent with the findings of Nelson (1982) and Viveros and Balakrishnan (1994).

8.7.2 Example 2: Two-parameter exponential case

Spinelli and Stephens (1987) reported data with 32 observations on measurements of modulus of repute (a measure of the breaking strength of lumber) of wood beams. For the purpose of illustrating the test procedure outlined in Section 8.5, a progressively Type-II right censored sample of size $m = 20$ has been randomly generated from the $n = 32$ observations in Table 3 of Spinelli and Stephens (1987). The observations, the removal pattern applied and the corresponding spacings computed from Eq. (8.19) are as follows:

Progressively censored sample generated from the measurements of modulus of repute of wood beams data by Spinelli and Stephens (1987)

i	1	2	3	4	5	6	7	8	9	10
$x_{i:m:n}$	43.19	49.44	51.55	56.63	67.27	78.47	86.59	90.63	94.38	98.21
R_i	0	2	0	0	2	0	0	0	0	0
S_i^*	—	193.75	59.08	137.16	276.64	257.60	178.64	84.84	75.00	72.77

i	11	12	13	14	15	16	17	18	19	20
$x_{i:m:n}$	98.39	99.74	100.22	103.48	105.54	107.13	108.14	108.94	110.81	116.39
R_i	2	2	0	0	0	0	1	1	0	2
S_i^*	3.24	20.25	5.76	35.86	20.60	14.31	8.08	4.80	7.48	16.74

Spinelli and Stephens (1987) studied tests based on regression and the empirical distribution function for testing the null hypothesis of exponentiality using the complete sample. They found that all the test statistics were highly significant (with p-value < 0.01) and rejected the null hypothesis that the data are exponentially distributed with p.d.f. (8.18).

The test statistic in (8.22) for testing the validity of a two-parameter exponential distribution is computed as

$$T^* = \frac{\sum_{i=2}^{m-1}(m-i)S_i}{(m-2)\sum_{i=2}^{m} S_i} = \frac{19983.72}{26506.8} = 0.75391,$$

and the p-value is

$$2\left[1-\Phi\left(\frac{0.75391-0.5}{\sqrt{1/216}}\right)\right]=2\times 0.0000951=0.00019026.$$

From this p-value, we observe that the data provide enough evidence to reject the null hypothesis that the progressively censored sample comes from a two-parameter exponential distribution, which agrees with the conclusion of Spinelli and Stephens (1987) drawn from the complete sample.

8.8 Multi-Sample Extension

To test that k independent progressively censored samples $X_{1:m_i:n_i}^{(R_{1i},\ldots,R_{m_ii})}, \ldots, X_{m_i:m_i:n_i}^{(R_{1i},\ldots,R_{m_ii})}, i=1,2,\ldots,k$, come from exponential populations $E(\mu_i,\sigma_i)$, we can generalize the test statistic T^* in (8.22) as follows:

$$T^*=\frac{\sum_{i=1}^k(m_i-2)T_i^*}{\sum_{i=1}^k(m_i-2)} \tag{8.23}$$

where T_i^* is the test statistic computed from the ith sample. Small and large values of T^* indicate the non-exponentiality of at least one of the k samples.

If we wish to test that the samples come from one-parameter exponential populations $E(\sigma_i)$, we can generalize the test statistic T in (8.5) as follows:

$$T_*=\frac{\sum_{i=1}^k(m_i-1)T_i}{\sum_{i=1}^k(m_i-1)} \tag{8.24}$$

where T_i is the test statistic computed from the ith sample. Small and large values of T_* indicate the non-exponentiality of at least one of the k samples. Note that, in both cases we may have the censoring schemes ($R'_{ij}s$), sample sizes (n_i), and effective sample sizes (m_i) for the k samples to be different.

The null distribution of T^* (T_*) may once again be shown to be equivalent to the distribution of the average of $\sum_{i=1}^k(m_i-2)$ ($\sum_{i=1}^k(m_i-1)$) Uniform $U(0,1)$ random variables. To compute the power under different alternatives, we may use a similar approximation to the one discussed in Section 8.4. In this case, however, we will not be able to write $P(T^*>c)$ in terms of a probability involving linear combinations of progressively Type-II right censored order statistics from each sample. We rely on the expressions in (8.16) and (8.17) to compute the moments of T^* and T_*, and the corresponding normal approximations to the probabilities.

Table 8.7 presents some simulation results for $k=2,3$ in the case of the one-parameter exponential model. The approximate values of power are reasonably close to the simulated values for most cases considered. It is of interest to note that combinations of censoring schemes for the k samples provide distinctly different power values.

8.9 Conclusions

In this article, we have proposed goodness-of-fit tests for the one- and two-parameter exponential models under general progressive Type-II censoring. These tests are based on normalized spacings, generalizing tests proposed by Tiku (1980). The exact and asymptotic null distribution of the test statistics have been derived. Further, two approximations to compute the power under different alternatives have been suggested.

Results of the simulation study for a wide range of sample sizes and censoring schemes show that the test performs well in detecting departures from exponentiality. If the alternative model is distinctly different from exponential, the power values are close to 1. The approximations for the power are very close to the values obtained through simulations. The proposed test procedures are illustrated on some real data for the one- and two-parameter exponential models. The conclusions drawn from these tests are consistent with those drawn by other authors using different procedures. Finally, some extensions to the multi-sample case have been suggested.

There are several theoretical aspects that still need to be looked at carefully. In particular, it would be useful to develop limit theorems for linear combinations of progressively Type-II censored order statistics. This would provide theoretical justification for the normal approximations suggested in this paper. Finally, it would also be interesting to develop analogous goodness-of-fit tests for the general location-scale family of distributions.

Table 8.1: Progressive censoring schemes used in the Monte Carlo simulation study

n	m	$(R_1, R_2, \ldots, R_m)$	Scheme No.
20	8	$R_1 = 12,\ R_i = 0$ for $i \neq 1$	[1]
		$R_8 = 12,\ R_i = 0$ for $i \neq 8$	[2]
		$R_1 = R_8 = 6,\ R_i = 0$ for $i \neq 1, 8$	[3]
	12	$R_1 = 8,\ R_i = 0$ for $i \neq 1$	[4]
		$R_{12} = 8,\ R_i = 0$ for $i \neq 12$	[5]
		$R_3 = R_5 = R_7 = R_9 = 2,\ R_i = 0$ for $i \neq 3, 5, 7, 9$	[6]
	16	$R_1 = 4,\ R_i = 0$ for $i \neq 1$	[7]
		$R_{16} = 4, R_i = 0$ for $i \neq 16$	[8]
		$R_5 = 4,\ R_i = 0$ for $i \neq 5$	[9]
40	10	$R_1 = 30,\ R_i = 0$ for $i \neq 1$	[10]
		$R_{10} = 30,\ R_i = 0$ for $i \neq 10$	[11]
		$R_1 = R_5 = R_{10} = 10,\ R_i = 0$ for $i = 1, 5, 10$	[12]
	20	$R_1 = 20,\ R_i = 0$ for $i \neq 1$	[13]
		$R_{20} = 20,\ R_i = 0$ for $i \neq 20$	[14]
		$R_i = 1$, for $i = 1, 2, \ldots, 20$	[15]
	30	$R_1 = 10,\ R_i = 0$ for $i \neq 1$	[16]
		$R_{30} = 10,\ R_i = 0$ for $i \neq 30$	[17]
		$R_1 = R_{30} = 5,\ R_i = 0$ for $i \neq 1, 30$	[18]
60	20	$R_1 = 40,\ R_i = 0$ for $i \neq 1$	[19]
		$R_{20} = 40,\ R_i = 0$ for $i \neq 20$	[20]
		$R_1 = R_{20} = 10,\ R_{10} = 20\ ,\ R_i = 0$ for $i \neq 1, 10, 20$	[21]
	40	$R_1 = 20,\ R_i = 0$ for $i \neq 1$	[22]
		$R_{40} = 20,\ R_i = 0$ for $i \neq 40$	[23]
		$R_{2i-1} = 1,\ R_{2i} = 0$ for $i = 1, 2, \ldots, 20$	[24]
	50	$R_1 = 10,\ R_i = 0$ for $i \neq 1$	[25]
		$R_{50} = 10,\ R_i = 0$ for $i \neq 50$	[26]
		$R_1 = R_{50} = 5,\ R_i = 0$ for $i \neq 1, 50$	[27]

Table 8.2: Monte Carlo power estimates for Weibull distribution at 10% and 5% levels of significance

	Weibull(0.5)									
	10%					5%				
c.s.	T	App(L)	App(W)	A^2	W_E	T	App(L)	App(W)	A^2	W_E
[1]	0.71672	0.79033	0.70467	0.60746	0.55796	0.61452	0.69872	0.59668	0.52883	0.45663
[2]	0.51847	0.66998	0.57311	0.37713	0.33930	0.39990	0.55197	0.45694	0.29668	0.24614
[3]	0.57377	0.72616	0.63247	0.43776	0.39414	0.45581	0.61273	0.51555	0.35465	0.29455
[4]	0.83379	0.83901	0.81327	0.66873	0.63183	0.76001	0.77399	0.72796	0.59327	0.53617
[5]	0.69796	0.78209	0.73798	0.47934	0.44554	0.59111	0.68995	0.63459	0.39233	0.34245
[6]	0.85449	0.85515	0.83683	0.68503	0.65829	0.79102	0.80064	0.76174	0.61718	0.56993
[7]	0.90389	0.87568	0.88596	0.71755	0.68749	0.85230	0.82815	0.82318	0.64605	0.59762
[8]	0.84024	0.85959	0.86140	0.59537	0.56385	0.76609	0.80119	0.78752	0.51014	0.46244
[9]	0.92164	0.89050	0.90996	0.74722	0.72349	0.87656	0.84947	0.85511	0.68152	0.63910
[10]	0.80630	0.83480	0.79047	0.69662	0.64007	0.72293	0.76423	0.69720	0.61928	0.54240
[11]	0.58011	0.69950	0.62607	0.39267	0.36022	0.46239	0.58494	0.51114	0.31052	0.26233
[12]	0.70747	0.81095	0.75718	0.51728	0.48377	0.60430	0.72898	0.65652	0.43550	0.38432
[13]	0.95650	0.91312	0.94612	0.80383	0.77610	0.92708	0.88138	0.90818	0.74337	0.69770
[14]	0.86101	0.87312	0.88381	0.56856	0.54223	0.78663	0.81655	0.81548	0.48023	0.43650
[15]	0.94605	0.91715	0.95127	0.75547	0.73113	0.91204	0.88713	0.91625	0.68957	0.64844
[16]	0.99036	0.95235	0.98797	0.87649	0.86136	0.98199	0.93601	0.97634	0.83067	0.80121
[17]	0.97063	0.94666	0.97797	0.74441	0.72546	0.94722	0.92471	0.95751	0.67077	0.63439
[18]	0.97969	0.95122	0.98410	0.79352	0.77495	0.96253	0.93245	0.96856	0.72763	0.69231
[19]	0.95994	0.91637	0.95033	0.82456	0.79270	0.93151	0.88588	0.91435	0.76548	0.71624
[20]	0.83761	0.85969	0.86431	0.53713	0.50973	0.75680	0.79580	0.78887	0.44811	0.40481
[21]	0.94038	0.92336	0.95982	0.72449	0.70207	0.90314	0.89470	0.92723	0.65344	0.61401
[22]	0.99840	0.97427	0.99792	0.93655	0.92493	0.99638	0.96575	0.99528	0.90704	0.88510
[23]	0.98985	0.97129	0.99328	0.80242	0.78826	0.97958	0.95890	0.98529	0.73386	0.70346
[24]	0.99819	0.97611	0.99834	0.92981	0.91833	0.99601	0.96820	0.99615	0.89855	0.87707
[25]	0.99970	0.98529	0.99961	0.96329	0.95653	0.99916	0.98049	0.99901	0.94347	0.92864
[26]	0.99872	0.98636	0.99921	0.90278	0.89330	0.99691	0.98104	0.99797	0.85950	0.83900
[27]	0.99924	0.98660	0.99949	0.92795	0.92069	0.99818	0.98174	0.99866	0.89362	0.87540

	Weibull(2.0)									
	10%					5%				
c.s.	T	App(L)	App(W)	A^2	W_E	T	App(L)	App(W)	A^2	W_E
[1]	0.81945	0.89734	0.93811	0.25905	0.23759	0.68849	0.81533	0.84627	0.17926	0.14879
[2]	0.49956	0.55783	0.55755	0.14362	0.13726	0.34470	0.39873	0.39878	0.08634	0.07574
[3]	0.60582	0.68462	0.68882	0.16607	0.15746	0.44619	0.52574	0.52552	0.10594	0.09065
[4]	0.91172	0.94854	0.97258	0.30676	0.25619	0.82316	0.89889	0.92478	0.21770	0.16360
[5]	0.72826	0.78185	0.78672	0.19354	0.16742	0.58402	0.64632	0.64752	0.12369	0.09740
[6]	0.87935	0.92728	0.94502	0.24654	0.21070	0.76253	0.85062	0.86532	0.16674	0.12852
[7]	0.95772	0.97482	0.98732	0.34094	0.27326	0.90236	0.94434	0.96109	0.24467	0.17720
[8]	0.89272	0.92350	0.93404	0.26365	0.21288	0.79847	0.84933	0.85772	0.17796	0.13134
[9]	0.96362	0.97831	0.98915	0.33296	0.27218	0.91216	0.94911	0.96429	0.23641	0.17514
[10]	0.93802	0.95734	0.99257	0.42401	0.35878	0.87462	0.92656	0.97241	0.32530	0.24949
[11]	0.58036	0.63285	0.63318	0.15917	0.14561	0.42602	0.47993	0.47986	0.09810	0.08196
[12]	0.73598	0.79830	0.80971	0.19894	0.17756	0.58615	0.66508	0.66851	0.12964	0.10439
[13]	0.99378	0.99339	0.99923	0.53410	0.40719	0.98161	0.98593	0.99659	0.42401	0.29094
[14]	0.91227	0.93098	0.93719	0.28490	0.22011	0.83312	0.86453	0.86957	0.19431	0.13708
[15]	0.98015	0.98502	0.99220	0.37401	0.29222	0.94816	0.96359	0.97446	0.27154	0.19144
[16]	0.99956	0.99932	0.99994	0.61063	0.45501	0.99800	0.99806	0.99964	0.49531	0.33160
[17]	0.99342	0.99455	0.99691	0.45004	0.32440	0.98097	0.98533	0.98956	0.33607	0.21861
[18]	0.99723	0.99742	0.99910	0.51169	0.37044	0.99111	0.99282	0.99636	0.39531	0.25746
[19]	0.99673	0.99372	0.99978	0.63063	0.48642	0.98981	0.98830	0.99890	0.52833	0.36543
[20]	0.89434	0.91226	0.91695	0.26949	0.20967	0.80681	0.83526	0.83849	0.18194	0.12919
[21]	0.97705	0.98067	0.98895	0.37580	0.29183	0.94247	0.95524	0.96683	0.27360	0.19119
[22]	1.00000	0.99994	1.00000	0.76301	0.57640	0.99993	0.99982	0.99999	0.66509	0.45036
[23]	0.99878	0.99896	0.99953	0.54811	0.38065	0.99612	0.99674	0.99805	0.42790	0.26786
[24]	1.00000	0.99994	1.00000	0.67704	0.50448	0.99984	0.99977	0.99997	0.56469	0.37931
[25]	1.00000	1.00000	1.00000	0.81038	0.61641	1.00000	0.99999	1.00000	0.71880	0.49141
[26]	0.99999	0.99996	0.99999	0.69585	0.49341	0.99988	0.99985	0.99996	0.58462	0.36861
[27]	0.99999	0.99998	1.00000	0.74535	0.54158	0.99997	0.99994	0.99999	0.64051	0.41547

Table 8.3: Monte Carlo power estimates for Lomax distribution at 10% and 5% levels of significance

	Lomax(0.5)									
	10%					5%				
c.s.	T	App(L)	App(W)	A^2	W_E	T	App(L)	App(W)	A^2	W_E
[1]	0.82303	0.93658	0.79763	0.68983	0.68715	0.77238	0.91681	0.73874	0.64701	0.63292
[2]	0.21607	0.31514	0.23585	0.12015	0.11848	0.13322	0.23460	0.15311	0.06864	0.06345
[3]	0.35568	0.48739	0.37394	0.17834	0.17364	0.25550	0.37055	0.27518	0.11914	0.10764
[4]	0.93605	0.96461	0.92302	0.82777	0.81768	0.91141	0.95555	0.89321	0.79650	0.77966
[5]	0.50119	0.61830	0.51626	0.22339	0.20745	0.39396	0.50200	0.41055	0.15846	0.13678
[6]	0.87040	0.93725	0.83274	0.73201	0.71268	0.83437	0.92106	0.78946	0.69530	0.66573
[7]	0.97759	0.97571	0.97297	0.90291	0.89338	0.96663	0.97001	0.95990	0.88231	0.86585
[8]	0.84985	0.88405	0.85203	0.54634	0.51649	0.79228	0.84954	0.79410	0.47630	0.43396
[9]	0.97492	0.97467	0.97028	0.89796	0.88618	0.96332	0.96873	0.95639	0.87614	0.85794
[10]	0.89205	0.95408	0.87313	0.76905	0.76184	0.85680	0.94141	0.83023	0.73191	0.71581
[11]	0.15319	0.22198	0.16594	0.10395	0.10412	0.08547	0.15746	0.09761	0.05345	0.05197
[12]	0.30073	0.38308	0.31154	0.14420	0.13878	0.20851	0.28029	0.22031	0.08717	0.07877
[13]	0.99203	0.98155	0.99126	0.94559	0.93672	0.98753	0.97742	0.98616	0.93202	0.91870
[14]	0.51911	0.57796	0.53041	0.17396	0.15807	0.40612	0.45920	0.41917	0.11173	0.09385
[15]	0.87879	0.88714	0.84641	0.63073	0.57919	0.83686	0.85845	0.80122	0.57361	0.51157
[16]	0.99947	0.98876	0.99965	0.98655	0.98275	0.99906	0.98639	0.99934	0.98181	0.97631
[17]	0.95434	0.91384	0.95475	0.61017	0.55853	0.92672	0.88995	0.92805	0.53408	0.46811
[18]	0.98875	0.94983	0.98856	0.83422	0.80025	0.98088	0.93788	0.98073	0.79008	0.74271
[19]	0.99196	0.98152	0.99125	0.94550	0.93663	0.98750	0.97739	0.98614	0.93192	0.91861
[20]	0.27506	0.31672	0.29002	0.11468	0.10877	0.18182	0.21508	0.19538	0.06215	0.05715
[21]	0.67153	0.72728	0.67173	0.29673	0.26005	0.58254	0.64608	0.58271	0.22258	0.18253
[22]	0.99998	0.99205	0.99999	0.99707	0.99574	0.99997	0.99041	0.99998	0.99578	0.99347
[23]	0.95153	0.90840	0.95330	0.49868	0.43662	0.92107	0.87992	0.92351	0.41043	0.33957
[24]	0.99844	0.97332	0.99211	0.96215	0.94623	0.99740	0.96775	0.98879	0.95045	0.92816
[25]	1.00000	0.99392	1.00000	0.99937	0.99884	1.00000	0.99269	1.00000	0.99906	0.99835
[26]	0.99923	0.96646	0.99920	0.91593	0.88434	0.99846	0.95875	0.99839	0.88664	0.84007
[27]	0.99997	0.97647	0.99992	0.98098	0.97197	0.99987	0.97147	0.99982	0.97274	0.95854

	Lomax(2.0)									
	10%					5%				
c.s.	T	App(L)	App(W)	A^2	W_E	T	App(L)	App(W)	A^2	W_E
[1]	0.30907	0.28451	0.21716	0.17958	0.16996	0.22452	0.21132	0.14050	0.12285	0.10901
[2]	0.11154	0.18329	0.12257	0.09971	0.09993	0.05553	0.13127	0.06576	0.05001	0.05019
[3]	0.12760	0.19612	0.13394	0.10209	0.10199	0.06827	0.14152	0.07401	0.05164	0.05101
[4]	0.40318	0.34265	0.28542	0.21513	0.19097	0.31774	0.24740	0.19945	0.15517	0.12782
[5]	0.14348	0.18665	0.14423	0.10234	0.10098	0.07840	0.12598	0.08144	0.05257	0.05097
[6]	0.34351	0.26750	0.22272	0.19374	0.17011	0.26190	0.18905	0.14520	0.13441	0.11023
[7]	0.48634	0.41182	0.35544	0.24258	0.20713	0.39831	0.30515	0.26333	0.17987	0.14128
[8]	0.24019	0.25206	0.21992	0.11499	0.11005	0.15713	0.16879	0.14033	0.06245	0.05625
[9]	0.47963	0.40199	0.34740	0.24079	0.20562	0.39190	0.29672	0.25631	0.17800	0.13977
[10]	0.36079	0.30957	0.25018	0.19619	0.18124	0.27491	0.22425	0.16867	0.13741	0.11851
[11]	0.10710	0.16402	0.11660	0.09873	0.10080	0.05239	0.11186	0.06150	0.04911	0.04951
[12]	0.12625	0.17748	0.12956	0.10021	0.10212	0.06505	0.12231	0.07084	0.05093	0.05065
[13]	0.55863	0.47640	0.42100	0.27115	0.22498	0.46974	0.36730	0.32598	0.20418	0.15610
[14]	0.13945	0.16562	0.14154	0.10118	0.09983	0.07666	0.10342	0.07927	0.05174	0.05031
[15]	0.28328	0.23982	0.21548	0.13414	0.12020	0.20092	0.15828	0.13758	0.07975	0.06556
[16]	0.69864	0.60886	0.56620	0.32850	0.26260	0.61953	0.51160	0.47313	0.25704	0.18883
[17]	0.28986	0.29036	0.27554	0.11381	0.10897	0.19623	0.19331	0.18472	0.06078	0.05647
[18]	0.40740	0.39444	0.37317	0.13461	0.12268	0.30562	0.28362	0.27218	0.07802	0.06664
[19]	0.55835	0.47602	0.42058	0.27110	0.22499	0.46945	0.36690	0.32560	0.20421	0.15603
[20]	0.11311	0.14384	0.11934	0.09941	0.09869	0.05773	0.08832	0.06338	0.05016	0.04953
[21]	0.17575	0.19212	0.16926	0.10483	0.10222	0.10522	0.12260	0.10054	0.05452	0.05210
[22]	0.79492	0.70105	0.67852	0.37994	0.29782	0.72959	0.62078	0.59503	0.30652	0.21959
[23]	0.26916	0.26786	0.25884	0.10682	0.10408	0.17633	0.17449	0.16974	0.05611	0.05310
[24]	0.59787	0.49216	0.46968	0.23909	0.18454	0.50840	0.38396	0.36923	0.17237	0.11847
[25]	0.86092	0.76694	0.76334	0.42547	0.33020	0.80912	0.70166	0.69209	0.34902	0.24792
[26]	0.51287	0.49757	0.48655	0.13395	0.12176	0.40291	0.38241	0.37653	0.07531	0.06508
[27]	0.65010	0.61974	0.60888	0.17696	0.15089	0.55128	0.51589	0.50570	0.11144	0.08636

Table 8.4: Monte Carlo power estimates for Lognormal distribution at 10% and 5% levels of significance

	Lognormal(0.5)									
	10%					5%				
c.s.	T	App(L)	App(W)	A^2	W_E	T	App(L)	App(W)	A^2	W_E
[1]	0.97317	0.99822	0.99953	0.57405	0.45279	0.93250	0.99138	0.99521	0.45287	0.32599
[2]	0.92186	0.95576	0.95264	0.36626	0.30817	0.82028	0.87119	0.86690	0.26666	0.20415
[3]	0.95509	0.98294	0.98435	0.44054	0.35978	0.88607	0.93814	0.93878	0.32760	0.24556
[4]	0.98369	0.99916	0.99951	0.61419	0.40693	0.95757	0.99580	0.99662	0.48638	0.28805
[5]	0.97790	0.99171	0.99046	0.49165	0.34131	0.93753	0.96818	0.96493	0.37131	0.23011
[6]	0.98599	0.99917	0.99931	0.53921	0.37542	0.95939	0.99506	0.99516	0.41481	0.25898
[7]	0.98804	0.99948	0.99947	0.62071	0.37729	0.96865	0.99725	0.99697	0.49201	0.26345
[8]	0.99050	0.99799	0.99747	0.57392	0.35703	0.97141	0.99078	0.98897	0.44477	0.24460
[9]	0.99008	0.99964	0.99960	0.60367	0.37816	0.97302	0.99787	0.99752	0.47486	0.26274
[10]	0.99883	0.99999	1.00000	0.91263	0.73497	0.99648	0.99993	1.00000	0.84945	0.61977
[11]	0.98880	0.99641	0.99603	0.58133	0.43831	0.96171	0.98239	0.98083	0.45922	0.31393
[12]	0.99844	0.99971	0.99988	0.70643	0.54012	0.99226	0.99791	0.99862	0.58682	0.40452
[13]	0.99968	1.00000	1.00000	0.93916	0.63184	0.99916	0.99999	1.00000	0.88579	0.50824
[14]	0.99983	0.99998	0.99998	0.83711	0.52330	0.99918	0.99985	0.99980	0.73616	0.39745
[15]	0.99995	1.00000	1.00000	0.88454	0.58418	0.99977	0.99999	1.00000	0.80170	0.45680
[16]	0.99987	1.00000	1.00000	0.94584	0.59018	0.99970	1.00000	1.00000	0.89330	0.46627
[17]	0.99999	1.00000	1.00000	0.92174	0.56214	0.99993	1.00000	0.99999	0.85183	0.43561
[18]	0.99999	1.00000	1.00000	0.93584	0.57814	0.99993	1.00000	1.00000	0.87549	0.45259
[19]	0.99998	1.00000	1.00000	0.99074	0.80103	0.99993	1.00000	1.00000	0.97817	0.70248
[20]	0.99997	1.00000	1.00000	0.90574	0.60691	0.99986	0.99998	0.99998	0.83300	0.48040
[21]	1.00000	1.00000	1.00000	0.95619	0.70913	1.00000	1.00000	1.00000	0.91205	0.58757
[22]	0.99999	1.00000	1.00000	0.99518	0.73613	0.99999	1.00000	1.00000	0.98564	0.62309
[23]	1.00000	1.00000	1.00000	0.98858	0.69547	1.00000	1.00000	1.00000	0.96990	0.57676
[24]	1.00000	1.00000	1.00000	0.99267	0.73544	0.99999	1.00000	1.00000	0.97930	0.62046
[25]	1.00000	1.00000	1.00000	0.99540	0.71991	1.00000	1.00000	1.00000	0.98692	0.60573
[26]	1.00000	1.00000	1.00000	0.99435	0.71123	1.00000	1.00000	1.00000	0.98352	0.59598
[27]	1.00000	1.00000	1.00000	0.99510	0.71781	1.00000	1.00000	1.00000	0.98578	0.60334

	Lognormal(1.0)									
	10%					5%				
c.s.	T	App(L)	App(W)	A^2	W_E	T	App(L)	App(W)	A^2	W_E
[1]	0.20500	0.29174	0.27413	0.11028	0.10874	0.12145	0.21294	0.18012	0.05800	0.05538
[2]	0.24330	0.27591	0.26382	0.10210	0.10222	0.13892	0.18235	0.16012	0.05253	0.05076
[3]	0.23727	0.28077	0.26728	0.10238	0.10244	0.13645	0.18965	0.16506	0.05240	0.05095
[4]	0.18549	0.24289	0.22190	0.11502	0.10925	0.11137	0.17729	0.14109	0.06320	0.05743
[5]	0.21975	0.25508	0.24096	0.09901	0.09883	0.12839	0.17135	0.14707	0.05157	0.04913
[6]	0.20836	0.27502	0.26113	0.11233	0.10829	0.12532	0.19362	0.16736	0.06123	0.05592
[7]	0.18059	0.20926	0.18788	0.12156	0.11289	0.11081	0.15041	0.11520	0.06750	0.05970
[8]	0.16825	0.21182	0.19219	0.10035	0.10005	0.09605	0.14480	0.11486	0.05062	0.04966
[9]	0.18455	0.21834	0.19881	0.12050	0.11267	0.11222	0.15713	0.12356	0.06690	0.05911
[10]	0.29002	0.41242	0.40955	0.13184	0.12372	0.19056	0.30891	0.29732	0.07706	0.06719
[11]	0.40977	0.43709	0.43629	0.11688	0.11248	0.27017	0.30055	0.29513	0.06533	0.05891
[12]	0.42399	0.46854	0.46800	0.11977	0.11430	0.28509	0.32841	0.32411	0.06761	0.05954
[13]	0.21872	0.28042	0.27187	0.14004	0.12703	0.13898	0.20647	0.18414	0.08305	0.06956
[14]	0.40377	0.42535	0.42453	0.11755	0.10848	0.27285	0.29961	0.29391	0.06296	0.05608
[15]	0.33082	0.40361	0.40213	0.11992	0.11126	0.21963	0.28892	0.28084	0.06599	0.05802
[16]	0.20384	0.21442	0.20436	0.15210	0.13327	0.13046	0.15171	0.12945	0.09250	0.07404
[17]	0.26370	0.29442	0.28827	0.11259	0.10725	0.16711	0.20260	0.18821	0.06001	0.05530
[18]	0.20548	0.24718	0.23741	0.11441	0.10902	0.12479	0.17094	0.15149	0.06116	0.05656
[19]	0.27206	0.37697	0.37584	0.16762	0.14182	0.18277	0.28612	0.27495	0.10273	0.08003
[20]	0.59741	0.61445	0.61335	0.14869	0.12681	0.45061	0.46773	0.46762	0.08561	0.06788
[21]	0.54944	0.59360	0.59258	0.14769	0.12547	0.41023	0.45363	0.45334	0.08552	0.06683
[22]	0.21695	0.22832	0.22276	0.18074	0.14920	0.14186	0.16210	0.14505	0.11487	0.08614
[23]	0.41991	0.44414	0.44395	0.13418	0.11751	0.29660	0.32185	0.31788	0.07462	0.06233
[24]	0.27574	0.35465	0.35287	0.15682	0.13262	0.18403	0.25781	0.24838	0.09450	0.07436
[25]	0.22026	0.19908	0.19353	0.18767	0.15237	0.14658	0.13547	0.12114	0.12045	0.08813
[26]	0.25109	0.28162	0.27666	0.12797	0.11442	0.16094	0.19489	0.18231	0.06991	0.05983
[27]	0.19203	0.22635	0.21912	0.13294	0.11846	0.11469	0.15524	0.13924	0.07404	0.06270

Table 8.5: Monte Carlo power estimates for Gamma distribution at 10% and 5% levels of significance

	Gamma(0.75)									
	10%					5%				
c.s.	T	App(L)	App(W)	A^2	W_E	T	App(L)	App(W)	A^2	W_E
[1]	0.18415	0.20186	0.14326	0.13881	0.13226	0.10768	0.14541	0.08019	0.08196	0.07277
[2]	0.17285	0.24856	0.18925	0.13105	0.12493	0.09870	0.18408	0.11582	0.07548	0.06725
[3]	0.17698	0.24464	0.18837	0.13428	0.12747	0.10133	0.17894	0.11471	0.07845	0.06911
[4]	0.20126	0.18905	0.15029	0.13406	0.12826	0.12070	0.12841	0.08523	0.07908	0.07091
[5]	0.19227	0.23571	0.20012	0.12946	0.12390	0.11468	0.16427	0.12364	0.07521	0.06779
[6]	0.22455	0.19843	0.15971	0.13788	0.13169	0.14011	0.13668	0.09233	0.08220	0.07366
[7]	0.21915	0.18877	0.16068	0.13048	0.12508	0.13728	0.12449	0.09287	0.07550	0.06775
[8]	0.21378	0.22985	0.20544	0.12798	0.12310	0.13250	0.15524	0.12734	0.07413	0.06615
[9]	0.23260	0.19744	0.17014	0.13328	0.12758	0.14705	0.13109	0.10003	0.07760	0.06965
[10]	0.20925	0.20162	0.15722	0.14563	0.13917	0.12582	0.13937	0.08970	0.08834	0.07907
[11]	0.19293	0.25732	0.21050	0.13239	0.12763	0.11360	0.18612	0.13223	0.07732	0.07132
[12]	0.21813	0.27227	0.23039	0.14202	0.13576	0.13313	0.19523	0.14796	0.08415	0.07732
[13]	0.25790	0.20852	0.18919	0.13505	0.12917	0.16655	0.13580	0.11353	0.07895	0.07154
[14]	0.24734	0.27371	0.25698	0.12966	0.12466	0.15974	0.18714	0.16810	0.07488	0.06859
[15]	0.28922	0.26257	0.24672	0.13907	0.13300	0.19353	0.17766	0.15884	0.08176	0.07444
[16]	0.30315	0.23667	0.22608	0.13627	0.13220	0.20500	0.15465	0.14195	0.07859	0.07327
[17]	0.29555	0.30082	0.29251	0.13444	0.13005	0.19685	0.20602	0.19660	0.07629	0.07161
[18]	0.29813	0.29032	0.28190	0.13498	0.13071	0.19872	0.19721	0.18755	0.07702	0.07198
[19]	0.26838	0.21659	0.19826	0.14554	0.13880	0.17535	0.14130	0.12004	0.08561	0.07798
[20]	0.25713	0.28799	0.27132	0.13654	0.13307	0.16713	0.19870	0.17981	0.07880	0.07305
[21]	0.30867	0.32241	0.30814	0.14807	0.14217	0.20850	0.22575	0.21093	0.08878	0.08112
[22]	0.35995	0.28396	0.27819	0.13843	0.13502	0.25210	0.19031	0.18315	0.08067	0.07448
[23]	0.35395	0.36277	0.35837	0.13581	0.13221	0.24655	0.25689	0.25220	0.07848	0.07290
[24]	0.39902	0.33282	0.32834	0.14345	0.13933	0.29018	0.23045	0.22488	0.08515	0.07792
[25]	0.40398	0.32213	0.31855	0.13907	0.13471	0.29295	0.22116	0.21665	0.07973	0.07439
[26]	0.39963	0.39085	0.38809	0.13739	0.13324	0.29200	0.28087	0.27803	0.07866	0.07312
[27]	0.40108	0.37789	0.37505	0.13829	0.13378	0.29301	0.26930	0.26625	0.07910	0.07378

	Gamma(2.0)									
	10%					5%				
c.s.	T	App(L)	App(W)	A^2	W_E	T	App(L)	App(W)	A^2	W_E
[1]	0.46065	0.62475	0.62450	0.12548	0.12036	0.31030	0.46339	0.46149	0.07321	0.06460
[2]	0.31230	0.35326	0.35024	0.10908	0.10780	0.18855	0.23024	0.22497	0.05885	0.05508
[3]	0.35993	0.42015	0.41776	0.11336	0.11036	0.22623	0.27977	0.27830	0.06245	0.05763
[4]	0.51783	0.67413	0.67629	0.13134	0.11943	0.37470	0.52369	0.52329	0.07698	0.06509
[5]	0.42380	0.46831	0.46754	0.11795	0.11029	0.28395	0.32517	0.32399	0.06544	0.05888
[6]	0.50328	0.64337	0.64437	0.12347	0.11446	0.35268	0.47912	0.47894	0.07051	0.06077
[7]	0.57175	0.71129	0.71348	0.13545	0.12040	0.42260	0.56653	0.56661	0.07713	0.06545
[8]	0.52100	0.58340	0.58317	0.12623	0.11473	0.37236	0.43239	0.43223	0.07143	0.06223
[9]	0.58871	0.72468	0.72717	0.13433	0.11948	0.43735	0.57774	0.57791	0.07667	0.06468
[10]	0.64455	0.79434	0.81290	0.18295	0.15373	0.50515	0.67882	0.68469	0.11693	0.09002
[11]	0.40708	0.44438	0.44338	0.12129	0.11380	0.27068	0.30636	0.30449	0.06812	0.06056
[12]	0.49898	0.56138	0.56073	0.13200	0.12107	0.34671	0.40423	0.40411	0.07695	0.06556
[13]	0.76435	0.86731	0.87863	0.19476	0.15400	0.64243	0.77634	0.78295	0.12190	0.08749
[14]	0.64708	0.67861	0.67864	0.15059	0.13037	0.50255	0.53621	0.53617	0.08910	0.07071
[15]	0.74438	0.81444	0.81902	0.16688	0.14117	0.60845	0.69161	0.69312	0.10085	0.07851
[16]	0.84495	0.91453	0.92076	0.19910	0.15700	0.74340	0.84214	0.84661	0.12385	0.08902
[17]	0.79250	0.82753	0.82892	0.17654	0.14476	0.67360	0.71617	0.71654	0.10737	0.08076
[18]	0.81585	0.85752	0.86045	0.18534	0.14961	0.70285	0.75773	0.75903	0.11441	0.08376
[19]	0.83133	0.90917	0.92771	0.24927	0.18278	0.73048	0.84521	0.86022	0.16829	0.11168
[20]	0.67630	0.70367	0.70371	0.16159	0.13807	0.53490	0.56499	0.56494	0.09688	0.07688
[21]	0.78997	0.83131	0.83647	0.18907	0.15461	0.66470	0.71712	0.71917	0.11754	0.08968
[22]	0.93598	0.96845	0.97375	0.25943	0.18517	0.87880	0.93434	0.94018	0.17081	0.11083
[23]	0.90123	0.91524	0.91700	0.21660	0.16524	0.82293	0.84274	0.84371	0.13600	0.09562
[24]	0.94513	0.96927	0.97321	0.24655	0.18208	0.89032	0.93273	0.93696	0.15958	0.10905
[25]	0.96211	0.98182	0.98469	0.27168	0.19690	0.92165	0.95854	0.96208	0.18012	0.11732
[26]	0.94769	0.95990	0.96196	0.24890	0.18512	0.89636	0.91673	0.91854	0.16102	0.10871
[27]	0.95310	0.96790	0.97045	0.25895	0.19005	0.90658	0.93157	0.93412	0.16873	0.11246

Table 8.6: Monte Carlo null probabilities of T for exponential distribution at levels 2.5 (2.5) 50%

m	8	10	12	16	20	30	40	50	60
2.5%	0.02253	0.02328	0.02371	0.02301	0.02429	0.02413	0.02476	0.02507	0.02462
5%	0.04890	0.04808	0.04835	0.04923	0.04911	0.04898	0.05009	0.04965	0.04871
7.5%	0.07512	0.07483	0.07351	0.07483	0.07470	0.07397	0.07582	0.07520	0.07412
10%	0.10062	0.10068	0.09847	0.09988	0.09974	0.10001	0.10157	0.10074	0.09971
12.5%	0.12675	0.12631	0.12446	0.12603	0.12449	0.12483	0.12693	0.12671	0.12484
15%	0.15306	0.15266	0.14985	0.15147	0.15003	0.15098	0.15134	0.15170	0.14946
17.5%	0.17857	0.17859	0.17529	0.17705	0.17558	0.17612	0.17663	0.17680	0.17457
20%	0.20468	0.20510	0.20045	0.20283	0.20115	0.20173	0.20129	0.20140	0.19986
22.5%	0.23018	0.23117	0.22611	0.22773	0.22700	0.22683	0.22697	0.22716	0.22475
25%	0.25656	0.25657	0.25116	0.25349	0.25133	0.25243	0.25266	0.25172	0.25019
27.5%	0.28182	0.28228	0.27619	0.27841	0.27594	0.27754	0.27726	0.27639	0.27496
30%	0.30769	0.30775	0.30175	0.30400	0.30082	0.30235	0.30175	0.30168	0.30020
32.5%	0.33301	0.33277	0.32711	0.32783	0.32573	0.32753	0.32655	0.32668	0.32499
35%	0.35814	0.35827	0.35161	0.35277	0.35141	0.35271	0.35235	0.35113	0.35056
37.5%	0.38273	0.38346	0.37726	0.37720	0.37649	0.37799	0.37732	0.37617	0.37520
40%	0.40781	0.40829	0.40274	0.40147	0.40152	0.40335	0.40204	0.40129	0.40072
42.5%	0.43366	0.43375	0.42809	0.42668	0.42666	0.42830	0.42760	0.42633	0.42546
45%	0.45883	0.45921	0.45420	0.45230	0.45167	0.45283	0.45339	0.45039	0.44989
47.5%	0.48439	0.48401	0.47919	0.47614	0.47604	0.47765	0.47838	0.47578	0.47557
50%	0.50943	0.50819	0.50365	0.50109	0.50159	0.50268	0.50383	0.50129	0.50035

Table 8.7: Simulated and approximate values of the power of T_* at 10% and 5% levels of significance

				Lognormal(1.0)				Lomax(2.0)			
				10%		5%		10%		5%	
k	n_j	m_j	c.s.	Sim.	App.	Sim.	App.	Sim.	App.	Sim.	App.
2	40	20	[13][14]	0.44296	0.51678	0.35086	0.39493	0.49618	0.37621	0.39880	0.27749
			[13][15]	0.42930	0.49896	0.33882	0.38034	0.59646	0.43641	0.50350	0.33563
			[14][14]	0.65778	0.68772	0.52172	0.55493	0.16830	0.17085	0.09696	0.10069
	60	40	[22][24]	0.25098	0.38508	0.16614	0.28199	0.81572	0.78496	0.76808	0.71048
			[23][24]	0.52032	0.61590	0.43032	0.49254	0.64546	0.53579	0.54524	0.42705
			[23][23]	0.66720	0.69734	0.54014	0.57330	0.39242	0.38314	0.28358	0.27471
3	40	20	[13][14][15]	0.53794	0.70703	0.44136	0.59081	0.57268	0.42501	0.47108	0.32097
			[14][14][15]	0.75916	0.82885	0.64342	0.72754	0.29140	0.24573	0.20196	0.15959
			[14][14][14]	0.81718	0.84408	0.71218	0.74542	0.19724	0.19967	0.12016	0.12231
	60	40	[22][23][24]	0.44540	0.60647	0.35110	0.49042	0.78820	0.81008	0.74146	0.73500
			[23][23][24]	0.65518	0.79441	0.56334	0.69244	0.70282	0.61208	0.60718	0.50176
			[23][23][23]	0.82042	0.84665	0.72174	0.75498	0.50552	0.49186	0.38716	0.37461

References

1. Balakrishnan, N. (1983). Empirical power study of a multi-sample test of exponentiality based on spacings, *Journal of Statistical Computation and Simulation*, **18**, 265–271.

2. Balakrishnan, N. and Aggarwala, R. (2000). *Progressive Censoring: Theory, Methods, and Applications*, Boston: Birkhäuser.

3. Balakrishnan, N. and Basu, A. P. (1995) (Eds.). *The Exponential Distribution: Theory, Methods and Applications*, Langhorne, Pennsylvania: Gordon and Breach.

4. Balakrishnan, N. and Rao, C. R. (1997). Large-sample approximations to the best linear unbiased estimation and best linear unbiased prediction based on progressively censored samples and some applications, In *Advances in Statistical Decision Theory and Applications* (Eds., S. Panchapakesan and N. Balakrishnan), pp. 431–444, Boston: Birkhäuser.

5. Cohen, A. C. (1963). Progressively censored samples in life testing, *Technometrics*, **5**, 327–329.

6. Cohen, A. C. (1966). Life testing and early failure, *Technometrics*, **8**, 539–549.

7. D'Agostino, R. B. and Stephens, M. A. (1986) (Eds.). *Goodness-of-Fit Techniques*, New York: Marcel Dekker.

8. David, H. A. (1981). *Order Statistics*, Second Edition, New York: John Wiley & Sons.

9. Johnson, N. L., Kotz, S., and Balakrishnan, N. (1994). *Continuous Univariate Distributions-Volume 1*, Second Edition, New York: John Wiley & Sons.

10. Kendall, M. G.and Stuart, A. (1969). *The Advanced Theory of Statistics*, Volume 1, London: Charles Griffin.

11. Nelson, W. (1982). *Applied Life Data Analysis*, New York: John Wiley & Sons.

12. Shapiro, S. S. (1995). Goodness-of-fit tests, In *The Exponential Distribution: Theory, Methods and Applications* (Eds., N. Balakrishnan and A. P. Basu), Chapter 13, Langhorne, Pennsylvania: Gordon and Breach.

13. Shapiro, S. S. and Wilk, M. B. (1972). An analysis of variance test for the exponential distribution, *Technometrics*, **14**, 355–370.

14. Spinelli, J. J. and Stephens, M. A. (1987). Tests for exponentiality when origin and scale parameters are unknown, *Technometrics*, **29**, 471–476.

15. Stephens, M. A. (1986). Tests for the exponential distribution, In *Goodness-of-Fit Techniques* (Eds., R. B. D'Agostino and M. A. Stephens), Chapter 10, New York: Marcel Dekker.

16. Tiku, M. L. (1980). Goodness of fit statistics based on the spacings of complete or censored samples, *Australian Journal of Statistics*, **22**, 260–275.

17. Viveros, R. and Balakrishnan, N. (1994). Interval estimation of parameters of life from progressively censored data, *Technometrics*, **36**, 84–91.

9

Goodness-of-Fit Statistics for the Exponential Distribution When the Data are Grouped

Sneh Gulati and Jordan Neus

Florida International University, Miami, Florida
State University of New York at Stony Brook, Stony Brook, New York

Abstract: In many industrial and biological experiments, the recorded data consist of the number of observations falling in an interval. In this paper, we develop two test statistics to test whether the grouped observations come from an exponential distribution. Following the procedure of Damianou and Kemp (1990), Kolmogrov-Smirnov type statistics are developed with the maximum likelihood estimator of the scale parameter substituted for the true unknown scale. The asymptotic theory for both the statistics is studied and power studies carried out via simulations.

Keywords and phrases: Empirical distribution function, exponential distribution, grouped data, Kolmogrov-Smirnov, goodness-of-fit, parametric bootstrap

9.1 Introduction

In a number of life testing experiments, it is impossible to monitor units continuously; instead one inspects the units intermittently or at prespecified times. Thus the data consists of the number of failures or deaths in an interval. For example, when testing a large number of inexpensive units for time to failure, it may be cost prohibitive to connect each one to a monitoring device. Thus an inspector may inspect them at predetermined time intervals and record the number of units that failed since the last inspection. Similarly in cancer follow up studies where the variable of interest is time to relapse, a patient may be monitored only at regular intervals or may seek help only after tangible symptoms of the disease appear. Thus the time to relapse cannot be specified exactly, but will only be known to lie between two successive clinic visits [see Yu *et al.* (2000) for details]. Grouped data also arise when it is not possible to measure units precisely due to the finite precision of the measuring instrument. As a result, one can only record the interval in which a measurement falls. See Steiner *et al.* (1994) for some excellent examples of how grouped data can arise naturally in the industry.

The first test to assess the goodness-of-fit of any model was developed by Karl Pearson and is the well-known chi-square test. The chi-square test is also the first test developed for grouped data since the test discretises any given data set and compares the observed cell counts to the expected cell counts. Next, came the empirical distribution function (EDF) tests, the Kolmogrov-Smirnov (KS) tests and the Cramér-von Mises statistics. Originally developed for complete data sets, they have also been extensively studied for testing goodness-of-fit for discrete and grouped data sets. The use of the KS test statistic for goodness-of-fit tests was by Schmid (1958), Conover (1972), Pettitt and Stephens (1977), Wood and Altavela (1978), among others. While Schmid (1958) studied the asymptotic distribution of the KS statistic for grouped data, Conover (1972) derived the exact null distribution of the test statistic as did Pettitt and Stephens (1974). A detailed study of the use of Cramér-von Mises statistics for studying goodness-of-fit of discrete data was done by Choulakian *et al.* (1994). They derived the asymptotic distribution of the W^2, U^2 and the A^2 statistics for a discrete distribution and showed that asymptotically all three test statistics can be written as a sum of independent non-central chi-square variates.

It is well known that the KS statistic is based on the maximum distance between the EDF and the hypothesized cumulative distribution function (CDF), while the Cramér-von Mises statistics are functions of the distance between the empirical CDF and the true CDF at all the observed data values (for continuous data the difference is measured at all data points, while for grouped data, the distance is measured at all the end points of the groups). Hence, the Cramér-von Mises are in general, more powerful than the KS statistics. As a result, a subclass of KS statistics which utilize the distance between the EDF and the hypothesized CDF at all data values or at certain quantiles have been proposed, among others by, Riedwyl (1967), Maag *et al.* (1973) and Green and Hegazy (1976) and more recently, Damianou and Kemp (1990). These test statistics are more powerful than the KS statistic, Watson's U-statistic and are comparable to the Cramér-von Mises statistics.

Most goodness-of-fit tests developed for grouped data so far have been for a completely specified null distribution, i.e., a simple null hypothesis. The purpose of this paper is to develop statistics to test whether the given grouped data comes from an exponential distribution with an unknown mean. We use the methodology of Damianou and Kemp (1990) to develop the test statistics. We develop the test statistics in Section 9.2. The asymptotic distribution of the statistics is studied in Section 9.3, and finally, in Section 9.4, we study the power of the test statistics via simulations. An example to show the practical applications of the test is also presented in Section 9.4.

9.2 The Model and the Test Statistics

Suppose that n independent observations are made on X, a lifetime with density $f(x)$. We assume that the units are observed at times $x_1, x_2, \ldots, x_{k-1}$, leading to observations in k groups $(0, x_1)$, (x_1, x_2), ..., (x_{k-1}, ∞). The recorded data then consists of $n_1, n_2, \ldots, n_k$, where n_i $(1 \leq i \leq k)$ is the number of observations falling in the ith interval. The purpose of this paper is to use the above grouped data to develop test statistics to test the null hypothesis

$$\text{H}_0 : f(x) = \theta e^{-x\theta}, \quad \theta \ \text{unknown}. \tag{9.1}$$

Since θ is unknown, our first step in testing the hypothesis involves the estimation of θ. From Kulldorff (1961) the maximum likelihood estimator, MLE, $\hat{\theta}$ of θ exists if and only if $n_1 < n$ and $n_k < n$, and is obtained by solving the following equation:

$$\sum_{i=1}^{k-1} \frac{n_i(x_i - x_{i-1})}{e^{\theta(x_i - x_{i-1})} - 1} - \sum_{i=2}^{k} n_i x_{i-1} = 0. \tag{9.2}$$

While the above equation can be solved easily by using iterative methods, note that if all the intervals are of the same length, then (9.2) has a closed form solution given as follows:

$$\hat{\theta} = \frac{1}{x_1} \ln \left(1 + \frac{n - n_k}{\sum_{i=1}^{k} (i-1) n_i} \right). \tag{9.3}$$

Let $\pi_i = e^{-\theta x_{i-1}} - e^{-\theta x_i}, 1 \leq i \leq n$, be the true probability under the null, of observing a value in the ith interval. Kulldorff (1961) has shown that $\hat{\theta}$ is consistent and asymptotically sufficient with asymptotic variance, $\sigma^2_{\hat{\theta}}$, given by $\left(n \sum \frac{1}{\pi_i} (d\pi_i/d\theta)^2\right)^{-1}$. From Nelson (1977), we also have that under the null hypothesis, $\hat{\theta}$ is asymptotically normally distributed with mean θ and asymptotic variance $\sigma^2_{\hat{\theta}}$.

Now in order to develop the test statistic, define the following quantities at the inspection times, x_i:

$F(x_i, \hat{\theta}) = 1 - e^{-x_i \hat{\theta}}$, the estimated null CDF,

$F_n(x_i) = \sum_{j=1}^{i} n_i/n$, the empirical CDF, and

$S_i = F_n(x_i) - F(x_i, \hat{\theta})$, the difference between the two estimates.

As in Damianou and Kemp (1990), we propose the following two test statistics to test the hypothesis (9.1) against the alternate hypothesis that the data do not come from an exponential distribution:

$$SW1 = \sqrt{n} \sum_{j=1}^{k-1} \frac{| S_j |}{\Psi_1(j)} \tag{9.4}$$

where $\Psi_1(j) = \sqrt{F(x_j, \hat{\theta})(1 - F(x_j, \hat{\theta}))}$, and

$$SW2 = \sqrt{n} \sum_{j=1}^{k-1} \{\Psi_2(j)\} | S_j | \tag{9.5}$$

where $\Psi_2(j) = (k/2 - j)^2+$, $x+ = 1$ for $x = 0$ and $x+ = x$ for $x \neq 0$.

Note that the weight function $\Psi_1(t)$ is the same weight function in the Anderson-Darling test and gives more weight to the tails of the distribution, while the weight function $\Psi_2(t)$ gives more weight to the center of the distribution.

To test the null hypothesis (9.1) against one-sided alternate hypothesis, one can also define the following one-sided test statistics:

$$SW1^* = \sqrt{n} \sum_{j=1}^{k-1} \frac{S_j}{\Psi_1(j)} \tag{9.6}$$

$$SW2^* = \sqrt{n} \sum_{j=1}^{k-1} \{\Psi_2(j)\} S_j \tag{9.7}$$

where the weights Ψ_1 and Ψ_2 are defined as in the two-sided statistics.

As is obvious, the small sample distribution of the test statistics will depend on the true value of θ as well as the cut points $x_1, x_2, \ldots, x_{k-1}$, hence the p-value of the test statistics is calculated based on their asymptotic distribution. This is not overly restrictive, since the test statistics approach their asymptotic distribution very quickly. Based on our simulation results and existing theory on the distribution of count data, we recommend that the methods presented in this paper can be used for n as small as 15.

Next we discuss the asymptotic theory of the test statistics.

9.3 Asymptotic Distribution

To study the asymptotic distribution of the test statistics, we define the following quantities:

1. $\hat{p} = \frac{n_i}{n}$, $1 \leq i \leq n$, the point estimate of the probability that an observation lies in the ith interval.

2. $\hat{\pi}_i = e^{-\hat{\theta}x_{i-1}} - e^{-\hat{\theta}x_i}$, $1 \leq i \leq n$, the estimated probability (under the null hypothesis) that an observation falls in the ith interval.

3. p_i, $1 \leq i \leq n$, the true probability of an observation falling in the ith interval.

4. $\pi_i = e^{-\theta x_{i-1}} - e^{-\theta x_i}$, $1 \leq i \leq n$, the true probability under the null, of observing a value in the ith interval.

Now from Bishop, Feinberg, and Holland (1975), we have that as $n \to \infty$: $\hat{p}_i \to p_i$ almost surely (w.p. 1) and $\hat{\pi}_i \to p_i = \pi_i$ w.p. 1 if the null is true.

As the next step, we define the following vectors and matrices based on the p_i's:

$$\vec{p} = (p_1, p_2, \ldots, p_k)' \;,$$

$$\frac{d\vec{p}}{d\theta} = \left(\frac{dp_1}{d\theta}, \frac{dp_2}{d\theta}, \ldots, \frac{dp_k}{d\theta}\right)'$$

and

$$D_p = \begin{pmatrix} p_1 & 0 & 0 & \cdots & 0 \\ 0 & p_2 & 0 & \cdots & 0 \\ & & \cdot & & \\ & & \cdot & & \\ & & \cdot & & \\ 0 & 0 & 0 & \cdots & p_k \end{pmatrix} .$$

Now if we let

$$\sigma^2 = (d\vec{p}/d\theta)'(d\vec{p}/d\theta) = \sum (dp_i/d\theta)^2/p_i$$

and

$$L = \frac{D_p^{-1}(d\vec{p}/d\theta)(d\vec{p}/d\theta)'}{\sigma^2} \;,$$

then, again from Bishop, Feinberg, and Holland (1975), we have the following:

Theorem 9.3.1 *Under the null hypothesis (9.1) and the regularity conditions defined in Chapter 14 of Bishop, Feinberg, and Holland (1975), the* $k \times 1$ *vector* $\widehat{\boldsymbol{W}}$ *defined as*

$$W = \begin{pmatrix} \sqrt{n}(\hat{p}_1 - \hat{\pi}_1) \\ \sqrt{n}(\hat{p}_2 - \hat{\pi}_2) \\ \cdot \\ \cdot \\ \cdot \\ \sqrt{n}(\hat{p}_k - \hat{\pi}_k) \end{pmatrix}$$

converges in distribution to a multivariate normal random $\boldsymbol{W}$ *vector with mean* $\mathbf{0}$ *and variance covariance matrix* $\Sigma = (D_p - \vec{p}\vec{p}')(I - L)$.

Since the one-sided test statistics defined in (9.5) and (9.6) are linear combinations of the vector $\widehat{\boldsymbol{W}}$, Theorem 9.3.1 then immediately gives us:

Theorem 9.3.2 *Assume that the null hypothesis defined in (9.1) is true and the aforementioned regularity conditions are satisfied. Then as $n \to \infty$, $SW1^*$ and $SW2^*$ converge in distribution to normal random variables with mean 0 and variance given by σ_1^2 and σ_2^2 respectively, where σ_1^2 and σ_2^2 are scalar functions of the matrix Σ.*

OUTLINE OF PROOF. To prove the theorem, we use techniques similar to those used by Choulakian *et al.* (1994) and define the following vectors and matrices:

Let $\boldsymbol{B}$ be the $1 \times k$ row vector $(1, 1, \ldots, 1)$, and C be the $k \times k$ matrix given by

$$C = \begin{pmatrix} 1 & 0 & 0 & \cdots & 0 \\ 1 & 1 & 0 & \cdots & 0 \\ & & \cdot & & \\ & & \cdot & & \\ & & \cdot & & \\ 1 & 1 & 1 & \cdots & 1 \end{pmatrix}.$$

We also define the $k \times k$ matrices of the weight functions as follows: Q_{Ψ_1} is the $k \times k$ diagonal matrix with its kth diagonal entry 0 and for $1 \leq j \leq k-1$, the jth diagonal entry is $\dfrac{1}{\Psi_1(j)} = \dfrac{1}{\sqrt{F(x_j, \hat{\theta})(1 - F(x_j, \hat{\theta}))}}$ and Q_{Ψ_2} is defined similarly for the weight function Ψ_2. That is, Q_{Ψ_2} is the $k \times k$ diagonal matrix with its kth diagonal entry 0 and for $1 \leq j \leq k-1$, the jth diagonal entry is $\Psi_2(j)$.

Now note that

$$SW1^* = \sum_{j=1}^{k-1} \left(\sum_{i=1}^{j} \frac{\widehat{W}_j}{\Psi_1(j)} \right) = (\boldsymbol{B} Q_{\Psi_1} C)\widehat{\boldsymbol{W}} \tag{9.8}$$

$$SW2^* = \sum_{j=1}^{k-1} \left(\sum_{i=1}^{j} \frac{\widehat{W}_j}{\Psi_2(j)} \right) = (\boldsymbol{B} Q_{\Psi_2} C)\widehat{\boldsymbol{W}}. \tag{9.9}$$

From Kulldorff (1961), under the null hypothesis (9.1), $\hat{\theta} \to \theta$ w.p. 1, $F(x_j, \theta)$ and from Theorem 9.3.1, $\widehat{\boldsymbol{W}}$ converges in distribution to the multivariate normal vector $\boldsymbol{W}$. The theorem now follows from the asymptotic theory described in Section 14.9 of Bishop, Feiberg, and Holland (1975) and the properties of the multivariate normal distribution with σ_1^2 and σ_2^2 defined appropriately. ■

While a one tailed test is not commonly used to test hypotheses of the form (9.1), Theorem 9.3.2 provides the foundation for the distribution theory of the two-sided test statistics. To test the hypothesis (9.1) against a general omnibus alternate hypothesis, that is, H$_a$: the data do not come from an exponential distribution, we use the test statistics defined in 9.4 and 9.5. Note that we can write the statistics $SW1$ and $SW2$ as follows:

$$SW1 = \sum_{j=1}^{k-1} \left| \sum_{i=1}^{j} \frac{\widehat{W}_j}{\Psi_1(j)} \right| \quad \text{and} \quad SW2 = \sum_{j=1}^{k-1} \left| \sum_{i=1}^{j} \frac{\widehat{W}_j}{\Psi_2(j)} \right| .$$

From 9.8 and 9.9, we have that

$$SW1 = (BQ_{\Psi_1}C)|\widehat{\boldsymbol{W}}| \qquad \text{and} \qquad SW2 = (BQ_{\Psi_2}C)|\widehat{\boldsymbol{W}}|.$$

Thus from Theorem 9.3.2, $SW1$ and $SW2$ converge in distribution to the sum of the absolute values of the components of a multivariate normal random vector. Note that while the asymptotic distribution of the test statistics is not known in closed form, with the proliferation of high-speed computers, it can be simulated quite easily and enable us to calculate "bootstrapped p-values" for the test.

Finally, the testing procedure is given as follows. From the given data set, calculate the test statistic $SW1$ (or $SW2$), henceforth referred to as the data test statistic SW_{dat}. As mentioned previously, the distribution theory outlined above allows us to calculate the p-value of the test through the following parametric bootstrap technique. Using the estimate, $\hat{\theta}$, of θ calculated from the data, generate 5,000 samples of size n from the density $f(x, \hat{\theta}) = \hat{\theta}e^{-x\hat{\theta}}$. Each sample is then grouped into the intervals defined by $(0, x_1), (x_1, x_2), \ldots, (x_{k-1}, \infty)$, and $\hat{\theta}$, the "bootstrapped" test statistic $SW1$ (or $SW2$) are all calculated. The p-value of the test is defined to be the proportion of "bootstrapped" test statistics less than or equal to the data test statistic, SW_{dat}. The test is rejected for small p-values. A FORTRAN program to calculate the p-value of the test is available from the authors upon request.

9.4 Power Studies

We studied the power of the test statistics $SW1$ and $SW2$ in testing the null hypothesis (9.1) with θ equal to 1 against the omnibus alternate hypothesis, H$_a$: the data do not come from an exponential distribution. To calculate the power of the test, we used the parametric bootstrap technique as mentioned earlier. The steps of the technique are outlined below:

1. Generate a single grouped sample from the alternate.

2. Estimate θ and calculate $SW1$ and $SW2$.

3. Generate 5000 "grouped" exponential samples with scale parameter $\hat{\theta}$. We call these samples bootstrapped samples.

4. For the above samples, calculate the bootstrapped p-value.

5. Repeat the above steps 5000 times.

The power of the test is then given by the proportion of bootstrapped p-values which are less than or equal to α.

For the alternate distributions, considered the following distributions: Weibull (shape $= 1.5$), Weibull (shape $= 1.5$, scale $= 0.886$), Half-Normal $(0, 1)$, Half-Normal $(0, 2)$, Chi Square (4), Uniform $(0, 2)$, Uniform $(0, 3)$, Gamma (shape $= 1.5$), Chi Square (1), Weibull (shape $= 0.8$), and Lognormal $(0, 1)$.

For each alternate distribution mentioned above, we studied the power for a sample size $n = 25$, 100(25) and significance levels $\alpha = 0.2$, 0.1, 0.05, 0.025, 0.01 and 0.005. We used 4, 5, 6, and 10 intervals with varying (but equal) lengths. In general, we found that the test does very well against most alternate distributions, except for the Weibull, with shape parameter equal to 0.8, and the lognormal distribution. As expected, the power increases as the sample size increases, and depends on the number of intervals and the distance between them. Also as expected, if the length of the intervals is held constant, the power increases as the number of intervals increases. When comparing the two test statistics, it was found that $SW2$ always did better than the Anderson-Darling type statistic. Some of these results are presented in Tables 9.1 and 9.2 and in Figure 9.1.

Finally, as an example, we decided to study the performance of the test on the air traffic data described in Hsu (1979). The data consist of 213 aircraft arrival times (in a particular sector of the Newark Airport) from noon till 8 PM, on April 30, 1969. Hsu has shown through various tests that the distribution of the arrival times is best described by the Poisson distribution and that of the interarrival times is best described by the exponential distribution. We divided both the arrival times and the interarrival times into 10 intervals and calculated the p-value of the test statistics $SW1$ and $SW2$ on the two data sets. For the arrival times, both tests reject exponentiality with a p-value of 0.0 and for the interarrival times, both tests accept exponentiality with the p-value for $SW1$ being 0.6788 and that for $SW2$ equal to 0.7756.

Table 9.1: Power comparisons, $n = 50$, 5 cutpoints @ 0.4, 0.8, 1.2, 1.6, 2.0

Alternate Distribution	α	Power of $SW1$	Power of $SW2$
Gamma (1.5)	0.10	0.4428	0.4776
Gamma (1.5)	0.05	0.3324	0.3516
Weibull (1.5)	0.10	0.8490	0.8886
Weibull (1.5)	0.05	0.7616	0.8084
Weibull (2)	0.10	0.9990	0.9994
Weibull (2)	0.05	0.9976	0.9988
Half Normal $(0, 2)$	0.10	0.3046	0.2874
Half Normal $(0, 2)$	0.05	0.2066	0.1882
Log-Normal	0.10	0.2510	0.2884
Log-Normal	0.05	0.1536	0.1810

Table 9.2: Power comparisons, $n = 50$, 9 cutpoints @ 0.25, 0.5, 0.75, 1.0, 1.25, 1.5, 1.75, 2.0, 2.25

Alternate Distribution	α	Power of $SW1$	Power of $SW2$
Gamma (1.5)	0.10	0.7476	0.8516
Gamma (1.5)	0.05	0.6418	0.7690
Weibull (1.5)	0.10	0.3250	0.4034
Weibull (1.5)	0.05	0.2228	0.2938
Weibull (2)	0.10	1.0	0.9986
Weibull (2)	0.05	0.9998	0.9962
Half Normal $(0, 2)$	0.10	0.6404	0.6428
Half Normal $(0, 2)$	0.05	0.5156	0.5124
Log-Normal	0.10	0.8912	0.9284
Log-Normal	0.05	0.8140	0.8684

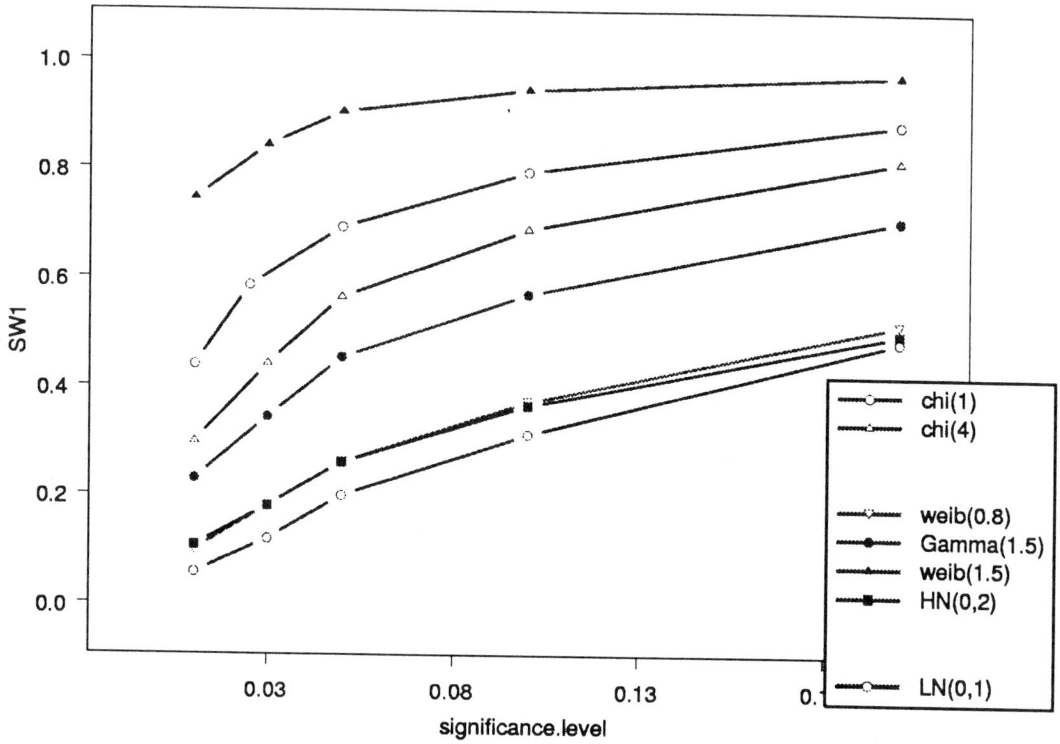

Figure 9.1: Power comparisons for $SW1$, $k = 6$, distance $= 0.4$

References

1. Bishop, Y. M., Feinberg, S. E., and Holland, P. W. (1975). *Discrete Multivariate Analysis*, MIT Press: Cambridge.

2. Choulakian, V., Lockhart, R. A., and Stephens, M. A. (1994). Cramér-von Mises statistics for discrete distributions, *Canadian Journal of Statistics*, **22**, 125–137.

3. Conover, W. J. (1972). A Kolmogrov goodness-of-fit test for discontinuous distributions, *Journal of the American Statistical Association*, **67**, 591–596.

4. Damianou, C. and Kemp, A. W. (1990). New goodness of statistics for discrete and continuous data, *American Journal of Mathematical and Management Sciences*, **10**, 275–307.

5. Green, J. R. and Hegazy, Y. A. S. (1976). Powerful modified-EDF goodness-of-fit tests, *Journal of the American Statistical Association*, **71**, 204–209.

6. Hsu, D. A. (1979). Detecting shifts of parameter in gamma sequences with applications to stock prices and air traffic flow analysis, *Journal of the American Statistical Association*, **74**, 31–40.

7. Kulldorff, G. (1961). *Estimation from Grouped and Partially Grouped Samples*, John WIley & Sons: New York.

8. Maag, U. R., Streit, P., and Drouilly, P. A. (1973). Goodness-of-fit test for grouped data, *Journal of the American Statistical Association*, **68**, 462–465.

9. Nelson, W. (1977). Optimum demonstration tests with grouped inspection data from an exponential distribution, *IEEE Transactions in Reliability*, **23**, 226–231.

10. Pettitt, A. N. and Stephens, M. A. (1977). The Kolmogrov-Smirnov goodness-of-fit statistic with discrete and grouped data, *Technometrics*, **19**, 205–210.

11. Riedwyl, H. (1967). Goodness-of-fit, *Journal of the American Statistical Association*, **62**, 390–398.

12. Schmid, P. (1958). On the Kolmogrov and Smirnov limit theorems for discontinuous distribution functions, *Annals of Mathematical Statistics*, **29**, 1011–1027.

13. Steiner, S. H., Geyer, P. L., and Wesolowsky, G. O. (1994). Control charts based on grouped data, *International Journal of Production Research*, **32**, 75–91.

14. Wood, C. L. and Altavela, M. M. (1978). Large sample results for the Kolmogrov-Smirnov statistics for discrete distributions, *Biometrika*, **65**, 235–239.

15. Yu, Q., Li, L., and Wong, G. (2000). On consistency of the self-consistent estimator of survival functions with interval censored data, *Scandinavian Journal of Statistics*, **27**, 35–44.

10

Characterization Theorems and Goodness-of-Fit Tests

Carol E. Marchetti and Govind S. Mudholkar

Rochester Institute of Technology, Rochester, New York
University of Rochester, Rochester, New York

Abstract: Karl Pearson's chi-square goodness-of-fit test of 1900 is considered an epochal contribution to the science in general and statistics in particular. Regarded as the first objective criterion for agreement between a theory and reality, and suggested as "beginning the prelude to the modern era in statistics," it stimulated a broadband enquiry into the basics of statistics and led to numerous concepts and ideas which are now common fare in statistical science. Over the decades of the twentieth century the goodness-of-fit has become a substantial field of statistical science of both theoretical and applied importance, and has led to development of a variety of statistical tools. The characterization theorems in probability and statistics, the other topic of our focus, are widely appreciated for their role in clarifying the structure of the families of probability distributions. The purpose of this paper is twofold. The first is to demonstrate that characterization theorems can be natural, logical and effective starting points for constructing goodness-of-fit tests. Towards this end, several entropy and independence characterizations of the normal and the inverse gaussian (IG) distributions, which have resulted in goodness-of-fit tests, are used. The second goal of this paper is to show that the interplay between distributional characterizations and goodness-of-fit assessment continues to be a stimulus for new discoveries and ideas. The point is illustrated using the new concepts of IG symmetry, IG skewness and IG kurtosis, which resulted from goodness-of-fit investigations and have substantially expanded our understanding of the striking and intriguing analogies between the IG and normal families.

Keywords and phrases: Goodness-of-fit, characterization

10.1 Introduction and Summary

Karl Pearson's classic "The Grammar of Science" (1892), his definition of the product moment correlation coefficient with Filon (1898), and his chi-square goodness-of-fit (1900) are major landmarks in the history of scientific and statistical decision making. The chi-square goodness-of-fit (GOF) test in particular contributed the first objective criterion for agreement between a theory and reality, and has been considered among the "20 Discoveries That Changed Our Lives"; see Hacking (1984). Arguably, it became the launching platform for the statistical science; or in the words of Bingham (2000), it "may be regarded as beginning the prelude to the modern era in statistics." The chi-square GOF test led to tabulation of the incomplete gamma function in Pearson's Statistical Laboratory, the subsequent square root transformation due to Fisher, the cube root transformation by Wilson and Hilferty for approximating the chi-square distribution, and numerous developments surrounding these topics. It is the corner stone of the important field of methods to verify distributional assumptions used in statistical modeling and data analysis. Early contributors to this area include Pearson (1933), Greenwood (1946), Moran (1951), Kolmogorov (1933), and many others; for references, see D'Agostino and Stephens (1986). Much of this earlier work was focused on the simple GOF hypothesis, with possible use of the plug-in approach for large sample testing of the composite GOF hypotheses. In the early 1960's, when asked by I. Vincze (1984), A. N. Kolmogorov named testing the composite GOF hypothesis of normality as the most important statistical problem of the period. Soon the well-known and widely used Shapiro-Wilk test of normality (1965) appeared, and was followed by many alternatives.

Characterization theorems are generally well appreciated for their aesthetic appeal, mathematical completeness and the light they shed on the structures of the probability distributions. Although logically self evident, but not well recognized, is the fact that they can be natural and effective bases for constructing GOF tests, useful for assessing the validity of distributional models such as normal, exponential, inverse gaussian (IG) etc., commonly used in statistical analyses of data.

To marshal this point, in Section 10.2 we describe some characterization theorems based on either maximum entropy properties of probability distributions or independence of sample statistics. Goodness-of-fit tests based on the maximum entropy properties are outlined in Section 10.3. In Section 10.4, we describe how some of the independence characterizations have been used to develop GOF tests for the normal, multivariate normal and IG composite hypotheses.

The IG distribution, introduced by Schrödinger (1915) and Smoluchovsky

(1915) in the context of Brownian motion, entered statistical applications via the works of Tweedie (1945) and Wald (1947). The IG distribution was independently discovered by E. Halphen in the context of hydrology, but for political reasons, was published under the authorship of Dugué (1941); see Seshadri (1999). It is now widely used for modeling and analyzing non-negative, right-skewed data; see Seshadri (1993, 1999). In a review paper, Folks and Chhikara (1978) highlighted some remarkable similarities between the gaussian (G) and the IG distributions; see also Iyengar and Patwardhan (1988) and Mudholkar and Natarajan (2000). As expressed by Dawid, a discussant of the Folks and Chhikara review paper, these G-IG analogies "intrigued and baffled" many and still remain a curiosity. In Section 10.5 we show how development of the IG goodness-of-fit tests based on maximum entropy and independence characterization properties led to the notions of IG symmetry, IG skewness and IG kurtosis, and how they substantially expanded the list of G-IG analogies; see Mudholkar and Natarajan (1999) and Mudholkar and Tian (2000). This supports the proposition that, as Pearson's chi-square GOF test did, goodness-of-fit investigations today continue to generate new statistical notions and knowledge.

10.2 Characterization Theorems

One of the earliest and prettiest of the characterization results is Cramér's (1936) convolution characterization of the normal distribution. It states that if X_1 and X_2 are independent random variables such that $X_1 + X_2$ has a normal distribution, then X_1 and X_2 are normally distributed. The body of literature on characterization results for probability distributions now forms a well-developed area of probability and statistics; see the monographs by Lukacs and Laha (1964) and Kagan, Linnik, and Rao (1973), the Proceedings of the 1974 Calgary Conference, and, more recently, the monograph by Rao and Shanbhag (1986). In this section we describe some characterizations of relevance to our present objective.

10.2.1 Entropy characterizations

The notion of entropy and the maximum entropy principle have their roots in the 19th and early 20th century physics in the formulation of disorder and chaos, e.g. by Boltzman, in thermodynamics and statistical mechanics. It is closely related to Kullback's (1959) notion of divergence, and Kullback-Leibler information measures; see Kullback and Leibler (1951), Kullback (1959) and Csiszar (1975).

The entropy, $H(f)$, of a random variable X with p.d.f. $f(x)$ is given by $H(f) = E[-\log f(X)]$. The following characterizations of some familiar probability distributions will be used to illustrate a theme of this paper, namely the role of the characterization results in GOF tests.

1. The Uniform Distribution. This is the simplest, the earliest and the best-known entropy characterization. Among all random variables taking values in $[0, 1]$, the $U(0, 1)$ variate has maximum entropy.

2. The Normal Distribution. This famous result was introduced by Shannon (1949) in the context of information theory, i.e. the mathematical theory of communication. It states that among all real-valued random variables with given fixed values of $E(X)$ and $E(X^2)$, the normal variate with these expectations attains the maximum value $\sigma(2\pi(e)^{1/2}$ of the entropy. The result may alternatively be expressed as: Among all real valued random variables with variance equal to σ^2, the normal variates with variance σ^2 have the maximum possible entropy.

3. The Exponential Distribution. Among all nonnegative random variables with $E(X) = \mu$, the exponential random variable attains the maximum value $(1 + \log \mu)$ of the entropy.

4. The Gamma Distribution. Among all nonnegative random variables with fixed values of $E(X)$ and $E(\log X)$, the gamma variate has maximum entropy.

The above examples are particular cases of a general theorem characterizing members of Koopman's multi-parameter exponential family; see Kagan, Linnik, and Rao (1973), and Csiszar (1975). Seshadri (1993) gives an extensive treatment of the IG as an exponential family, which implies the corresponding IG entropy characterization. However, this characterization is not expedient for testing goodness-of-fit. Instead, the following characterization due to Mudholkar and Tian (2000) can be used to construct a maximum entropy GOF test for IG models.

5. The Inverse Gaussian Distribution. Let X be a non-negative random variable $Y = \sqrt{X}$ and let $E(Y^2) - 1/E(Y^2) = \xi^2$ have a given fixed value. Then Y attains the maximum possible value $\log(\sqrt{2\pi e}\xi/2)$ when X is appropriate IG random variable.

10.2.2 Statistical independence

The independence of the mean and the variance of a normal sample is perhaps the best known and most commonly used result involving independence of sample statistics. Actually, this is a characterization of a random sample from a normal population, which stimulated a substantial body of results characterizing populations in terms of the independence of statistics of random samples from them. For an excellent account of such results, see Kagan, Linnik, and Rao (1973). The following are a few of these results which have been used as starting points for developing GOF tests for parametric statistical models.

1. Sample Mean and Variance. The mean and variance of a random sample from a population are independent if and only if the population is normal. This result is attributed to Geary (1936), Lukacs (1942) and Zinger (1951).

2. Sample Mean and Third Central Moment. If $X_1, X_2, \ldots, X_n$ is a random sample from a normal population then it is a simple exercise involving characteristic functions to show that the sample mean $\bar{X}$ and the vector $(X_1 - \bar{X}, X_2 - \bar{X}, \ldots, X_n - \bar{X})$ are independently distributed. Consequently, it follows that $\bar{X}$ and the third central moment $m_3 = \Sigma(X_i - \bar{X})^3/n$ are independently distributed. Actually, Kagan, Linnik, and Rao (1973), as a corollary of a general result characterizing the normal distribution in terms of the independence of linear and polynomial statistics of a random sample, show that the sample mean and the third central sample moment are independent if and only if the population is normal.

3. Multivariate Mean and Covariance Matrix. The multivariate version of the characterization result 1 above states that for a random sample $\boldsymbol{X}_1, \boldsymbol{X}_2, \ldots, \boldsymbol{X}_n$ from a p-variate population, the sample mean $\boldsymbol{X}$ and the sample covariance matrix $\boldsymbol{S}$ are independent if and only if the population is a p-variate normal population.

4. Mean and Difference from Harmonic Mean. From the classical arithmetic, geometric and harmonic inequality it follows that $E(X^{-1}) - E(X) \geq 0$, with equality if and only if X is degenerate. Actually, the above difference is a legitimate scale parameter for the distribution of the reciprocal of the IG(μ, λ) variate. It is well known that the maximum likelihood estimates $\hat{\mu} = \bar{X}$ and $1/\hat{\lambda} = V = \Sigma(1/X_i - 1/\bar{X})$ based on a random sample from the IG population are independently distributed. Khatri (1962) has shown that X and V are independently distributed if and only if the population is inverse gaussian.

5. Mean and Coefficient of Variation. It is well-known that if random variables X and Y are i.i.d., then $X+Y$ and X/Y are independent if and only if the common distribution of X and Y is gamma. A recent result by Hwang and Hu (1999) states that the mean and coefficient of variation of a random sample are independent if and only if the population is gamma.

These simple characterization results have been extended in a variety of ways. For an overview account of such extensions, see Kagan, Linnik, and Rao (1973) and Rao and Shanbhag (1986).

10.3 Maximum Entropy Tests

The earliest explicit use of a characterization theorem for constructing a goodness-of-fit test is by Vasicek (1976), who used Shannon's maximum entropy characterization to construct a test for the composite hypothesis of normality. However, some earlier tests of uniformity and exponentiality can now be interpreted as maximum entropy tests. Vasicek's approach involved constructing a nonparametric estimate of entropy and rejecting normality for its small values.

1. Vasicek's Test. Let $X_1, X_2, \ldots, X_n$ be a random sample from a population with density function $f.(.)$. Then Vasicek's estimator of the population entropy $H(f) = E(-\log f(x))$ is

$$\hat{H}(f) = H_{mn} = \frac{1}{n}\sum_{i=1}^{n} \log\left\{\frac{n}{2m}(X_{(i+m)} - X_{(i-m)})\right\}, \tag{10.1}$$

where $X_{(1)} \leq X_{(2)} \leq \cdots \leq X_{(n)}$, m is a positive integer less than $n/2$, $X_{(i)} = X_{(1)}$ for $i < 1$ and $X_{(i)} = X_{(n)}$ for $i > n$. Vasicek then proposed rejecting the null hypothesis H_0 that the population is normal if is small, or equivalently if

$$K_{mn} = \frac{n}{2ms}\left\{\prod_{i=1}^{n}(X_{(i+m)} - X_{(i-m)})\right\}^{1/n} \leq C_\alpha. \tag{10.2}$$

Vasicek gave an empirical tabulation of the 5% points of the null distribution of the K_{mn} test, and presented results of a study of its power properties. He also argued consistency of the test by showing that under H_0, as $n \to \infty$, K_{mn} converges in probability to $\sigma^{-1}\exp\{H(f)\} < \sqrt{2\pi e}$.

Vasicek's approach can be used to interpret Greenwood's (1946) test of uniformity and it's extension by Cressie (1976) in the entropy terms; e.g. see Dudewicz and van der Meulen (1981). Furthermore, it has also been used to construct GOF tests for a variety of parametric models; e.g. see Gokhale (1983) and Mudholkar and Lin (1984) for the maximum entropy tests of exponentiality.

Now there exists a substantial body of literature on the distributional properties of Vasicek's entropy estimator, the sum-log of generalized spacings; e.g. see Kashimov (1989), Hall (1984, 1986), van Es (1992), Beirlant *et al.* (1997), and references therein. However, the null distribution of K_{mn}, which involves an estimate of the nuisance parameter, is still intractable.

The entropy measure $H(f)$ is closely related to the measures of divergence and Kullback-Leibler information. Nonparametric estimation of entropy and information, and use of these estimates for testing goodness-of-fit against simple and restricted composite alternatives, e.g. $H_0 : N(\mu, \sigma)$ versus $H_1 : N(0, \sigma)$ or

$H_1 : N(0,1)$, has been considered by Ebrahimi, Habibullah, and Soofi (1992) and Soofi, Ebrahimi, and Habibullah (1995).

2. Inverse Gaussian Model. The IG distribution of non-negative right-skewed random variables is increasingly used for modeling and analysis of data in applied research; see Seshadri (1999). Mudholkar and Tian (2000) have developed an IG analog of Vasicek's test, i.e. a maximum entropy test of the composite IG hypothesis using the entropy characterization for the root-reciprocal IG distribution given in Section 10.2.1.

Specifically, if $X_{(1)}, X_{(2)}, \ldots, X_{(n)}$ are the order statistics of a random sample and $Y_{(i)} = 1/\sqrt{X_{(n-i+1)}}$, for $i = 1, 2, \ldots, n$, then the maximum entropy test rejects the IG character of the population if

$$K(Y_1, \ldots, Y_n) = K_{m,n}(f_Y) = \exp\{H_{mn}(f_Y)\}/(w/2), \tag{10.3}$$

where H_{mn} is the sample entropy of the $Y_{(i)}$'s as defined by Vasicek and w is found by

$$w^2 = \sum_{i=1}^{n} Y_i^2/(n-1) - n^2 \left(\sum_{i=1}^{n} Y_i^{-2} \right)^{-1} \Big/ (n-1) \tag{10.4}$$

to be small. As in the case of the entropy test of normality, even the asymptotic null distribution of this test is analytically intractable. Hence an empirical table of the 5% points was constructed and compared with the similar table in Vasicek (1976). Interestingly, the values in the two tables are remarkably close, but not close enough to be considered identical. We shall return to this point in Section 10.5. Mudholkar and Tian have also considered the use of the Kullback-Leibler information measure for testing the composite IG hypothesis against simple or restricted composite alternatives.

10.4 Four Z Tests

In this section, we outline the use of independence characterizations mentioned in Section 10.3 as the starting points for developing goodness-of-fit tests. Specifically, four tests for assessing model appropriateness are outlined. Two are for normality, one each for multivariate normality and for the composite inverse gaussian hypothesis.

1. The Z_2-Test. This test due to Lin and Mudholkar (1980), then labeled the Z test, used the characteristic independence of the sample mean and variance of random samples from normal populations. For obvious reasons, Mudholkar, Marchetti, and Lin (2000) later relabeled it.

Let $\bar{X}$ and s^2 denote respectively the mean and the variance of a random sample from a normal population. Obviously, a single replication $(\bar{X}, s^2)$ is insufficient to test the independence. To obtain a measure of dependence between $\bar{X}$ and s^2 Lin and Mudholkar start with the jackknife pseudovalues, or equivalently, $(\bar{X}_{-i}, s^2_{-i})$, $i = 1, 2, \ldots, n$, obtained by deleting one observation at a time.

It is well-known that the rank tests of independence, e.g. due to Höeffding (1948) or Blum, Kiefer, and Rosenblatt (1961), which are consistent against all bivariate dependence alternatives, suffer from very low power. For this and the practical reason of simplicity, Lin and Mudholkar first approximately symmetrize s^2 by the well-known Wilson-Hilferty (1931) cube-root transformation and propose the correlation coefficient $r = Corr(\bar{X}_{-i}, (s^2_{-i})^{1/3})$ as a measure of dependence. Furthermore, because of the normality and robustness properties of Fisher's arctanh^{-1} transform, they propose

$$Z_2 = \frac{1}{2} \log \left(\frac{1+r}{1-r} \right), \tag{10.5}$$

as a test statistic for normality. Under normality, as $n \to \infty$, $\sqrt{n} Z_2 \to N(0, 3)$. For use with small samples, $n \geq 5$, they empirically obtain approximations for $Var(Z_2)$ and $Kurtosis(Z_2)$ and recommend use of Edgeworth or Cornish-Fisher corrections to the null distribution. Furthermore, Lin and Mudholkar show that the Z_2 test is consistent against and appropriate for detecting all skewed alternatives to normality.

2. The Z_3-Test. The characterization of normality in terms of the independence between $\bar{X}$ and m_3 was used, as in the Z_2 test, by Mudholkar, Marchetti, and Lin (2000) to construct the Z_3 test of normality. However, because of the symmetry of m_3 a transformation was considered unnecessary. Specifically, for testing normality of a population based on a random sample $X_1, X_2, \ldots, X_n$, they proposed the test statistic

$$Z_3 = \frac{1}{2} \log \left(\frac{1+r_3}{1-r_3} \right), \tag{10.6}$$

where $r_3 = Corr(\bar{X}_{-i}, m_{3,-i})$. To distinguish between the two Z statistics, they named Lin and Mudholkar's statistic Z_2 and the new statistic Z_3. They show that, as $n \to \infty$ under the composite hypothesis of normality, $\sqrt{n} Z_3 \to N(0, 4)$. They also show that Z_2 and Z_3 are independent asymptotically, and empirically for samples as small as $n = 5$. Furthermore, Z_3 is shown to be consistent against all alternatives with non-normal kurtosis.

In summary, as $n \to \infty$, $\sqrt{n}(Z_2, Z_3)$ are asymptotically bivariate normal with mean $\mathbf{0}$ and covariance matrix $\mathbf{\Sigma}$ with diagonal elements $Var(Z_2) = 3$,

$Var(Z_3) = 4$ and $Cov(Z_2, Z_3) = 0$. Mudholkar, Lin, and Marchetti use the two Z-tests to detect four targeted skewness-kurtosis alternatives: right-skew heavy-tail, right-skew light-tail, left skew heavy-tail and left skew light-tail. This is done by combining the one-tail versions of the two Z-tests, which are for all practical purposes independent, using the Fisher (1932) classical method of combining independent p-values.

3. The Multivariate Z_p-Test. Mudholkar, McDermott, and Srivastava (1992) developed a p-dimensional adaptation of Lin and Mudholkar's test of normality. Let $\boldsymbol{X}_1, \boldsymbol{X}_2, \ldots, \boldsymbol{X}_n$ be a random sample from a p-variate population, with sample mean $\bar{\boldsymbol{X}}$ and sample covariance matrix $\boldsymbol{S}$. Then it is well known that, if the population is normal, the Mahalonobis distances

$$D_i^2 = (\boldsymbol{X}_i - \bar{\boldsymbol{X}})'\boldsymbol{S}^{-1}(\boldsymbol{X}_i - \bar{\boldsymbol{X}}) \tag{10.7}$$

are asymptotically independently distributed chi square variates. Hence, for large n they may be transformed into approximately i.i.d. normal variates, and the multivariate normality of the population may be tested by testing univariate normality of the transforms. Towards this end, Mudholkar, McDermott, and Srivastava (1992), empirically refine the Wilson-Hilferty transformation, and claim that, for $p \geq 2$,

$$Y_i = (D_i^2)^h, \quad i = 1, 2, \ldots, n, \text{ where } h = 1/3 - 0.11/p \tag{10.8}$$

may be considered approximately i.i.d. normal variates. They propose testing p-variate normality of the X's using

$$Z_p = Z_2(Y_1, Y_2, \ldots, Y_n) \tag{10.9}$$

as the test statistic, i.e. by applying the Z_2 test to the Y's. They derive the asymptotic null distribution of Z_p, and offer its empirical refinement, which is applicable for $n \geq 10$.

4. The Z(IG) Test. Mudholkar, Natarajan, and Chaubey (2000) have employed Khatri's (1962) characterization of the inverse gaussian distribution, as in the examples above, to construct the Z(IG) statistic for testing the composite IG hypothesis. They find that asymptotically, under the null hypothesis, as $n \to \infty$, $\sqrt{n}Z(\text{IG})$ is normally distributed with zero mean and variance 3, and present a small sample refinement of the distribution. It is interesting that the asymptotic null distribution of Z(IG) is exactly same as that of the Z_2 statistic of normality. We shall return to amplify this point in the next section.

We close this section by noting the paucity of the GOF tests for the important composite gamma distributional assumption and report that a Z test based on the characterization of the gamma distribution stated in the previous section is under development.

10.5 Byproducts: The G-IG Analogies

This section is given to supporting the proposition, suggested in the introduction, that construction of GOF tests and related issues can generate new statistical notions and knowledge. The parallelism between the characterization results for the normal and the IG families and the related GOF tests are used to illustrate the point.

It is well-known that the approximately 100 years old IG family is strikingly similar to the nearly three centuries old gaussian family in terms of its analytical simplicity and remarkably parallel statistical properties and methods; e.g. see discussion of Folks and Chhikara (1978). Arguably, the IG and the gaussian distributions may be regarded as fraternal twins, or the IG and the G families may be said to form parallel universes. Work on the GOF tests for the composite IG hypothesis described in the earlier sections resulted in a substantial expansion of the IG inference methodology and of the list of the G-IG analogies. Some of these results and connections are now outlined.

1. IG-Skewness and IG-Kurtosis. Lin and Mudholkar's Z test resulting from the characteristic independence of the sample mean and variance is targeted to detect skew alternatives. For the parallel IG test, based on the independence of the maximum likelihood estimators of the two IG parameters, the nature of its target alternatives was raised by Mudholkar, Natarajan, and Chaubey (2000). The investigation into an answer led to the following definitions of the coefficient δ_1 of IG-skewness and δ_2 of IG-kurtosis

$$\begin{aligned}\delta_1 &= \frac{E(X^2)/E^2(X) - E(X^{-1})}{(E(X)E(X^{-1})-1)\sqrt{E(X^2)/E^2(X)-1}},\\ \delta_2 &= \left[\frac{Var(V)}{E^2(V)}\right]+1, \end{aligned} \tag{10.10}$$

where $V = \Sigma(1/X_i - 1/\bar{X})/n$. Interestingly, the asymptotic joint distribution of d_1 and d_2, the estimators of δ_1 and δ_2, obtained by substituting sample moments of IG samples for the population moments, is exactly the same as that of the Pearson coefficients b_1 and b_2 based on normal samples. That is, $\delta_1 \sim N(0, 6/n)$ and $\delta_2 \sim N(0, 24/n)$ and the two are asymptotically independent; see Mudholkar and Natarajan (1999). A development of the GOF tests of the IG composite hypothesis based on d_1 and d_2, along the lines of the tests of normality based on b_1 and b_2 [D'Agostino and Stephens (1986)], is in progress.

2. IG Symmetry. The concept of IG-skewness was developed as a measure of departure from IG-symmetry, where X is a nonnegative random variable. X

is said to be IG-symmetric about its expectation μ, if

$$E\left[\left(\frac{X}{\mu}\right)^{-r}\right] = E\left[\left(\frac{X}{\mu}\right)^{r+1}\right], \qquad r = 1, 2, \dots . \tag{10.11}$$

Mudholkar and Natarajan (1999) show that the contaminated IG distributions, i.e. the IG scale mixtures, the lognormal distributions and mixtures of these are IG-symmetric, and suggest that these families may have a role in future developments of robust IG methods.

3. The ANORE Alternatives. The procedure for testing homogeneity of IG means is known as the analysis of reciprocals (ANORE), and, as in the normal theory ANOVA, uses the F distribution. The best known contrast-based normal theory alternatives to the ANOVA test are Tukey's studentized range and Dunnet's comparison with control tests. Mudholkar and Tian (2000) define the IG analogs of these tests for comparing IG means and show that the null distributions of these statistics for comparing IG means and show that their null distributions are the same as those of their normal theory counterparts. Thus these procedures can be implemented without new tables or software.

4. Robust Tests. In her dissertation, Natarajan (1998) began with the IG-GOF tests, then developed and studied IG analogs of the classical tests for equality of variances due to Bartlett, Cochran, Hartley and others, and the IG analogs of the order constrained versions of these tests due to Fujino (1979). She also developed IG analogs of the robust tests for homogeneity of variances in Mudholkar, McDermott, and Aumont (1993) and Mudholkar, McDermott, and Mudholkar (1995). Also considered in her dissertation is an IG analog of the transformation methods of Box and Cox (1964). The motivation for the entire investigation was the similarity between the normal theory and the IG theory originally stimulated by the GOF problem.

5. Order Constrained Inference. The theory for statistical inference subject to a priori order constraints goes back to Chernoff (1954) and Bartholomew (1959a,b). An excellent account of a substantial part of the growing body of literature on the subject appears in Barlow, Bartholomew, Bremner and Brunk (1971) and Robertson, Wright, and Dykstra (1988). Mudholkar and Tian (2000) have developed and studied the likelihood ratio test for comparing IG means subject to linear order constraints, and have shown that the null distributions of these tests correspond well with the normal theory results. They have also studied, in the IG context, a simple approach for general order restrictions proposed in Mudholkar and McDermott (1989) and in McDermott and Mudholkar (1993), and concluded that the results are essentially analogous to those in the normal theory.

6. Extreme Value Distributions. Extreme value and extreme spacings distributions are elegant and important parts of statistical theory and practice. Friemer *et al.* (1989) showed that the asymptotic distributions of the extremes and extreme spacings of random samples can be simply derived using elementary methods, such as Taylor expansions, with quantile functions. They also showed that if $a_n X_{n:n} + b_n$, where $X_{i:n}$ denotes the ith largest in a random sample of size n, has an asymptotic distribution, then the high probability order of magnitude of the tail length as measured by the extreme spacings is given by $X_{n:n} - X_{(n-1):n} = O_p(a_n^{-1})$. However, their work did not cover the pedagogically important normal and gamma populations, which lack closed form expressions for their quantile functions. Mudholkar and Tian (2000) fill in this gap and additionally consider the IG, the reciprocal IG (RIG) and the reciprocal root IG (RRIG) populations.

7. Simulation Considerations. Because the null distribution of Vasicek's maximum entropy test statistic is intractable, he had to tabulate it empirically. The IG test statistic by Mudholkar and Tian (2000) had the same problem and consequently had to resort to the same approach. A comparison of their table with Vasicek's showed the two to be remarkably close, but not close enough to be considered identical. However, it was noticed that in his Monte Carlo study, as recommended by IBM at the time, Vasicek used the mean of twelve uniform random variables as the generator for normal deviates. When Vasicek's experiment was repeated by Mudholkar and Tian using the current IMSL default generator, the difference between the percentiles for the entropy tests of normality and the IG hypothesis were seen to be statistically very insignificant. This observation has obvious implications for the results of early Monte Carlo experiments.

8. Conclusion. Statistical measurements are generally nonnegative and positively skewed. Hence, the IG family of such distributions may often be better suited for modeling and analysis of data than its older and better-known gaussian twin. The normal theory for statistical analysis is substantial and very well developed, whereas the theory for the IG inference may be considered to be in its infancy. There are indications that the two methodologies may be strongly analogous, and that much of the effort expended in the development of normal theory may have served the cause of statistical analysis of IG models. We emphasize that these findings were part of investigations stimulated by the interplay between the areas of characterization theorems and goodness-of-fit tests.

References

1. Barlow, R. E., Bartholomew, D. J., Bremner, J. M., and Brunk, H. D. (1971). *Statistical inference under order restrictions*, New York: John Wiley & Sons.

2. Bartholomew, D. J. (1959a). A test of homogeneity for ordered alternatives, *Biometrika*, **46**, 36–38.

3. Bartholomew, D. J. (1959b). A test of homogeneity for ordered alternatives II, *Biometrika*, **46**, 328–335.

4. Beirlant, J., Dudewicz, E. J., Györfi, L., and van der Meulen, E. C. (1997). Nonparametric entropy: An overview, *International Journal of Mathematical and Statistical Sciences*, **6**, 17–39.

5. Bingham, N. H. (2000). Studies in the history of probability and statistics XLVI. Measure into probability: From Lebesgue to Kolmogorov, *Biometrika*, **87**, 145–156.

6. Blum, J. R., Kiefer, J., and Rosenblatt, M. (1961). Distribution free tests of independence based on the sample distribution function, *Annals of Mathematical Statistics*, **32**, 485–498.

7. Box, G. E. P. and Cox, D. R. (1964). An analysis of transformations, *Journal of the Royal Statistical Society, Series B*, **26**, 211–252.

8. Chernoff, H. (1954). Testing homogeneity against ordered alternatives, *Annals of Mathematical Statistics*, **34**, 945–956.

9. Cramér, H. (1936). Über eine Eigenschaft der normalen Verteilungsfunktion, *Math. Zeitschrift*, **41**, 405–411.

10. Cressie, N. (1976). On the logarithms of high-order spacings, *Biometrika*, **63**, 343–355.

11. Csiszar, I. (1975). I-divergence geometry of probability distributions an minimization problems, *Annals of Probability*, **3**, 146–158.

12. D'Agostino, R. B. and Stephens, M. A. (Eds.) (1986). *Goodness-of-fit Techniques*, Marcel Dekker: New York.

13. Darling, D. A. (1953). On a class of problems related to the random division of an interval, *Annals of Mathematical Statistics*, **24**, 239–253.

14. Dawid, A. P. (1978). Comments on "The inverse Gaussian distribution and its statistical application: A review," *Journal of the Royal Statistical Society, Series B*, **40**, 280.

15. Dudewicz, E. and van der Meulen, E. C. (1981). Entropy-based tests of uniformity, *Journal of the American Statistical Association*, **76**, 967–974.

16. Dugué, D. (1941), Sur un nouveau type de courbe de frequence, *Comptes Rendus de l'Academie des Sciences Paris*, **213**, 634–635.

17. Ebrahimi, N., Habibullah, M., and Soofi, E. S. (1992). Testing for exponentiality based on Kullback-Leiber information, *Journal of the Royal Statistical Society, Series B*, **54**, 739–748.

18. Fisher, R. A. (1932). *Statistical Methods for Research Workers*, London: Oliver and Boyd,.

19. Folks, J. L. and Chhikara, R. S. (1978). The inverse gaussian distribution and its statistical application - a review, *Journal of the Royal Statistical Society, Series B*, **40**, 263–289.

20. Freimer, M., Kollia, G., Mudholkar, G. S., and Lin, C. T. (1989). Extremes, extreme spacings and outliers in the Tukey and Weibull families, *Communications in Statistics—Theory and Methods*, **18**, 4261–4274.

21. Fujino, Y. (1979). Tests for the homogeneity of a set of variances against ordered alternatives, *Biometrika*, **66**, 133–140.

22. Geary, R. C. (1936). The distribution of 'Student's' ratio for non-normal samples, *Supplement to the Journal of the Royal Statistical Society*, **3**, 178–184.

23. Gokhale, D. V. (1983). On entropy-based goodness-of-fit tests, *Computational Statistics and Data Analysis*, **1**, 157–165.

24. Greenwood, M. (1946). The statistical study of infectious disease, *Journal of the Royal Statistical Society, Series B*, **109**, 85–110.

25. Hacking, I. (1984). Trial by numbers, *Science-84*, **5**, 69–73.

26. Hall. P. (1984). Limit theorems for sums of general functions of m-spacings, *Mathematical Statistics and Data Analysis*, **1**, 517–532.

27. Hall, P. (1986). On powerful distributional tests on sample spacings, *Journal of Multivariate Analysis*, **19**, 201–255.

28. Höeffding, W. (1948). A nonparametric test of independence, *Annals of Mathematical Statistics*, **19**, 546.

29. Hwang, T-Y. and Hu, C-Y. (1999). On a characterization of the gamma distribution: The independence of the sample mean and the sample coefficient of variation, *Annals of the Institute of Statistical Mathematics*, **51**, 749–753.

30. Iyengar, S. and Patwardhan, G. (1988). Recent developments in the inverse Gaussian distribution, *Handbook of Statistics Volume 7*, 479–490.

31. Kagan, A. M., Linnik, Y. V., and Rao, B. (1973). *Characterization Problems in Mathematical Statistics*, New York: John Wiley & Sons.

32. Kashimov, S. A. (1989). Asymptotic properties of functions of spacings, *Theory of Probability and its Applications*, **34**, 298–307.

33. Khatri, C. G. (1962). A characterization of the inverse gaussian distribution, *Annals of Mathematical Statistics*, **33**, 800–803.

34. Kolmogorov, A. (1933). Sulla determinazione empirica di una legge di distribuzione, *Gior. Ist. Ital. Attuari*, **4**, 83–91.

35. Kullback, S. (1959). *Information Theory and Statistics*, p. 15, New York: John Wiley & Sons.

36. Kullback, S. and Leibler, R. A. (1951). On information and sufficiency, *Annals of Mathematical Statistics*, **22**, 79–86.

37. Lin, C. C. and Mudholkar, G. S. (1980). A simple test for normality against asymmetric alternatives, *Biometrika*, **67**, 455–461.

38. Lukacs, E. (1942). A characterization of the normal distribution, *Annals of Mathematical Statistics*, **13**, 91–93.

39. Lukacs, E. and Laha, R. G. (1964). *Applications of Characterizations*, New York: Hafner.

40. McDermott, M. P. and Mudholkar, G. S. (1993). A simple approach to testing homogeneity of order-constrained means, *Journal of the American Statistical Association*, **88**, 1371–1379.

41. Moran, P. A. P. (1951). The random division of an interval - Part II, *Journal of the Royal Statistical Society, Series B*, **9**, 92–98.

42. Mudholkar, G. S. and Lin, C. T. (1984). On two applications of characterization theorems to goodness of fit, *Colloquia Mathematica Societatis Janos Bolyai*, **45**, 395–414.

43. Mudholkar, G. S., Marchetti, C. E., and Lin, C. T. (2000). Independence characterizations and testing normality, *Journal of Statistical Planning and Inference* (to appear).

44. Mudholkar, G. S. and McDermott, M. P. (1989). A class of tests for equality of ordered means, *Biometrika*, **76**, 161–168.

45. Mudholkar, G. S., McDermott, M. P., and Aumont, J. (1993). Testing homogeneity of ordered variances, *Metrika*, **40**, 271–281.

46. Mudholkar, G. S., McDermott, M. P., and Mudholkar, A. (1995). Robust finite-intersection tests for homogeneity of ordered variances, *Journal of Statistical Planning and Inference*, **43**, 185–195.

47. Mudholkar, G. S., McDermott, M. P., and Srivastava, D. K. (1992). A test of p-variate normality, *Biometrika*, **79**, 850–854.

48. Mudholkar, G. S. and Natarajan, R. (1999). The inverse gaussian analogs of symmetry, skewness and kurtosis, *Annals of the Institute of Statistical Mathematics* (to appear).

49. Mudholkar, G. S., Natarajan, R., and Chaubey, Y. P. (2000), Independence characterization and inverse gaussian goodness of fit composite hypothesis, *Sankhyā* (to appear).

50. Mudholkar, G. S. and Tian, L. (2000). An entropy characterization of the inverse gaussian distribution and related goodness of fit test, *Technical Report*, University of Rochester, Rochester, NY. Submitted for publication.

51. Natarajan, R. (1998). An investigation of the inverse Gaussian distribution with an emphasis on Gaussian analogies, *Ph.D. Thesis*, University of Rochester, Rochester, NY.

52. Pearson, K. (1892). *The Grammar of Science*, London: W. Scott.

53. Pearson, K. (1900). On the criterion that a given system of deviations from the probable in the case of a correlated system of variables is such that it can reasonably be supposed to have risen from random sampling, *Phil. Mag.*, **5**, 157–175.

54. Pearson, K. (1933). On a method of determining whether a sample of given size n supposed to have been drawn from a parent population having a known probability integral has probably been drawn at random, *Biometrika*, **25**, 379–410.

55. Pearson, K. and Filon, L. N. G. (1898). *Philosophical Transactions of the Royal Society of London*, **191**, 229–311.

56. Rao, C. R. and Shanbhag, D. N. (1986). Recent results on characterization of probability distributions: A unified approach through extensions of Deny's theorem, *Advances in Applied Probability*, **18**, 660–678.

57. Robertson, T., Wright, F. T., and Dykstra, R. L. (1988). *Order Restricted Statistical Inference*, New York: John Wiley & Sons.

58. Schrödinger, E. (1915). Zur Theorie der Fall-und-Steigversuche an Teilchenn mit Brownsche Bewegung, *Physikalische Zeitschrift*, **16**, 289–295.

59. Seshadri, V. (1993). *The Inverse Gaussian Distribution: A Case Study in Exponential Families*, Oxford: Clarendon Press.

60. Seshadri, V. (1999). *The Inverse Gaussian Distribution: Statistical Theory and Applications*, New York: Springer-Verlag.

61. Shannon, C. E. (1949). *The Mathematical Theory of Communication*, p. 55, University of Illinois Press.

62. Shapiro, S. S. and Wilk, M. B. (1965). An analysis of variance test for normality (complete samples), *Biometrika*, **52**, 591–611.

63. Smoluchovsky, M. V. (1915). Notiz uber die berechnung der Browschen molekular-bewegung bei der ehrenhaft-millikanschen versuchsanordnung, *Phy. Z.*, **16**, 318–321.

64. Soofi, E. S., Ebrahimi, N., and Habibullah, M. (1995). Information distinguishability with application to analysis of failure data, *Journal of the American Statistical Association*, **90**, 657–668.

65. Tweedie, M. C. K. (1945). Inverse statistical variance, *Nature*, **155**, 453.

66. van Es, B. (1992). Estimating functionals related to a density by a class of statistics based on spacings, *Scandinavian Journal of Statistics*, **19**, 61–72.

67. Vasicek, O. (1976). A test for normality based on the sample entropy, *Journal of the Royal Statistical Society, Series B*, **38**, 54–59.

68. Vincze, I. (1984). *Colloquia Mathematica Societatis Janos Bolyai*, **45**, 395–414.

69. Wald, A. (1947). *Sequential Analysis*, New York: John Wiley & Sons.

70. Wilson, E. B. and Hilferty, M. M. (1931). The distribution of chi-square, *Proceedings of the National Academy of Sciences*, **17**, 684–688.

71. Zinger, A. A. (1951). On independent samples from normal populations, *Uspeki Mat. Nauk*, **6**, 172–175.

11

Goodness-of-Fit Tests Based on Record Data and Generalized Ranked Set Data

Barry C. Arnold, Robert J. Beaver, Enrique Castillo, and Jose Maria Sarabia

University of California, Riverside, California
University of California, Riverside, California
University of Cantabria, Santander, Spain
University of Cantabria, Santander, Spain

Abstract: Assume that observations have common distribution function F. We wish to test $H : F = F_0$ where F_0 is a completely specified distribution. Two kinds of data are considered: (i) The first $k+1$ record values $X^{(0)}, X^{(1)}, \ldots, X^{(k)}$ or possibly several independent sets of records based on observations with distribution F. (ii) Generalized ranked set data, i.e., J independent order statistics $X_{i_j:n_j}$; with common parent distribution F. Several appropriate goodness-of-fit tests are described and evaluated by simulation studies. The more general problem dealing with the composite hypothesis $H : F \in \{F(\cdot;\boldsymbol{\theta}) : \boldsymbol{\theta} \in \Theta\}$ is also discussed.

Keywords and phrases: Chi-squared test, Watson statistic, missing data, order statistics

11.1 Introduction

The classic problem of goodness-of-fit involves determining whether a set of i.i.d. observations can be reasonably supposed to have common distribution function F_0, a completely specified distribution. It is often assumed, and is assumed here, that F_0 is continuous. Thus, via a straightforward transformation, we reduce the problem to one of testing goodness-of-fit to either a uniform or an exponential distribution, whichever is deemed convenient. The non-standard feature of the analysis in this paper rests in the nature of the data assumed available. We assume that our data will consist either of record values from observations with common distribution F, or will consist of independent order statistics with common parent distribution F (i.e. generalized ranked set samples [Kim and Arnold (1999)]. In both cases we will wish to test $H : F = F_0$.

It is natural to also consider the problem of testing a composite hypothesis $H : F \in \{F_{\boldsymbol{\theta}} : \boldsymbol{\theta} \in \Theta\}$ using record and ranked set data configurations. In such a situation the first step will be to use the data to estimate $\boldsymbol{\theta}$.

Simulation based power studies are provided for the proposed tests in the simple hypothesis case. The major emphasis will be on the ranked set data case. As we shall see in Section 11.2, record value data is readily analysed by taking advantage of characteristic properties of record spacings for exponentially distributed data.

11.2 Record Data

If our data consists of a set of $k+1$ records $X_{(0)}, X_{(1)}, \ldots, X_{(k)}$ (or perhaps J independent sets of records) then our problem is readily transformed to a standard exponential goodness-of-fit problem. This is true because the spacings of the transformed records $Y_{(i)} = -\log(1 - F_0(X_{(i)}))$ will, under $H : F = F_0$, be i.i.d. exponential variables.

If more generally, the hypothesis to be tested is $H : F \in \{F_{\boldsymbol{\theta}} : \boldsymbol{\theta} \in \Theta\}$, then it will be necessary to obtain record based estimates of $\boldsymbol{\theta}$ to be used in transforming the records before testing the spacings for exponentiality. See Arnold, Balakrishnan, and Nagaraja (1998) for discussion of record based parameter estimates.

11.3 Generalized Ranked Set Data

Here our data will consist of J independent order statistics $X_{i_1:n_1} X_{i_2,n_2}, \ldots, X_{i_J:n_J}$, with common parent distribution F. To test $H : F = F_0$, we may consider

$$Y_{i_j:n_j} = F_0(X_{i_j:n_j}) \tag{11.1}$$

and ask whether these are reasonably supposed to be uniform order statistics. A Pearson-like goodness-of-fit statistic for this is of the form

$$T = \sum_{j=1}^{J} \frac{(Y_{i_j:n_j} - \frac{i_j}{n_j+1})^2}{i_j(n_j - i_j + 1)(n_j+1)^{-2}(n_j+2)^{-1}} \,. \tag{11.2}$$

Large values of T will be cause for rejection of H. The null distribution of T would be expected to be approximately χ^2_J if J is large, the n_j's are large and if the ratios i_j/n_j are not too extreme. In practice however, the n_j's will be small. If J is large a χ^2_J approximation may be adequate. If J is small then a more accurate evaluation of the null distribution of T will be needed. Balanced rank set samples are most commonly used. These consist of m independent replicates of a complete set of n independent order statistics $X_{1:n}, X_{2:n}, \ldots, X_{n:n}$ where n is small and m is usually not so small. Simulation based upper 90, 95 and 99th percentiles of the statistic T for such balanced ranked set samples are provided in Table 11.1 for an array of choices of values for m and n. These simulations based on 200,000 replications for some representative choices of m and n can be expected to provide two figure accuracy and often three figure accuracy. More extensive tables will be published elsewhere.

The discrepancies between the percentiles displayed in Table 11.1 and the corresponding χ^2_{mn} approximation can be quite large. Some representative comparisons are given in Table 11.2. Note that the percentage error is in the range -7.8% to 6.3%.

It is evident from Table 11.2, that only for large values of mn (say $mn > 100$) is it reasonable to use a χ^2 approximation for the 90th percentile of T. Even larger values of mn are required if we wish to accurately approximate the 95th and 99th percentiles. In general recourse should be made to the simulated values in Table 11.1.

Of course, one could instead have transformed to get exponential order statistics instead of uniform ones. Thus we might define

$$Z_{i_j:n_j} = -\log(1 - F_0(X_{i_j:n_j})) \tag{11.3}$$

and consider the related Pearson type statistic

$$\tilde{T} = \sum_{i=1}^{J} (Z_{i_j:n_j} - \mu_{i_j:n_j})^2 / \sigma^2_{i_j:n_j} \tag{11.4}$$

where

$$\mu_{i_j:n_j} = E(Z_{i_j:n_j}) = \sum_{k=0}^{i_j-1} \frac{1}{n_j - k}$$

and

$$\sigma^2_{i_j:n_j} = var(Z_{i_j:n_j}) = \sum_{k=0}^{i_j-1} \frac{1}{(n_j - k)^2} \,.$$

Table 11.3 includes simulation based percentiles for the statistics $\tilde{T}$ for balanced ranked set samples for an array of choices of m and n. These simulations are based on 200,000 replications for each choice of m and n. More extensive tables will be published elsewhere.

Table 11.4 provides indications of the discrepancies between the entries of Table 11.3 and the corresponding χ^2_{mn} approximation for the distribution of $\tilde{T}$. Here too large values of mn are required if we wish to reliably approximate the

Table 11.1: Simulation based upper 90, 95 and 99th percentiles of the statistic T for different values of n and m

n	m	$T_{0.90}$	$T_{0.95}$	$T_{0.99}$
2	1	4.2774	5.5352	7.6272
3	1	5.8970	7.6109	11.3275
5	1	9.1312	11.3177	16.8318
10	1	16.2907	19.3516	26.9705
1	3	5.1017	5.7119	6.8890
3	3	14.4149	16.6576	21.2247
5	3	22.5614	25.7760	32.6179
10	3	41.3461	45.9678	56.7119
2	5	15.0453	16.8732	20.4360
3	5	21.9053	24.4662	29.8155
5	5	34.7321	38.5275	46.5919
10	5	64.8435	70.3943	82.6211
1	10	13.7028	14.8317	16.9072
3	10	39.7595	43.1229	49.9132
5	10	63.8748	68.7633	78.7493
10	10	120.8672	128.2020	143.1316
2	25	60.9863	64.4107	71.3143
3	25	90.0779	95.0251	104.6656
5	25	146.7982	154.0079	168.1895
10	25	283.1010	293.8117	315.2286
1	100	111.5231	114.8760	121.2865
3	100	329.8593	338.9327	356.5544
5	100	543.0533	556.2458	581.1844
10	100	1065.0612	1084.8261	1122.8240

Table 11.2: Accuracy of chi-square approximations for percentiles of T

n	m	Percentile	Simulated value	Chi-square approximation	%Error
2	5	.90	15.0453	15.9872	6.26
5	5	.95	38.5275	37.6525	−2.27
10	5	.99	82.6211	76.1539	−7.83
3	10	.90	39.7595	40.2560	1.24
6	10	.95	80.9052	79.0819	−2.25
9	10	.99	118.4600	124.1163	4.77
4	25	.90	118.7316	118.4980	−0.19
7	25	.95	210.4848	206.8668	−1.72
10	25	.99	315.2286	304.9369	−3.26
2	50	.90	115.4174	118.4980	2.67
5	50	.95	290.2510	287.8815	−0.84
10	50	.99	589.0752	576.4928	−2.14

90th percentile of $\tilde{T}$ by the corresponding χ^2_{mn} percentile. Even larger values are needed to accurately approximate the 95th and 99th percentiles.

Predictably, the test procedures based on T and $\tilde{T}$ will be more powerful against different alternatives.

A small simulation study shows for example that, for testing $F(x) = e^x/1+e^x$ (a logistic distribution) based on a balanced ranked set sample with $m = 10, n = 3$, the test based on T is more powerful against a standard normal alternative while a test based on $\tilde{T}$ is more powerful against a standard Cauchy alternative. Such power considerations will be discussed further in Section 11.4.

An attractive alternative approach in the generalized ranked set sampling scenario is to treat the problem as one involving a considerable amount of missing data. Thus we have one observation $X_{i_1:n_1}$ from the first set of n_1 observations, $n_1 - 1$ of which are missing. Similarly $X_{i_2:n_2}$ has $n_2 - 1$ missing observations associated with it. Using the notation $N = \sum_{j=1}^{J} n_j$ we have J observations and $N - J$ missing values. Our approach then is to simulate the missing data assuming $H : F = F_0$ and then use standard goodness-of-fit procedures applied to the augmented samples. In practice we will transform using (11.1) or (11.3) to uniform or exponential order statistics and readily simulate the missing uniform or exponential order statistics. For example if we transform to uniform order statistics then the missing data corresponding to $Y_{i_j:n_j}$ can be simulated by generating:

$$(i_j - 1) \text{ i.i.d. uniform } (0, Y_{i_j:n_j}) \text{ variates}$$

Table 11.3: Simulation based upper 90, 95 and 99th percentiles of the statistic $\tilde{T}$ for different values of n and m

n	m	$T_{0.90}$	$T_{0.95}$	$T_{0.99}$
1	1	1.7345	4.0986	13.1464
3	1	6.5532	10.2042	21.5125
5	1	10.2478	14.3793	26.2311
10	1	17.5866	22.1089	34.1520
1	3	6.7389	10.9627	24.0590
3	3	17.4239	23.1078	38.3326
5	3	25.8054	31.9416	47.2485
10	3	44.2939	50.8899	66.8919
1	5	11.0281	16.2471	31.3853
3	5	26.6323	33.3458	50.9238
5	5	39.4436	46.7528	63.7632
10	5	68.6998	76.6287	94.9124
1	10	19.9969	26.7644	45.2539
3	10	47.4434	56.0884	76.7791
5	10	71.1083	80.2365	101.6653
10	10	126.6812	137.1765	159.6729
1	25	42.5964	52.0486	75.1563
3	25	103.2947	115.2631	141.7282
5	25	158.6502	171.6593	199.9937
10	25	292.0612	306.7315	337.4402
1	100	136.7013	151.6927	184.0668
3	100	356.4998	376.3445	417.3805
5	100	566.5579	589.0772	634.3068
10	100	1083.0568	1109.6497	1162.5362

Table 11.4: Accuracy of chi-square approximations for percentiles of $\tilde{T}$

n	m	Percentile	Simulated value	Chi-squared approximation	% Error
2	5	.90	19.3748	15.9872	−17.48
5	5	.95	46.7528	37.6252	−19.46
10	5	.99	94.9124	76.1539	−19.76
3	10	.90	47.4434	40.2560	−15.15
6	10	.95	91.8737	79.0819	−13.73
9	10	.99	147.7456	124.1163	−15.99
4	25	.90	131.1570	118.4980	−9.65
7	25	.95	226.4419	206.8668	−8.64
10	25	.99	337.4402	304.9396	−9.63
2	50	.90	134.5036	118.4980	−11.90
5	50	.95	313.9644	287.8815	−8.31
10	50	.99	618.3149	576.4928	−6.76

and

$$(n_j - i_j) \text{ i.i.d. uniform } (Y_{i_j:n_j}, 1) \text{ variates .}$$

Simulations will enable us to determine the power of these procedures against a variety of alternatives. Currently our simulations are based on balanced ranked set samples, but obviously unbalanced data can be accommodated.

Assuming that we transform to uniform order statistics using (11.1) a reasonable test statistic to apply to the augmented sample of size $N = \sum_{j=1}^{k} n_j$ is Stephens' (1970) modified version of the Watson (1961) U^2 statistic. For background see D'Agostino and Stephens (1986, p. 248–249). If we denote the ordered N observations by $Y_{(1)}, \ldots, Y_{(N)}$ then

$$U^2 = \frac{1}{12N} + \sum_{i=1}^{N} (\frac{2i-1}{2N} - Y_{(i)})^2 - N(\bar{Y} - 0.5)^2 \tag{11.5}$$

and the modified statistic is given by

$$U^2_{MOD} = \{U^2 - \frac{0.1}{N} + \frac{0.1}{N^2}\}\{1 + \frac{0.8}{N}\}. \tag{11.6}$$

Critical values for U^2_{MOD} when $N > 10$ were supplied by Stephens. They are: 90th percentile = 0.152, 95th percentile = 0.187 and 99th percentile = 0.267.

For values of $N \leq 10$, Quesenberry and Miller (1977) provide simulated critical values that differ only slightly from the values corresponding to the case $N > 10$.

Under the null hypothesis $F = F_0$, our augmented sample of Y's will be distributed as a sample of size N from a uniform $(0, 1)$ distribution and so the relevant critical value of U^2_{MOD} is the customary value for a random sample of size N.

11.4 Power

In Section 11.3, we introduced 3 different tests for a goodness-of-fit of $H : F = F_0$ where F_0 is a completely specified distribution. Critical values for the test statistics $T, \tilde{T}$ and U^2_{MOD}, based on simulation studies, were provided. We turn now to consider how well the tests perform. A priori it is not easy to visualize which of the 3 tests will be best for particular situations. A small power simulation study, reported in Section 11.3, indicated that sometimes T is more powerful than $\tilde{T}$ and sometimes the situation is reversed.

Simulated power studies can provide some guidance in selection of a test from the 3 available. They provide only limited information since the results obtained may well be specific to the particular alternatives considered and the particular sample sizes used, etc. The simulation studies to be reported in this section are based on balanced rank set samples with a spectrum of choices of values of m and n. In all cases a test of size .05 was used. The null hypothesis was that F is a standard normal distribution. Four alternative hypotheses were considered: Normal (0,4), Normal (2,1), Logistic (0,1) and Logistic (0,4).

The simulated power determinations are based on 10,000 replications for T and $\tilde{T}$ and on 20,000 replications for U^2_{MOD}. The results for a selection of values of m and n are displayed in Tables 11.5–11.7. More extensive tables will be presented elsewhere.

Comparison of Tables 11.5–11.7 reveals that, almost uniformly over the range of values of m and n considered, the $\tilde{T}$ test is more powerful than the T test which itself is more powerful than the U^2_{MOD} test. It must be emphasized that this may be specific to the choice of null hypothesis (normal) and the choices of alternatives. We know that for a logistic null hypothesis, as reported in Section 11.3, $\tilde{T}$ is not uniformly more powerful than T. More extensive and detailed power simulations will be required to resolve the issue. For the moment, however, for a standard normal null hypothesis the test based on $\tilde{T}$ seems to be the one to choose.

The reader will have noticed from Tables 11.5–11.7 that none of the tests is really able to distinguish standard normal data from standard logistic data. This is especially true for the test based on T which actually appears to be biased since the test of size .05 actually rejects normality less than 5% of the

Table 11.5: Power of the T test of size .05 with a standard normal null hypothesis

n	m	95th Percentile	N(0,1)	N(0,4)	N(2,1)	L(0,1)	L(0,4)
1	1	2.709000	0.049300	0.319200	0.514400	0.057200	0.296400
3	1	7.610900	0.055000	0.191400	0.925600	0.039900	0.156300
5	1	11.317700	0.055100	0.286700	0.999700	0.042000	0.242700
10	1	19.351601	0.049500	0.638600	1.000000	0.048200	0.491700
1	3	5.711900	0.051100	0.446000	0.813000	0.040000	0.374200
3	3	16.657600	0.051100	0.452200	1.000000	0.034600	0.366800
5	3	25.775999	0.050300	0.636800	1.000000	0.036900	0.508400
10	3	45.967800	0.048500	0.975200	1.000000	0.046800	0.909700
1	5	8.461500	0.049900	0.576100	0.939900	0.038300	0.481800
3	5	24.466200	0.056000	0.640900	1.000000	0.029200	0.538100
5	5	38.527500	0.048300	0.842100	1.000000	0.032700	0.721600
10	5	70.394302	0.046800	0.999300	1.000000	0.047000	0.986100
1	10	14.831700	0.050300	0.794400	0.997100	0.030300	0.695800
3	10	43.122898	0.050700	0.896600	1.000000	0.025800	0.810600
5	10	68.763298	0.051300	0.985800	1.000000	0.026100	0.947600
10	10	128.201996	0.046000	1.000000	1.000000	0.042600	1.000000
1	25	32.535400	0.047500	0.984100	1.000000	0.021000	0.946800
3	25	95.025101	0.054900	0.999000	1.000000	0.018600	0.991200
5	25	154.007904	0.045500	0.999900	1.000000	0.020800	1.000000
10	25	293.811707	0.045700	1.000000	1.000000	0.037900	1.000000
1	100	114.875999	0.053200	1.000000	1.000000	0.006500	1.000000
3	100	338.932709	0.053900	1.000000	1.000000	0.004900	1.000000
5	100	556.245789	0.047800	1.000000	1.000000	0.007400	1.000000
10	100	1084.826050	0.046200	1.000000	1.000000	0.026900	1.000000

Table 11.6: Power of the $\tilde{T}$ test of size .05 with a standard normal null hypothesis

n	m	crit0.95	N(0,1)	N(0,4)	N(2,1)	L0,1)	L(0,4)
1	1	4.098600	0.050300	0.196600	0.631800	0.048300	0.185500
3	1	10.204200	0.054000	0.417400	0.993100	0.049300	0.370300
5	1	14.379300	0.051800	0.586000	1.000000	0.052700	0.530800
10	1	22.108900	0.048900	0.906300	1.000000	0.060800	0.848100
1	3	10.962700	0.050200	0.383100	0.890400	0.059300	0.345100
3	3	23.107800	0.048100	0.721300	1.000000	0.058300	0.663200
5	3	31.941601	0.050500	0.904600	1.000000	0.070200	0.851700
10	3	50.889900	0.045300	0.999300	1.000000	0.067400	0.997000
1	5	16.247101	0.049300	0.494200	0.969600	0.071100	0.443300
3	5	33.345798	0.054800	0.857000	1.000000	0.068800	0.807100
5	5	46.752800	0.046500	0.981300	1.000000	0.067600	0.959300
10	5	76.628700	0.043800	1.000000	1.000000	0.077400	0.999900
1	10	26.764400	0.045400	0.694100	0.999100	0.080400	0.643800
3	10	56.088402	0.048200	0.978100	1.000000	0.074000	0.958400
5	10	80.236504	0.047700	0.999300	1.000000	0.073600	0.997000
10	10	137.176498	0.041500	1.000000	1.000000	0.085100	1.000000
1	25	52.048599	0.044500	0.933300	1.000000	0.101000	0.894900
3	25	115.263100	0.048900	1.000000	1.000000	0.090900	0.999700
5	25	171.659302	0.046400	1.000000	1.000000	0.088600	1.000000
10	25	306.731506	0.049800	1.000000	1.000000	0.099400	1.000000
1	100	151.692703	0.043400	1.000000	1.000000	0.152000	0.999900
3	100	376.344513	0.043500	1.000000	1.000000	0.113000	1.000000
5	100	589.077209	0.044700	1.000000	1.000000	0.103400	1.000000
10	100	1109.649658	0.040900	1.000000	1.000000	0.134300	1.000000

Table 11.7: Power of the U^2_{MOD} test of size .05 with a standard normal null hypothesis

n	m	N(0,1)	N(0,4)	N(2,1)	L(0,1)	L(0,4)
1	1					
3	1	0.04755	0.14545	0.71445	0.04755	0.11595
5	1	0.04970	0.21365	0.96800	0.05045	0.16185
10	1	0.04675	0.46550	1.00000	0.05090	0.34700
1	3	0.05505	0.18225	0.57390	0.05660	0.14675
3	3	0.04970	0.33290	0.99520	0.05195	0.25010
5	3	0.05220	0.52980	1.00000	0.05140	0.39555
10	3	0.05075	0.8948	1.00000	0.05775	0.76895
1	5	0.05015	0.27310	0.82135	0.05220	0.20155
3	5	0.04915	0.51920	1.00000	0.05210	0.39370
5	5	0.04925	0.75505	1.00000	0.05860	0.59155
10	5	0.05265	0.98940	1.00000	0.06780	0.94410
1	10	0.04995	0.50465	0.99230	0.05895	0.37085
3	10	0.04755	0.82990	1.00000	0.06070	0.67680
5	10	0.05050	0.97180	1.00000	0.06060	0.89150
10	10	0.04845	1.00000	1.00000	0.08535	0.99875
1	25	0.05010	0.89535	1.00000	0.06230	0.76280
3	25	0.05135	0.99740	1.00000	0.07750	0.97630
5	25	0.05215	1.00000	1.00000	0.09565	0.99960
10	25	0.04715	1.00000	1.00000	0.14555	1.00000
1	100	0.04935	1.00000	1.00000	0.12535	0.99990
3	100	0.04675	1.00000	1.00000	0.17100	1.00000
5	100	0.04680	1.00000	1.00000	0.25155	1.00000
10	100	0.05215	1.00000	1.00000	0.48115	1.00000

Table 11.8: Ranked set sample of shrub sizes

	Initial data				
Rank: 1	0.79	0.20	0.57	0.35	0.75
Rank: 2	1.45	0.97	0.97	0.98	1.50
Rank: 3	0.52	0.62	2.54	2.12	1.86
Rank: 1	-0.235722	-1.609440	-0.562119	-1.049820	-0.287682
Rank: 2	0.371564	-0.030459	-0.030459	-0.020203	0.405465
Rank: 3	-0.653926	-0.478036	0.932164	0.751416	0.620576

time when the data has a standard logistic distribution. The tests based on $\tilde{T}$ and U^2_{MOD} do better but an embarrassingly low power is achieved for a standard logistic alternative even for large values of m and n. It has been observed by many authors that the normal and the logistic densities are not easy to distinguish. The current study reinforces that observation.

11.5 Composite Null Hypotheses

If an hypothesis involves a parametric family of distributions then, prior to transforming to uniform or exponential order statistics, it will be necessary to utilize a ranked set based estimate of the unknown parameters. Either an EM algorithm implementation of maximum likelihood or a Gibbs sampler based diffuse prior Bayesian estimation procedure [Kim and Arnold (1999)] can be used for this part of the analysis. Simulation based evaluations of the power of such procedures in balanced ranked set settings are currently underway.

To illustrate the approach consider the following data set from Muttlak and McDonald (1990). They report an application in which interest centers on the size of shrubs. In the application the sample contained 46 shrubs. From the initial 46 shrubs a balanced ranked set sample was taken with $n = 3$ and $m = 5$. The data are in Table 8. In this example we test whether or not the sample is lognormally distributed. Thus, we take logarithms and test for normality.

To test for normality we use the following steps:

Step 1: We obtain the initial crude estimates of μ and σ using the estimates

$$\bar{\mu} = \sum_{i=1}^{k} b_j X_{i_j,n_j} \tag{11.7}$$

and

$$\bar{\sigma} = \sum_{i=1}^{k} c_j X_{i_j,n_j} \tag{11.8}$$

where

$$b_j = \left\{ \frac{1}{n} - \frac{\bar{\alpha}(\alpha_j - \bar{\alpha}}{\sum_{i=1}^{k}(\alpha_i - \bar{\alpha})^2} \right\} \tag{11.9}$$

$$c_j = \frac{\bar{\alpha}(\alpha_j - \bar{\alpha})}{\sum_{i=1}^{k}(\alpha_i - \bar{\alpha})^2} \tag{11.10}$$

and

$$\alpha_j = E(X_{i_j, n_j}). \tag{11.11}$$

These estimates for the data in Table 11.8 are:

$$\bar{\mu} = -0.125112; \quad \bar{\sigma} = 0.283775.$$

Step 2: We complete the sample by simulating the missing data using the actual estimates of μ and σ. To this end we first transform the data to a uniform sample using the cdf associated with these values of μ and σ, we simulate the uniform missing data, and finally, we return to our normal sample.

Step 3: We calculate $\bar{x} = \sum_{i=1}^{N} x_i/N$ and $s^2 - \sum_{i=1}^{N}(x_i - \bar{x})^2(N-1)$ using the actual completed sample.

Step 4: We simulate σ_i^2, an inverted gamma $IG(\frac{N-1}{2}, \frac{s^2}{2})$ random value.

Step 5: We simulate μ_i, a normal $N(\bar{x}, \frac{\sigma^2}{N})$ random value.

Step 6: We repeat Steps 2 to 5 $N_1 + N_2$ times (in the example we have used $N_1 = 500, N_2 = 500$).

Step 7: We disregard the first N_1 iterations and then estimate the parameters using

$$\hat{\mu} = \frac{1}{N_2} \sum_{i=N_1+1}^{N_1+N_2} \mu_i = -0.125202; \quad \hat{\sigma}^2 = \frac{1}{N_2} \sum_{i=N_1+1}^{N_1+N_2} \sigma_i^2 = 0.502888 \; .$$

Step 8: We complete the sample, as in Step 2, but using the estimates for μ and σ from Step 7.

Step 9: We obtain an iid uniform sample transforming the sample using the transformation

$$u_i = \Phi\left(\frac{x_i - \hat{\mu}}{\hat{\sigma}}\right),$$

where $\Phi()$ is the cdf of the standard $N(0,1)$ distribution and simulating missing uniform observations. This sample of size 45 after being sorted becomes:

0.00158157, 0.0315106, 0.0329857, 0.0552611, 0.0624795, 0.0880001, 0.100829, 0.101379, 0.112136, 0.121759, 0.146543, 0.192474, 0.214823, 0.217738, 0.24146, 0.252666, 0.292152, 0.311323, 0.373312, 0.395422, 0.413025, 0.448858, 0.536143, 0.539685, 0.569218, 0.574718, 0.574718, 0.582695, 0.599063, 0.623866, 0.624577, 0.771671, 0.803022, 0.838381, 0.841018, 0.854342, 0.895774, 0.900308, 0.930962, 0.942142, 0.954493, 0.959348, 0.980614, 0.982249, 0.993978

Step 10: We test the uniformity of this sample using the U^2 statistic given by

$$U^2 = \frac{1}{12N} + \sum_{i=1}^{N} \left\{ \frac{2i-1}{2N} - u_{(i)} \right\}^2 - N(\bar{u} - 0.5)^2 = 0.141958, \qquad (11.12)$$

where $u_{(i)}$ is the ith order statistic from the transformed sample of size N. The value of the test statistic is modified as follows prior to entering the table of critical values

$$U^2_{MOD} = \left\{ U^2 - \frac{0.1}{N} + \frac{0.1}{N^2} \right\} \left\{ 1 + \frac{0.8}{N} \right\} = 0.14227.$$

Observe that we get a value that is smaller than the critical value 0.187 at the 0.05 significance level. Thus, we cannot reject the assumption that the sample comes from a lognormal population.

11.6 Remarks

(i) Only minor modifications of our ranked set procedures will be required if more than one unit in each ranked set is measured, i.e. if some of the $X_{i_j:n_j}$'s are dependent, coming from the same sample.

(ii) In spite of our expectations, we found that the χ^2-approximations for both T and $\tilde{T}$ were not very accurate. As a result, it is important that more extensive tables for both statistics be made available as planned.

References

1. Arnold, B. C., Balakrishnan, N., and Nagaraja, H.N. (1998). *Records*, New York: John Wiley & Sons.

2. D'Agostino, R. B. and Stephens, M. A. (1986). *Goodness-of-Fit Techniques*, New York: Marcel Dekker.

3. Kim, Y. H. and Arnold, B. C. (1999). Parameter estimation under generalized ranked set sampling, *Statistics and Probability Letters*, **42**, 353–360.

4. Mattlak, H. A. and McDonald, L. L. (1990). Ranked set sampling with size-biased probability of selection, *Biometrics*, **46**, 435–445.

5. Quesenberry, C. P. and Miller, F. L., Jr. (1977). Power studies of some test of uniformity, *Journal of Statistical Computation and Simulation*, **5**, 169–191.

6. Stephens, M. A. (1970). Use of the Kolmogorov-Smirnov, Cramer-von Mises and related statistics without extensive tables, *Journal of the Royal Statistical Society, Series B*, **32**, 115–122.

7. Watson, G. S. (1961). Goodness-of-fit tests on a circle, *Biometrika*, **48**, 109–114.

PART IV

Regression and Goodness-of-Fit Tests

12

Gibbs Regression and a Test for Goodness-of-Fit

Lynne Seymour

University of Georgia, Athens, Georgia

Abstract: We explore a model for social networks that may be viewed either as an extension of logistic regression or as a Gibbs distribution on a complete graph. The model was developed for data from a mental health service system which includes a neighborhood structure on the clients in the system. This neighborhood structure is used to develop a Markov chain Monte Carlo goodness-of-fit test for the fitted model, with pleasing results.

Keywords and phrases: Gibbs distribution, Markov chain Monte Carlo, Pearson's goodness-of-fit statistic, dependent binary data

12.1 Introduction

Researchers in the social sciences require an understanding of the social network within which individuals act, as well as the individual interactions within that network. In an attempt to capture the global and local interactions simultaneously, spatial models, in which the spatial adjacency matrix is replaced by a matrix of social interdependencies, were considered [Doreian (1980, 1982, 1989)] with some success [e.g., Gould (1991)]. Another modeling effort looks at log-linear models. These models, in which the social interdependency is the observed random variable, have also been successful in modeling social networks [Strauss and Ikeda (1990), Galaskiewicz and Wasserman (1993) and Wasserman and Pattison (1996)]. Such logistic regression models are called Markov random graphs in the social science literature.

In statistical image analysis, the Gibbs distribution — which was originally introduced by Gibbs (1902) to model particle interactions in statistical mechanics — models categorical responses at each point of a regular grid; specifically, the model is known as a Gibbs random field [Besag (1974) and Geman and Geman (1984)], and may be interpreted intuitively as a distribution describing an image — a collection of (stochastic) colors at each pixel on a computer screen. Within the random field, however, "neighboring pixels" are those which are within some fixed distance of one another, which is not necessarily suitable for modeling a social network. This research article explores using a Gibbs distribution on a complete graph (i.e., one in which each point is a neighbor of every other point) as a model for social networks — a model which was initially proposed by Seymour *et al.* (2000) and which we have dubbed a Gibbs regression. The social network interpretation of this model sees individuals as vertices on a complete graph, with categorical responses at each vertex.

This paper proceeds in Section 12.2 with a brief description of the data and of the Gibbs regression model; more details may be found in Seymour *et al.* (2000). Section 12.3 fits the model, and develops and implements a goodness-of-fit test based on the neighborhood structure necessary for Gibbs distributions. Section 12.4 concludes with discussion.

12.2 The Motivation and the Model

This implementation of Gibbs regression was motivated by data from a study which investigates how the continuity of an individual client's case management (i.e., the supervision of the client's care within the service system) is affected by the management structure of a city's social services [Lehman *et al.* (1994) and Morrissey *et al.* (1994)]. Continuity of case management was considered by the experts involved to be an indicator of the stability of the service system.

In that data, there is w_{ij}, the number of organizations visited in common by clients i and j, which will be used as a measure of dependence between those individuals. In addition, there are traditional covariates: for the ith client, we have

x_{i1} indicating sex (0 if female; 1 if male),

x_{i2} indicating age (1 if 18-27; 2 if 28-33; 3 if 34-39; 4 if 40+),

x_{i3} indicating level of education (1 if did not complete high school; 2 if completed high school but did not attend college; 3 if some college education),

x_{i4} indicating schizophrenia diagnosis (0 if not schizophrenic; 1 otherwise), and

x_{i5} indicating marital status (0 if not married; 1 otherwise).

The dependent variable, Y_i, is a discrete measure of whether or not the ith client received case management.

For this study, Y_i is taken to be the simplest case – a symmetric binary variable (-1 or 1) – but in general Y_i may be a multinomial response. For example, the response of interest here is the number of case managers a client has had. Ideally, the responses should be reflect whether client i has had no case manager, has had one case manager, or has had more than one case manager. Due to model complexity and the available data, however, Y_i is herein considered as a binary variable.

Initial considerations imply a model of the form

$$P(Y_i = 1) = \frac{\exp\left\{\vec{\beta}'\vec{x}_i\right\}}{\exp\left\{\vec{\beta}'\vec{x}_i\right\} + \exp\left\{-\vec{\beta}'\vec{x}_i\right\}}, \tag{12.1}$$

where

$Y_i \in \{-1, 1\}$ is the response of the ith client;

$\vec{\beta} = \{\beta_0, \ldots, \beta_5\}'$ is a vector of regression coefficients ($'$ denotes transpose); and

$\vec{x}_i = \{1, x_{i1}, \ldots, x_{i5}\}'$ is the vector of covariates for the ith client.

If there is no interdependence, then the model in (12.1) — the logistic regression model — is adequate. However, an extension of the logistic regression model which can account for client interdependence is required. The description of and estimation strategies for such a model follows, the bulk of which may also be found in Seymour *et al.* (2000).

A general solution to this problem was given by Besag (1974). Assuming that there are no interactions of orders greater than two, and that the second-order interactions are determined by the w_{ij}'s, the general formula in Seymour *et al.* (2000) yields

$$P\{Y_1 = y_1, \ldots, Y_N = y_N\} \propto \exp\left\{\sum_i y_i \vec{\beta}'\vec{x}_i + \frac{1}{2}\sum_i\sum_j \rho(w_{ij})y_i y_j\right\}, \tag{12.2}$$

in which the interdependencies are captured by combining $\{w_{ij}\} = W$ with a real-valued function $\rho(\cdot)$ governing the strength of dependence for a given value of w_{ij}. An elementary version of this sort of interaction has in fact been considered for network autocorrelation models by Doreian (1980), specifically with $\rho(w_{ij}) = \xi w_{ij}$ where ξ is a real-valued parameter to be estimated. The conditional form of (12.2) is

$$P\{Y_i = 1|Y_j = y_j, j \neq i\}$$
$$= \frac{\exp\left\{\vec{\beta}'\vec{x}_i + \sum_{j\neq i} \rho(w_{ij})y_j\right\}}{\exp\left\{\vec{\beta}'\vec{x}_i + \sum_{j\neq i} \rho(w_{ij})y_j\right\} + \exp\left\{-\vec{\beta}'\vec{x}_i - \sum_{j\neq i} \rho(w_{ij})y_j\right\}}, \quad (12.3)$$

which is a Gibbs distribution on the network of clients, which we call Gibbs regression. Note that (12.3) reduces cleanly to (12.1) if $\rho(w_{ij}) \equiv 0$ for all values of w_{ij}.

The Gibbs regression model bears similarities to both the spatial and random graph approaches from the social sciences. While the interactions employed in (12.3) are similar to those used in the spatial approach, the responses in that approach are not categorical but are typically continuous; e.g., the spatial approach might use a model of the form $Y_i = \vec{\beta}'\vec{x}_i + \sum_{j\neq i} \rho(w_{ij})Y_j$. The random graph approach is based on log-linear models such as (12.1) and (12.3), with much of the current methodology coming from Markov random field theory. Indeed, Gibbs and Markov random fields are equivalent if the client has a fixed and finite set of neighbors which is not dependent upon the number of clients in the network Besag (1974). But the Markov property clearly does not hold for the Gibbs regression, since all clients are neighbors no matter how many clients are within the service system. More importantly, the random graph approach considers the client interactions as random responses, whereas the Gibbs regression in (12.3) assumes that all clients interact in a deterministic – and possibly zero-valued – way.

A major difficulty with (12.2) is that the normalizing constant required to make it a proper likelihood function requires summation over all possible outcomes, which is a prohibitively large set. Though this makes the direct calculation of the likelihood function impossible, there are two methods of circumventing the problem.

One such method is to use the pseudolikelihood function of Besag (1975), which is calculated by multiplying the probabilities in (12.3) over all clients. Indeed, Strauss and Ikeda (1990) use the pseudolikelihood for estimation in the random graph approach. Though the maximum pseudolikelihood estimate (MPLE) is consistent; see, Comets (1992), there are at least two problems with using the MPLE, particularly for social network models. The first of these is that the variance of the MPLE is unknown and/or intractable, depending on the strength of dependence in the random field [Possolo (1991), Cressie (1993) and Seymour (2000)]. Without the variance, there is no statistical inference for parameters based on the MPLE. The second problem is that the MPLE is unstable for "small" random fields, where the concept of "small" depends on the strength of dependence in the random field; see, Seymour (2000) . If the dependence is very weak, then a "small" sample size is 100 clients; the sample size considered "small" increases as the strength of dependence increases.

A computational technique which circumvents the intractable likelihood function in a more statistically satisfying way is the Markov chain Monte Carlo (MCMC) approximation to the likelihood function derived by Geyer and Thompson (1992). The value that maximizes that MCMC approximation with respect to θ is called a Monte Carlo MLE (MCMLE) and converges almost surely to the true MLE as the length of the chain goes to infinity. In addition, since this technique estimates the true log-likelihood function (to within a multiplicative constant), the approximation of standard errors using the observed information matrix [approximated numerically via quasi-Newton methods; see Georgii (1988)] is valid.

In principle, θ_0 may be any value in the parameter space, but in practice, it is known that the procedure works best if θ_0 is not too far from the MLE. For this purpose, the current demonstration uses the MLE under independence as θ_0 since it is easily implemented. In order to get values of θ_0 which are closer to the dependence MLE, subsequent values of the Monte Carlo MLE are iteratively assigned to θ_0 as the Monte Carlo procedure is run again. This procedure was suggested by Geyer and Thompson (1992), and results obtained using this procedure appear to be numerically stable.

12.3 Application and Evaluation of the Model

The data set described above is from a metropolitan area in the Midwestern United States. There are 34 clients who have changed case management (response is $+1$) within this mental health service system, and 63 who have retained the same case manager (response is -1) throughout the demonstration.

Table 12.1 lists information about the interaction matrix W. For example, there are 908 pairs of clients who share no organizations in common. Of those, 436 share a positive response, so that 48.02% of those pairs who have no organizations in common have each changed case management. For the sake of parsimony, it is desirable to combine or "lump" some of the values of w_{ij} together. Indeed, this was also indicated as different versions of the model were fitted to the data. The optimal lumping chosen is also given in Table 12.1. Notice that by combining the values of 2, 3, and 4, the percentage of positive responses for 2 without lumping and for 2^+ with lumping has not changed significantly (and, in an off-the-cuff comparison, a p-value of 0.7114 fails to reject the null hypothesis that these two percentages are equal).

Table 12.1: Interaction profile

	Without lumping			With lumping		
w_{ij}	$\#w_{ij}$	$\#\{Y_i = Y_j = 1\}$	%age	$\#w_{ij}$	$\#\{Y_i = Y_j = 1\}$	%age
0	908	436	0.4802	908	436	0.4802
1	2618	1079	0.4121	2618	1079	0.4121
2	994	393	0.3954	1130	438	0.3876
3	121	43	0.3554			
4	15	2	0.1333			

In order to avoid degeneracy in (12.2), the function $\rho(\cdot)$ must be restricted in some way. There are several such restrictions which could make sense in this context, such as forcing $\sum_k \rho(k)$ to be 0 or 1. In the current application, we impose the restriction that $\rho(0) = 0$, which indicates that clients who share no organizations in common are not dependent, and we force $|\rho(\cdot)|$ to be small (specifically, less than 1), which limits the dependence structure in a reasonable way given the relatively small number of individuals in the study. Indeed, there is currently no understanding of how such a strong dependence structure – called phase transition in the random field literature – will affect inference for a random field on a complete graph, though there is some understanding of inference for strongly dependent random fields on the integer lattice [Georgii (1988) and Comèts (1992)].

Table 12.2 compares the fit of three similar models to the data: the logistic (independence) regression; an extension of the logistic regression, called "logistic(+)", in which $x_{i6} = \sum_j w_{ij}$ is taken to be a covariate; and the Gibbs (dependence) regression. The logistic and logistic(+) regressions were fit using straightforward maximum likelihood parameter estimation. The Gibbs regression was fit using the MCMC approximation to the log-likelihood, as described in Section 12.2. A backward elimination procedure was used to eliminate the parameters which were not significantly different from zero. Estimates of the standard errors for the remaining parameter estimates are given in parentheses.

All three models indicate that a client's sex and level of education (variables 1 and 3) are potentially significant predictors of whether the client changes case management (males are more likely to change case management; the likelihood of changing case management increases as educational level increases), whereas a client's age, schizophrenia diagnosis, and marital status (variables 2, 4, and 5) are nowhere near significant. An interesting phenomenon is that inclusion of the W information forces the intercept to be zero in both the logistic(+) and Gibbs regressions. Particularly for Gibbs regression, this makes some sense since an intercept should depend on the neighboring responses of a given client. Using

Table 12.2: Parameter estimates

Parameters	Logistic	Logistic(+)	Gibbs
β_0	-0.5660 (0.2997)	0	0
β_1	0.3431 (0.2166)	0.4836 (0.2307)	0.4080 (0.2546)
β_2	0	0	0
β_3	0.4477 (0.1649)	0.4773 (0.1179)	0.4534 (0.1476)
β_4	0	0	0
β_5	0	0	0
β_6	—	-0.0066 (0.0020)	—
$\rho(0)$	—	—	0
$\rho(1)$	—	—	-0.0312 (0.0141)
$\rho(2)$	—	—	-0.0224 (0.0220)
NLLH	57.4805	57.6425	48.9700

the negative log-likelihoods and the standard errors of the parameter estimates as a guide, the Gibbs regression appears to contribute something significant towards explaining the relationships involved.

In order to further assess the fit of the Gibbs regression, an MCMC version of Pearson's goodness-of-fit statistic may be calculated for the "contingency table" interaction profile in Table 12.1 (with lumping). In a traditional contingency table setting, let c be the number of categories into which N responses uniquely fall. Let O_i be the observed number of responses in category $i \in \{1, \ldots, c\}$, $N = O_1 + \ldots + O_c$. Let E_i be the expected number of responses in category $i \in \{1, \ldots, c\}$, under an assumed model. Then Pearson's goodness-of-fit statistic is given by

$$\sum_{i=1}^{c} \frac{(O_i - E_i)^2}{E_i}, \tag{12.4}$$

which has a $\chi^2(c-1)$ distribution under the null hypothesis that the assumed model is the true model, assuming that the responses making up the contingency table are independent. The following development is the first goodness-of-fit test developed for Gibbs distributions.

In the current setting, the categories are the numbers of organizations shared in common (as in Table 12.1), a response is whether both of a given pair of clients have changed case management, the observed counts are given in the sixth column of Table 12.1, and the expected counts must be estimated via MCMC methods. In addition, the responses are not independent. Hence, the goodness-of-fit statistic (12.4) may not have a χ^2 distribution.

In order to evaluate (12.4), we first generate a Markov chain of social networks via the Metropolis algorithm [Metropolis *et al.* (1953)], using the candidate model with both the W matrix and the covariate information held constant. "Expected" counts of shared positive responses for each value of w_{ij} are aggregated from the chain, and the statistic (12.4) is then calculated using the number of shared positive responses in Table 12.1 as "observed" values. In a traditional contingency table setting, since $c = 3$, the appropriate distribution for this statistic is $\chi^2(2)$. In this situation, however, the MCMC Pearson statistic for the Gibbs regression model in Table 12.2 appears to be distributed Gamma(α, θ) as in Table 12.3, where α and θ depend on the length of the Markov chain. (N. B. $\chi^2(2)$ =Gamma$(1, 2)$.) We did not explore this distribution under the logistic or logistic(+) regression models.

Table 12.3: Gamma parameters for MCMC Pearson statistics

Chain Length	α	θ
10	0.8636	18.1036
100	0.8922	15.8216
1000	0.9004	14.9529

The MCMC Pearson statistics for each of the models are shown in Table 12.4; all used a Markov chain of length 1000. The expected counts and Pearson statistics of the logistic and logistic(+) regressions were calculated simply for comparison; in fact, it is expected that their MCMC Pearson statistics will be distributed differently from that of the Gibbs regression. Nevertheless, note that the expected counts match up best with the observed counts under the Gibbs regression.

Table 12.4: Results for MCMC Pearson statistic

w_{ij}	O	E-Logistic	E-Logistic(+)	E-Gibbs
0	436	399.315	679.513	428.423
1	1079	1087.619	1927.686	1045.687
2^+	438	460.915	828.139	423.341
	Statistic	4.578	644.706	1.703

Table 12.5 gives the percentiles from the simulated distribution (sample size of 5000) for the MCMC Pearson statistic under the chosen Gibbs regression for the chain lengths shown in Table 12.5. One can easily determine that one cannot reject the null hypothesis that the Gibbs regression fits this data.

Table 12.5: Percentiles under Gibbs regression

Percentiles	$M = 10$	$M = 100$	$M = 1000$
1st	0.43	0.50	0.58
5th	1.39	1.19	1.13
10th	2.31	1.98	1.80
25th	4.95	4.60	4.56
50th	10.28	9.28	8.66
75th	20.16	17.58	16.94
90th	35.87	31.60	29.59
95th	48.30	43.08	42.99
99th	80.17	72.90	69.52

12.4 Discussion

Though the Gibbs regression model may be very effective in modeling social networks, there are some difficulties with the data and with the model.

The fit of a Gibbs regression will almost surely be improved by weighting the individual organizations according to their expected impact upon the response. Unfortunately, the data described herein gave no information about the individual organizations within the service system.

There is an abundance of model diagnostic tools in the logistic regression literature which may be extended to Gibbs regression, some of which were used in Seymour *et al.* (2000); however, there are no such diagnostic tools in the Gibbs random field literature. For model selection, an ad hoc kind of backwards selection from classical multiple regression was used to choose the models in Table 12.2; some other ad hoc selection criterion could easily have been used. Again, there are numerous criteria from the logistic regression literature that could be extended. In addition, there are two criteria from the Gibbs-Markov random field literature that could be used for Gibbs regression: one MCMC-based Bayesian information criterion [Seymour and Ji (1996)], and one pseudolikelihood criterion [Ji and Seymour (1996)].

Unfortunately, there are no theoretical properties for a Gibbs distribution on a complete graph. Hopefully, the potential of the model demonstrated herein will motivate such results.

References

1. Besag, J. E. (1974). Spatial interaction and the statistical analysis of lattice systems (with discussion), *Journal of Royal Statistical Society, Series B*, **36**, 192–236.

2. Besag, J. E. (1975). Statistical analysis of non-lattice data, *The Statistician*, **24**, 179–195.

3. Comêts, F. (1992). On consistency of a class of estimators for exponential families of Markov random fields on the lattice, *Annals of Statistics*, **20**, 455–468.

4. Cressie, N. (1993). *Statistics for Spatial Data*, New York: John Wiley & Sons.

5. Doreian, P. (1980). Linear models with spatially distributed data: spatial disturbances or spatial effects, *Sociological Methods & Research*, **9**, 29–61.

6. Doreian, P. (1982). Maximum likelihood methods for linear models: spatial effect and spatial disturbance terms, *Sociological Methods & Research*, **10**, 243–269.

7. Doreian, P. (1989). Network autocorrelation models: problems and prospects. Paper presented at 1989 Symposium *"Spatial Statistics: Past, Present, Future"*, Department of Geography, Syracuse University.

8. Galaskiewicz, J. and Wasserman, S. (1993). Social network analysis: Concepts, methodology, and directions for the 1990s, *Sociological Methods & Research*, **22**, 3–22.

9. Gibbs, J. W. (1902). *Elementary Principles of Statistical Mechanics*, Yale University Press.

10. Geman, S. and Geman, D. (1984). Stochastic relaxation, Gibbs distributions, and Bayesian restoration of images, *IEEE Transactions on Pattern Analysis and Machine Intelligence*, **6**, 721–741.

11. Georgii, H. O. (1988). *Gibbs Measures and Phase Transitions*, Berlin: Walter de Gruyter.

12. Geyer, C. J. and Thompson, E. A. (1992). Constrained Monte Carlo maximum likelihood for dependent data (with discussion), *Journal of Royal Statistical Society, Series B*, **54**, 657–699.

13. Gould, R. (1991). Multiple networks and mobilization in the Paris Commune, 1871, *American Sociological Review*, **56**, 716–29.

14. Ji, C. and Seymour, L. (1996). A consistent model selection procedure for Markov random fields based on penalized pseudolikelihood, *Annals of Applied Probability*, **6**, 423–443.

15. Lehman, A., Postrado, L., Roth, D., McNary, S., and Goldman, H. (1994). An evaluation of continuity of care, case management, and client outcomes in the Robert Wood Johnson Program on chronic mental illness, *The Milbank Quarterly*, **72**, 105–122.

16. Metropolis, N., Rosenbluth, A. W., Rosenbluth, M. N., Teller, A. H., and Teller, E. (1953). Equations of state calculations by fast computing machines, *Journal of Chemical Physics*, **21**, 1087–1092.

17. Morrissey, J. P., Calloway, M., Bartko, W. T., Ridgley, S., Goldman, H., and Paulson, R. I. (1994). Local mental health authorities and service system change: Evidence from the Robert Wood Johnson Foundation Program on Chronic Mental Illness, *The Milbank Quarterly*, **72**, 49–80.

18. Nash, J. C. (1990). *Compact Numerical Methods for Computers - Linear Algebra and Function Minimisation* (2nd edition), Bristol: Adam Hilger.

19. Possolo, A. (1991). Subsampling a random field, In *Spatial Statistics and Imaging* (Ed., A. Possolo), Vol. 20, pp. 286–294, IMS Lecture Notes - Monograph Series.

20. Seymour, L. (2000). A note on the variance of the maximum pseudolikelihood estimator, Submitted to *Proceedings of the Symposium on Stochastic Processes*, Athens, Georgia.

21. Seymour, L. and Ji, C.(1996). Approximate Bayes model selection criteria for Gibbs-Markov random fields, *Journal of Statistical Planning and Inference*, **51**, 75–97.

22. Seymour, L., Smith, R., Calloway, M., and Morrissey, J. P. (2000). Lattice models for social networks with binary data, *Technical Report 2000-24*, Department of Statistics, University of Georgia.

23. Strauss, D. and Ikeda, M. (1990). Pseudolikelihood estimation for social networks. *Journal of the American Statistical Association*, **85**, 204–212.

24. Wasserman, S. and Pattison, P. (1996). Logit models and logistic regressions for social networks: I. An introduction to Markov graphs and p^*, *Psychometrika*, **61**, 401–425.

13

A CLT for the L_2 Norm of the Regression Estimators Under α-Mixing: Application to G-O-F Tests

Cheikh A. T. Diack

University of Warwick, Coventry, UK

Abstract: We establish a central limit theorem for integrated square error of least squares splines estimators based on α-mixing. The new theorem is used to study the behavior of an asymptotic goodness-of-fit test.

Keywords and phrases: G-O-F tests, B-splines, mixing

13.1 Introduction

The local and global properties of commonly used nonparametric estimators on the basis of i.i.d. observations are now well known and allow powerful methods of statistical inference such as goodness-of-fit tests. However, much less is known in the case of dependent observations. Whereas there are many papers in nonparametric curve estimation under mixing, only local properties are usually established.

In this paper, we consider the problem of estimating a regression function when the design points are nonrandom and the errors are dependent. We estimate the regression function using splines. The rate of convergence for such estimators are derived by Burman (1991). Our objective is to obtain a global measure of quality for the least squares spline as an estimate of the regression function. Specifically, we derive the central limit theorem for the integrated square error of the least squares splines estimator. We apply this new result to validating an asymptotic goodness-of-fit test. We also discuss the consistency of the proposed tests.

We consider the following regression model

$$Y_i = g(x_i) + Z_i, i = 1, \cdots, n. \tag{13.1}$$

The design points $\{x_i\}_{i=1}^n$ are deterministic. Without loss of generality, we assume that $x_i \in [0,1]$. We also assume that $\{Z_k, k \in \mathbb{Z}\}$ is the two-sided moving average

$$Z_t = \sum_{j=-\infty}^{+\infty} \psi_j X_{t-j}, \tag{13.2}$$

where $X_t \sim IID\left(0, \sigma^2\right)$ and the sequence $\{\psi_j\}$ is absolute summable. Let

$$\gamma_k = \mathbb{E} Z_i Z_{i+k} \tag{13.3}$$

be its covariance sequence. Let $\sigma(Z_i, i \leq 0)$ and $\sigma(Z_i, i \geq j)$ be the σ-fields generated by $\{Z_i, i \leq 0\}$ and $\{Z_i, i \geq j\}$, respectively. We assume that the sequence $\{Z_k, k \in \mathbb{Z}\}$ is α-mixing, that is:

$$\alpha_j = \sup_{\substack{A \in \sigma(Z_i, i \leq 0) \\ B \in \sigma(Z_i, i \geq j)}} |P(AB) - P(A) P(B)| \to 0 \text{ as } j \to +\infty.$$

We also introduce the maximal coefficient of correlation

$$\rho_j^* = \sup_{\substack{A \in L_2(\sigma(Z_i, i \leq 0)) \\ B \in L_2(\sigma(Z_i, i \geq j))}} |\mathrm{corr}(A, B)| \to 0 \text{ as } j \to +\infty.$$

We assume that the spectral density of Z is bounded away from zero and infinity.

13.2 Estimators

To estimate the function g, we use a least squares spline estimator. Let $\eta_0 = 0 < \eta_1 < ... < \eta_{k+1} = 1$ be a subdivision of the interval $[0,1]$ by k distinct points. We define $S(k,d)$ as the collection of all polynomial splines of order d (i.e., degree $\leq d-1$) having a sequence of knots $\eta_1 < ... < \eta_k$. The class $S(k,d)$ of such splines is a linear space of functions with dimension $(k+d)$. A basis for this linear space is provided by the B-splines [see Schumaker (1981)]. Let $\{N_1, .., N_{k+d}\}$ denote the set of normalized B-splines. The least squares spline estimator of g *is* defined by

$$\hat{g}(x) = \sum_{p=1}^{k+d} \hat{\theta}_p N_p(x),$$

where

$$\hat{\theta} = \left(\hat{\theta}_1, \cdots, \hat{\theta}_{k+d}\right)' = \arg\min_{\theta \in \mathbb{R}^{k+d}} \sum_{i=1}^{n} \left\{ Y_i - \sum_{p=1}^{k+d} \theta_p N_p(x_i) \right\}^2 . \tag{13.4}$$

We need to specify some conditions. For any two sequences of positive real numbers $\{a_n\}$ and $\{b_n\}$, we write $a_n \sim b_n$ to mean that a_n/b_n stays bounded between two positive constants.

We assume that the sequence of knots is generated by $p(x)$, a positive continuous density on $[0,1]$ such that

$$\int_0^{\eta_i} p(x)\, dx = \frac{i}{k+1}, i = 0, ..., k+1.$$

We set $\delta_k = \max_{0 \le i \le k} (\eta_{i+1} - \eta_i)$, then it is easy to see that

$$\delta_k \sim k^{-1}. \tag{13.5}$$

We assume that

$$\sup_{x \in [0,1]} |H_n(x) - H(x)| = o\left(k^{-1}\right) \tag{13.6}$$

where $H_n(x)$ is the empirical distribution function of $\{x_i\}_{i=1}^n$ and $H(x)$ is the limit distribution with positive density $h(x)$.

13.3 A Limit Theorem

In this section we study the global properties of our estimates. The asymptotic distribution of the functional $\int \{\hat{g}(x) - g(x)\}^2 h(x)\, dx$ is evaluated under appropriate conditions as the sample size $n \to +\infty$. We set

$$T = \int \{\hat{g}(x) - g(x)\}^2 h(x)\, dx. \tag{13.7}$$

It is convenient to introduce the following notation. Let $N(x)$ be the vector of $N_p(x), p = 1, ..., k+d$ and

$$F = (N(x_1), ..., N(x_n)) \text{ and } M_n = \frac{1}{n} \sum_{i=1}^{n} N(x_i) N(x_i)'.$$

We define the $(k+d) \times (k+d)$ matrix M_h by

$$M_h = \int N(x) N(x)' h(x)\, dx. \tag{13.8}$$

We denote the $n \times n$ matrices with (i,j)th element $\Gamma_{ij} = \gamma_{|i-j|}$ and $\Gamma_{ij}^{+} = \gamma_{i+j}$ by Γ and Γ^{+} respectively. We assume that the spectral density of $\{Z_k, k \in \mathbb{Z}\}$ is bounded away from zero and infinity. A classical result on Toeplitz matrices [see Grenander and Szegö (1984)] proves that $2\pi\lambda_{\min}\Gamma$ and $2\pi\lambda_{\max}\Gamma$ (where $\lambda_{\min}\Gamma$ and $\lambda_{\max}\Gamma$ are the smallest and the largest eigenvalues of Γ respectively) converge, respectively, to the minimum and the maximum of the spectral density of Z. Hence, the assumption on the spectral density of $\{Z_k, k \in \mathbb{Z}\}$ guarantees that the eigenvalues of Γ are bounded away from zero and infinity.

Let π_{pq} be the (p,q)th element of the $n \times n$ matrix $F'M_n^{-1}M_hM_n^{-1}F$ and

$$\pi_0 = \frac{1}{n}\mathrm{tr}\left(F'M_n^{-1}M_hM_n^{-1}F\right), \text{for } |p| = 1, ..., n-1 : \pi_p = 1/n \sum_{q=1}^{n-|p|} \pi_{|p|+q,q}. \tag{13.9}$$

We set $\Lambda = (\pi_1, ..., \pi_{n-1})'$.

Theorem 13.3.1 *Suppose that $\sum_p^{\infty} \alpha_p^{1-2/\epsilon} < \infty$ and $E|Z_1|^{2\epsilon} < \infty$ for some $\epsilon > 2$. Assume that (13.5) and (13.6) hold and $\lim_{n\to\infty} \rho_n^* < 1$. We also assume that $\{Z_t\}$ is the two-sided moving average*

$$Z_t = \sum_{j=-\infty}^{+\infty} \psi_j X_{t-j},$$

where $X_t \sim IID(0, \sigma^2)$ and $\sum_{j=-\infty}^{+\infty} |j\psi_j| < +\infty$. Then, if $k = o(n)$ and $EX_1^4 = \eta\sigma^4 < \infty$,

$$\frac{nT - \frac{1}{n}tr\left(F'M_n^{-1}M_hM_n^{-1}F\Gamma\right) - \frac{nB_{2d}}{(2d)!k^{2d}} \int \left\{g^{(d)}(x)/p^d(x)\right\}^2 h(x)\,dx}{\sqrt{2\gamma_0\Lambda'(\Gamma+\Gamma^+)\Lambda + \pi_0^2\gamma_0^2 + (\eta-3)\pi_0^2\gamma_0\sum_{p=0}^{\infty}\gamma_p + \left(\sum_{|p|<n}\pi_p\gamma_p\right)^2}}$$

$$\rightsquigarrow \mathcal{N}(0,1).$$

The above theorem says that

$$IMSE \approx \frac{1}{n^2}tr\left(F'M_n^{-1}M_hM_n^{-1}F\Gamma\right) + \frac{B_{2d}}{(2d)!k^{2d}} \int \left\{g^{(d)}(x)/p(x)^d\right\}^2 h(x)\,dx. \tag{13.10}$$

When the errors in the models are uncorrelated, $\Gamma = \gamma_0 I$ (with I the identity matrix) and $\frac{1}{n}tr\left(F'M_n^{-1}M_hM_n^{-1}F\Gamma\right) \sim k\gamma_0$. Therefore, (13.10) agrees with the results of Agarwal and Studden (1980).

Actually, we believe that Theorem 13.3.1 is new even for the case of uncorrelated errors when the variance is $2\gamma_0^2\sum_{p=0}^{n-1}\pi_p^2 + (\eta-1)\pi_0^2\gamma_0^2$.

13.4 Inference

In this section we use Theorem 13.3.1 to construct consistent nonparametric tests. We prove that the tests have asymptotic powers for some local alternatives.

Goodness-of-fit tests

The null hypothesis is that $H_\circ : g = g_0$. Against an unrestricted alternative, it is natural to use the L_2 distance between the estimator $\hat{g}$ and g_0. Therefore, the statistic of the test is given by

$$T = \int \left\{\hat{g}(x) - g_0(x)\right\}^2 h(x)\, dx.$$

Using Theorem 13.3.1, we see that the null hypotheses can be rejected at asymptotic level α if

$$nT \geq q_\alpha \sqrt{V} + \frac{1}{n} \mathrm{tr}\left(F' M_n^{-1} M_h M_n^{-1} F \Gamma\right) + \frac{n B_{2d}}{(2d)! k^{2d}} \int \left\{g_0^{(d)}(x)\right\}^2 h(x)\, dx \tag{13.11}$$

where

$$V = 2\gamma_0 \Lambda' \left(\Gamma + \Gamma^+\right) \Lambda + \pi_0^2 \gamma_0^2 + (\eta - 3)\, \pi_0^2 \gamma_0 \sum_{p=0}^{\infty} \gamma_p + \left(\sum_{|p|<n} \pi_p \gamma_p\right)^2 \tag{13.12}$$

and q_α is the upper 100α percentile of the standard normal distribution.

Specification test

Under some assumptions, the same cutoff point for the goodness-of-fit test may be used for testing composite hypotheses of the form $H_\circ : g = g_0(., \beta)$ where $\beta \in \Theta$, is an unknown parameter. However, we must use the statistic T by substituting an estimate $\hat{\beta}$ for the unknown parameter β. We need the following assumption:

$$\sup_{t \in [0,1]} \left| g(t, \beta) - m\left(t, \hat{\beta}\right) \right| = o_p\left(\left\{n^{-1} k\right\}^{1/2}\right).$$

Under some mild regularity conditions, estimators such as the least squares, generalized method for moments or the adaptive efficient weighted estimators satisfy the required assumption. Hence, the specification test has the same properties as the goodness-of-fit test.

Asymptotic power

To make a local power calculation for the tests described above, we need to consider the behavior of different statistics (calculated under a fixed but unknown point $g_0 \in H_o$) for a sequence of alternatives of the form

$$g_n(x) = g_0(x) + \tau_n \varphi(x), \tag{13.13}$$

where g_n lies in the alternative hypothesis, $\varphi(.)$ is a known function and τ_n is a sequence of real variables converging to zero.

Theorem 13.4.1 *We suppose that the assumptions of Theorem 13.3.1 hold and that*

$$nk^{-1}\tau_n^2 \to +\infty. \tag{13.14}$$

Then T has a power equal to one under the local alternatives.

Discussion

We have proposed an asymptotic goodness-of-fit test and a specification test based on least squares spline estimators. The tests are consistent and have power against some local alternatives.

In applications, the covariance matrix Γ is unknown. Therefore, we must estimate it. The estimators which we shall use for γ_r, $r \geq 0$ are

$$\hat{\gamma}_r = \frac{1}{n}\sum_{i=1}^{n-h}\left(Y_i - \bar{Y}\right)\left(Y_{i+r} - \bar{Y}\right), r = 0, ..., n-1,$$

where $\bar{Y}$ is the sample mean. The estimators $\hat{\gamma}_r, r = 0, ..., n-1$, have the desirable property that for each $n \geq 1$, the matrix $\hat{\Gamma}$ with elements $\hat{\Gamma}_{ij} = \hat{\gamma}_{|i-j|}$, is non-negative definite [cf. Brockwell and Davis (1991)]. However, plug-in $\hat{\Gamma}$ in order to estimate the variance does not guarantee that we have a consistent estimator. This is an open problem which is under study.

Using regression splines may be advantageous when we want to impose properties such as monotonicity and/or convexity. We could then test the shape of a regression function by using the functional T defined in (13.7) when we substitute an constrained estimator for g. This problem is also under study.

13.5 Proofs

Proof of Theorem 13.3.1. We can write $T = T_1 + T_2 + T_3$ where

$$T_1 = \int \left\{\hat{g}(x) - \mathbb{E}\hat{g}(x)\right\}^2 h(x)\, dx,$$

$$T_2 = \int \{g(x) - \mathbb{E}\hat{g}(x)\}^2 h(x)\,dx$$

and

$$T_3 = \int \{\hat{g}(x) - \mathbb{E}\hat{g}(x)\}\{\mathbb{E}\hat{g}(x) - g(x)\} h(x)\,dx.$$

From Theorem 3.1 in Agarwal and Studden (1980), we have

$$T_2 \approx \frac{B_{2d}}{(2d)!k^{2d}} \int \left\{g^{(4)}(x)/p(x)^d\right\}^2 h(x)\,dx. \tag{13.15}$$

On the other hand, $ET_3 = 0$ and $\text{var}(T_3) = o(\text{var}(T))$. Therefore, to prove Theorem 3, it is enough to prove that

$$T_1 \rightsquigarrow \mathcal{N}(U, V), \tag{13.16}$$

where $U = \frac{1}{n^2}\text{tr}\left(F'M_n^{-1}M_h^{(j)}M_n^{-1}F\Gamma\right)$ and

$$V = 1/n^2\left(2\gamma_0\Lambda'\left(\Gamma + \Gamma^+\right)\Lambda + \pi_0^2\gamma_0^2 + (\eta - 3)\,\pi_0^2\gamma_0\sum_{p=0}^{\infty}\gamma_p + \left(\sum_{|p|<n}\pi_p\gamma_p\right)^2\right).$$

We have

$$T_1 = \frac{1}{n^2}(FZ)'\,M_n^{-1}M_hM_n^{-1}(FZ).$$

It follows easily that $ET_1 = U$. Now, since $\{Z_k, k \in \mathbb{Z}\}$ is a strictly stationary sequence, straightforward calculations prove that

$$\text{var}\left(T_1^{(j)}\right) = \text{var}\left[\frac{1}{n}\sum_{|p|<n}\pi_p Z_0 Z_p\right],$$

where the π_p are defined by (1.9). Hence,

$$\text{var}(T_1) = \frac{1}{n^2}\sum_{|p|<n}\sum_{|q|<n}\pi_p\pi_q\text{cov}(Z_0Z_p, Z_0Z_q).$$

This can be rewritten in the following form

$$\text{var}(T_1) = \frac{1}{n^2}\sum_{|p|<n}\sum_{|q|<n}\pi_p\pi_q\left\{\mathbb{E}Z_0^2Z_pZ_q - \gamma_p\gamma_q\right\}.$$

Now $Z_t = \sum_{j=-\infty}^{\infty}\psi_j X_{t-j}, \{X_t\} \sim IID\left(0, \sigma^2\right)$ with $EX_t^4 = \eta\sigma^4$. Hence, straightforward calculations show that

$$\mathbb{E}Z_0^2Z_pZ_q = (\eta - 3)\,\sigma^4\sum_{i=-\infty}^{\infty}\psi_i^2\psi_{i+p}\psi_{i+q} + \gamma_0\gamma_{p-q} + 2\gamma_p\gamma_q.$$

It follows that

$$\operatorname{var}(nT_1) = \sum_{|p|<n}\sum_{|q|<n} \pi_p \pi_q \left\{ (\eta - 3)\,\sigma^4 \sum_{i=-\infty}^{\infty} \psi_i^2 \psi_{i+p} \psi_{i+q} + \gamma_0 \gamma_{p-q} + \gamma_p \gamma_q \right\}. \tag{13.17}$$

One can show easily the following equality

$$\sum_{|p|<n}\sum_{|q|<n} \pi_p \pi_q \gamma_0 \gamma_{p-q} = 2\gamma_0 \Lambda' \left(\Gamma + \Gamma^+\right) \Lambda + \pi_0^2 \gamma_0^2.$$

The last term of (13.17) is equal to $\left(\sum_{|p|<n} \pi_p \gamma_p\right)^2$.

On the other hand,

$$\begin{aligned} |\pi_p - \pi_0| \;\; &\leq \;\; \frac{1}{n} \left| \sum_{q=1}^{n-p} \sum_{r,s} N_r(x_q) \left(N_s(x_q) - N_s(x_{p+q}) \right) m_{rs}^* \right| \\ &\quad + \frac{1}{n} \left| \sum_{q=1}^{n-p} \sum_{r,s} N_r(x_q)\, N_s(x_q)\, m_{rs}^* \right| \end{aligned} \tag{13.18}$$

where m_{rs}^* is the (r,s)th element of the matrix $M_n^{-1} M_h M_n^{-1}$. Using equation 6.22 in Agarwal and Studden (1980) and Lemma 6.3 in Zhou, Shen, and Wolf (1998), we see that $|m_{rs}^*| = O\left(k\nu^{|r-s|}\right)$ with $\nu \in (0,1)$. Therefore, it is easy to see that the second term of the right hand side of (13.18) is $O(kp/n)$. Besides,

$$|N_s(x_q) - N_s(x_{p+q})| \;\leq\; |x_{p+q} - x_q| \sup_x |N'(x)|.$$

A classical result on B-splines proves that $\sup_x |N'(x)| = O(k)$. Moreover, using (13.6), we see that $|x_{p+q} - x_q| = \circ\left(k^{-1}p\right)$. Finally, we obtain

$$|\pi_p - \pi_0| = \mathcal{O}\left(\frac{kp}{n} \{1 + n\varepsilon_n\} \right) \tag{13.19}$$

with $\varepsilon_n \to 0$. It is worth noting that $\pi_0 \sim k$. Using this, we can write

$$\begin{aligned} &\sum_{|p|<n}\sum_{|q|<n} \pi_p \pi_q \sum_{i=-\infty}^{\infty} \psi_i^2 \psi_{i+p} \psi_{i+q} \\ &= \;\; \pi_0^2 \sum_{|p|<n}\sum_{|q|<n}\sum_{i=-\infty}^{\infty} \psi_i^2 \psi_{i+p} \psi_{i+q} \\ &\quad + \pi_0 \sum_{|p|<n}\sum_{|q|<n}\sum_{i=-\infty}^{\infty} (\pi_q - \pi_0 + \pi_p - \pi_0)\, \psi_i^2 \psi_{i+p} \psi_{i+q} \\ &\quad + \sum_{|p|<n}\sum_{|q|<n}\sum_{i=-\infty}^{\infty} (\pi_q - \pi_0)(\pi_p - \pi_0)\, \psi_i^2 \psi_{i+p} \psi_{i+q}. \end{aligned} \tag{13.20}$$

Interchanging the order of summation we find that

$$\lim_{n\to\infty} \pi_0^2\sigma^4 \sum_{|p|<n}\sum_{|q|<n}\sum_{i=-\infty}^{\infty} \psi_i^2\psi_{i+p}\psi_{i+q} = \pi_0^2\gamma_0 \sum_{p=0}^{\infty}\gamma_p \sim k^2.$$

We will show that the two other terms of (1.20) are $\circ\left(k^2\right)$. In what follows, we will denote the generic constants by $c_1, c_2, \ldots$ Using the absolute summability of $\{j\psi_j\}$ we have

$$\begin{aligned}
&\pi_0 \left| \sum_{|p|<n}\sum_{|q|<n}\sum_{i=-\infty}^{\infty} (\pi_q - \pi_0 + \pi_p - \pi_0)\, \psi_i^2\psi_{i+p}\psi_{i+q} \right| \\
&\le c_1\pi_0 \frac{k}{n}\{1+n\varepsilon_n\} \sum_{|p|<n}\sum_{|q|<n}\sum_{i=-\infty}^{\infty} (|p|+|q|) \left|\psi_i^2\psi_{i+p}\psi_{i+q}\right| \\
&\le c_2\pi_0 \frac{k}{n}\{1+n\varepsilon_n\} \sum_{|p|<n}\sum_{i=-\infty}^{\infty} |p| \left|\psi_i^2\psi_{i+p}\right| \sum_{q=-\infty}^{\infty} |\psi_{i+q}| \\
&\le c_3\pi_0 \frac{k}{n}\{1+n\varepsilon_n\} \sum_{p=-\infty}^{\infty}\sum_{i=-\infty}^{\infty} (|p+i|+|i|) \left|\psi_i^2\psi_{i+p}\right| \\
&\le c_4\pi_0 \frac{k}{n}\{1+n\varepsilon_n\} \sum_{p=-\infty}^{\infty}\sum_{i=-\infty}^{\infty} (|p+i|+|i|) \left|\psi_i\psi_{i+p}\right| \\
&\le 2c_4\pi_0 \frac{k}{n}\{1+n\varepsilon_n\} \left(\sum_{i=-\infty}^{\infty} |i|\,|\psi_i|\right) \sum_{p=-\infty}^{\infty} |\psi_p| \\
&= \; o\left(k^2\right).
\end{aligned}$$

Reasoning as above, we see that

$$\begin{aligned}
&\left| \sum_{|p|<n}\sum_{|q|<n}\sum_{i=-\infty}^{\infty} (\pi_q - \pi_0)(\pi_p - \pi_0)\, \psi_i^2\psi_{i+p}\psi_{i+q} \right| \\
&\le c_5 \frac{k^2}{n^2}\{1+n\varepsilon_n\}^2 \sum_{|p|<n}\sum_{|q|<n}\sum_{i=-\infty}^{\infty} |p|\,|q| \left|\psi_i^2\psi_{i+p}\psi_{i+q}\right| \\
&\le c_5 \frac{k^2}{n^2}\{1+n\varepsilon_n\}^2 \sum_{i=-\infty}^{\infty} \left(\sum_{p=-\infty}^{\infty} |p| \left|\psi_i\psi_{i+p}\right|\right)^2 \\
&\le c_6 \frac{k^2}{n^2}\{1+n\varepsilon_n\}^2 \sum_{i=-\infty}^{\infty} \psi_i^2 \left(\sum_{p=-\infty}^{\infty} (|p+i|+|i|) \left|\psi_{i+p}\right|\right)^2 \\
&\le c_7 \frac{k^2}{n^2}\{1+n\varepsilon_n\}^2 \sum_{i=-\infty}^{\infty} \psi_i^2 \left(1+i^2\right) \\
&= o\left(k^2\right).
\end{aligned}$$

Hence we have

$$\begin{aligned}\operatorname{var}(nT_1) &= 2\gamma_0\Lambda'\left(\Gamma+\Gamma^+\right)\Lambda+\pi_0^2\gamma_0^2+(\eta-3)\,\pi_0^2\gamma_0\sum_{p=0}^{\infty}\gamma_p \\ &\quad+\left(\sum_{|p|<n}\pi_p\gamma_p\right)^2+o\left(k^2\right). \qquad (13.21)\end{aligned}$$

Next, we show that T_1 is Gaussian. But we first show that $var\,(nT_1) \sim k^2$. Using Lemma 6.5 in Zhou, Shen, and Wolf (1998), we see that we just need to show that $\|\Lambda\|^2 \sim k^2$. We have

$$\|\Lambda\|^2=\sum_{p=0}^{n-1}\pi_p^2,$$

and

$$\begin{aligned}|\pi_p| &= \frac{1}{n}\left|\sum_{q=1}^{n-p}\sum_{r,s}N_r\left(x_q\right)N_s\left(x_{p+q}\right)m_{rs}^*\right| \\ &\le \frac{c_8k}{n}\sum_{q=1}^{n-p}\sum_{r,s}N_r\left(x_q\right)N_s\left(x_{p+q}\right)\nu^{|r-s|}\end{aligned}$$

Noting that $N_r(x_q)=0$ when $x_q\notin(t_r,t_{r+d})$ and since $|x_{p+q}-x_q|=\varepsilon_n k^{-1}p$, we have

$$|\pi_p|\le\frac{c_9k}{n}(n-p)\left(\nu^{\varepsilon_n k^{-1}}\right)^p.$$

Hence $\|\Lambda\|^2\sim k^2$. Using the definition of π_{pq} in (1.9) , we can write

$$T_1=\frac{1}{n^2}\sum_{p=1}^{n}\sum_{q=1}^{n}\pi_{pq}Z_pZ_q=T_{1,1}+T_{1,2},$$

with

$$T_{1,1}=\frac{1}{n^2}\sum_{p=1}^{n}\pi_{pp}Z_p^2+\sum_{p\neq q}\pi_{pq}\mathbb{E}Z_pZ_q$$

and

$$T_{1,2}=\frac{1}{n^2}\sum_{p\neq q}\pi_{pq}\left(Z_pZ_q-\mathbb{E}Z_pZ_q\right).$$

We have $ET_{1,1}=ET_1$. Besides,

$$\operatorname{var}\left(T_{1,1}\right)=\frac{1}{n^4}\sum_{p,q}\pi_{pp}\pi_{qq}cov\left(Z_p^2,Z_q^2\right).$$

Using Lemma 4.1 in Burman (1991) and since $|\pi_{pp}| \leq k^2$ we have for some $\epsilon > 2$

$$var\,(T_{1,1}) \leq \frac{c_{10}k^2}{n^4} \sum_{p,q} \alpha_{|p-q|}^{1-2/\epsilon} \leq \frac{c_{10}k^2}{n^3} \sum_{p} \alpha_p^{1-2/\epsilon} = \mathrm{o}\left(k^2\right).$$

Therefore it is enough to prove that $T_{1,2}$ is Gaussian. We have var$(T_{1,2})$ $\approx$var(T_1) According to Corollary 2.1 in Peligrad (1996), it is enough to prove

$$\max_{p,q} \frac{|\pi_{pq}|}{n^2\sqrt{var\,(T_1)}} \to 0 \text{ as } n \to \infty, \tag{13.22}$$

$$\sup_n \frac{1}{n^4 var\,(T_1)} \sum_{p,q} \pi_{pq}^2 < \infty, \tag{13.23}$$

and for every $\zeta > 0$

$$\frac{1}{\sigma_n^2} \sum_{p\neq q} \mathbb{E}\left(Z_pZ_q - \mathbb{E}Z_pZ_q\right)^2 \mathbb{I}\left(|Z_pZ_q - \mathbb{E}Z_pZ_q| > \zeta\sigma_n\right) \to 0 \text{ as } n \to \infty, \tag{13.24}$$

where $\sigma_n^2 = var\left(\sum_{p\neq q} Z_pZ_q - \mathbb{E}Z_pZ_q\right)$. Eqs. (13.22) and (13.23) are trivial. Reasoning as above one can show that $\sigma_n^2 \geq n^3$ hence, (13.24) follows easily and Theorem 13.3.1 is proven. ■

Proof of Theorem 13.4.1. The theorem follows quite readily from Theorem 13.3.1. ■

Acknowledgement. This research was carried out while the author was a Fellow of EURANDOM in Eindhoven, The Netherlands.

References

1. Agarwal, G. G. and Studden, W. J. (1980). Asymptotic integrated mean square error using least squares and bias minimizing splines, *The Annals of Statistics*, **8**, 1307–1325.

2. Brockwell, P. J. and Davis, R. A. (1991). *Times Series: Theory and Methods*, Second Edition, New York: Springer-Verlag.

3. Burman, P. (1991). Regression function estimation from dependent observations, *Journal of Multivariate Analysis*, **36**.

4. Grenander, U. and Szegö, (1984). *Toeplitz Forms and their Applications*, New York: Chelsea Publishing Company.

5. Peligrad, M. (1996). On the asymptotic normality of sequences of weak dependent random variables, *Journal of Theoretical Probability*, **9**, 703–715.

6. Schumaker, L. (1981). *Spline Function: Basic Theory*, New York: John Wiley & Sons.

7. Zhou, X., Shen, S., and Wolf, D. A. (1998). Local asymptotics for regression splines and confidence regions, *The Annals of Statistics*, **26**, 1760–1782.

14

Testing the Goodness-of-Fit of a Linear Model in Nonparametric Regression

Zaher Mohdeb and Abdelkader Mokkadem

Université Mentouri, Constantine, Algéria
University of Versailles-Saint-Quentin, Versailles, France

Abstract: We construct a linear hypothesis test on the regression function f in nonparametric regression model; more precisely, we test that f is an element of U, where U is a finite dimensional vector space. The test statistic is easy to compute and we give the asymptotic level and the asymptotic power of the test. Even if the procedure is based on large sample behaviour, simulation experiments reveal that, for small samples, the proposed statistic is close to the asymptotic distribution and the test has good power properties.

Keywords and phrases: Linear hypothesis, nonparametric regression, nonparametric test

14.1 Introduction

We consider the following regression model

$$Y_i = f(t_i) + \varepsilon_i\,, \quad i = 1, \ldots, n\,, \tag{14.1}$$

where f is an unknown real function, defined on the interval $[0, 1]$ and $t_1 = 0 < t_2 < \cdots < t_n = 1$, is a fixed sampling on $[0, 1]$. The errors ε_i are independent and identically distributed random variables with zero mean and variance σ^2.

Our aim is to construct a linear hypothesis test on the regression function f. Let $g_1(t), \ldots, g_p(t)$ be linearly independent functions in $[0, 1]$ and let U_p be the vector space spanned by $g_1, \ldots, g_p$; we want to test the hypothesis

$$\mathrm{H}_0: \; f \in U_p \text{ against } \mathrm{H}_1: \; f \notin U_p\,. \tag{14.2}$$

The use of nonparametric regression methods for developing tests in regression analysis has been the subject of several works. A considerable amount of literature is based on the spline method: Cox and Koh (1989), Eubank and Spiegelman (1990) derived goodness-of-fit tests for a linear model based on test statistics constructed from nonparametric regression fits to residuals from linear regression; Jayasuriya (1996) proposed a test based on nonparametric fits to the residuals from kth order polynomial regression. A method based on a kernel estimate for f is proposed by Müller (1992). Härdle and Mammen (1993) suggested a test statistic based on a weighted L^2-distance between the parametric fit and a nonparametric fit based on a kernel estimator. The use of empirical Fourier coefficients of f to construct a hypothesis test in the model (14.1) is developed by Eubank and Spiegelman (1990) and Eubank and Hart (1992).

In order to test the hypothesis H_0 in (14.2), Dette and Munk (1998) proposed a procedure based on the large sample behaviour of an empirical L^2-distance between f and the subspace U_p. Their test statistic makes use of weights. In the present paper, we propose a test statistic based on a similar approach but without weight and show that it has the same asymptotic behaviour.

The remainder of this paper is organized as follows. In Section 14.2 a test is derived for H_0 and its asymptotic distribution under the null and alternative hypotheses is given. In Section 14.3 we employ insights from Section 14.2 to derive a practical test for H_0 using Monte Carlo techniques for small samples.

14.2 The Test Statistic

We assume that the design $\{t_1, \ldots, t_n\}$ is associated to a positive density function h on the interval $[0,1]$. We denote $L^2(d\mu)$, where $d\mu(t) = h(t)\,dt$, the space of square integrable functions, equipped with the usual inner product $\langle .,. \rangle$. The distance between f and U_p is denoted by $T(f)$. We assume that the following assumptions are satisfied:

(A1) $\displaystyle \max_{i=2,\ldots,n} \left| \int_{t_{i-1}}^{t_i} h(t)\,dt - \frac{1}{n} \right| = o\left(\frac{1}{n}\right)$;

(A2) h, f, g_k, , $k = 1, \ldots, p$, satisfy the Hölder condition with order $\gamma > 1/2$;

(A3) $\forall n$, $\varepsilon_1, \ldots, \varepsilon_n$ are independent and $\exists C \in \mathbb{R}^+$ such that $E(\varepsilon_i^4) < C$, $\forall i$.

Set

$$(\widehat{a}_1, \ldots, \widehat{a}_p)' = \underset{(\alpha_1,\ldots,\alpha_p)\in\mathbb{R}^p}{\operatorname{argmin}} \left\{ \frac{1}{n} \sum_{i=1}^{n} \left| Y_i - \sum_{k=1}^{p} \alpha_k g_k(t_i) \right|^2 \right\}. \tag{14.3}$$

Since the hypothesis "$f \in U_p$" is equivalent to "$\phi = f - \sum_{k=1}^{p} \widehat{a}_k g_k \in U_p$," as a measure of discrepancy between the regression function and the subspace U_p, we use the square distance

$$T^2(\phi) = \min_{v \in U_p} \int_0^1 |\phi(t) - v(t)|^2 \, h(t)dt \, .$$

This square distance can be represented also as

$$T^2(\phi) = \frac{G(\phi, g_1, \ldots, g_p)}{G(g_1, \ldots, g_p)} \, , \tag{14.4}$$

where $G(v_1, \ldots, v_k)$ denotes the Gramian determinant $\left|\Big(\langle v_i, v_j \rangle\Big)\right|_{i,j=1,\ldots,k}$ for $v_1, \ldots, v_k$ in $L^2(d\mu)$.

We thus need to estimate $T^2(\phi)$; for this, we introduce the observations $X = (X_1, \ldots, X_n)'$, where

$$X_i = Y_i - \sum_{k=1}^{p} \widehat{a}_k g_k(t_i) \, , \quad i = 1, \ldots, n \, .$$

We follow the procedure of Dette and Munk (1998), but applied to ϕ and X_i, and without use of weights. Let $\Delta_i = t_i - t_{i-1}$ $(i = 2, \ldots, n)$, $\Delta_1 = \Delta_2$, $W = \operatorname{diag}\Big(\Delta_i h(t_i)\Big)_{i=1,\ldots,n}$ and $g_{k,n} = \Big(g_k(t_1), \ldots, g_k(t_n)\Big)'$, $k = 1, \ldots, n$. Let $U_{p,n}$ be the vector subspace of $\mathbb{R}^n$ spanned by $(g_{1,n}, \ldots, g_{p,n})$ and $\Pi_n^\perp$ the projection matrix on the orthogonal of $U_{p,n}$.

We define $G_n(X, g_1, \ldots, g_p)$ as the determinant obtained by replacing in (14.4) the inner products $\langle \phi, \phi \rangle$ and $\langle \phi, g_k \rangle$, $(k = 1, \ldots, p)$ respectively by

$$X'WX = \sum_{i=1}^{n} \Delta_i h(t_i) X_i^2$$

and

$$X'W g_{k,n} = \sum_{i=1}^{n} \Delta_i h(t_i) g_k(t_i) X_i \quad k = 1, \ldots, p \, .$$

Hence, we estimate $T^2(\phi)$ by

$$\widehat{T}_n^2 = \frac{G_n(X, g_1, \ldots, g_p)}{G(g_1, \ldots, g_p)} - S_\varepsilon^2 \operatorname{tr}\left(W \Pi_n^\perp\right) \, , \tag{14.5}$$

where S_ε^2 is the estimator of Gasser, Sroka, and Jennen-Steinmetz (1986), defined by

$$S_\varepsilon^2 = \frac{1}{6(n-2)} \sum_{i=2}^{n-1} (Y_{i+1} + Y_{i-1} - 2Y_i)^2 \, . \tag{14.6}$$

We reject H_0 if $\widehat{T}_n^2 > \tau$. We have the following asymptotic behaviour of our test statistic.

Theorem 14.2.1 *If the assumptions (A1) – (A3) are satisfied, then*

$$\sqrt{n}\left(\widehat{T}_n^2 - T^2(f)\right) \xrightarrow{d} \mathcal{N}\left(0, \frac{17}{9}\sigma^4 + 4\sigma^2 T^2(f)\right), \quad \text{as } n \to +\infty,$$

where $\xrightarrow{d}$ *denotes the convergence in distribution.*

Steps of the proof

First, we show that $G_n(X, g_1, \ldots, g_p)$ is invariant with respect to change of basis in U_p. Using then a fixed orthonormal basis $\{v_1, \ldots, v_p\}$ of U_p, the test statistic may be written

$$\begin{aligned} \widehat{T}_n^2 &= G_n(X, v_1, \ldots, v_p) - S_\varepsilon^2 \operatorname{tr}\left(W\Pi_n^\perp\right) \\ &= \sum_{i=1}^n \Delta_i h(t_i) X_i^2 - \sum_{k=1}^p \left|\sum_{i=1}^n \Delta_i h(t_i) v_k(t_i) X_i\right|^2 - S_\varepsilon^2 \operatorname{tr}\left(W\Pi_n^\perp\right). \end{aligned}$$

Then, we use analysis arguments and Lemma A.1 in Dette and Munk (1998) for the weights $w_{i,n} = 1$, $i = 1, \ldots, n$, to obtain

$$\widehat{T}_n^2 - T^2(f) = \frac{1}{n}\sum_{i=1}^{n-2} \eta_i + o_P\left(\frac{1}{\sqrt{n}}\right), \tag{14.7}$$

where $\eta_i = \eta_{i,n}$, $i = 1, \ldots, n$ are random variables which form a centered row-wise 2-dependent array.

We show that

$$\lim_{n\to\infty} \operatorname{Var}\left(\frac{1}{\sqrt{n}}\sum_{i=2}^n \eta_i\right) = \frac{17}{9}\sigma^4 + 4\sigma^2 T^2(f).$$

Now, the convergence of the distribution of $\frac{1}{\sqrt{n}}\sum_{i=2}^n \eta_i$ to normal law $\mathcal{N}\left(0, \frac{17}{9}\sigma^4 + 4\sigma^2 T^2(f)\right)$ follows from Orey (1958). Theorem 14.2.1 is obtained from (14.7).

The test procedure

Since $T^2(f) = 0$ when $f \in U_p$, the theorem provides a very simple test of the hypothesis H_0; let $\widehat{\sigma}^2$ be any consistent estimator of σ^2. We reject H_0 at the level α if the test statistic $(9n/17)^{1/2}\widehat{T}_n^2/\widehat{\sigma}^2$ exceeds the $(1-\alpha)$-quantile $c_{1-\alpha}$ of the standard normal distribution. $\widehat{\sigma}^2$ may be replaced by S_ε; however, from our simulations, it appears that the use of S_ε does not always behave well; for this reason, we suggest the use of the empirical estimator of σ^2:

$$\widehat{\widehat{\sigma}}^2 = \frac{1}{n}\sum_{i=1}^n \left\{Y_i - \sum_{k=1}^p \widehat{a}_k g_k(t_i)\right\}^2, \tag{14.8}$$

which is consistent under H_0 and gives a better approximation by the normal distribution.

14.3 Simulations

In order to investigate both the power and level of our test (14.2), we conducted a small-scale simulation using the model $Y_i = f(t_i) + \varepsilon_i$, $i = 1, \ldots, n$ ($n = 50$) with $t = (i-1)/(n-1)$, $i = 1, \ldots, n$ and the uncorrelated normal random errors with variance σ^2. In our simulations, we study the test of the hypothesis $\mathrm{H}_0 : f \in U_2$, where U_2 is the subspace of $L^2(d\mu)$ spanned by $g_1(t) = t$ and $g_2(t) = 1$. The Monte Carlo study, for small sample size $n = 50$, turns on the comparison of the statistic $\widehat{T}_n^2$ and the statistic $\hat{M}_n^2$ proposed by Dette and Munk (1998). The empirical quantiles, denoted by Q_T and Q_{DM} of the test statistics $\widehat{T}$ and $\hat{M}_n^2$ respectively, are given in Table 14.1 and Table 14.2. As shown in Table 14.1, use of the $\widehat{\sigma}^2 = S_\varepsilon^2$ estimator of Gasser, Sroka, and Jennen-Steinmetz (1986), in the statistics reveals that the normal approximation is not satisfactory. In Table 14.2, $\widehat{\sigma}^2$ is replaced by $\widehat{\widehat{\sigma}}^2 = (1/n)\sum_{i=1}^n |Y_i - \widehat{a}_1 t_i - \widehat{a}_2|^2$; it appears, that the normal law is better approximated with $\widehat{T}_n^2$ than $\hat{M}_n^2$.

In order to study the power, we consider two different forms for f, namely $f_1(t) = -1.5t + 0.5 + \beta t e^{-2t}$ and $f_2(t) = -1.5t + 0.5 + \beta t^2$ with several choices on the interval $[0, 2]$ for β. The empirical powers of $\widehat{T}$ and $\hat{M}_n^2$ are denoted by P_T and P_{DM}. As shown in Table 14.3 and Table 14.4, where the level is taken at $\alpha = 5\%$, the results of the empirical power study give a little advantage to the statistic $\widehat{T}_n^2$. We note also that the test has good power properties.

Table 14.1: Empirical quantiles, when σ^2 is estimated by S_ε^2; (theoretical values at level 1%, 5%, 10%, are 2.33, 1.65, 1.28 respectively)

	$\alpha = 1\%$		$\alpha = 5\%$		$\alpha = 10\%$	
σ	Q_T	Q_{DM}	Q_T	Q_{DM}	Q_T	Q_{DM}
0.05	3.7457	3.5102	2.3354	2.1856	1.7829	1.6124
0.10	3.6813	3.4737	2.2886	2.1561	1.7751	1.6080
0.20	3.8228	3.5927	2.3632	2.1640	1.7832	1.6098
0.50	3.7958	3.5486	2.4075	2.2063	1.7798	1.6192
1.00	3.5249	3.2360	2.2697	2.0983	1.6775	1.5229
2.00	3.5239	3.3212	2.2464	2.0923	1.6716	1.5190

Table 14.2: Empirical quantiles, when σ^2 is estimated by $\widehat{\widehat{\sigma}}^2$; (theoretical values at level 1%, 5%, 10%, are 2.33, 1.65, 1.28 respectively)

	$\alpha = 1\%$		$\alpha = 5\%$		$\alpha = 10\%$	
σ	Q_T	Q_{DM}	Q_T	Q_{DM}	Q_T	Q_{DM}
0.05	2.2711	2.1428	1.7099	1.5900	1.3495	1.2362
0.10	2.2250	2.0973	1.6540	1.5408	1.3419	1.2198
0.20	2.1796	2.0346	1.6846	1.5678	1.3870	1.2530
0.50	2.2001	2.1122	1.6989	1.5707	1.3855	1.2514
1.00	2.2982	2.1976	1.7188	1.6167	1.4083	1.2708
2.00	2.2778	2.1173	1.6877	1.5558	1.3519	1.2311

Table 14.3: Proportion of rejections in 1000 samples of size $n = 50$; with two examples of alternatives: $f_1(t) = a_1 t + a_2 + \beta t e^{-2t}$ and $f_2(t) = a_1 t + a_2 + \beta t^2$, ($\sigma^2$ estimated by S_ε^2)

		$f_1(t)$		$f_2(t)$	
	β	P_T	P_{DM}	P_T	P_{DM}
$\sigma = 0.05$	0.0	0.052	0.053	0.041	0.042
	0.1	0.060	0.058	0.067	0.065
	0.5	0.160	0.155	0.715	0.670
	1.0	0.666	0.616	0.999	0.999
	1.5	0.972	0.964	1.000	1.000
	2.0	0.999	0.999	1.000	1.000
$\sigma = 0.10$	0.0	0.057	0.059	0.054	0.052
	0.1	0.047	0.044	0.055	0.051
	0.5	0.071	0.072	0.182	0.169
	1.0	0.163	0.136	0.687	0.651
	1.5	0.355	0.327	0.990	0.986
	2.0	0.644	0.599	0.977	0.965
$\sigma = 0.20$	0.0	0.047	0.045	0.049	0.047
	0.1	0.060	0.057	0.048	0.045
	0.5	0.059	0.059	0.055	0.054
	1.0	0.075	0.072	0.200	0.190
	1.5	0.109	0.097	0.415	0.393
	2.0	0.179	0.158	0.375	0.350
$\sigma = 0.50$	0.0	0.065	0.062	0.055	0.050
	0.1	0.056	0.055	0.038	0.037
	0.5	0.050	0.047	0.053	0.048
	1.0	0.048	0.055	0.064	0.066
	1.5	0.057	0.055	0.090	0.086
	2.0	0.066	0.064	0.088	0.087

Table 14.4: Proportion of rejections in 1000 samples of size $n = 50$; with two examples of alternatives: $f_1(t) = a_1 t + a_2 + \beta t e^{-2t}$ and $f_2(t) = a_1 t + a_2 + \beta t^2$, ($\sigma^2$ estimated by $\widehat{\widehat{\sigma}}^2$)

		$f_1(t)$		$f_2(t)$	
	β	P_T	P_{DM}	P_T	P_{DM}
$\sigma = 0.05$	0.0	0.052	0.049	0.055	0.047
	0.1	0.056	0.054	0.059	0.060
	0.5	0.160	0.145	0.732	0.671
	1.0	0.647	0.576	1.000	1.000
	1.5	0.966	0.944	1.000	1.000
	2.0	1.000	0.998	1.000	1.000
$\sigma = 0.10$	0.0	0.052	0.050	0.045	0.044
	0.1	0.054	0.051	0.049	0.044
	0.5	0.067	0.066	0.170	0.166
	1.0	0.165	0.146	0.715	0.667
	1.5	0.359	0.314	0.981	0.966
	2.0	0.640	0.568	1.000	1.000
$\sigma = 0.20$	0.0	0.043	0.046	0.052	0.057
	0.1	0.047	0.053	0.043	0.041
	0.5	0.060	0.056	0.048	0.047
	1.0	0.077	0.074	0.183	0.167
	1.5	0.104	0.092	0.384	0.343
	2.0	0.173	0.155	0.693	0.635
$\sigma = 0.50$	0.0	0.050	0.048	0.028	0.030
	0.1	0.054	0.050	0.061	0.058
	0.5	0.056	0.058	0.065	0.055
	1.0	0.038	0.037	0.063	0.064
	1.5	0.051	0.052	0.097	0.093
	2.0	0.068	0.064	0.117	0.106

References

1. Cox, D. and Koh, E. (1989), A smoothing spline based test of model adequacy in polynomial regression, *Annals of the Institute of Statistical Mathematics*, **41**, 383–400.

2. Dette, H. and Munk, A. (1998). Validation of linear regression models, *Annals of Statistics*, **26**, 778–800.

3. Eubank, R. L. and Hart, J. D. (1992). Testing goodness-of-fit in regression using nonparametric via order selection criteria, *Annals of Statistics*, **20**, 1412–1425.

4. Eubank, R. L. and Spiegelman, C. H. (1990). Testing the goodness-of-fit of a linear model via nonparametric regression techniques, *Journal of the American Statistical Association*, **85**, 387–392.

5. Gasser, T., Sroka, L., and Jennen-Steinmetz, C. (1986). Residual variance and residual pattern in nonlinear regression, *Biometrika,* **73**, 625-633.

6. Härdle, W. and Mammen, E. (1993). Comparing nonparametric versus parametric regression fits, *Annals of Statistics*, **21**, 1926–1947.

7. Jayasuriya, B. R. (1996). Testing for polynomial regression using nonparametric regression techniques, *Journal of the American Statistical Association*, **91**, 1626–1631.

8. Müller, H. G. (1992). Goodness of fit diagnostics for regression models, *Scandinavian Journal of Statistics*, **19**, 157–172.

9. Orey, S. (1958). A central limit theorem for m-dependant random variables, *Duke Mathematics Journal*, **52**, 543–546.

15

A New Test of Linear Hypothesis in Regression

Y. Baraud, S. Huet, and B. Laurent

Ecole Normale Supérieure, Paris, France
INRA, Jouy-en-Josas, France
Université Paris-Sud, Orsay, France

Abstract: In the Gaussian regression model, we propose a new test, based on model selection methods, for testing that the regression function F belongs to a linear space. The test is free from any prior assumption on F and on the variance σ^2 of the errors. The procedure is rate optimal over both smooth and directional alternatives and the simulations studies show that it is also robust with respect to the non-Gaussianity of the errors.

Keywords and phrases: Adaptive test, model selection, linear hypothesis, minimax hypothesis testing, nonparametric alternative, goodness-of-fit, nonparametric regression, Fisher test

15.1 Introduction

We consider the regression model

$$Y_i = F(x_i) + \sigma\varepsilon_i, \quad i = 1, ..., n \tag{15.1}$$

where the x_i's are known deterministic points belonging to $\mathbf{R}^k$, F is an unknown function from $\mathbf{R}^k$ into $\mathbf{R}$, σ is an unknown positive number and $(\varepsilon_1, \ldots, \varepsilon_n)$ is an unobservable sample of i.i.d. standard Gaussian random variables, i.e. $\varepsilon_i \sim \mathcal{N}(0, 1)$. Let V^* be some finite dimensional subspace of the set of functions $\mathcal{F}(\mathbf{R}^k, \mathbf{R}) = \{F : \mathbf{R}^k \to \mathbf{R}\}$, that is, $V^* = \{\sum_{j=1}^{d} \theta_j \psi_j, \theta_1, \ldots, \theta_d \in \mathbf{R}\}$ where the ψ_j's are given functions in $\mathcal{F}(\mathbf{R}^k, \mathbf{R})$.

The aim of this paper is to propose a test of the null hypothesis "F belongs to V^*" against that it does not under no prior information on F and σ.

The testing procedure can be described in the following way. We consider a finite collection of linear subspaces of $\mathcal{F}(\mathbf{R}^k, \mathbf{R})$, $\{S_m^*,\ m \in \mathcal{M}\}$. We associate to the collection $\{S_m^*,\ m \in \mathcal{M}\}$ a suitable collection of numbers $\{\alpha_m,\ m \in \mathcal{M}\}$

belonging to $]0,1[$ and consider the family of Fisher tests of level α_m of the null hypothesis $F \in V^*$ against the alternative $F \in V^* + S_m^*$. Our procedure rejects the null hypothesis if one of these Fisher tests does.

Let us give some examples. If F is defined on $I = [0,1]$ one can take for $m \in \mathbf{N}$, S_m^* as the space of trigonometric polynomials of degree not larger than m, the space of piecewise polynomials with fixed degree based on the grid $\{k/m,\ k = 0,...,m\}$ or the space spanned by the Haar basis, $\{\varphi_{j,k},\ j \in \{0,...,m\},\ k \in \{1,...,2^j\}\}$. Moreover, it is possible to mix several kinds of these spaces to constitute the collection $\{S_m^*,\ m \in \mathcal{M}\}$. If F is defined on $I = [0,1]^2$, one can take for example the space of constant functions on a finite partition of $[0,1]^2$ in rectangles.

Under the Gaussian assumption on the errors, the results given in this paper are nonasymptotic. For each n, the test has the desired level and we characterize a set of functions over which our test is powerful. Under the posterior assumption that F belongs to the Hölderian ball

$$H_s(R) = \left\{F : [0,1] \to \mathbf{R}, \forall (x,y) \in [0,1]^2, |F(x) - F(y)| \le R|x-y|^s\right\}, \quad (15.2)$$

for some $R > 0$ and $s \in]0,1]$, we show that the test is rate optimal among the adaptive tests (that is among the tests which have no prior knowledge of R nor s). Such a result has been obtained by Horowitz and Spokoiny (2000) for a different procedure that will be described in Section 15.4. In addition our test recovers the parametric rate of testing over directional alternatives. A similar result has been obtained by Eubank and Hart (1992).

Finally, we present a simulation study to evaluate the performances of our testing procedure both when the errors are Gaussian and non-Gaussian.

15.2 The Testing Procedure

Let $\{S_m^*\ , m \in \mathcal{M}\}$ be a collection of subspaces of $\mathcal{F}(\mathbf{R}^k, \mathbf{R})$. For each $m \in \mathcal{M}$, we denote by S_m the linear subspace of $\mathbf{R}^n$ defined as the orthogonal projection of

$$\{(F(x_1),...,F(x_n))^T,\ F \in S_m^*\}$$

onto $V^\perp$ where

$$V = \{(F(x_1),...,F(x_n))^T,\ F \in V^*\},$$

and set $V_m = V + S_m$. We make the following assumption:

Condition ($\mathbf{H}_\mathcal{M}$): For all $m \in \mathcal{M}$ $S_m \neq \{0\}$ and $V_m \neq \mathbf{R}^n$.

This condition is usually met in practice if with regard to the design points x_i's, the collection $\{S_m^*, m \in \mathcal{M}\}$ is chosen judiciously. It implies that the dimensions of S_m and $V_m^\perp$, respectively denoted by D_m and N_m, are both different from 0.

We denote by $\mathbf{P}_F$ the law of the observation $Y = (Y_1, \ldots, Y_n)^T$ drawn from model (15.1) and by $\|.\|$ the Euclidean norm in $\mathbf{R}^n$.

For each $m \in \mathcal{M}$, we denote by Π_m and Π_{V_m} the orthogonal projectors onto S_m and V_m respectively.

15.2.1 Description of the procedure

Let us fix some $\alpha \in]0,1[$ and assume that the collection $\{S_m^*\ m \in \mathcal{M}\}$ satisfies $(\mathrm{H}_{\mathcal{M}})$. Let T_α be the test statistic defined by

$$T_\alpha = \sup_{m \in \mathcal{M}} \left\{ \frac{N_m \|\Pi_m Y\|^2}{D_m \|Y - \Pi_{V_m} Y\|^2} - \bar{F}^{-1}_{D_m,N_m}(\alpha_m) \right\} \tag{15.3}$$

where for each $u \in]0,1[$, $\bar{F}^{-1}_{D_m,N_m}(u)$ is the $1-u$ quantile of a Fisher random variable with D_m and N_m degrees of freedom and where $\{\alpha_m, m \in \mathcal{M}\}$ is a collection of numbers in $]0,1[$ satisfying

$$\forall F \in V^*,\ \mathbf{P}_F\left(T_\alpha > 0\right) \leq \alpha. \tag{15.4}$$

We reject the null hypothesis when T_α is positive. In the sequel, we choose the collection $\{\alpha_m,\ m \in \mathcal{M}\}$ in accordance with one of the following procedures:

P1 For all $m \in \mathcal{M}$, $\alpha_m = a_n$ where a_n is the largest number a in $]0,1[$ satisfying

$$\mathbf{P}\left(\sup_{m \in \mathcal{M}} \left\{ \frac{N_m \|\Pi_m \varepsilon\|^2}{D_m \|\varepsilon - \Pi_{V_m} \varepsilon\|^2} - \bar{F}^{-1}_{D_m,N_m}(a) \right\} > 0 \right) \leq \alpha. \tag{15.5}$$

where ε is a standard Gaussian vector of dimension n.

P2 The α_m's verify the following equality

$$\sum_{m \in \mathcal{M}} \alpha_m = \alpha.$$

Comments: For each $m \in \mathcal{M}$, the Fisher test of level α_m of the hypothesis $F \in V^*$ against $F \notin V^*$, $F \in V^* + S_m^*$ rejects the null hypothesis if the test statistic

$$T_m = \frac{N_m \|\Pi_m Y\|^2}{D_m \|Y - \Pi_{V_m} Y\|^2} - \bar{F}^{-1}_{D_m,N_m}(\alpha_m),$$

is positive. Since our test statistic T_α is defined as the supremum of those statistics, clearly our test rejects the null if one of those Fisher tests does.

15.2.2 Behavior of the test under the null hypothesis

The test associated to the procedure P1 has the property to be of size α. More precisely, we have

$$\forall F \in V^*, \quad \mathbf{P}_F\left(T_\alpha > 0\right) = \alpha. \tag{15.6}$$

This result follows from the definition of a_n noticing that $S_m \subset V^\perp$ implies that $\Pi_m Y = \Pi_m \varepsilon$ and $V \subset V_m$ implies that $Y - \Pi_{V_m} Y = \varepsilon - \Pi_{V_m}\varepsilon$. The drawback of this procedure is that it requires heavy computations to obtain the value of a_n. In contrast, the choice of the α_m's as proposed in the procedure P2 does not require any computation at all. A test associated to the procedure P2 leads to a Bonferonni test for which we only have (15.4) by arguing as follows:

$$\begin{aligned}\mathbf{P}_F(T_\alpha > 0) &\leq \sum_{m \in \mathcal{M}} \mathbf{P}_F\left(\frac{N_m \|\Pi_m Y\|^2}{D_m \|Y - \Pi_{V_m} Y\|^2} > \bar{F}^{-1}_{D_m, N_m}(\alpha_m)\right) \\ &\leq \sum_{m \in \mathcal{M}} \alpha_m = \alpha.\end{aligned}$$

15.2.3 A toy framework: The case of a known variance

For the sake of simplicity and to clarify the proofs as much as possible, we only state our main results in the simpler case where the variance σ^2 is known. This means that we content ourselves with proving those results for the test statistic $\hat{T}_\alpha$ (instead of T_α) defined as

$$\hat{T}_\alpha = \sup_{m \in \mathcal{M}} \left\{ \|\Pi_m Y\|^2 - \bar{\chi}^{-1}_{D_m}(\alpha_m)\sigma^2 \right\}, \tag{15.7}$$

where for each $u \in]0,1[$, $\bar{\chi}^{-1}_{D_m}(u)$ denotes the $1-u$ quantile of a χ^2 random variable with D_m degrees of freedom (note that in this case the assumption that $N_m \neq 0$ can be relaxed). For a complete proof of those results in the case of an unknown variance, i.e. for the test statistic T_α, we refer the reader to the paper by Baraud, Huet, and Laurent (2000).

15.3 The Power of the Test

15.3.1 The main result

The aim of this section is to describe a subset of $\mathcal{F}(\mathbf{R}^k, \mathbf{R})$ over which the test defined in Section 15.2.3 is powerful. For any $F \in \mathcal{F}(\mathbf{R}^k, \mathbf{R})$ and $\mathcal{G} \subset \mathcal{F}(\mathbf{R}^k, \mathbf{R})$, we define

$$d_n^2(F, \mathcal{G}) = \inf_{G \in \mathcal{G}} \frac{1}{n} \sum_{i=1}^{n} \left(F(x_i) - G(x_i)\right)^2.$$

Theorem 15.3.1 *Let α and β be two numbers in $]0,1[$. Let $\hat{T}_\alpha$ be the test statistic defined by (15.7). Let*

$$v_m^2(\beta) = \frac{\sigma^2}{n}\left(2.5\sqrt{2D_m \log(1/(\alpha_m\beta))} + 12.5\log(1/(\alpha_m\beta))\right).$$

Then, $\mathbf{P}_F(\hat{T}_\alpha > 0) \geq 1-\beta$ for all F belonging to the set

$$\mathcal{F}_n(\beta) = \left\{F \in \mathcal{F}(\mathbf{R}^k, \mathbf{R}),\ d_n^2(F, V^*) \geq \inf_{m\in\mathcal{M}}\left(d_n^2(F, V^* + S_m^*) + v_m^2(\beta)\right)\right\}. \tag{15.8}$$

Comment: The definition of $\mathcal{F}_n(\beta)$ given in (15.8) shows that we would take advantage of a collection of spaces $\{S_m^*, m \in \mathcal{M}\}$ containing spaces with good approximation properties, that is spaces S_m^*'s such that both $d_n^2(F, S_m^*)$ and the dimension of S_m^* are small.

15.3.2 Rates of testing

By evaluating the order of magnitude of the infimum in (15.8) under the assumption that F belongs to some class of smoothness, we derive from Theorem 15.3.1 an upper bound for the rate of testing achieved by our procedure over this class.

We take full advantage of the possibility to mix several linear spaces in the collection $\{S_m^*, m \in \mathcal{M}\}$, each of them being appropriate for detecting specific alternatives. We show that for an adequate choice of the collection, we obtain a test that is both rate optimal (in some sense) over the s-Hölderian balls of radius R defined by (15.2), simultaneously for all $s \in]0,1]$ and $R > 0$, and that is able to detect local alternatives at the parametric rate of testing. For this aim, we introduce the following collections of α_m's and S_m^*'s.

- Let
$$\mathcal{M}_1 = \left\{k \in \{1, ..., n\}, k \in \{2^j,\ j \geq 0\} \cup \{n\}\right\}.$$
We set $\alpha_{(k,1)} = \alpha/(4|\mathcal{M}_1|)$ if $k \neq n$ and $\alpha_{(n,1)} = \alpha/4$. Let $S_{(k,1)}^*$ be the linear subspace of $\mathcal{F}([0,1], \mathbf{R})$ spanned by the piecewise constant functions $\left(\mathbb{1}_{]\frac{j-1}{k}, \frac{j}{k}]},\ j = 1, ..., k\right)$ if $k \neq n$ and $S_{(n,1)}^* = \mathcal{F}([0,1], \mathbf{R})$.

- Let $(\phi_j)_{j\geq 1}$ be a Hilbert basis of $\mathcal{F}([0,1], \mathbf{R})$. For each $k \in \mathcal{M}_2 = \{1, ..., n\}$ we define $\alpha_{(k,2)} = 3\alpha/(\pi^2 k^2)$ and $S_{(k,2)}^*$ as the linear space spanned by the function ϕ_k.

In the sequel, we set $\mathcal{M} = \{(k,1),\ k \in \mathcal{M}_1\} \cup \{(k,2),\ k \in \mathcal{M}_2\}$ and consider the collections $\{S_m^*,\ m \in \mathcal{M}\}$ and $\{\alpha_m, m \in \mathcal{M}\}$. Note that $\sum_{m\in\mathcal{M}} \alpha_m \leq \alpha$, and therefore the inequality (15.4) holds.

Corollary 15.3.1 *Let α and β be two numbers in $]0,1[$. Let $\hat{T}_\alpha$ be the test statistic defined by (15.7), then the following holds:*

(i) Let us denote by L_n the quantity $\log\log(n)$, and assume that $R^2 \geq \dfrac{\sqrt{L_n}\sigma^2}{n}$. There exists a constant $C(\alpha,\beta)$ depending only on α and β such that for all $s \in]0,1]$, for all $R>0$ and for all $F \in H_s(R)$ such that

$$
\begin{aligned}
& d_n^2(F, V^*) \\
& \geq \; C(\alpha,\beta)\left\{\left(R^{2/(1+4s)}\left(\frac{\sqrt{L_n}}{n}\sigma^2\right)^{4s/(1+4s)} + \frac{L_n}{n}\sigma^2\right) \wedge \left(\frac{\sigma^2}{\sqrt{n}}\right)\right\},
\end{aligned}
$$

we have $\mathbf{P}_F\left(\hat{T}_\alpha > 0\right) \geq 1-\beta$.

(ii) Assume that $F \in \mathbf{L}_2([0,1], dx)$ and that the x_i's satisfy for all $k \geq 1$,

$$
\lim_{n\to+\infty} \frac{1}{n}\sum_{i=1}^n F(x_i)\phi_k(x_i) = \int_0^1 F(x)\phi_k(x)dx, \quad \lim_{n\to+\infty} \frac{1}{n}\sum_{i=1}^n \phi_k^2(x_i) = 1.
$$

If for some $k_0 \geq 1$,

$$
\left(\int_0^1 F(x)\phi_{k_0}(x)dx\right)^2 > -4\sigma^2\log(2\alpha_{(k_0,2)}\beta),
$$

then for n large enough,

$$
\mathbf{P}_{F/\sqrt{n}}\left(\hat{T}_\alpha > 0\right) \geq 1-\beta.
$$

Comments: In Case (i), the rate of testing is of order $(\sqrt{\log\log(n)}/n)^{2s/(1+4s)}$ provided that $s > 1/4$. In the Gaussian white noise model, the rate $n^{-2s/(1+4s)}$ is known to be minimax over s-Hölderian balls [we refer to Ingster (1993)]. The loss of efficiency of a $\log\log(n)$ factor is due to the fact that the test is adaptive, which means that it does not use any prior knowledge on s and R [for further details we refer to the work of Spokoiny (1996)]. When $s < 1/4$, the rate of testing is of order $n^{-1/4}$. This rate is known to be minimax as proved in Baraud (2000). Lastly, let us emphasize that the result given in (i) holds under no assumption on the x_i's.

The result presented in (ii) shows that we obtain the parametric rate $1/\sqrt{n}$ for directional alternatives.

15.4 Simulations

We carry out a simulation study in order to evaluate the performances of our procedure both when the errors are normally distributed and when they are not, and to compare its performances with the testing procedure proposed recently by Horowitz and Spokoiny (2000). This section is organized as follows: we present the simulation experiment, then we describe how our testing procedure is performed. Finally we give the results of the simulation study.

15.4.1 The simulation experiment

We evaluate the performances of our test by using the same simulation experiment as Horowitz and Spokoiny (2000). They considered three distributions of the errors ε_i.

1. The Gaussian distribution: $\varepsilon_i \sim \mathcal{N}(0, 4)$
2. The mixture of Gaussian distributions: ε_i is distributed as $\pi X_1 + (1-\pi)X_2$ where π is distributed as a Bernoulli variable with expectation 0.9, X_1 and X_2 are centered Gaussian variables with variance respectively equal to 2.43 and 25, π, X_1 and X_2 are independent. In that case the variance of ε_i equals 3.9. This distribution has heavy tails.
3. The Type I distribution: ε_i has density $(s/2)f_X(\mu+(s/2)x)$ where $f_X(x) = \exp\{-x - \exp(-x)\}$, and where μ and s^2 are the expectation and the variance of a variable X with density f_X. These ε_i's are centered variables with variance 4. This distribution is asymmetrical.

By considering the distributions 2 and 3, we evaluate the robustness of our procedure to discrepancies to the assumption of normality.

Three regression function F are considered:

$$\begin{aligned} F_0(x) &= 1 + x \\ F_\tau(x) &= 1 + x + \frac{5}{\tau}\phi\left(\frac{x}{\tau}\right) \quad \text{with } \tau = 0.25 \text{ and } \tau = 1 \end{aligned}$$

where ϕ is the density of a standard Gaussian variable. When $\tau = 0.25$ the regression function F_τ presents a peak, when $\tau = 1$, F_τ presents a small bump. We will test the linearity of the function F at level $\alpha = 5\%$.

The number of observations $n = 250$. The x_i's are simulated once and for all, as centered Gaussian variables with variance equal to 25 and are constrained to lie in the interval $[\Phi^{-1}(0.05), \Phi^{-1}(0.95)]$, where Φ is the distribution function of a standard Gaussian variable.

15.4.2 The testing procedure

The testing procedure depends on the choice of the collections $\{S_m^*, m \in \mathcal{M}\}$ and $\{\alpha_m, m \in \mathcal{M}\}$.

The collection $\{S_m^*, m \in \mathcal{M}\}$. We consider the spaces S_m^* based on piecewise functions and trigonometric polynomials. More precisely, for each $k = 1, \ldots n$, we consider the spaces $S_{(k,1)}^*$ of regular histograms based on the regular grid $\{l/k, l = 0, \ldots k\}$ and $S_{(k,2)}^*$ the space of trigonometric polynomials of degree not larger than k. For each $\delta \in \{1, 2\}$, and for each $k = 1, \ldots n$ let us recall than $S_{(k,\delta)}$ is the linear subspace of $\mathbf{R}^n$ defined as the orthogonal projection of $\{(F(x_1), ..., F(x_n))^T, F \in S_{(k,\delta)}^*\}$ onto $V^\perp$, where V is the space of $\mathbf{R}^n$ with dimension $d = 2$ spanned by the vectors $(1, \ldots, 1)^T$ and $(x_1, \ldots, x_n)^T$. Now let us set

$$\mathcal{M} = \{(k, \delta) \in \{1, \ldots, n\} \times \{1, 2\},\ \dim(S_{(k,\delta)}) \in \Big\{\{2^j, j \geq 0\} \cap \{1, \ldots, n/2\}\Big\}.$$

The collection $\{\alpha_m, m \in \mathcal{M}\}$. We consider the procedure **P2** (see Section 15.2.1) where the α_m's equal $\alpha/|\mathcal{M}|$ where $|\mathcal{M}|$ denotes the cardinality of $\mathcal{M}$ and $\alpha = 5\%$. Note that, under the assumption of Gaussian errors, the test based on the procedure **P1** is of size α and is more powerful than the test based on the procedure **P2** since a_n is larger than the α_m's. But the calculation of a_n makes the procedure P1 complicated to deal with, a_n depending on n, on V, and on the collection $\{S_m, m \in \mathcal{M}\}$. Therefore we decided to evaluate the performances of the procedure **P2** that is very easily implemented.

The test statistic is equal to T_α defined by Equation 15.3, and the null hypothesis is rejected if $T_\alpha > 0$. The results of the simulation study are reported in Table 15.1 under the column T_α. They are based on 4000 simulations.

15.4.3 The test proposed by Horowitz and Spokoiny (2000)

They proposed an adaptive testing procedure for testing that the regression function belongs to some parametric family of functions. Their procedure rejects the null hypothesis if for some bandwidth among a grid, the distance between the nonparametric kernel estimator and the kernel smoothed parametric estimator of F under the null hypothesis is large. The quantiles of their test statistic are estimated by a bootstrap method.

The results of the simulation experiment are reported in Table 15.1, under the column **HS**-test. They used 1000 simulations for estimating the level of the test and 250 simulations for estimating the power.

Table 15.1: Percentage of rejection

	Null hypothesis is true		Null hypothesis is false					
	HS-test	T_α	τ	**HS**-test	T_α	τ	**HS**-test	T_α
Normal	0.07	0.03	1	0.79	0.93	0.25	0.92	1
Mixture	0.05	0.05	1	0.80	0.93	0.25	0.93	1
Type I	0.05	0.03	1	0.82	0.94	0.25	0.97	1

15.4.4 Results of the simulation study

As expected, when the distribution of the errors is Gaussian, the test based on T_α is of level 5% (see Section 15.2.2). For the distributions "mixture" and "Type I", it has also the desired level showing the robustness of the procedure with respect to non-Gaussianity in these cases. In addition it turns out to be powerful over the alternatives considered by Horowitz and Spokoiny (2000).

15.5 Proofs

15.5.1 Proof of Theorem 15.3.1

For the sake of simplicity and to keep our formulae as short as possible we assume that $\sigma^2 = 1$. By definition of $\hat{T}_\alpha$, for any $F \in \mathcal{F}(\mathbf{R}^k, \mathbf{R})$,

$$\begin{aligned}\mathbf{P}_F\left(\hat{T}_\alpha \le 0\right) &= \mathbf{P}_F\left(\forall m \in \mathcal{M}, \|\Pi_m Y\|^2 \le \bar{\chi}_{D_m}^{-1}(\alpha_m)\right) \\ &\le \inf_{m \in \mathcal{M}} \mathbf{P}_F\left(\|\Pi_m Y\|^2 \le \bar{\chi}_{D_m}^{-1}(\alpha_m)\right).\end{aligned}$$

Let $f = (F(x_1), ..., F(x_n))^T$. Let us denote by $\chi_D^{-1}(a, u)$ the u quantile of a noncentral χ^2 variable with D degrees of freedom and noncentrality parameter a. For each $m \in \mathcal{M}$, the random variable $\|\Pi_m Y\|^2$ is distributed as a noncentral χ^2 variable with D_m degrees of freedom and noncentrality parameter $\|\Pi_m f\|^2$. It follows that if for some m in $\mathcal{M}$,

$$\chi_{D_m}^{-1}(\|\Pi_m f\|^2, \beta) \ge \bar{\chi}_{D_m}^{-1}(\alpha_m), \tag{15.9}$$

then $\mathbf{P}_F(\hat{T}_\alpha \le 0) \le \beta$. We shall use the two following inequalities, respectively due to Laurent and Massart (2000) and to Birgé (2000). For all $u \in]0, 1[$:

$$\begin{aligned}\bar{\chi}_D^{-1}(u) &\le D + 2\sqrt{D\log(1/u)} + 2\log(1/u), &(15.10)\\ \chi_D^{-1}(a, u) &\ge D + a - 2\sqrt{(D + 2a)\log(1/u)}. &(15.11)\end{aligned}$$

In the following, we set $L_m = \log(1/\alpha_m)$ and $L = \log(1/\beta)$. By using the inequalities $\sqrt{u+v} \leq \sqrt{u} + \sqrt{v}$, $2\sqrt{uv} \leq \theta u + \theta^{-1} v$ which hold for all positive numbers u, v, θ we derive from (15.11) that

$$\chi_{D_m}^{-1}(\|\Pi_m f\|^2, \beta) \geq D_m + \frac{4}{5}\|\Pi_m f\|^2 - 2\sqrt{D_m L} - 10L.$$

Therefore, using (15.10) the inequality (15.9) holds if

$$\frac{4}{5}\|\Pi_m f\|^2 \geq 2\sqrt{D_m}\left(\sqrt{L_m} + \sqrt{L}\right) + 2L_m + 10L.$$

Since $\sqrt{u} + \sqrt{v} \leq \sqrt{2}\sqrt{u+v}$ for any positive numbers u, v, inequality (15.9) holds as soon as

$$\|\Pi_m f\|^2 \geq 2.5\sqrt{2D_m(L_m + L)} + 12.5(L_m + L).$$

For any linear space $W \subset \mathbf{R}^n$, we denote by Π_W the orthogonal projector onto W. Using the fact that $S_m \subset V^\perp$, by the Pythagorean inequality, $\|\Pi_m f\|^2 = \|f - \Pi_V f\|^2 - \|\Pi_{V^\perp} f - \Pi_m f\|^2$. Noting that $\|\Pi_{V^\perp} f - \Pi_m f\|^2 = \|f - \Pi_{V+S_m} f\|^2$, we get $\|\Pi_m f\|^2/n = d_n^2(F, V^*) - d_n^2(F, V^* + S_m^*)$, which concludes the proof of Theorem 15.3.1.

15.5.2 Proof of Corollary 15.3.1

PROOF OF CASE (i). All along the proof, C denotes some constant that may vary from line to line. The dependency of C with respect to various quantities is specified by the notation $C(.)$. Let

$$\rho_n^2(F) = \inf_{m \in \mathcal{M}} \left(d_n^2(F, V^* + S_m^*) + v_m^2(\beta)\right).$$

It follows from Theorem 15.3.1 that $\mathbf{P}_F(\hat{T}_\alpha > 0) \geq 1 - \beta$ for all F such that $d_n^2(F, V^*) \geq \rho_n^2(F)$. Let us therefore give an upper bound for $\rho_n^2(F)$.

Note that since $S_{(n,1)}^* = \mathcal{F}([0,1], \mathbf{R})$, we have $d_n^2(F, S_{(n,1)}^*) = 0$ and since $D_{(n,1)} \leq n$ (we do not assume the design points to be distinct) and $\alpha_{(n,1)} = \alpha/4$ we have

$$\rho_n^2(F) \leq C(\alpha, \beta)\frac{\sigma^2}{\sqrt{n}}. \tag{15.12}$$

Noting that $d_n^2(F, V^* + S_{(k,1)}^*) \leq d_n^2(F, S_{(k,1)}^*)$, the statement of the first part of the corollary follows from the two following inequalities: for all $F \in H_s(R)$ and for all $k \in \mathcal{M}_1$,

$$d_n^2(F, S_{(k,1)}^*) \leq R^2 k^{-2s} \tag{15.13}$$

$$v_{(k,1)}^2(\beta) \leq \frac{\sigma^2}{n} C(\alpha, \beta)\left(\sqrt{kL_n} + L_n\right). \tag{15.14}$$

Let us prove (15.13). For any $k \in \mathcal{M}_1$, we define the function F^k on $[0,1]$ as follows: for $x \in](j-1)/k, j/k]$, $F^k(x) = F(j/k)$ for $j = 1, ..., k$ and $F^k(0) = F(1/k)$. Clearly $F^k \in S^*_{(k,1)}$ and under the assumption that $F \in H_s(R)$ the following inequalities hold

$$d_n(F, S^*_{(k,1)}) \leq d_n(F, F^k) \leq \|F - F^k\|_\infty \leq Rk^{-s}.$$

The inequality (15.14) follows easily by noting that for all $k \in \mathcal{M}_1$,

$$\log\left(1/\alpha_{(k,1)}\right) \leq C(\alpha, \beta) \log\log(n),$$

and $D_{(k,1)} = k$.

Therefore, we have

$$C^{-1}(\alpha, \beta)\rho_n^2(F) \leq \inf_{k \in \mathcal{M}_1} \left(R^2 k^{-2s} + \sqrt{kL_n}\frac{\sigma^2}{n}\right) + L_n\frac{\sigma^2}{n}. \tag{15.15}$$

Note that $R^2 k^{-2s} \leq \sqrt{kL_n}\sigma^2/n$ if and only if

$$k \geq k^* = \left(\frac{R^2 n}{\sigma^2 \sqrt{L_n}}\right)^{2/(1+4s)}.$$

Under the assumption on R we know that $k^* \geq 1$. Let us now distinguish between two cases:

1. If $k^* \leq n$ then there exists some $k' \in \mathcal{M}_1$ such that $k^* \leq k' \leq 2k^*$. It follows that

$$\inf_{k \in \mathcal{M}_1} \left(R^2 k^{-2s} + \sqrt{kL_n}\frac{\sigma^2}{n}\right) \leq 2\sqrt{k' L_n}\frac{\sigma^2}{n} \leq 2\sqrt{2k^* L_n}\frac{\sigma^2}{n}.$$

   Hence, by (15.15) we get

$$C^{-1}(\alpha, \beta)\rho_n^2(F) \leq \sqrt{k^* L_n}\frac{\sigma^2}{n} + L_n\frac{\sigma^2}{n}. \tag{15.16}$$

2. If $k^* > n$ then by (15.12)

$$C^{-1}(\alpha, \beta)\rho_n^2(F) \leq \frac{\sigma^2}{\sqrt{n}} \leq \sqrt{k^* L_n}\frac{\sigma^2}{n} + L_n\frac{\sigma^2}{n}. \tag{15.17}$$

Now, by collecting (15.12), (15.16) and (15.17), one obtains that

$$C^{-1}(\alpha, \beta)\rho_n^2(F) \leq \left(\sqrt{k^* L_n}\frac{\sigma^2}{n} + L_n\frac{\sigma^2}{n}\right) \wedge \left(\frac{\sigma^2}{\sqrt{n}}\right)$$

which concludes the proof in case (i) by replacing k^* by its value. ■

PROOF OF CASE *(ii)*. Let us set $f^n = (F(x_1), \ldots, F(x_n))^T/\sqrt{n}$, and assume that $\sigma^2 = 1$ for simplicity.

Following the proof of the theorem, we have to show that for some $k \in \mathcal{M}_2$

$$\chi_1^{-1}\left(\|\Pi_{(k,2)} f^n\|^2, \beta\right) \geq \bar{\chi}_1^{-1}(\alpha_{(k,2)}). \tag{15.18}$$

Using that $2P(X > u) \leq \exp(-u^2/2)$ for all $u > 0$, where $X \sim \mathcal{N}(0,1)$, it is easy to show that for all $t \in]0,1]$

$$\bar{\chi}_1^{-1}(t) \leq -2\log(t).$$

In the same way using that for all $\mu \in \mathbf{R}$ and $0 < u < \mu$,

$$P((X+\mu)^2 \leq u^2) \leq P(X \geq \mu - u) \leq (1/2)\exp\left(-(\mu-u)^2/2\right),$$

we get

$$\chi_1^{-1}\left(\mu^2, t\right) \geq \left(\mu - \sqrt{-2\log(2t)}\right)^2.$$

Then (15.18) is verified if

$$\|\Pi_{(k,2)} f^n\| \geq \sqrt{-2\log(\alpha_{(k,2)})} + \sqrt{-2\log(2\beta)}.$$

This holds as soon as $\|\Pi_{(k,2)} f^n\|^2 \geq -4\log(2\alpha_{(k,2)}\beta)$.

Let $r_n^2(k_0) = \sum_{i=1}^n \phi_{k_0}^2(x_i)/n$. We conclude by noticing that

$$\|\Pi_{(k_0,2)} f^n\|^2 = \frac{1}{r_n^2(k_0)}\left(\frac{1}{n}\sum_{i=1}^n F(x_i)\phi_{k_0}(x_i)\right)^2.$$

■

References

1. Baraud, Y. (2000). Non asymptotic minimax rates of testing in signal detection, *Technical Report 00.25*, Ecole Normale Supérieure, Paris.

2. Baraud, Y., Huet, S., and Laurent, B. (2000). Adaptive tests of linear hypotheses by model selection, *Technical Report 99-13*, Ecole Normale Supérieure, Paris.

3. Birgé, L. (2000). An alternative point of view on Lepski's method, In *State of the Art in Probability and Statistics; Festschrift for Willem R. van Zwet* (Eds., Mathisca C. M. de Gunst, Chris A. J. Klaassen, and Aad W. van der Vaart), Institute of Mathematical Statistics, Lecture Notes–Monograph Series, (to appear).

4. Eubank, R. L. and Hart, J. D. (1992). Testing goodness-of-fit in regression via order selection criteria, *Annals of Statistics*, **20**, 1412–1425.

5. Horowitz, J. L. and Spokoiny, V. G. (2000). An adaptive, rate-optimal test of a parametric model against a nonparametric alternative, *Econometrica*, (to appear).

6. Ingster, Yu. I. (1993). Asymptotically minimax testing for nonparametric alternatives I, *Mathematical Methods of Statistics*, **2**, 85–114.

7. Laurent, B. and Massart, P. (2000). Adaptive estimation of a quadratic functional by model selection, *Annals of Statistics*, (to appear).

8. Spokoiny, V. G. (1996). Adaptive hypothesis testing using wavelets, *Annals of Statistics*, **24**, 2477–2498.

PART V

Goodness-of-Fit Tests in Survival Analysis and Reliability

16

Inference in Extensions of the Cox Model for Heterogeneous Populations

Odile Pons

INRA Biométrie, Jouy-en-Josas, France

Abstract: The analysis of censored time data in heterogeneous populations requires extensions of the Cox model to describe the distribution of duration times when the conditions may change according to various schemes. New estimators and tests are presented for a model with a non-stationary baseline hazard depending on the time at which the observed phenomenon starts, and a model where the regression coefficients are functions of an observed variable.

Keywords and phrases: Asymptotic distribution, censored data, Cox model, non-parametric estimation, non-stationarity

16.1 Introduction

The distribution of a survival time T^0 conditionally on a vector Z of explanatory variables or processes is characterized by the hazard function

$$\lambda(t \mid Z) = \lim_{\Delta t \downarrow 0} \frac{1}{\Delta t} Pr(t \leq T^0 < t + \Delta t | T^0 \geq t, Z(s), 0 \leq s \leq t).$$

Cox's model (1972) is widely used for the analysis of censored survival data under controlled experimental conditions, it expresses the hazard function of T^0 in the form $\lambda(t \mid Z) = \lambda(t) e^{\beta^T Z(t)}$ where β is a vector of unknown regression parameters and λ is an unknown non-parametric baseline hazard function. If the survival time is right censored at a random time C, the observations are the censored time $T = T^0 \wedge C$, the indicator $\delta = 1_{\{T^0 \leq C\}}$ and the covariate Z. The regression parameter is estimated by the value that maximizes the "partial likelihood" [Cox (1972)] and an estimator of the cumulative hazard function $\Lambda(t) = \int_0^t \lambda(s)\, ds$ was defined by Breslow (1972).

In the fields of epidemiology or ecology, heterogeneous populations are observed and more complex models are required to describe the distribution of event times when the conditions may change in time or according to the value of a variable. This paper presents results for two extensions of the classical Cox model involving a non-parametric baseline hazard function and a regression on a p-dimensional process,

- a model for the duration $X = T^0 - S$ of a phenomenon with a non-stationary baseline hazard depending non-parametrically on the time S at which the observed phenomenon starts,

$$\lambda_{X|S,Z}(x \mid S, Z) = \lambda_{X|S}(x; S)\, e^{\beta^T Z(S+x)}, \tag{16.1}$$

- a model where the regression coefficients are smooth functions of an observed variable X,

$$\lambda(t \mid X, Z) = \lambda(t) e^{\beta(X)^T Z(t)}. \tag{16.2}$$

These models have been discussed and studied in the literature by several authors, in particular by Keiding (1991) for (16.1), by Brown (1975) and Zucker and Karr (1990) for time-varying coefficients and by Hastie and Tibshirani (1993) for (16.2). In Pons and Visser (2000) and Pons (1999), I proposed new estimators of the regression parameters and the cumulative baseline hazard function, and I studied their asymptotic properties. They are based on the local likelihood estimation method introduced by Tibshirani and Hastie (1987) for non-parametric regression models and adapted with kernels by Staniswalis (1989). They are defined using a kernel K that is a continuous and symmetric density with support $[-1, 1]$ and $K_{h_n}(s) = h_n^{-1} K(h_n^{-1} s)$, where the bandwidth h_n tends to zero at a convenient rate. The asymptotic properties of the estimators $\widehat{\beta}_n$ and $\widehat{\Lambda}_n$ of β and of the cumulative baseline hazard function will be presented for both models. They follow the classical lines but the kernel estimation requires modifications. In model (16.1), the convergence rate of $\widehat{\beta}_n$ is not modified by the kernel smoothing, it is $n^{-1/2}$ as in the Cox model and $\widehat{\Lambda}_n$ converges at the non-parametric rate for a kernel estimator as expected. In model (16.2), $\widehat{\Lambda}_n$ only involves kernel terms through the regression functions but both $\widehat{\beta}_n$ and $\widehat{\Lambda}_n$ have the same non-parametric rate of convergence as it was also the case for splines estimators in Zucker and Karr (1990). Goodness-of-fit test statistics for the classical Cox model against the alternatives of models (16.1) or (16.2) are deduced from these results.

16.2 Non-Stationary Cox Model

On a probability space $(\Omega, \mathcal{F}, P_0)$, let S and T^0 be the times at which a phenomenon starts and ends respectively, and let $X = T^0 - S > 0$ denote the duration of the phenomenon. Let Z be a p-dimensional vector of left-continuous process of covariates with right-hand limits. We assume that the conditional hazard function of X given (S, Z) follows (16.1) with the parameters $(\beta_0, \lambda_{0,X|S})$, that T^0 may be right-censored at a random time C independent of (S, T^0) conditionally on Z and non informative for β_0 and $\lambda_{0,X,S}$, and that S is uncensored. We observe a sample of n observations $(S_i, T_i, \delta_i, Z_i)_{1\le i\le n}$ drawn from the distribution of (S, T, δ, Z), where $T = T^0 \wedge C$ and $\delta = 1_{\{T^0 \le C\}}$ is the censoring indicator. The data are observed on a finite time interval $[0, \tau]$ strictly included in the support of the distributions of the variables S, T^0 and C, and (S, T^0) belongs to the triangle $I_\tau = \{(s, x) \in [0, \tau] \times [0, \tau]; s + x \le \tau\}$. Up to additive terms constant in $(\beta, \lambda_{X|S})$, the log-likelihood of $(S_i, T_i, \delta_i, Z_i)$ under (16.1) is

$$\begin{aligned} l_{(i)} &= \delta_i\{\log \lambda_{X|S}(X_i; S_i) + \beta^T Z_i(T_i)\} \\ &\quad - \int_0^\tau Y_i(y) \exp\{\beta^T Z_i(S_i + y)\}\lambda_{X|S}(y; S_i)\, dy. \end{aligned}$$

For S_i in a neighborhood of s, the baseline hazard $\lambda_{X|S}(\cdot; S_i)$ is approximated by $\lambda_{X|S}(\cdot; s)$, which yields an approximation $l_{(i)}(s)$ for $l_{(i)}$, and the local log-likelihood at $s \in [h_n, \tau - h_n]$ is defined as $l_n(s) = \sum_i K_{h_n}(s - S_i) l_{(i)}(s)$. The estimator of $\Lambda_{0,X|S}(x; s) = \int_0^x \lambda_{0,X|S}(y; s)\, dy$ is then defined for $(s, x) \in I_{n,\tau} = \{(s, x); s \in [h_n, \tau - h_n], x \in [0, \tau - s]\}$ by $\hat\Lambda_{n,X|S}(x; s) = \hat\Lambda_{n,X|S}(x; s, \hat\beta_n)$ with

$$\hat\Lambda_{n,X|S}(x; s, \beta) = \sum_i \frac{K_{h_n}(s - S_i) 1_{\{S_i \le C_i, X_i \le x \wedge (C_i - S_i)\}}}{n S_n^{(0)}(X_i; s, \beta)}$$

where $S_n^{(0)}(x; s, \beta) = n^{-1} \sum_j K_{h_n}(s - S_j) Y_j(x) \exp\{\beta^T Z_j(S_j + x)\}$ and $Y_i(x) = 1_{\{T_i^0 \wedge C_i \ge S_i + x\}} = 1_{\{\widetilde X_i \ge x\}}$ with $\widetilde X_i = X_i \wedge (C_i - S_i)$. The estimator $\widehat\beta_n$ of the regression coefficient maximizes the following partial likelihood

$$l_n(\beta) = \sum_i \delta_i \left[\beta^T Z_i(T_i^0) - \log\{n S_n^{(0)}(X_i; S_i, \beta)\}\right] \varepsilon_n(S_i)$$

where $\varepsilon_n(s) = 1_{[h_n, \tau - h_n]}(s)$. In this section, we shall use the following notations and assumptions. We assume that under P_0, S has a density f_S such that for some $\eta > 0$, the supports of the distributions of S and C contain $[-\eta, \tau + \eta]$ and we define $I_{\tau,\eta} = \{(s, x) : x \in [0, \tau], s \in [-\eta, \tau + \eta], s + x \in [-\eta, \tau + \eta]\}$. For a vector z in $\mathbb{R}^p$, let $z^{\otimes 0} = 1$, $z^{\otimes 1} = z$ and $z^{\otimes 2} = zz^T$. The norms in $\mathbb{R}^p$ and in $(\mathbb{R}^p)^{\otimes 2}$ are denoted $\|\cdot\|$ and the uniform norm of a function or a process on a set J is denoted $\|\cdot\|_J$. For $k = 0, 1, 2$, (s, x) in $I_{n,\tau}$ and under the next conditions let

$$\begin{aligned} S_n^{(k)}(x; s, \beta) &= n^{-1} \sum_i K_{h_n}(s - S_i) Y_i(x) Z_i^{\otimes k}(S_i + x) e^{\beta^T Z_i(S_i + x)} \\ s^{(k)}(x; s, \beta) &= \mathbb{E}[Y(x) Z^{\otimes k}(s + x) e^{\beta^T Z(s + x)} | S = s]\, f_S(s), \end{aligned}$$

$$\begin{aligned}
s_n^{(k)}(x;s,\beta) &= \mathbb{E}[S_n^{(k)}(x;s,\beta)] = \int_0^\tau K_{h_n}(s-u)s^{(k)}(x;u,\beta)\,du,\\
s^{(k,l)}(x,y;s) &= \mathbb{E}[Y(y)Z^{\otimes k}(s+x)e^{\beta_0^T Z(s+x)}(Z^{\otimes l}(s+y))^T\\
&\qquad \times e^{\beta_0^T Z(s+y)}|S=s]\, f_S(s),\ k,l=0,1,\\
I(\beta) &= \int_{I_\tau}\{s^{(2)}s^{(0)-1}-[s^{(1)}s^{(0)-1}]^{\otimes 2}\}(x;s,\beta)\\
&\qquad \times s^{(0)}(x;s,\beta_0)\lambda_{0,X|S}(x;s)\,dx\,ds,\\
\widehat{H}_n(s,x) &= n^{-1}\sum_i \delta_i 1_{\{S_i\le s\}}1_{\{X_i\le x\}},\\
\widehat{H}_{1n}(x;s) &= n^{-1}\sum_i K_{h_n}(s-S_i)\delta_i 1_{\{X_i\le x\}},\\
H(s,x) &= P_0(S\le s, X\le x, \delta=1),\\
H_{1n}(x;s) &= \int_0^\tau K_{h_n}(s-u)H(du,x),\\
H_1(x;s) &= \int_0^x s^{(0)}(y;s)\lambda_{0,X|S}(y;s)\,dy,\\
W_i^{(k)}(s,x) &= Z_i^{\otimes k}(S_i+x)e^{\beta_0^T Z_i(S_i+x)}1_{\{S_i\le s\}}1_{\{\widetilde{X}_i\ge x\}},\quad i=1\dots,n,
\end{aligned}$$

$W^{(k)}=\mathbb{E}W_i^{(k)}$ and $\bar{W}_n^{(k)}=n^{-1}\sum_i W_i^{(k)}$. We also denote $\gamma_2(K)=\int_{-1}^1 K^2(z)\,dz$ and we omit β_0 in the notations. We assume that the next conditions hold,

C1 β_0 and $(\widehat{\beta}_n)_n$ belong to the interior of B a compact and convex subset of $\mathbb{R}^p$, $\|\lambda_{0,X|S}\|_{I_{\tau,\eta}}<\infty$ and $\|f_S\|_{[-\eta,\tau+\eta]}<\infty$.

C2 The process Z has sample paths in a subset $\mathcal{Z}$ of the space of left-continuous functions with right-hand limits, the matrix $\mathrm{Var}Z(t)$ is positive definite for any $t\in[0,\tau]$, $\inf_{\beta\in B}\inf_{(s,x)\in I_{\tau,\eta}} s^{(0)}(x;s,\beta)>0$, and there exists a constant M_1 such that

$$\mathbb{E}\sup_{t\in[-\eta,\tau+\eta]}\sup_{\beta\in B}[\|Z(t)\|^{2k+1}e^{2\beta^T Z(t)}]\le M_1,\ k=0,1,2.$$

C3 The functions $s\mapsto\lambda_{0,X|S}(x;s)$ and $s\mapsto s^{(k)}(x;s,\beta)$, $k=0,1$, are twice differentiable on $\{s:(s,x)\in I_{\tau,\eta}\}$, with second derivatives such that $\sup_x\sup_{|s_1-s_2|\le h}|\ddot{\lambda}_{0,X|S}(x;s_1)-\ddot{\lambda}_{0,X|S}(x;s_2)|$ and $\sup_x\sup_{\beta\in B}\sup_{|s_1-s_2|\le h}\|\ddot{s}^{(k)}(x;s_1,\beta)-\ddot{s}^{(k)}(x;s_2,\beta)\|$ tend to zero as $h\to 0$, and the functions $s^{(k,l)}(x,y;\cdot)$ are continuous, uniformly in x and y.

C4 $\int_{-1}^1|dK(z)|<\infty$, $nh_n^2\to\infty$ and $h_n=o(n^{-1/4})$.

C5 $\|S_n^{(k)}-s^{(k)}\|_{I_{n,\tau},B}\xrightarrow{P_0}0$, for $k=0,1,2$.

Using an integration by parts, C5 is verified if $n^{1/2}\|\bar{W}_n^{(k)}-W^{(k)}\|_{I_\tau,B}$ converges weakly. In particular, this convergence holds under C1–C3 if Z is a random variable and if there exist conditional densities $f_{X|S,Z}$ and $g_{C|S,Z}$ for X and C and a constant M_2 such that

$$\sup_{\beta\in B} \sup_{(s,x_1,x_2):(s,x_1),(s,x_2)\in I_{\tau,\eta}} \mathbb{E}[\|Z^{\otimes k}(s+x_1)e^{\beta^T Z(s+x_1)}\|^2\{g_{C|S,Z}(s+x_2;s,Z) \\ +f_{X|S,Z}(x_2;s,Z)\}|S=s] \le M_2.$$

The weak consistency of $\widehat{\beta}_n$ and the asymptotic normality of $n^{1/2}(\widehat{\beta}_n - \beta_0)$ are established by the classical arguments in maximum likelihood estimation. For large n, an expansion of the score process gives

$$\widehat{\beta}_n - \beta_0 = I_n^{-1}(\beta_n)U_n,$$

where β_n is between β_0 and $\widehat{\beta}_n$, $I_n(\beta) = -n^{-1}d^2 l_n(\beta)/d\beta^{\otimes 2}$ satisfies

$$\sup_{\|\beta-\beta_0\|\le\varepsilon} \|I_n(\beta) - I(\beta_0)\| \xrightarrow{P_0} 0$$

as $\varepsilon \to 0$ and $n \to \infty$,

$$U_n = n^{-1}(dl_n/d\beta)(\beta_0) = n^{-1}\sum_i \varepsilon_n(S_i)\delta_i\left\{Z_i(T_i) - [S_n^{(1)}S_n^{(0)-1}](X_i;S_i,\beta_0)\right\}$$

is approximated using a statistic of the form $\sum_{i\ne j}\varphi_n(\xi_i,\zeta_j)$ with $\xi_i = (S_i, X_i, \delta_i)$ and $\zeta_j = (S_j, X_j \wedge (C_j - S_j), Z_j)$. Denoting $\psi_n((\xi_i,\zeta_i),(\xi_j,\zeta_j)) = \frac{1}{2}\{\varphi_n(\xi_i,\zeta_j) + \varphi_n(\xi_j,\zeta_i)\}$, we obtain a U-statistic of order 2, $\sum_{i\ne j}\psi_n((\xi_i,\zeta_i),(\xi_j,\zeta_j))$, and the weak convergence of $n^{1/2}U_n$ follows from a Hoeffding decomposition [Serfling (1980)].

Theorem 16.2.1 *$n^{1/2}(\widehat{\beta}_n - \beta_0)$ converges weakly to a Gaussian variable $\mathcal{N}(0, I^{-1}(\beta_0))$ and $I^{-1}(\beta_0)$ is the minimal variance for a regular estimator of β_0.*

For fixed s in $[h_n, \tau - h_n]$, let $D([0,\tau - s])$ be the space of right continuous real functions with left-hand limits on $[0, \tau - s]$. In Pons and Visser (2000), conditions are given for weak convergence of the process $\widehat{L}_n(\cdot;s) = (nh_n)^{1/2}(\widehat{\Lambda}_{n,X|S} - \Lambda_{0,X|S})(\cdot;s)$ in the Skorohod topology on $D[0,\tau - s]$. Its limit is a continuous Gaussian process L with mean zero and covariance $\sigma(x' \wedge x; s)$ at $(x;s)$ and $(x';s)$ in $I_{n,\tau}$, where

$$\sigma^2(x;s) = \gamma_2(K)\int_0^x s^{(0)-1}(y;s)\lambda_{0,X|S}(y;s)\,dy. \tag{16.3}$$

The asymptotic variance of $\widehat{L}_n(x;s)$ is consistently estimated by

$$\widehat{\sigma}_n^2(x;s) = \gamma_2(K)\int_0^x S_n^{(0)-2}(y;s,\widehat{\beta}_n)\widehat{H}_{1n}(dy;s) \tag{16.4}$$

which provides a point-wise asymptotic confidence interval for $\Lambda_{0,X|S})(x;s)$. The weak convergence of $\widehat{L}_n$ can be extended on $I_{n,\tau}$: Let $\alpha_n^{(0)} = n^{1/2}(\bar{W}_n^{(0)} - W^{(0)})$, $\alpha_n^H = n^{1/2}(\widehat{H}_n - H)$ and $\alpha_n = (\alpha_n^{(0)}, \alpha_n^H)$. For every n the variables $\alpha_n(s,x)$ and $\alpha_n(s',x')$ have the same covariance $\Sigma_\alpha(s,s',x,x')$ defined by

$$\begin{aligned}
\mathbb{E}\{\alpha_n^H(s,x)\alpha_n^H(s',x')\} &= H(s\wedge s', x\wedge x') - H(s,x)H(s',x'),\\
\mathbb{E}\{\alpha_n^{(0)}(s,x)\alpha_n^{(0)}(s',x')\} &= \mathbb{E}[1_{\{S_i\le s\wedge s'\}}1_{\{\widetilde{X}_i\ge x\vee x'\}}e^{\beta_0^T\{Z_i(s+x)+Z_i(s'+x')\}}]\\
&\quad - W^{(0)}(s,x)W^{(0)}(s',x'),\\
\mathbb{E}\{\alpha_n^{(0)}(s,x)\alpha_n^H(s',x')\} &= \mathbb{E}[1_{\{S_i\le s\wedge s'\}}1_{\{x\le\widetilde{X}_i\le x'\}}e^{\beta_0^T Z_i(s+x)}]\\
&\quad - W^{(0)}(s,x)H(s',x') \text{ if } x\le x',\\
\mathbb{E}\{\alpha_n^{(0)}(s,x)\alpha_n^H(s',x')\} &= -W^{(0)}(s,x)H(s',x') \text{ if } x\ge x', \qquad (16.5)
\end{aligned}$$

thus $\Sigma_\alpha(s,s',x,x')$ is written in the form

$$C_\alpha(s\wedge s', x, x') - E_\alpha(s,x)\{E_\alpha(s',x')\}^T.$$

Theorem 16.2.2 *If there exists a sequence of centered Gaussian processes B_n on $I_{n,\tau}$ with covariance matrix Σ_α (16.5) such that $\|\alpha_n - B_n\|_{I_{n,\tau}} = o_p(h_n^{1/2})$ and if $s^{(0)}(\cdot,s)$ has a derivative $\dot{s}^{(0)}(\cdot,s)$ such that $\dot{s}^{(0)}$ is continuous on $I_{\tau,\eta}$, then $(nh_n)^{1/2}(\widehat{\Lambda}_{n,X|S} - \Lambda_{0,X|S})$ converges weakly on $I_{n,\tau}$ to a centered Gaussian process L.*

PROOF. On $I_{n,\tau}$, $\widehat{L}_n$ is written

$$\widehat{L}_n(x;s) = (nh_n)^{1/2}\Big\{\int_0^x S_n^{(0)-1}(y;s,\widehat{\beta}_n)\,\widehat{H}_{1n}(dy;s) - \int_0^x s^{(0)-1}(y;s)\,H_1(dy;s)\Big\}$$

and it develops as $\widehat{L}_n = L_n + L_{1n} + L_{2n}$, with

$$L_n(x;s) = (nh_n)^{1/2}\Big\{\int_0^x s_n^{(0)-1}(y;s)\,H_{1n}(dy;s) - \int_0^x s^{(0)-1}(y;s)\,H_1(dy;s)\Big\}$$

which is a uniform $O((nh_n^5)^{1/2})$ under the regularity conditions, and

$$L_{1n}(x;s) = (nh_n)^{1/2}\int_0^x \{S_n^{(0)-1}(y;s,\widehat{\beta}_n) - S_n^{(0)-1}(y;s)\}\,\widehat{H}_{1n}(dy;s),$$

$$L_{2n}(x;s) = (nh_n)^{1/2}\Big\{\int_0^x S_n^{(0)-1}(y;s)\,\widehat{H}_{1n}(dy;s) - \int_0^x s_n^{(0)-1}(y;s)\,H_{1n}(dy;s)\Big\}.$$

By an expansion of $S_n^{(0)-1}(\widehat{\beta}_n)$ for $\widehat{\beta}_n$ in a neighborhood of β_0 and by Theorem 16.2.1,

$$L_{1n}(x;s) + (nh_n)^{1/2}(\widehat{\beta}_n - \beta_0)^T\int_0^x [s^{(1)}s^{(0)-2}](y;s)\,H_1(dy;s)$$

and therefore L_{1n} are $o_p(1)$ uniformly on $I_{n,\tau}$.

Finally,

$$\begin{aligned}L_{2n}(x;s) &= s^{(0)-1}(x;s)\widetilde{\alpha}_n^H(x;s) - \int_0^x s^{(0)-2}(y;s)\widetilde{\alpha}_n^H(y;s)\dot{s}^{(0)}(y;s)\,dy \\ &\quad - \int_0^x s^{(0)-2}(y;s)\widetilde{\alpha}_n^{(0)}(y;s)H_1(dy;s)\end{aligned}$$

with $\widetilde{\alpha}_n(x;s) = \int_0^\tau K_{h_n}(s-u)\,\alpha_n(x,du)$ for $(s,x) \in I_{n,\tau}$. An integration by parts entails

$$\|h_n^{1/2}\int_0^\tau K_{h_n}(s-u)(\alpha_n - B_n)(x;du)\| \le h_n^{-1/2}\|\alpha_n - B_n\|_{I_\tau}\,(\int_{-1}^1 |dK| + \|K\|),$$

and it tends to zero in probability. The processes B_n^* defined on I_τ as the stochastic integral $B_n^*(x;s) = h_n^{1/2}\int_0^\tau K_{h_n}(s-u)B_n(du,x)$ are centered Gaussian processes with the same limit covariance since $\int K^2(z)C_\alpha(dz,x,x')$ is finite for every x and x', from the expression (16.5). Moreover, under the above assumptions $s_n^{(0)}(\cdot,s)$ has a derivative $\dot{s}_n^{(0)}(\cdot,s)$ and $\dot{s}_n^{(0)}$ converges to $\dot{s}^{(0)}$ on I_τ, therefore $\|L_{2n} - L_{2n}^*\|_{I_{n,\tau}}$ converges to zero in probability, with

$$\begin{aligned}L_{2n}^*(x;s) &= s^{(0)-1}(x;s)B_n^{H*}(x;s) - \int_0^x s^{(0)-2}(y;s)B_n^{H*}(y;s)\dot{s}^{(0)}(y;s)\,dy \\ &\quad - \int_0^x s^{(0)-2}(y;s)B_n^{(0)*}(y;s)H_1(dy;s)\end{aligned}$$

and $\widehat{L}_n$ converges weakly to a centered process L whose covariance is the limiting covariance of L_{2n}^*.

If Z is a variable with values in a bounded subset $\mathcal{Z}$ of $\mathbb{R}$, let ν_n be the empirical process of the variables $(S_i,\widetilde{X}_i,Z_i)_{i\le n}$ and ν_n^δ be the empirical process of the variables $(S_i,\widetilde{X}_i,Z_i)$ with $\delta_i = \delta$, $i \le n$. By Theorem 1 in Massart (1989), there exist a constant c and a sequence of Gaussian processes G_n such that $P_0(\|\nu_n - G_n\|_{I_\tau\times\mathcal{Z}} > c(\log n)^{3/2}n^{-1/6})$ tends to zero as $n\to\infty$. As in Burke, Csörgő, and Horváth (1981) and in Castelle (1991) for the empirical process of a one-dimensional variable associated with a discrete variable, it is then expected that (ν_n,ν_n^0,ν_n^1) may be approximated by Gaussian processes at the same rate, so that the rate of the Gaussian approximation of (ν_n,ν_n^0,ν_n^1) is an $o_p(h_n^{1/2})$ if $h_n = O(n^{-d})$ with $d < 1/3$. As

$$\alpha_n^{(0)}(s,x) = \int_{\mathcal{Z}} e^{\beta_0^T z}\{\int_0^\tau \nu_n(s,dt,dz) - \nu_n(s,x^-,dz)\}$$

and

$$\alpha_n^H(s,x) = \int_{\mathcal{Z}} \nu_n^1(s,x,dz),$$

the approximation $\|\alpha_n - B_n\|_{I_\tau} = o_p(h_n^{1/2})$ assumed in Theorem 16.2.2 is then satisfied.

A goodness-of-fit test of a Cox model for the duration X against the alternative of a non-stationary model (16.1), may be viewed as a test for stationarity, with the hypothesis H_0: $\lambda_{X|S}(x; s) = \lambda_X(x)$ for every $(s, x) \in I_\tau$, hence

$$\lambda_{X|S,Z}(x; S, Z) = \lambda_X(x) \exp\{\beta^T Z(S + x)\}.$$

Tests for H_0 can be based on the difference between the estimated hazard functions under the hypothesis and under the alternative, i.e. on the process D_{1n} defined on $I_{n,\tau}$ by

$$D_{1n}(x; s) = (nh_n)^{1/2}\{\hat{\Lambda}_{n,X|S}(x; s) - \hat{\Lambda}_{n,X}(x)\}$$

where $\hat{\Lambda}_{n,X}$ is Breslow's estimator of Λ_X under the null hypothesis. Under H_0 and under conditions C1–C5, $\hat{\Lambda}_{n,X} - \Lambda_{0,X|S} = O_p(n^{-1/2})$, therefore

$$D_{1n} = (nh_n)^{1/2}(\hat{\Lambda}_{n,X|S} - \Lambda_{0,X|S}) + o_p(1)$$

and it converges weakly to a Gaussian process under the conditions of Theorem 16.2.2. Under the alternative, $\|D_{1n}\|_{I_{n,\tau}}$ tends in probability to infinity.

As for the comparison of the hazard functions of several groups [Andersen *et al.* (1993)], nonparametric tests for comparing $\Lambda_{X|S}$ and Λ_X can be based on processes of the form

$$T_n(s) = \int_0^{\tau - s} B_n(x) D_{1n}(dx; s), \qquad s \in [h_n, \tau - h_n],$$

where B_n is a predictable process with respect to the filtration generated by the observations in the duration scale under H_0. For instance a test statistic can be based on a discretization of D_{1n} on a finite grid: Let $(x_j)_{0 \le j \le J}$ be an increasing sequence of $[0, \tau]$ with $x_0 = 0$ and let $V_{1n}(s)$ be the vector of dimension J with components the variables

$$D_{1n}(s, x_j) - D_{1n}(s, x_{j-1}), \qquad 1 \le j \le J, \ s \in [h_n, \tau - h_n].$$

By (16.3) and (16.4), the components of $V_{1n}(s)$ are asymptotically independent and their asymptotic variance $v_j(s) = \sigma^2(x_j; s) - \sigma^2(x_{j-1}; s)$ is estimated by $\widehat{v}_{nj}(s) = \widehat{\sigma}^2(x_j; s) - \widehat{\sigma}^2(x_{j-1}; s)$. Let $A_{1n}(s)$ be the diagonal matrix with elements $\widehat{v}_{nj}(s)$, under the conditions of Theorem 16.2.2 the process $A_{1n}^{-1/2} V_{1n}$ tends to a Gaussian process with mean zero and variance identity. However, its asymptotic distribution under H_0 depends on the unknown parameters Λ_0 and β_0 through its covariances. Bootstrap tests with statistics

$$\int_{h_n}^{\tau - h_n} V_{1n}^T(s) A_{1n}^{-1}(s) V_{1n}(s)\, ds$$

or

$$\int_{h_n}^{\tau-h_n} V_{1n}^T(s)A_{1n}^{-1}(s)V_{1n}(s)\,d\widehat{F}_{n,S}(s)$$

may be used, with $\widehat{F}_{n,S}$ the empirical distribution function of $(S_i)_{i\leq n}$.

A test for stationarity of the baseline hazard function against an alternative of differences at fixed times s_i, of $[h_n, \tau - h_n]$ can very simply be performed if $(s_i)_{i\leq I}$ is an increasing sequence with $s_i - s_{i-1} > 2h_n$. Let V_n be the vector of dimension IJ with components the variables

$$D_{1n}(s_i, x_j) - D_{1n}(s_i, x_{j-1}), \qquad 1 \leq i \leq I,\ 1 \leq j \leq J$$

and let A_n be the diagonal matrix with elements

$$A_{n,ij} = \widehat{\sigma}^2(x_j; s_i) - \widehat{\sigma}^2(x_{j-1}; s_i)$$

defined by (16.4). From the expression of Σ_α, the covariance of

$$\int_0^\tau K_{h_n}(s-u)\,\alpha_n(x, du) \text{ and } \int_0^\tau K_{h_n}(s'-u)\,\alpha_n(x', du),$$

(s, x) and $(s', x') \in I_{n,\tau}$, is asymptotically equivalent to

$$\int_{-1}^1 K(u)K(u + h_n^{-1}(s-s'))\,C_\alpha(s + h_n\,du, x, x'),$$

and therefore to zero if $|s - s'| > 2h_n$ since K is zero outside $[-1, 1]$. Then under C1–C5, the statistic $V_n^T A_n^{-1} V_n$ has an asymptotic χ^2 distribution with IJ degrees of freedom under H_0 and it tends in probability to infinity if $\Lambda_{X|S}(x_j; s_i) \neq \Lambda_X(x_j)$ for some $i \leq I$ and $j \leq J$.

16.3 Varying-Coefficient Cox Model

On a probability space $(\Omega, \mathcal{F}, P_0)$ we observe a sample $(T_i, \delta_i, X_i, Z_i)_{1\leq i\leq n}$ of n independent and identically distributed observations where $T_i = T_i^0 \wedge C_i$ and $\delta_i = 1_{\{T_i^0 \leq C_i\}}$ are defined as in the introduction , X_i is a real random variable and Z_i is a p-dimensional vector of processes. We assume that conditionally on (X_i, Z_i), T_i^0 has a hazard function defined by (16.2) with parameters (β_0, λ_0) and C_i is independent of T_i^0 conditionally on (X_i, Z_i). We suppose that the time observations T_i belong to a finite time interval $[0, \tau]$ and that the X_i's have a density f_X. The problem is to estimate the function β_0 on a compact subset J_X of the support of f_X, so that $\beta_0(x)$ belongs to a compact set B for every x in J_X. In Pons (1999), the estimator $\widehat{\beta}_n(x)$ is defined as the value of β which maximizes

$$\begin{aligned} l_{n,x}(\beta) &= \sum_{i\le n} \delta_i K_{h_n}(x - X_i) \\ &\quad \times \left[\beta^T Z_i(T_i) - \log\left\{ \sum_{j\le n} K_{h_n}(x - X_j) Y_j(T_i) e^{\beta^T Z_j(T_i)} \right\} \right] \end{aligned}$$

where $Y_i(t) = 1_{\{T_i \ge t\}}$ is the risk indicator for individual i at t, and an estimator of the integrated baseline hazard function follows,

$$\hat{\Lambda}_n(t) = \sum_i \frac{\delta_i 1_{\{T_i \le t\}}}{\sum_j Y_j(T_i) 1_{\{X_j \in J_X\}} e^{\widehat{\beta}_n(X_j)^T Z_j(T_i)}}.$$

We assume that J_X and $J_{X,\eta} = \{X : \inf_{y\in J_X} |x - y| \le \eta\}$ are strictly included in $\mathcal{X} = \{x \in \mathbb{R} : f_X(x) > 0\}$ for some $\eta > 0$ and we modify the notations and conditions of Section 16.2. Let

$$\bar{S}_n^{(k)}(t, \beta) = \sum_i 1_{\{X_i \in J_X\}} Y_i(t) Z_i^{\otimes k}(t) e^{\beta(X_i)^T Z_i(t)},$$

where β is a function satisfying the conditions below. For $\beta \in \mathbb{R}^p$, $x \in J_X$, let

$$\begin{aligned} S_n^{(k)}(t, \beta, x) &= \sum_i K_{h_n}(x - X_i) Y_i(t) Z_i^{\otimes k}(t) e^{\beta^T Z_i(t)}, \\ \widetilde{S}_n^{(k)}(t, x) &= \sum_i K_{h_n}(x - X_i) Y_i(t) Z_i^{\otimes k}(t) e^{\beta_0(X_i)^T Z_i(t)}, \\ m^{(k)}(t, \beta, x) &= \mathbb{E}\{Y_i(t) Z_i^{\otimes k}(t) e^{\beta^T Z_i(t)} \mid X_i = x\}, \\ s^{(k)}(t, \beta, x) &= m^{(k)}(t, \beta, x) f_X(x), \\ \widetilde{s}^{(k)}(t, x) &= m^{(k)}(t, \beta_0(x), x) f_X(x), \\ I(\beta, x) &= \int_0^\tau \{s^{(2)} - s^{(1)\otimes 2} s^{(0)-1}\}(t, \beta, x)\, d\Lambda_0(t), \\ s_n^{(k)}(t, \beta, x) &= \mathbb{E} n^{-1} S_n^{(k)}(t, \beta, x) \\ &= \int K_{h_n}(x - y) m^{(k)}(t, \beta, y) f_X(y)\, dy, \\ \widetilde{s}_n^{(k)}(t, x) &= \mathbb{E} n^{-1} \widetilde{S}_n^{(k)}(t, x) \\ &= \int K_{h_n}(x - y) m^{(k)}(t, \beta_0(y), y) f_X(y)\, dy, \\ m^{(k,l)}(s, t, x_1, x_2, x) &= \mathbb{E}\{Y_i(t) Z_i^{\otimes k}(s) (Z_i^{\otimes l}(t))^T \\ &\quad \times e^{\beta_0(x_1)^T Z_i(s) + \beta_0(x_2)^T Z_i(t)\}} \mid X_i = x\}, \\ &\qquad x_1, x_2, x \in J_{X,\eta},\ k, l = 0, 1. \end{aligned}$$

Now, we denote $U_n(\beta, x)$ and $-I_n(\beta, x)$ the first two derivatives of $n^{-1}l_{n,x}(\beta)$ with respect to β, and we simply write

$$\begin{aligned} U_n(x) &= U_n(\beta_0(x), x), \\ S_n^{(k)}(t,x) &= S_n^{(k)}(t, \beta_0(x), x), \\ s_n^{(k)}(t,x) &= s_n^{(k)}(t, \beta_0(x), x), \\ s^{(k)}(t,x) &= s^{(k)}(t, \beta_0(x), x). \end{aligned}$$

We throughout assume that the next conditions C1–C5 are satisfied,

C1 For every x in $J_{X,\eta}$, $f_X(x) > 0$ and $\beta_0(x)$ belongs to the interior of a compact and convex subset B of $\mathbb{R}^p$.

C2 The process Z has sample paths in a subset $\mathcal{Z}$ of the space of left-continuous functions with right-hand limits, the matrix $\mathrm{Var}\{Z(t)|X = x\}$ is positive definite for any t in $[0, \tau]$ and x in $J_{X,\eta}$, $\Lambda_0(\tau) < \infty$, $\|f_X\|_{J_X} < \infty$,

$$\inf_{t\in[0,\tau]} \inf_{\beta\in B} \inf_{x\in J_{X,\eta}} s^{(0)}(t, \beta, x) > 0$$

and there exists a constant M_1 such that for $k = 0, 1, 2$,

$$\mathbb{E} \sup_{t\in[0,\tau]} \sup_{\beta\in B} [\|Z(t)\|^{2k+1} e^{2\beta^T Z(t)}] \le M_1.$$

C3 The functions β_0 and f_X are twice continuously differentiable on $J_{X,\eta}$. For $k, l = 0, 1$ and for every s and t in $[0, \tau]$ and β in B, the functions $m^{(k,l)}(s, t, \cdot, \cdot, \cdot)$ are continuous on $J^3_{X,\eta}$, the functions $m^{(k)}(t, \cdot, \cdot)$ are twice continuously differentiable on $B \times J_{X,\eta}$, with second derivatives such that

$$\lim_{\varepsilon,\varepsilon'\to 0} \sup_{|x_1-x_2|\le\varepsilon} \sup_{\|\beta_1-\beta_2\|\le\varepsilon'} \sup_{t\in[0,\tau]} \|\ddot{m}^{(k)}(t, \beta_1, x_1) - \ddot{m}^{(k)}(t, \beta_2, x_2)\| = 0.$$

C4 $nh \to \infty$ and $h_n = o(n^{-1/5})$.

C5 The variables $\|n^{-1}S_n^{(k)} - s^{(k)}\|_{[0,\tau]\times B\times J_X}$ and $\|n^{-1}\widetilde{S}_n^{(k)} - \widetilde{s}^{(k)}\|_{[0,\tau]\times J_X}$ converge in probability to zero, $k = 0, 1, 2$.

By classical arguments, $\widehat{\beta}_n(x)$ and $I_n(\beta, x)$ converge in probability to $\beta_0(x)$ and $I(\beta, x)$ for any x in J_X and β in B, and this point-wise weak consistency is extended as a uniform convergence under C5: The variables

$$\sup_{x\in J_X} \|\widehat{\beta}_n(x) - \beta_0(x)\|$$

and

$$\sup_{x\in J_X} \sup_{\|\beta-\beta'\|\le\|\widehat{\beta}_n(x)-\beta_0(x)\|} \|I_n(\beta, x) - I(\beta', x)\|$$

converge in probability to zero.

Let

$$U_n(x) = n^{-1} \sum_i K_{h_n}(x - X_i)\delta_i \left\{Z_i(T_i) - (S_n^{(1)} S_n^{(0)-1})(T_i, x)\right\},$$

x in J_X. The asymptotic normality of $(nh_n)^{1/2}(\widehat{\beta}_n(x) - \beta_0(x))$ relies on a uniform expansion of the score process, which implies

$$U_n(x) = \{I(\beta_0, x) + r_n(x)\}(\widehat{\beta}_n(x) - \beta_0(x))$$

with $\|r_n\|_{J_X} = o_p(1)$. Using a Hoeffding decomposition of $U_n(x)$ we obtain the approximation $(nh_n)^{1/2}U_n(x) = n^{1/2}U_n^*(x)\{1 + o_p(1)\}$ with a uniform $o_p(1)$ on J_X and

$$U_n^*(x) = n^{-1} \sum_i \{V_{1,n,i}(x) - \mathbb{E}V_{1,n,i}(x)\} + n^{-1} \sum_i \{V_{2,n,i}(x) - \mathbb{E}V_{2,n,i}(x)\}, \tag{16.6}$$

where

$$\begin{aligned} V_{1,n,i}(x) &= h_n^{1/2} K_{h_n}(x - X_i)\delta_i\{Z_i(T_i) - \frac{s^{(1)}}{s^{(0)}}(T_i, x)\}, \\ V_{2,n,i}(x) &= h_n^{1/2} K_{h_n}(x - X_i) \int_0^\tau Y_i(t) e^{\beta_0(x)^T Z_i(t)} \left\{\frac{s^{(1)}}{s^{(0)}}(t, x) - Z_i(t)\right\} \\ &\quad \times \frac{\widetilde{s}^{(0)}}{s^{(0)}}(t, x)\, d\Lambda_0(t). \end{aligned}$$

We deduce a point-wise weak convergence result.

Theorem 16.3.1 *The variable $(nh_n)^{1/2}(\widehat{\beta}_n - \beta_0)(x)$ converges weakly to a Gaussian variable $\mathcal{N}(0, \gamma_2(K)I_0^{-1}(x))$ for every x in J_X.*

For the weak convergence of the process $(nh_n)^{1/2}(\widehat{\beta}_n - \beta_0)$, let us define for $t \in [0, \tau]$, x and $y \in J_X$ the following processes:

$$\begin{aligned} W_i^{(1)}(t, y) &= \delta_i \, 1_{\{T_i \le t\}} \, 1_{\{X_i \le y\}}, \\ W_i^{(2)}(t, y) &= \delta_i Z_i(T_i) \, 1_{\{X_i \le y\}}, \\ W_i^{(3)}(t, x, y) &= e^{\beta_0(x)^T Z_i(t)} \, 1_{\{T_i \ge t\}} \, 1_{\{X_i \le y\}} \end{aligned}$$

and

$$W_i^{(4)}(t, x, y) = Z_i(t) e^{\beta_0(x)^T Z_i(t)} 1_{\{T_i \ge t\}} 1_{\{X_i \le y\}}, \ 1 \le i \le n.$$

Let $W^{(k)} = \mathbb{E}W_i^{(k)}$, $\alpha_n^{(k)} = n^{1/2}(n^{-1}\sum_i W_i^{(k)} - W^{(k)})$, $k = 1, \ldots, 4$, and $\alpha_n = (\alpha_n^{(1)}, \ldots, \alpha_n^{(4)})^T$. For every n the variables $\alpha_n(t, x)$ and $\alpha_n(t', x')$ have the same covariance matrix $\Sigma_\alpha(t, t', x, x')$ which is again of the form

$$C_\alpha(t, t', x \wedge x') - E_\alpha(t, x)\{E_\alpha(t', x')\}^T.$$

Theorem 16.3.2 *If there exists a sequence of centered Gaussian processes B_n on $[0,\tau]\times J_X$ with covariance matrix Σ_α such that $\|\alpha_n - B_n\|_{[0,\tau]\times J_X} = o_p(h_n^{1/2})$ and if the functions $s^{(k)}(t,x)$, $k=0,1$, have a continuous derivative with respect to $t \in [0,\tau]$, then $(nh_n)^{1/2}(\widehat{\beta}_n - \beta_0)$ converges weakly to a centered Gaussian process on $[0,\tau]$.*

This result is based on the approximation of the process U_n given by (16.6) which implies

$$\begin{aligned} U_n^*(x) &= h_n^{1/2} \int K_{h_n}(x-y) \Bigg[\alpha_n^{(2)}(dy) - \int_0^\tau \frac{s^{(1)}}{s^{(0)}}(t,x)\alpha_n^{(1)}(dt,dy) \\ &\quad + \int_0^\tau \left\{ \frac{s^{(1)}}{s^{(0)}}(t,x)\alpha_n^{(3)}(t,x,dy) - \alpha_n^{(4)}(t,x,dy) \right\} \frac{\widetilde{s}^{(0)}}{s^{(0)}}(t,x)\, d\Lambda_0(t) \Bigg] \end{aligned}$$

and the proof is similar to the convergence of the process L_{2n} for Theorem 16.2.2.

The asymptotic behavior of the process $(\widehat{\Lambda}_n - \Lambda_0)$ relies on an expansion of $\bar{S}_n^{(0)}(\widehat{\beta}_n)$ for $\widehat{\beta}_n$ close to β_0, and therefore on the behavior of $\bar{S}_n^{(0)}(\widehat{\beta}_n) - \bar{S}_n^{(0)}(\beta_0)$. From a development of this sum and due to the convergence rate $(nh_n)^{-1/2}$ of the estimators $\widehat{\beta}_n(X_i)$, the process

$$G_n = (nh_n)^{1/2}\{\bar{S}_n^{(0)}(\widehat{\beta}_n) - \bar{S}_n^{(0)}(\beta_0)\}$$

converges weakly to a Gaussian process G, and its mean is a $O((nh_n^5)^{1/2})$ under a Lipschitz condition for $m^{(1)}(\cdot, 2\beta_0, x)$ and from the convergence in probability to zero for the variables

$$\|n^{-1}S_n^{(k)} - s^{(k)}\|_{[0,\tau]\times B\times J_X} \text{ and } \|n^{-1}\widetilde{S}_n^{(k)} - \widetilde{s}^{(k)}\|_{[0,\tau]\times J_X},$$

$k = 0,1,2$ [Pons (1999)]. The asymptotic distribution of $\widehat{\Lambda}_n$ is deduced from

$$(nh_n)^{1/2}(\widehat{\Lambda}_n - \Lambda_0) = -\int_0^\cdot \frac{G_n}{n^{-2}\bar{S}_n^{(0)}(\widehat{\beta}_n)\bar{S}_n^{(0)}(\beta_0)}\, d\Lambda_0 + o_p(1).$$

Theorem 16.3.3 *The process $(nh_n)^{1/2}(\widehat{\Lambda}_n - \Lambda_0)$ converges weakly to $-\int_0^\cdot G(t)\{\int_{J_X} s^{(0)}(t,y)\,dy\}^{-2}\,d\Lambda_0(t)$.*

A goodness-of-fit test of a Cox model for the survival time T^0 against the alternative of a model (16.2) where the regression coefficients vary with the values of the variable X on J_X is a test for the hypothesis $H_0 : \beta(x) = \beta \in B$ for every x in J_X. Tests for H_0 can be based on the process

$$D_{2n}(x) = (nh_n)^{1/2}\{\hat{\beta}_n(x) - \hat{\beta}_{n,0}\},$$

x in J_X, where $\widehat{\beta}_{n,0}$ is the Cox estimator of the regression parameter β_0 under the null hypothesis. Under H_0, $\widehat{\beta}_{n,0}$ converges to β_0 at the rate $n^{-1/2}$ and

$$D_{2n}(x) = (nh_n)^{1/2}(\hat{\beta}_n(x) - \beta_0) + o_p(1) = I_0^{-1}(x)\,(nh_n)^{1/2}U_n(x) + o_p(1).$$

Under C1–C5 and under the conditions of Theorem 16.3.3, it converges weakly under H_0 to a Gaussian process with mean zero and variance $I_0^{-1}(x)$ and $\|D_{2n}\|_{J_X}$ tends to infinity under the alternative. A bootstrap test based on the process $D_{2n}^T(x)I_n^{-1}(x)D_{2n}(x)$ may be used, where $I_n(x) = I_n(\hat{\beta}_n(x), x)$ is a consistent estimator of $I_0(x)$.

If we restrict the alternative to possible differences between values $\beta(x_i)$ on a finite subset $(x_i)_{i\leq I}$ of J_X such that $(x_i)_{i\leq I}$ is an increasing sequence in J_X with $x_i - x_{i-1} > 2h_n$. Let V_{2n} be the vector of dimension I with components the variables $D_{2n}(x_i)$. From the expression of the covariances of α_n, the covariance of

$$\int_0^T K_{h_n}(x-u)\,\alpha_n(du) \text{ and } \int_0^T K_{h_n}(x'-u')\,\alpha_n(du')$$

tends to zero for every s and $s' \in J_X$ such that $|s - s'| > 2h_n$. The asymptotic variance of the variable V_{2n} is therefore a block-diagonal matrix of dimension pI and with the sub-matrices $I_0^{-1}(x_i), i \leq I$, as diagonal blocks, it is consistently estimated by the block-diagonal matrix A_{2n} with sub-matrices $I_n(\widehat{\beta}_n(x_i))$. Then a simple test statistic for constant regression coefficients is given by $V_{2n}^T A_{2n}^{-1} V_{2n}$, under conditions C1–C5 it has an asymptotic χ^2 distribution with I degrees of freedom under H_0 and it tends to infinity under the alternative if $\beta_0(x_i) \neq \beta_0(x_j)$ for some i and $j \leq I$.

References

1. Andersen, P. K., Borgan, O., Gill, R. D., and Keiding, N. (1993). *Statistical Models Based on Counting Processes*, New York: Springer-Verlag.

2. Breslow, N. E. (1972). Discussion of the paper by Cox, D. R., *Journal of the Royal Statistical Society, Series B*, **34**, 216–217.

3. Breslow, N. E. and Crowley, J. J. (1974). A large sample study of the life table and product limit estimates under random censorship, *Annals of Statistics*, **2**, 437–453.

4. Brown, C. (1975). On the use of indicator variables in studying the time-dependence of parameters in a response-time model, *Biometrics*, **31**, 863–872.

5. Burke, M. D. and Csörgő, M., and Horváth, L. (1981). Strong approximations of some biometric estimates under random censorship, *Zeitschrift für Wahrscheinlichkeitstheorie und Verwandte Gebiete*, **56**, 87–112.

6. Castelle, N. (1991). *Principes d'invariance et application à la statistique de modèles gaussiens*, Thesis Orsay: Université Paris-Sud.

7. Cox, D. R. (1972). Regression model and life tables (with discussion), *Journal of the Royal Statistical Society, Series B*, **34**, 187–220.

8. Dabrowska, D. M. (1987). Non-parametric regression with censored survival time data, *Scandinavian Journal of Statistics*, **14**, 181–197.

9. Hastie, T. and Tibshirani, R. (1993). Varying-coefficient models, *Journal of the Royal Statistical Society, Series B*, **55**, 757–779.

10. Keiding, N. (1991). Age-specific incidence and prevalence: A statistical perspective, *Journal of the Royal Statistical Society, Series A*, **154**, 371–412.

11. Massart, P. (1989). Strong approximations for multidimensional empirical and related processes via KMT constructions, *Annals of Probability*, **17**, 266–291.

12. Pons, O. and Visser, M. (2000). A non-stationary Cox model, *Scandinavian Journal of Statistics*, **27**, 619–639.

13. Pons, O. (1999). Nonparametric estimation in a varying-coefficient Cox model, *Mathematical Methods of Statistics, to appear.*

14. Serfling, R. J. (1980). *Approximation Theorems of Mathematical Statistics*, New York: John Wiley & Sons.

15. Staniswalis, J. G. (1989). The kernel estimate of a regression function in likelihood-based models, *Journal of the American Statistical Society*, **84**, 276–283.

16. Tibshirani, R. and Hastie, T. (1987). Local likelihood estimation, *Journal of the American Statistical Society*, **82**, 559–567.

17. Tsiatis, A. A. (1981). A large sample study of Cox's regression model, *Annals of Statistics*, **9**, 93–108.

18. Zucker, D. and Karr, A. (1990). Nonparametric survival analysis with time-dependent covariate effects: A penalized partial likelihood approach, *Annals of Statistics*, **18**, 329–353.

17

Assumptions of a Latent Survival Model

Mei-Ling Ting Lee and G. A. Whitmore

Harvard University, Cambridge, Massachusetts
McGill University, Montreal, Quebec, Canada

Abstract: Whitmore, Crowder, and Lawless (1998), henceforth WCL, consider a model for failure of engineering systems in which the physical degradation process is latent or unobservable but a time-varying marker, related to the degradation process, is observable. Lee, DeGruttola, and Schoenfeld (2000), henceforth LDS, extend the WCL model and investigate the relationship between a disease marker and clinical disease by modeling them as a bivariate stochastic process. The disease process is assumed to be latent or unobservable. The time to reach the primary endpoint or failure (for example, death, disease onset, etc.) is the time when the latent disease process first crosses a failure threshold. The marker process is assumed to be correlated with the latent disease process and, hence, tracks disease, albeit imperfectly perhaps. The general development of this latent survival model does not require the proportional hazards assumption. The Wiener processes assumptions of the WCL model and the extended model by LDS, however, must be verified in actual applications to have confidence in the validity of the findings in these applications. In this article, we present a suite of techniques for checking assumptions of this model and discuss a number of remedies that are available to make the model applicable.

Keywords and phrases: First hitting time, inverse Gaussian distribution, latent status, marker, Wiener processes

17.1 Introduction

Whitmore, Crowder, and Lawless (1998), henceforth WCL, consider a model for failure of engineering systems in which the physical degradation process is latent or unobservable but a time-varying marker, related to the degradation process, is observable. Lee, DeGruttola, and Schoenfeld (2000), henceforth LDS, extend the WCL model and investigate the relationship between a disease marker and clinical disease by modeling them as a bivariate stochastic process. The disease process is assumed to be latent or unobservable. The time to reach the primary endpoint or failure (for example, death, disease onset, etc.) is the time when the latent disease process first crosses a failure threshold. The marker process is assumed to be correlated with the latent disease process and, hence, tracks disease, albeit imperfectly perhaps. The general development of this latent survival model does not require the proportional hazards assumption. The Wiener processes assumptions of the original WCL model and the extended model by LDS, however, must be verified in actual applications to have confidence in the validity of the findings in these applications. In this article, we present a suite of techniques for checking assumptions of this model and discuss a number of remedies that are available to make the model applicable.

17.2 Latent Survival Model

We shall follow the extended model adapted by LDS in considering the latent process as a disease process, rather than an engineering degradation process as originally proposed by WCL.

Let $\{X(t)\}$ denote a latent *disease process* which represents the level of deterioration of the health of a subject. Likewise, let $\{Y_w(t)\}$ represent an observed *marker process* that is correlated with the disease process $\{X(t)\}$ and tracks its progress. For example, in an AIDS setting, CD4 count tracks the latent disease. We consider the two-dimensional Wiener diffusion process $\{X(t), Y_w(t)\}$, for $t \geq 0$, with initial values $\{X(0), Y_w(0)\}$. The vector $\{X(t), Y_w(t)\}$ has a bivariate normal distribution with mean vector $\{X(0), Y_w(0)\} + t\mu$, where $\mu = (\mu_x, \mu_y)$, and covariance matrix $t\Sigma$, where

$$\Sigma = \begin{pmatrix} \sigma_{xx} & \sigma_{xy} \\ \sigma_{yx} & \sigma_{yy} \end{pmatrix}. \tag{17.1}$$

We assume that the subject's initial disease level is some negative number $X(0) < 0$. This initial level is unknown and will be estimated. We set a failure threshold at zero on the disease scale. The closer the value of $X(t)$ is to zero, the more diseased is the subject. A subject *fails*, i.e, reaches a primary endpoint, when the disease level first reaches the origin. The distance from the initial level to the origin will be denoted by parameter δ, where $\delta = |X(0)|$. Also, we find it convenient to consider a modified marker process that measures changes in the marker process from its initial level, i.e., we consider the marker change process $\{Y(t)\}$ where $Y(t) = Y_w(t) - Y_w(0)$. The initial marker level $Y_w(0) = y_{w0}$ is used as a baseline covariate. We denote the first passage time from the initial disease level to the failure threshold by random variable S. To relate our notation to that of WCL replace their a and x by δ and $x + \delta$, respectively.

17.3 Data and Parameter Estimation

As the disease process is assumed to be latent, inferences about the disease process must be based on observations on the marker process, supplemented by observed survival times for subjects who fail. No direct observation is made of the latent disease process, although one can infer that the X process has reached the failure threshold at the moment a subject fails and that X has not reached that level at any time in the interval $(0, t]$ if the subject has survived to time t. In the following presentation, we will consider the available data to have the following form.

We consider m independent subjects ($i = 1, \dots, m$). For the ith subject, we observe the marker process $\{Y_i(t)\}$ at $n_i + 1$ times or ages: $0 = t_{i0} < t_{i1} < \cdots < t_{in_i}$. The realized initial marker level for subject i is $Y_{wi}(t_{i0}) = Y_{wi}(0) = y_{wi0}$. The realized marker increments over the subsequent time intervals $\Delta t_{ij} = t_{ij} - t_{i,j-1}$ are denoted by $\Delta Y_{ij} = Y_{wi}(t_{ij}) - Y_{wi}(t_{i,j-1}) = \Delta y_{ij}$, for $j = 1, \dots, n_i$ and $i = 1, \dots, m$. We let $n = \sum_i n_i$. The corresponding realized increments of the disease process $\{X_i(t)\}$ for subject i are denoted by $\Delta X_{ij} = X_i(t_{ij}) - X_i(t_{i,j-1}) = \Delta x_{ij}$ and the initial disease level is $X_i(t_{i0}) = X_i(0) = x_{i0}$. These disease increments and the initial degradation level are unobservable by assumption.

We denote the first passage time for subject i by random variable S_i with realization s_i. If the ith subject is observed for a period of length b_i then all subjects with $S_i > b_i$ will have censored failure times. Subjects that fail before the end of their observation periods b_i are given subscripts $i = 1, \dots, p$, where $p \leq m$. We assume that $t_{in_i} = s_i$ for failing subjects and $t_{in_i} = b_i$ for subjects whose failures are censored. This last statement implies that a marker reading is available for each subject either at the time of failure or at the end of the observation period b_i, as the case may be.

The termination time t_{in_i} for the sample path of any subject i is a random variable in advance of data collection because it is uncertain if the sample path will end in survival, in which case $t_{in_i} = b_i$, or in failure, in which case $t_{in_i} = S_i$. Thus, the index n_i of the last increment in a sample path is random, as well as the length of the last time increment Δt_{in_i}.

As demonstrated by both WCL and LDS, maximum likelihood can be used to estimate the model parameters. As the disease process is unobservable, one of its parameters must be fixed arbitrarily. For example, the initial disease level x_{i0} or the variance σ_{xx} might be set to unity. LDS implemented their estimation procedure and demonstrated how the model parameters may be related to relevant baseline covariates of the subjects using generalized link functions. In the following discussion, we do not take explicit account of possible individual variation in model parameters that arises because of different baseline covariates. Modifications in the methods that are needed to take account of individual parametric variability will be apparent.

17.4 Model Validation Methods

The latent survival model makes a number of assumptions that would require verification in any application to ensure confidence in scientific findings based on it. These assumptions center mainly on the bivariate Wiener process for the joint latent disease and marker processes and the use of a first passage time as a survival time. As a Wiener process requires independent normally distributed increments, checks are required for these two features. The first passage time to a fixed barrier in a Wiener process with drift towards the barrier follows an inverse Gaussian distribution so the censored survival data provide a check on this feature. In this section we describe methods for examining the available data to check these assumptions.

LDS illustrated the application of a latent survival model to data from the AIDS Clinical Trial Group (ACTG) 116A Study. In brief, this study was a multicenter, randomized double-blind trial that compared the clinical efficacy of zidovudine (at varying dosage levels) with two dosage levels of didanosine (500 and 750 mg/d) for patients with AIDS, AIDS-related complex or asymptomatic HIV. Death was the primary study endpoint. Their data file contained records for 787 patients. The logarithm of CD4 cell count was taken as the marker. LDS have illustrated several of the model checking methods listed below. We give a cross-reference to LDS in each case so the reader can see an application of the checking method.

1. Checking the Wiener Property of the Latent Disease Process:

   The censored survival data can help in verifying the form of the latent disease process. As already noted, the first passage time to a fixed barrier in a Wiener process follows an inverse Gaussian (IG) distribution. Hence, a probability plot can be used to compare the censored survival data to this distribution. Alternatively, a Kaplan-Meier (KM) nonparametric survival function plot can be compared with the IG survival curves implied by the model. LDS illustrate a comparison of IG and KM survival plots for their case study – see Figure 3, p. 757 of Lee, DeGruttola, and Schoenfeld (2000).

2. Checking the Wiener Property of the Observable Marker Process:

The marker increments Δy_{ij} are distributed as $N(\mu_x \Delta t_{ij}, \sigma_{xx} \Delta t_{ij})$ under the latent survival model. Therefore, given the parameter estimates $\hat{\mu}_x$ and $\hat{\sigma}_{xx}$, the Δy_{ij} may be standardized using the known time increments Δt_{ij} as follows:

$$w_{ij} = \frac{\Delta y_{ij} - \hat{\mu}_x \Delta t_{ij}}{(\hat{\sigma}_{xx} \Delta t_{ij})^{1/2}} \tag{17.2}$$

A normal probability plot of these standardized values should confirm approximate normality.

The Wiener process also requires that both the mean and variance of the marker increment Δy_{ij} should vary in direct proportion to the time increment Δt_{ij} used in computing the standardized values in (17.2). Therefore, a scatter plot of the w_{ij} against the Δt_{ij} should show no systematic pattern.

LDS illustrate both of these plots – see Figure 2, p. 755, and Figure 1, p. 754, respectively, of Lee, DeGruttola, and Schoenfeld (2000).

3. Checking the Assumption of Independent Marker Increments:

   With longitudinal marker data, the assumption of independent increments can be checked by examining the autocorrelation structure of the marker increments. Specifically, consecutive marker increments of the form Δy_{ij} and $\Delta y_{i,j-1}$, $j = 1, \ldots, n_i$, for subject i should be uncorrelated under the independent increments assumption. As a practical matter, we note that since both of these marker increments share the same measured marker value at time t_{j-1}, i.e., they share the reading for $Y_{wi}(t_{i,j-1})$, there will be some negative autocorrelation if the marker reading is subject to independent measurement error. If the measurement error is not large, the negative correlation will be relatively small, although possibly statistically significant.

   LDS did not report that they examined this model feature but we have used their data to perform a small demonstration check. We have computed the correlation coefficient for the last two marker increments for subjects in their study who had three or more longitudinal measurements of the marker. The coefficient is -0.178. It is small and negative, but significant, indicating that some measurement error is present. We note, however, that the squared correlation coefficient is only 0.03 which suggests that the measurement error variance is small relative to the marker process variance. CD4 cell counts are known to have some measurement error so the finding is not unexpected. Thus, the marker increments in this study are reasonably consistent with the assumption of independent increments, except for the presence of small (but material) measurement error. In later discussion, we explain how measurement error can be modelled if it is judged to be great enough to warrant explicit attention.

4. Goodness-of-fit Test for Survival Time:

   The following chi-square test provides a predictive check on model validity. Partition survival time into a set of time intervals. Assign each subject to an interval based on his or her actual survival time. The total counts in the intervals provide the observed frequencies for the test. Compute the survival probability for each interval for each subject from the fitted model, taking account of individual covariates. Sum these probabilities for each interval for all subjects. These probability sums provide the expected frequencies for the test. A chi-square test is then used to compare the observed and expected frequencies. A low chi-square value would indicate a valid model. The test is more discriminating if the survival probabilities are computed from a model that is fitted without the reference subject (i.e., on a remove-one basis). The test is modified in a standard way for censored survival data.

   LDS do not perform this model check as it is somewhat redundant if the methods described under item 1 above have already been applied. We also will not illustrate the method as its application is straightforward in principle, albeit somewhat tedious to apply.

5. Consistency with Established Science:

   The latent survival model must always meet validation checks that arise from subject matter knowledge and the established science of the field of application. The marker, for example, should be either a known cause or a known effect of the disease. The robustness of the latent survival model has yet to be tested. Experience with other statistical methods based on normal theory suggests that parametric inferences from this model are probably more robust than predictive inferences. Some departures from the model assumptions are not likely to be a major concern. For example, a marker that must have a monotonic sample path (for example, lifetime cumulative tobacco smoking as a marker for lung disease) may still be reasonably well represented by a Wiener process. In this case, the fitted process parameters are likely to be consistent with a relatively smooth sample path, having a small coefficient of variation $(\sigma_{xx})^{1/2}/\mu_x$.

   LDS provide a strong defense of the validity of the latent survival model in their case study based on medical scientific understanding of the disease course of AIDS and its relationship to CD4 cell count.

17.5 Remedies to Achieve a Better Model Fit

A number of remedies are available that frequently can help to bring the latent survival model into line with the data in a particular application and thus achieve an acceptable degree of model fit. We itemize some of these remedies below, although many others are available that may be appropriate in particular studies but not for general use.

1. Transformation:

   Appropriate transformations may bring an application context into conformity with the model. Engineering applications of this kind of model show that monotonic transformations of the time scale are often needed so a Wiener process has a constant drift over time [see, for example, Whitmore and Schenkelberg (1997)]. Some disease or marker processes will tend to accelerate or decelerate with time. A monotonic transformation of the calendar time scale r, such as $t = r^{\gamma}$ for some $\gamma > 0$, may produce a constant process mean parameter on the transformed time scale t. Parameters in this time transformation (such as γ) would then require estimation. Transformations may be suggested by scientific knowledge rather than by the data. Longitudinal data would be useful for checking on the suitability of a transformation for a marker but there are some checks that don't require this. For example, a monotonic transformation of the time scale may bring the observed survival times into conformity with an inverse Gaussian distribution. Nonstationarity in both the marker and disease processes may be an issue. As the marker is expected to track the disease, a similar transformation may be required for both processes.

2. Measurement Errors:

   If independent measurement errors are present in the *increments* of the marker process then the associated measurement bias and variability will already be incorporated in the parameters μ_y and σ_{yy}. The marker increments are then interpreted as 'marker increments measured with error' and no modifications are required in the analysis or the model. If measurement error is an independent disturbance term that appears in each marker reading (as indicated by a negative correlation of marker increments) then the true marker Y becomes latent or unobservable. The jth measurement on the marker process then becomes disguised by the presence of a measurement error term ϵ_j as follows (here we suppress the subject index i in the notation).

   $$O_{wj} = Y_{wj} + \epsilon_j \tag{17.3}$$

Here O_{w_j} is the observed reading and Y_{w_j} is the true marker value. In this situation, we might assume that the ϵ_j are independently distributed as $N(0, \nu)$. The likelihood function of the model can then be extended accordingly and parameter ν estimated together with the process parameters. See Whitmore (1995) for a similar extension of the Wiener model to include measurement error.

It should be noted that an independent measurement error of the type shown in (17.3) will introduce a negative correlation between the baseline marker reading (with error) $O_{w0} = Y_{w0} + \epsilon_0$ and the observed marker increment $\Delta O_{w1} = O_{w1} - O_{w0}$ because they share the measurement error ϵ_0 in the initial reading O_{w0}. The same negative correlations would appear in the marker increments for longitudinal marker data, as we noted earlier. This kind of correlation is found in longitudinal blood pressure readings for example. We note, however, that the presence of measurement error can be confounded with the presence of other dependencies in the increments of a marker process.

3. Risks Associated with Competing Disease Processes:

The investigation of any disease process must take account of competing diseases. A marker may reflect the progress of one or more diseases simultaneously. One can view the course of several diseases as a multidimensional latent process $\{\mathbf{X}(t)\}$ with failure being triggered when the process enters a failure region $\mathcal{F}$ for the first time. The failure time would then be defined as $S = \min_t (\mathbf{X}(t) \in \mathcal{F})$. To be more specific, the set $\mathcal{F}$ may be defined by a separate failure threshold in each dimension of $\mathbf{X}(t)$ and the time of failure would be determined by the first threshold which the subject crosses.

Competing risks can be handled easily within the current latent survival model as follows, using an approach that is quite standard in the literature. In this approach, the model and methods are applied to each mode of failure separately, treating failures from competing modes as censored observations. The separate analysis by failure mode can provide important insights into the relationship between the marker and the different disease processes, as well as the effects of covariates if these effects are included in the model.

Acknowledgements. We acknowledge with thanks the financial support provided for this research by the National Institutes of Health grants HL40619-09 and NIGMS 55326-02 (Lee) and by a grant from the Natural Sciences and Engineering Research Council of Canada (Whitmore).

References

1. Lee, M.-L.T., DeGruttola, V., and Schoenfeld, D. (2000). A model for marker and latent health status, *Journal of Royal Statistical Society, Series B*, **62**, 747–762.

2. Whitmore, G. A. (1995). Estimating degradation by a Wiener diffusion process subject to measurement error, *Lifetime Data Analysis*, **1**, 307–319.

3. Whitmore, G. A. and Schenkelberg, F. (1997). Modelling accelerated degradation data using Wiener diffusion with a time scale transformation, *Lifetime Data Analysis*, **3**, 1–19.

4. Whitmore, G. A., Crowder, M. J., and Lawless, J. F. (1998). Failure inference from a marker process based on a bivariate Wiener model, *Lifetime Data Analysis*, **4**, 229–251.

18

Goodness-of-Fit Testing for the Cox Proportional Hazards Model

Karthik Devarajan and Nader Ebrahimi

Northern Illinois University, DeKalb, Illinois

Abstract: For testing the validity of the Cox proportional hazards model, a goodness-of-fit test of the null proportional hazards assumption is proposed based on a semi-parametric generalization of the Cox model, whereby the hazard functions can cross for different values of the covariates, using Kullback-Leibler distance. The proposed method is illustrated using some real data. Our test was compared to that of previously described tests by using simulation experiments and found to perform very well.

Keywords and phrases: Cox proportional hazards, hazard function, cumulative hazard function, Kullback-Leibler discrimination information measure, directed divergence measure, goodness-of-fit test, Weibull distribution, Chi-square distribution

18.1 Introduction

The Cox proportional hazards (PH) model [Cox (1972)] offers a method for exploring the association of covariates with the failure time variable often seen in medical and engineering studies. It is a widely used tool in the analysis of survival data and hence testing its validity is a matter of prime importance.

For a given vector of observed covariates $\boldsymbol{z} = (1, z_1, \ldots, z_p)'$, the hazard time t is modeled as

$$\lambda(t|\boldsymbol{z}) = \lambda_0(t)\exp(\boldsymbol{\beta}'\boldsymbol{z}), \tag{18.1}$$

where $\boldsymbol{\beta} = (\beta_0, \beta_1, \ldots, \beta_p)'$ is a $(p+1)$ vector of regression coefficients and $\lambda_0(t)$ is an unspecified function of t and is referred to as the baseline hazard function.

Over the years, numerous graphical and analytical procedures have been developed to test the assumption of proportional hazards. The graphical methods include plotting the logarithm of survival [Kalbfleisch and Prentice (1980)] and methods based on the definitions of different residuals, among others. Schoenfeld (1982) and Lin and Wei (1991) recommended plotting the elements of the Schoenfeld residuals against failure times. Pettitt and Bin Daud (1990) suggested smoothing the Schoenfeld residuals to consider time dependent effects. Wei (1984) and Therneau, Grambsch and Fleming (1990) recommended plotting the cumulative sums of martingale residuals. Lagakos (1980) proposed a graphical method for assessing covariates in the Cox PH model based on the cumulative hazard transformation and the score from the partial likelihood. Thaler (1984) proposed nonparametric estimation and plotting of the hazards ratio to check for non-proportionality. Arjas (1988) proposed a graphical method based on comparisons between observed and expected frequencies of failures as estimated from the Cox PH model. Departure from the proportional hazards assumption is indicated by an imbalance between such frequencies as shown by the graphs. Other graphical methods include those by Kay (1977), Crowley and Hu (1977), Cox (1979) and Crowley and Storer (1983). In general, graphical procedures give a first-hand idea about the departure from proportionality but are quite subjective.

The analytical methods include, among others, tests for equality sensitive to crossing hazards [Fleming *et al.* (1980)] and tests of proportionality for grouped data [Gail (1981)]. A number of tests have been proposed based on the time-weighted score tests of the proportional hazards hypothesis. These include models in which the parameter vector $\boldsymbol{\beta}$ is defined as a given function of time as considered by Cox (1972) and Stablein *et al.* (1981), tests in which $\boldsymbol{\beta}$ varies as a step function according to a given partition of the time axis [Moreau, O'Quigley, and Mesbah (1985)], tests in which $\boldsymbol{\beta}$ has defined trends along the time intervals [O'Quigley and Pessione (1989)] and tests in which $\boldsymbol{\beta}$ varies as a step function according to a given partition of the covariate space [Schoenfeld (1980) and Andersen (1982)], among others. Wei (1984) and Gill and Schumaker (1987) developed tests using time-weighted score tests based on the two-sample hazards ratio. In the multiple regression setting, similar tests were developed for the parameters using a rank transformation of time by Harrell (1986) and Harrell and Lee (1986). Nagelkerke, Oosting, and Hart (1984) considered testing the global validity of the proportional hazards hypothesis without reference to any alternative. Horowitz and Neumann (1992) and Lin, Wei, and Ying (1993) proposed global tests based on the cumulative sums of martingale residuals. Most of these tests have been shown to be special cases of the methods developed by Therneau and Grambsch (1994). These methods are applicable when a prespecified form is given for departures from proportionality.

Lin and Wei (1991) extend the methods of White (1982) for detecting parametric model misspecification to the Cox partial likelihood. Kooperberg, Stone, and Truong (1995) introduced a model for the log-hazard function conditional on the covariates. The Cox PH model is a member of the class of models considered for the conditional log-hazard and hence they test the proportionality assumption. Hess (1994) considered cubic spline functions for assessing time by covariate interactions in the Cox PH model. Quantin *et al.* (1996) derived a global test of the proportional hazards hypothesis using the score statistic from the partial likelihood. Pena (1998) discussed smooth goodness-of-fit tests for the baseline hazard in the Cox PH model.

We define a semi-parametric generalization of the Cox PH model in which the hazard functions corresponding to different values of the covariates can cross. The cumulative hazard function corresponding to a covariate vector $\boldsymbol{z}$ is given by

$$\wedge(t|\boldsymbol{z}) = \{(\wedge_0(t))\}^{\exp(\boldsymbol{\gamma}'\boldsymbol{z})} \exp(\boldsymbol{\beta}'\boldsymbol{z}), \tag{18.2}$$

where $\wedge_0(t)$ is an arbitrary baseline cumulative hazard function and $\boldsymbol{\beta}$ and $\boldsymbol{\gamma}$ are unknown $(p+1)$ vectors of parameters.

In addition to being a semi-parametric generalization of the Cox PH model, the model (18.2) has several nice features. Below we only mention two. For other features and also more details about this model see Quantin *et al.* (1996) and Devarajan and Ebrahimi (2000).

i. For two different covariate vectors $\boldsymbol{z}_1$ and $\boldsymbol{z}_2$ we have $\log\left\{\frac{\wedge(t|\boldsymbol{z}_1)}{\wedge(t|\boldsymbol{z}_2)}\right\} = \alpha'(t)\ (\boldsymbol{z}_1 - \boldsymbol{z}_2)$, where $\alpha(t) = \beta + g(t)\gamma$ with $g(t) = \log \wedge_0(t)$. This is the Cox PH model with time dependent coefficients $\alpha(t)$ that allows for crossing of hazard curves;

ii. For two different values of covariates $\boldsymbol{z}_1$ and $\boldsymbol{z}_2$, we have $\frac{\lambda(t|\boldsymbol{z}_1)}{\wedge(t|\boldsymbol{z}_1)} = \frac{\lambda(t|\boldsymbol{z}_2)}{\wedge(t|\boldsymbol{z}_2)} \exp(\gamma'(\boldsymbol{z}_1 - \boldsymbol{z}_2))$.

That is, the corresponding ratios of the hazard function to the cumulative hazard function are proportional.

In this paper, our goal is to develop goodness-of-fit testing methods for the Cox PH model against the model (18.2) using Kullback-Leibler Discrimination (KLD) information measures [see Kullback (1978) and Ebrahimi *et al.* (1994)]. The structure of the paper is as follows.

The methods are described in Section 18.2. In Section 18.3, we compare the proposed test with existing tests based on empirical power estimates via simulation experiments. In Section 18.4, we illustrate our methods using real life data sets. Throughout this paper we assume that the data consists of independent observations on the triple $(X_i, \delta_i, \boldsymbol{z}_i), i = 1, \ldots, n$, where X_i is the minimum of a failure and censoring time pair $(T_i, C_i), \delta_i = I(T_i \leq C_i)$ is the indicator of the event that a failure has been observed, and $\boldsymbol{z}_i = (1, z_{i1}, \ldots, z_{ip})'$ is a $(p+1)$ vector of covariates for the i-th individual. The random variables T_i and C_i are assumed to be independent.

18.2 Goodness-of-Fit Testing for the Cox PH Model

In the context of goodness-of-fit testing for the Cox PH model against the alternative defined by (18.2), the null hypothesis of interest is

$$H_0 : \wedge(t|\boldsymbol{z}) = (\exp(\boldsymbol{\beta}'\boldsymbol{z}))\wedge_0(t), \tag{18.3}$$

where $\wedge_0(t)$ is an arbitrary baseline cumulative hazard function, $\wedge(t|\boldsymbol{z})$ is the conditional cumulative hazard function given the covariate vector $\boldsymbol{z}$ and $\boldsymbol{\beta}$ is the $(p+1)$ vector of regression parameters. The alternative is

$$H_1 : \wedge(t|\boldsymbol{z}) = \exp(\boldsymbol{\beta}'\boldsymbol{z})\{\wedge_0(t)\}^{\exp(\boldsymbol{\gamma}'\boldsymbol{z})}, \tag{18.4}$$

where $\boldsymbol{\gamma}$ is an additional $p+1$ vector of regression parameters. Hence, the null hypothesis corresponds to $\boldsymbol{\gamma} = 0$ and the alternative corresponds to $\boldsymbol{\gamma} \neq 0$.

Given that F and G are the cumulative distribution functions under H_0 and H_1 respectively, to discriminate between the two hypotheses (18.3) and (18.4), we use the KLD information between two distributions given by

$$I(F:G) = \int_0^\infty f(x)\log\left\{\frac{f(x)}{g(x)}\right\}dx. \tag{18.5}$$

Here, $I(F:G)$ measures the discrepency between two distributions F and G corresponding to the Cox model and the model (18.2) in the direction of H_0. It is well known that $I(F:G) \geq 0$ and the equality holds if and only if $F = G$. Similarly, one can also define

$$I(G;F) = \int_0^\infty g(x)\log\left\{\frac{g(x)}{f(x)}\right\}dx \tag{18.6}$$

which measures the discrepancy between the two distributions F and G in the direction of H_1.

We can derive the divergence measure

$$J(F:G) = I(F:G) + I(G:F), \tag{18.7}$$

measuring the difficulty of discriminating between H_0 and H_1. Both $I(F:G)$ and $I(G:F)$ are commonly referred to as the directed divergence measure and $J(F:G)$ is referred to as the divergence measure.

It is clear that the survival function and the density function corresponding to the Cox PH model given the covariate vector $\boldsymbol{z}$ are

$$\bar{F}(t|\boldsymbol{z}) = \exp(-\wedge_F(t|\boldsymbol{z})) = \exp\{-(\wedge_0(t))\exp(\boldsymbol{\beta}'\boldsymbol{z})\}$$

and

$$f(t|\boldsymbol{z}) = (\exp(\boldsymbol{\beta}'\boldsymbol{z}))\lambda_0(t)\exp\{-\wedge_F(t|\boldsymbol{z})\}. \tag{18.8}$$

Similarly, the survival function and the density function corresponding to the non-proportional hazards model of H_1, denoted by $\bar{G}(t|\boldsymbol{z})$ and $g(t|\boldsymbol{z})$ respectively, are

$$\begin{aligned}\bar{G}(t|\boldsymbol{z}) &= \exp(-\Lambda_G(t|\boldsymbol{z})) \\ &= \exp\Big(-\textstyle\int_0^t \lambda_0(y)\exp\{\boldsymbol{\beta}'\boldsymbol{z}+\boldsymbol{\gamma}'\boldsymbol{z} \\ &\times(e^{\boldsymbol{\gamma}'\boldsymbol{z}}-1)\log\Lambda_0(y)\}dy\Big),\end{aligned} \tag{18.9}$$

and

$$g(t|\boldsymbol{z}) = \exp\{-\Lambda_G(t|\boldsymbol{z})+\boldsymbol{\beta}'\boldsymbol{z}+\boldsymbol{\gamma}'\boldsymbol{z}\}\lambda_0(t)[\Lambda_0(t)]^{\exp(\boldsymbol{\gamma}'\boldsymbol{z}-1)}. \tag{18.10}$$

Thus using (18.8) and (18.10) the directed divergence $I(G:F|\boldsymbol{z})$ given the covariate vector $\boldsymbol{z}$, is given by

$$\begin{aligned}I(G:F|\boldsymbol{z}) &= (1-e^{-\boldsymbol{\gamma}'\boldsymbol{z}})\{-\boldsymbol{\beta}'\boldsymbol{z}+K_e\}+\boldsymbol{\gamma}'\boldsymbol{z}-1 \\ &+\exp(\boldsymbol{\beta}'\boldsymbol{z}-\boldsymbol{\gamma}'\boldsymbol{z}-\boldsymbol{\beta}'\boldsymbol{z}\bar{e}^{\boldsymbol{\gamma}'\boldsymbol{z}})\Gamma(\bar{e}^{\boldsymbol{\gamma}'\boldsymbol{z}}),\end{aligned} \tag{18.11}$$

and the directed divergence $I(F:G|\boldsymbol{z})$ given covariate vector $\boldsymbol{z}$, is given by

$$\begin{aligned}I(F:G|\boldsymbol{z}) &= (1-\bar{e}^{\boldsymbol{\gamma}'\boldsymbol{z}})\{-\boldsymbol{\beta}'\boldsymbol{z}+K_e\}-\boldsymbol{\gamma}'\boldsymbol{z} \\ &+\exp(\boldsymbol{\beta}'\boldsymbol{z}+\boldsymbol{\gamma}'\boldsymbol{z}-\boldsymbol{\beta}'\boldsymbol{z}e^{\boldsymbol{\gamma}'\boldsymbol{z}})\Gamma(\bar{e}^{\boldsymbol{\gamma}'\boldsymbol{z}})-1.\end{aligned} \tag{18.12}$$

Hence, given the covariate vector $\boldsymbol{z}$, the divergence measure $J(F:G|\boldsymbol{z})$ is

$$J(F:G|\boldsymbol{z}) = I(F,G|\boldsymbol{z})+I(G;F|\boldsymbol{z}). \tag{18.13}$$

Here $K_e = \int_0^\infty(\bar{e}^u\log u)du$ is Euler's constant and $\Gamma(\cdot)$ is the gamma function. An important feature of all three measures described in (18.11)–(18.13) is that they are all free from the baseline $\lambda_0(t)$.

Since evaluations of $I(G:F|\boldsymbol{z}), I(F:G|\boldsymbol{z})$ and $J(F:G|\boldsymbol{z})$ in (18.11)–(18.13) require complete knowledge of unknown parameters $\boldsymbol{\beta}$ and $\boldsymbol{\gamma}$, then these measures are not operational. We operationalize $I(G:F|\boldsymbol{z}), I(F:G|\boldsymbol{z})$ and $J(F:G|\boldsymbol{z})$ by developing discrimination information statistics $\hat{I}(G:F|\boldsymbol{z}), \hat{I}(F:G|\boldsymbol{z})$ and $\hat{J}(F:G|\boldsymbol{z})$, where $\boldsymbol{\beta}$ and $\boldsymbol{\gamma}$ are replaced by $\hat{\boldsymbol{\beta}}$ and $\hat{\boldsymbol{\gamma}}$. Here, $\hat{\boldsymbol{\beta}}$ and $\hat{\boldsymbol{\gamma}}$ are the maximum likelihood estimates obtained by approximating the baseline hazard function with a linear combination of cubic B spline basis functions . For more details about these estimates and their properties see Devarajan and Ebrahimi (2000). Thus, for a given value of $\boldsymbol{z}$, our goodness-of-fit test will be based on either $\hat{I}(G:F|\boldsymbol{z}), \hat{I}(F:G|\boldsymbol{z})$ or $\hat{J}(F:G|\boldsymbol{z})$.

Remark 18.2.1 Observing $\hat{I}(G:F|\boldsymbol{z}), \hat{I}(F:G|\boldsymbol{z})$ and $\hat{J}(F:G|\boldsymbol{z})$ we see that they all depend on the covariate vector $\boldsymbol{z}$. As a global measure of goodness-of-fit, we suggest averaging $\hat{I}(F:G|\boldsymbol{z}_i), \hat{I}(G:F|\boldsymbol{z}_i)$ and $\hat{J}(F:G|\boldsymbol{z}_i)$ over all individuals in the sample. That is considering $\frac{1}{n}\sum_{i=1}^{n}\hat{I}(F:G|\boldsymbol{z}_i), \frac{1}{n}\sum_{i=1}^{n}\hat{I}(G:F|\boldsymbol{z}_i)$ and $\frac{1}{n}\sum_{i=1}^{n}\hat{J}(F:G|\boldsymbol{z}_i)$. Another approach is taking $\hat{I}(F:G|\bar{\boldsymbol{z}}), \hat{I}(G:F|\bar{\boldsymbol{z}})$, and $\hat{J}(F:G|\bar{\boldsymbol{z}})$, where $\bar{\boldsymbol{z}}$ is the average covariate value over all the individuals.

Now, to implement test statistics $\hat{I}(F:G|\boldsymbol{z}), \hat{I}(G:F|\boldsymbol{z})$ and $\hat{J}(F:G|\boldsymbol{z})$ use the following steps:

Step 1: Use the Devarajan and Ebrahimi (2000) approach to estimate β and γ. Denote the estimates by $\hat{\beta}$ and $\hat{\gamma}$.

Step 2: Replace β and γ by $\hat{\beta}$ and $\hat{\gamma}$ in equations (18.11)–(18.13) to get $\hat{I}(F:G|\boldsymbol{z}), \hat{I}(G:F|\boldsymbol{z})$ and $\hat{J}(F:G|\boldsymbol{z})$.

Step 3: One can show that $2n\ \hat{I}(F:G|\boldsymbol{z})$, $2n\ \hat{I}(G:F|\boldsymbol{z})$ and $\hat{J}(F:G|\boldsymbol{z})$ have asymptotically chi-squared distribution with q degrees of freedom under H_0. [See Kullback (1978)]. Here q is the number of parameters under H_0. Therefore, if you are using $\hat{I}(F:G|\boldsymbol{z})$, then reject H_0 if $2n\ \hat{I}(F:G|\boldsymbol{z}) > \chi^2_{q,x}$, where $\chi^2_{q,x}$ is the chi-squared distribution with q degrees of freedom and α is the significance level of the test. If you are using $\hat{I}(G:F|\bar{\boldsymbol{z}})$, then reject H_0 if $2n\ \hat{I}(G:F|\boldsymbol{z}) > \chi^2_{q,x}$. Finally, if you are using $\hat{J}(F:G|\boldsymbol{z})$, then reject H_0 if $n\ \hat{J}(F:G|\boldsymbol{z}) > \chi^2_{q,x}$.

18.3 Comparison of the Proposed Goodness-of-Fit Test with Existing Methods

First, we compare our proposed goodness-of-fit test based on the divergence $\hat{J}(F:G|\boldsymbol{z})$ with existing tests for the Cox PH model using empirical power estimates based on 1000 simulations for different sample sizes and censoring patterns. The comparison is made with results from Quantin *et al.* (1996) to reduce the computational burden.

Simulations were performed based on a Weibull model for the two sample problem using hazard functions $\lambda_0(t) = 1$ in Group 0 (corresponding to the unit exponential distribution) and $\lambda_1(t) = \beta\alpha(\alpha t)^{\beta-1}$ in Group 1 (corresponding to a Weibull distribution with shape parameter β and scale parameter α). The experiment was repeated 1000 times for each of the following combinations: $\beta = 0.5, 2,\quad \alpha = 1, 2$ and $n = 30, 50, 100$ per group. Independent censoring samples were generated based on the uniform distribution over a fixed interval

(0,B) to result in 25% censoring. Computations for the proposed method were based on the full likelihood approach using B-spline approximations for the baseline hazard as described by Devarajan and Ebrahimi (2000).

The proposed method is compared with goodness-of-fit tests of Quantin *et al.* (1996), Breslow *et al.* (1984), Gill and Schumaker (1987), Nagelkerke *et al.* (1984), Wei (1984) and that of Cox (1972) incorporating time-dependent covariate effect. The critical values of the test statistics for the proposed test were computed based on the Chi-square distribution with 1 degree of freedom at a significance level of 0.05 as discussed in Section 18.2. Quantin *et al.* (1996) note that the critical values of all the test statistics in their comparison study were also based on the Chi-square distribution with 1 degree of freedom at a significance level of 0.05 except that of Wei (1984) whose critical values are based on the tables of Koziol and Byar (1975).

It was seen in the simulations that the proposed test achieves the specified significance level of 5% and hence it is a consistent test. Tables 18.1 through 18.6 present the results of the power study. From the results, there is clear evidence that the proposed test performs better than most of the existing tests for the Cox PH model. The empirical power of the proposed test is higher than all the other tests in most situations. Overall, the results are much better for the case of the Weibull distribution with shape parameter $\beta = 0.5$ for both choices of the scale parameter $\alpha = 1, 2$. The proposed test performs moderately well relative to the other tests for the case of $\beta = 2$ and $\alpha = 1$ in both uncensored and censored samples with group sizes 30 and 50.

It would be interesting to study how the proposed test performs in the case of other distributions such as a log-logistic and a lognormal distribution that allow the hazard functions corresponding to differing scale and shape parameters to cross but they do not satisfy the model (18.2). In order to get a first-hand idea on the performance of the proposed test in such situations, a small simulation study was performed. Simulations were performed based on a lognormal model for the two sample problem using hazard functions $\lambda_0(t)$ in Group 0 corresponding to the standard lognormal distribution with scale parameter $\mu = 0$ and shape parameter $\sigma = 1$ and $\lambda_1(t)$ in Group 1 corresponding to a lognormal distribution with scale parameter μ and shape parameter σ. The experiment was repeated 1000 times for uncensored samples for each of the following combinations: $\mu = 0, 1,\ \ \sigma = 0.5, 2$ and $n = 30, 50$ and 100 per group. The results are shown in Tables 18.7 through 18.9. From the tables, we see that the empirical power estimates are much better when $\sigma = 2$ relative to $\sigma = 1$. Even in the case of $\sigma = 1$, the empirical powers are higher for the case of $\mu = 1$ relative to $\mu = 0$. There is an indication that the proposed test is able to pick up crossing hazards even in situations where the underlying distributions do not belong to the family of models (i.e., Weibull) included in the non-proportional hazards model (18.2).

We also compare the goodness-of-fit statistics based on the directed divergence as given by $\hat{I}(F:G)$ and $\hat{I}(G:F)$ and the divergence $\hat{J}(F:G)$. The comparison is made for each combination of sample size, censoring percentage and Weibull distribution characteristics as given above. In each case, we observe that the directed divergence given by $\hat{I}(F:G)$ gives the highest empirical power among the three statistics. But $\hat{J}(F:G)$ measures the directed divergences in both directions as pointed out earlier. Overall, the power estimates based on these three measures are in proximity to each other.

Table 18.1: Simulation results: Power comparison of proposed test uncensored Weibull samples, Weibull sample size per group = 30

Test	$\beta = 0.5$		$\beta = 2$	
	$\alpha = 1$	$\alpha = 2$	$\alpha = 1$	$\alpha = 2$
Proposed test	1.000	1.000	0.657	0.899
Quantin *et al.*	0.884	0.828	0.842	0.823
Breslow *et al.*	0.786	0.734	0.787	0.672
Gill and Schumaker	0.492	0.537	0.754	0.807
Nagelkerke *et al.*	0.208	0.166	0.191	0.711
Wei	0.692	0.705	0.603	0.601
Harrell and Lee	0.788	0.794	0.888	0.728
Cox (time dependent)	0.704	0.677	0.825	0.716

Table 18.2: Simulation results: Power comparison of proposed test uncensored Weibull samples, sample size per group = 50

Test	$\beta = 0.5$		$\beta = 2$	
	$\alpha = 1$	$\alpha = 2$	$\alpha = 1$	$\alpha = 2$
Proposed test	1.000	1.000	0.832	0.948
Quantin *et al.*	0.986	0.979	0.978	0.951
Breslow *et al.*	0.955	0.923	0.957	0.898
Gill and Schumaker	0.673	0.772	0.953	0.933
Nagelkerke *et al.*	0.307	0.231	0.338	0.908
Wei	0.913	0.901	0.924	0.824
Harrell and Lee	0.957	0.949	0.990	0.908
Cox (time dependent)	0.892	0.863	0.958	0.915

Table 18.3: Simulation results: Power comparison of proposed test uncensored Weibull samples, sample size per group = 100

Test	$\beta = 0.5$		$\beta = 2$	
	$\alpha = 1$	$\alpha = 2$	$\alpha = 1$	$\alpha = 2$
Proposed test	1.000	1.000	0.968	0.991
Quantin *et al.*	1.000	1.000	1.000	0.999
Breslow *et al.*	0.996	0.998	0.999	0.997
Gill and Schumaker	0.939	0.971	1.000	0.999
Nagelkerke *et al.*	0.574	0.428	0.588	0.997
Wei	0.998	0.997	0.997	0.982
Harrell and Lee	0.957	0.949	0.990	0.908
Cox (time dependent)	0.993	0.986	1.000	0.997

Table 18.4: Simulation results: Power comparison of proposed test 25% censoring, Weibull samples, sample size per group = 30

Test	$\beta = 0.5$		$\beta = 2$	
	$\alpha = 1$	$\alpha = 2$	$\alpha = 1$	$\alpha = 2$
Proposed test	1.000	1.000	0.626	0.705
Quantin *et al.*	0.745	0.683	0.752	0.737
Breslow *et al.*	0.610	0.554	0.673	0.666
Gill and Schumaker	0.602	0.612	0.742	0.710
Nagelkerke *et al.*	0.098	0.166	0.123	0.290
Wei	0.686	0.662	0.696	0.641
Harrell and Lee	0.686	0.633	0.706	0.551
Cox (time dependent)	0.536	0.515	0.711	0.654

Table 18.5: Simulation results: Power comparison of proposed test 25% censoring, Weibull samples, sample size per group = 50

Test	$\beta = 0.5$		$\beta = 2$	
	$\alpha = 1$	$\alpha = 2$	$\alpha = 1$	$\alpha = 2$
Proposed test	1.000	1.000	0.787	0.946
Quantin *et al.*	0.934	0.891	0.937	0.912
Breslow *et al.*	0.829	0.775	0.741	0.861
Gill and Schumaker	0.820	0.797	0.909	0.896
Nagelkerke *et al.*	0.148	0.221	0.171	0.442
Wei	0.793	0.778	0.808	0.775
Harrell and Lee	0.904	0.888	0.917	0.815
Cox (time dependent)	0.694	0.719	0.908	0.872

Table 18.6: Simulation results: Power comparison of proposed test 25% censoring, Weibull samples, sample size per group = 100

Test	$\beta = 0.5$		$\beta = 2$	
	$\alpha = 1$	$\alpha = 2$	$\alpha = 1$	$\alpha = 2$
Proposed test	1.000	1.000	0.956	0.998
Quantin *et al.*	0.999	1.000	0.999	0.994
Breslow *et al.*	0.980	0.974	0.992	0.992
Gill and Schumaker	0.938	0.982	0.996	0.990
Nagelkerke *et al.*	0.196	0.208	0.304	0.723
Wei	0.986	0.980	0.991	0.973
Harrell and Lee	0.999	0.995	0.999	0.987
Cox (time dependent)	0.963	0.948	0.998	0.991

Table 18.7: Power comparison of directed divergence and divergence uncensored lognormal samples, sample size per group = 30

Test	$\sigma = 0.5$		$\sigma = 2$	
	$\mu = 0$	$\mu = 1$	$\mu = 0$	$\mu = 1$
$\hat{I}(F:G)$	0.584	0.876	1.000	1.000
$\hat{I}(G:F)$	0.540	0.851	1.000	1.000
$\hat{J}(F:G)$	0.561	0.865	1.000	1.000

Table 18.8: Power comparison of directed divergence and divergence uncensored lognormal samples, sample size per group = 50

Test	$\sigma = 0.5$		$\sigma = 2$	
	$\mu = 0$	$\mu = 1$	$\mu = 0$	$\mu = 1$
$\hat{I}(F:G)$	0.614	0.853	1.000	1.000
$\hat{I}(G:F)$	0.579	0.832	1.000	1.000
$\hat{J}(F:G)$	0.599	0.847	1.000	1.000

Table 18.9: Power comparison of directed divergence and divergence uncensored lognormal samples, sample size per group = 100

Test	$\sigma = 0.5$		$\sigma = 2$	
	$\mu = 0$	$\mu = 1$	$\mu = 0$	$\mu = 1$
$\hat{I}(F:G)$	0.646	0.849	1.000	1.000
$\hat{I}(G:F)$	0.629	0.834	1.000	1.000
$\hat{J}(F:G)$	0.641	0.841	1.000	1.000

Table 18.10: Power comparison of directed divergence and divergence uncensored Weibull samples, sample size per group = 30

Test	$\beta = 0.5$		$\beta = 2$	
	$\alpha = 1$	$\alpha = 2$	$\alpha = 1$	$\alpha = 2$
$\hat{I}(F:G)$	1.000	1.000	0.680	0.905
$\hat{I}(G:F)$	1.000	1.000	0.614	0.881
$\hat{J}(F:G)$	1.000	1.000	0.657	0.899

Table 18.11: Power comparison of directed divergence and divergence uncensored Weibull samples, sample size per group = 50

Test	$\beta = 0.5$		$\beta = 2$	
	$\alpha = 1$	$\alpha = 2$	$\alpha = 1$	$\alpha = 2$
$\hat{I}(F:G)$	1.000	1.000	0.842	0.952
$\hat{I}(G:F)$	1.000	1.000	0.810	0.943
$\hat{J}(F:G)$	1.000	1.000	0.832	0.948

Table 18.12: Power comparison of directed divergence and divergence uncensored Weibull samples, sample size per group = 100

Test	$\beta = 0.5$		$\beta = 2$	
	$\alpha = 1$	$\alpha = 2$	$\alpha = 1$	$\alpha = 2$
$\hat{I}(F:G)$	1.000	1.000	0.972	0.991
$\hat{I}(G:F)$	1.000	1.000	0.964	0.988
$\hat{J}(F:G)$	1.000	1.000	0.968	0.991

Table 18.13: Power comparison of directed divergence and divergence 25% censoring, Weibull samples, sample size per group = 30

Test	$\beta = 0.5$		$\beta = 2$	
	$\alpha = 1$	$\alpha = 2$	$\alpha = 1$	$\alpha = 2$
$\hat{I}(F:G)$	1.000	1.000	0.647	0.697
$\hat{I}(G:F)$	1.000	1.000	0.588	0.681
$\hat{J}(F:G)$	1.000	1.000	0.626	0.697

Table 18.14: Power comparison of directed divergence and divergence 25% censoring, Weibull samples, sample size per group = 50

Test	$\beta = 0.5$		$\beta = 2$	
	$\alpha = 1$	$\alpha = 2$	$\alpha = 1$	$\alpha = 2$
$\hat{I}(F:G)$	1.000	1.000	0.798	0.947
$\hat{I}(G:F)$	1.000	1.000	0.763	0.945
$\hat{J}(F:G)$	1.000	1.000	0.787	0.946

Table 18.15: Power comparison of directed divergence and divergence 25% censoring, Weibull samples, sample size per group = 100

Test	$\beta = 0.5$		$\beta = 2$	
	$\alpha = 1$	$\alpha = 2$	$\alpha = 1$	$\alpha = 2$
$\hat{I}(F:G)$	1.000	1.000	0.961	0.998
$\hat{I}(G:F)$	1.000	1.000	0.949	0.998
$\hat{J}(F:G)$	1.000	1.000	0.956	0.998

18.4 Illustration of the Goodness-of-Fit Test using Real-Life Data

To illustrate the usefulness of the proposed tests in applied work, we examine three data sets that have been considered in the literature.

Data 1: The data consists of times to remission (in weeks) of 42 leukemia patients from Kalbfleisch and Prentice (1980). The patients are randomized into two groups; a control group and a treatment group for which the drug 6-MP was administered. Each group consists of 21 patients. The covariate is group membership specified by the indicator variable z where $z = 0$ denotes the control group and $z = 1$ denotes the treatment group. For the complete data set see Kalbfleisch and Prentice (1980, p. 206).

Using $\hat{\beta} = 1.573$ and $\hat{\gamma} = -.111$ from Devarajan and Ebrahimi (2000), we have, $\hat{J}(F : G|z = 1) = 3.33X10^{-6}, n\widehat{J(F : G|z = 1)} = 0.00014$ which gives a p-value of 0.991 using an asymptotic χ^2 distribution with 1 degree of freedom under the null hypothesis. Since p-value $= 0.991 > \alpha = 0.05$, we do not reject H_0. This data set was also analyzed by Gill and Schumaker (1987). Their conclusion coincides with ours. However our p-value is larger.

Data 2: In this example, we analyze data from an animal experiment presented in Wei (1984) in which a group of male mice are subject to 300 rads of radiation and followed for cancer incidence. This is the placebo or the control group. The treatment group was placed in a germ free environment. Using $\hat{\beta} = -.414$ and $\hat{\gamma} = -.348$ from Devarajan and Ebrahimi (2000), we have, $\hat{J}(F : G|z = 1) = .317$. Consequently, $n\hat{J}(F : G|z = 1) = 16.167$ which gives a p-value of $(5.8) \times 10^{-5}$ using an asymptotic χ^2 distribution with 1 degree of freedom under the null hypothesis. Since p-value $= 5.8 \times 10^{-5}$, we do reject H_0 in favor of H_1. That is, the Cox model is not appropriate. This data set was analyzed by Wei (1984). He ended up with a similar conclusion. However, his p-value is larger.

Data 3: The data consists of the times to remission of 33 leukemia patients presented in Cox and Oakes (1984). There are two covariates, a binary variable indicating group membership (Control or Treatment group) and a continuous variable denoting the logarithm of White Blood Count (WBC). There are no censored observations.

Using $\hat{\beta}_1 = 13.941$, $\hat{\beta}_2 = -6.194$, $\hat{\gamma}_1 = 1.056$ and $\hat{\gamma}_2 = -0.595$ from Devarajan and Ebrahimi (2000) $\hat{J}(F : G|\mathbf{z}) = \hat{J}(F : G|z_1 = \bar{z}_1, z_2 = 1) = 2.032$. Now, $n\hat{J}(F : G|\mathbf{z}) = 67.056$ which gives a p-value of $2.78X10^{-15}$ using a χ^2 distribution with 2 degrees of freedom under the null hypothesis. Since p-value $= 2.78X10^{-15} < \alpha = 0.05$, we reject H_0 in favor of H_1. That is, the non-proportional hazards model is more appropriate.

This data was truncated before the observation value of 7 units. For the truncated data the parameter estimates are: $\hat{\beta}_1 = 1.336$, $\hat{\gamma}_1 = -1.404$, $\hat{\beta}_2 = -0.318$ and $\hat{\gamma}_2 = 1.007$; see Devarajan and Ebrahimi (2000). The value of our test statistic is $\hat{J}(F : G|\mathbf{z}) = 0.0168$. Thus, $n\hat{J}(F : G|\mathbf{z}) = 0.554$ which gives a p-value of 0.758 using a χ^2 distribution with 2 degrees of freedom under the null hypothesis. Since p-value $= 0.758 > \alpha = 0.05$, we do not reject H_0. That is for the truncated data the Cox PH model is appropriate.

It should be mentioned that mean value of $z_1 = \bar{z}_1$ and $z_2 = 1$ were used for computing $\hat{J}(F : G|\mathbf{z})$ in both cases above.

18.5 Concluding Remarks

A test for the Cox proportional hazards model based on an estimate of the KLD information between a non-proportional hazards model and the Cox PH model has been presented. The power of the proposed test was investigated and compared with the standard tests. The following results were obtained.

(a) Use of the information statistic enables us to test the Cox PH model against a non-proportional hazards model.

(b) Compared with all the tests, our proposed test is not dependent on the baseline hazard function $\lambda_0(t)$.

(c) The proposed test performs very well in terms of power compared with other leading tests against non-proportional hazards alternatives.

References

1. Andersen, K. (1982). Testing goodness-of-fit of Cox's regression and life model, *Biometrics*, **38**, 67–77.

2. Arjas, E. (1988). A graphical method for assessing goodness of fit in Cox's proportional hazards model, *Journal of the American Statistical Association*, **83**, 204–212.

3. Breslow, N. E. (1974). Covariance analysis of censored survival data, *Biometrics*, **30**, 89–100.

4. Breslow, N. E., Edler, L., and Berger, J. (1984). A two-sample censored-data rank test for acceleration, *Biometrics*, **40**, 1049–1062.

5. Cox, D. R. (1972). Regression models and life tables (with discussion), *Journal of the Royal Statistical Society, Series B*, **34**, 187–202.

6. Cox, D. R. (1979). A note on the graphical analysis of survival data, *Biometrika*, **66**, 188–190.

7. Cox, D. R. and Oakes, D. O. (1984). *Analysis of Survival Data*, London: Chapman Hall.

8. Crowley, J. and Hu, M. (1977). Covariance analysis of heart transplant survival data, *Journal of the American Statistical Association*, **72**, 27–36.

9. Crowley, J. and Storer, B. E. (1983). Comments on "A Reanalysis of the Standard Heart Transplant Data," *Journal of the American Statistical Association*, **78**, 277–281.

10. Devarajan, K. and Ebrahimi, N. (2000). Inference for a non-proportional hazards regression model and applications, submitted for publication.

11. Ebrahimi, N., Habibullah, M., and Soofi, E. (1994). Testing exponentiality based on Kullback-Leibler information, *Journal of the Royal Statistical Society, Series B*, **54**, 739–748.

12. Fleming, T. R., O'Fallon, J. R., O'Brien, P. C., and Harrington, D. P. (1980). Modified Kolmogorov-Smirnov test procedures with application to arbitrarily right-censored data, *Biometrics*, **36**, 607–625.

13. Gail, M. (1981). Evaluating serial cancer marker studies in patients at risk of recurring disease, *Biometrics*, **37**, 67–78.

14. Gill, R. and Schumaker, M. (1987). A simple test of the proportional hazards assumption, *Biometrika*, **74**, 289–300.

15. Harrell, F. E. (1986). *The PHGLM Procedure, SAS Supplemented Library User's Guide*, Version 5, Cary, North Carolina: SAS Institute Inc.

16. Harrell, F. E. and Lee, K. L. (1986). Verifying assumptions of the Cox proportional hazards model, In *Proceedings of the Eleventh Annual SAS Users' Group International Conference*, pp. 823–828, Cary, North Carolina: SAS Institute Inc.

17. Hess, K. R. (1994). Assessing time-by-covariate interactions in proportional hazards regression models using cubic-spline functions, *Statistics in Medicine*, **13**, 1045–1062.

18. Horowitz, J. L. and Neumann, G. R. (1992). A generalized moments specification test of the proportional hazards model, *Journal of the American Statistical Association*, **87**, 234–240.

19. Kalbfleisch, J. D. and Prentice, R. L. (1980). *The Statistical Analysis of Failure Time Data*, New York: John Wiley & Sons.

20. Kay, R. (1977). Proportional hazards regression models and the analysis of censored survival data, *Applied Statistics*, **26**, 227–237.

21. Kooperberg, C., Stone, C. J., and Truong, Y. K. (1995). Hazard regression, *Journal of the American Statistical Association*, **78**, 78–94.

22. Koziol, J. A. and Byar, D. P. (1975). Percentage points of the asymptotic distributions of one and two-sample K.S. tests for truncated or censored data, *Technometrics*, **17**, 507–510.

23. Kullback, S. (1978). *Information Theory and Statistics*, Dover Publications.

24. Kullback, S. and Leibler, R. A. (1951). On information and sufficiency, *Annals of Mathematical Statistics*, **22**, 79–86.

25. Kupperman, M. (1957). Further applications of information theory to multivariate analysis and statistical inference, *Ph.D. Dissertation*, Graduate Council of George Washington University.

26. Lagakos, S. W. (1980). The graphical evaluation of explanatory variables in proportional hazards regression models, *Biometrika*, **68**, 93–98.

27. Lin, D. Y., Wei, L. J., and Ying. Z. (1993). Checking the Cox model with cumulative sums of martingale residuals, *Biometrika*, **80**, 557–572.

28. Lin, D. Y. and Wei, L. J. (1991). Goodness-of-fit tests for the general Cox regression model, *Statistica Sinica*, **1**, 1–17.

29. Moreau, T., O'Quigley, J., and Mesbah, M. (1985). A global goodness-of-fit statistic for the proportional hazards model, *Applied Statistics*, **34**, 212–218.

30. Nagelkerke, N. J. D., Oosting, J., and Hart, A. A. M. (1984). A simple test for goodness-of-fit of Cox's proportional hazards model, *Biometrics*, **40**, 483–486.

31. O'Quigley, J. and Pessione, F. (1989). Score-tests for homogeneity of regression effect in the proportional hazards model, *Biometrics*, **45**, 135–145.

32. Pena, E. A. (1998). Smooth goodness-of-fit tests for the baseline hazard in Cox proportional hazards model, *Journal of American Statistical Association*, **93**, 673–692.

33. Pettitt, A. N. and Bin Daud, I. (1990). Investigating time-dependence in Cox's proportional hazards model, *Applied Statistics*, 313–329.

34. Quantin, C., Moreau, T., Asselain, B., Maccario, J., and Lellouch, J. (1996). A Regression model for testing the proportional hazards hypothesis, *Biometrics*, **52**, 874–885.

35. Schoenfeld, D. (1980). Chi-squared goodness-of-fit tests for the proportional hazards regression model, *Biometrika*, **67**, 145–153.

36. Schoenfeld, D. (1982). Partial residuals for the porportional hazards regression model, *Biometrika*, **69**, 239–241.

37. Stablein, D. H., Carter, W. H., and Novak, J. W. (1981). Analysis of survival data with nonproportional hazard functions, *Controlled Clinical Trials*, **2**, 149–159.

38. Thaler, H. T. (1984). Nonparametric estimation of the hazard ratio, *Journal of the American Statistical Association*, **79**, 290–293.

39. Therneau, T. R. and Grambsch, P. M. (1994). Proportional hazards tests and diagnostics based on weighted residuals, *Biometrika*, **81**, 515–526.

40. Therneau, T. R., Grambsch, P. M., and Fleming, T. R. (1990). Martingale-based residuals for survival models, *Biometrika*, **77**, 147–160.

41. Wei, L. J. (1984). Testing goodness-of-fit for the proportional hazards model with censored observations, *Journal of the American Statistical Association*, **79**, 649–652.

42. White, H. (1982). Maximum likelihood estimation of misspecified models, *Econometrica*, **50**, 1–26.

19

A New Family of Multivariate Distributions for Survival Data

Shulamith T. Gross and Catherine Huber-Carol

The City University of New York, New York
Université Paris V and U472 INSERM, Paris, France

Abstract: We introduce a family of multivariate distributions in discrete time which may be regarded as a multiple logistic distribution for discrete conditional hazards. In the independent case, the marginal laws are identical to those of the univariate logistic model for survival data discussed by Efron (1988). We present the analysis of a data set previously analyzed using frailty models.

Keywords and phrases: Discrete correlated survival data, censoring, frailty models, Markov assumption, log-odds, Akaike information criterion

19.1 Introduction

A large fraction of analyses of correlated survival data appearing in recent statistical publications is based on some form of frailty models. Frailty models have been used both to model population heterogeneity in univariate survival data and to model association in multiple survival data. The model we introduce in the present treatise is meant to offer an alternative to the latter only.

19.2 Frailty Models: An Overview

Frailty models, or random effect models, for multivariate survival times of d members of a family may be divided into two broad categories:

1. conditional frailty models in which the individual survivals within a cluster are assumed independent conditionally given the cluster frailty, and

2. marginal frailty models often referred to as copula models.

The former were first introduced by Vaupel, Manton, and Stallard (1979) and Clayton and Cuzick (1985). For recent references and an illuminating discussion of the two types of frailty models, we refer the reader to Scheike, Petersen, and Martinussen (1999), and for a rigorous study of the asymptotic properties of frailty models to Parner (1998). A Cox type conditional frailty model assumes that the data for a single cluster k, $k = 1, 2, \cdots, K$ consists of possibly censored observations $(T_{k1}, \cdots, T_{kd})$ and death indicators $(\delta_{k1}, \cdots, \delta_{kd})$ where

$$\begin{aligned} T_{kj} &= X_{kj} \wedge C_{kj} \\ \delta_{kj} &= 1_{\{X_{kj} \le C_{kj}\}}. \end{aligned}$$

The censoring variables $C_k = (C_{k1}, \cdots, C_{kd})$ are assumed independent and independent of the survival variables $X_k = (X_{k1}, \cdots, X_{kd})$. Their distributions are also assumed not to depend explicitly of the parameters β and γ characterizing the laws of X_k. The components of X_k are assumed independent given the frailty W_k and the covariates (Z, V) of cluster k with hazards (instantaneous in the continuous case and discrete in the discrete case)

$$\lambda_{kj}(t) = \lambda_0(t) e^{(Z_k'\beta + V_{kj}'\gamma_j)} W_k. \tag{19.1}$$

The frailties W_k are assumed independent identically distributed according to some convenient frailty law G. The W_k are also assumed independent of the covariates and the censoring variables. This conditional frailty model suffers from several important shortcomings. Chief among them is the lack of interpretability of the regression parameters. In general, one may no longer interpret single regression parameters, say a single e^{γ_j} as the effect of the covariate V_j on an individual hazard of failure keeping all other covariates fixed because one would need to condition on the unobservable frailty W_k of the corresponding cluster. Furthermore, the values of the parameters β and γ depend on the choice of the frailty distribution G, which is typically arbitrary and not easily estimated from the data. Copula or marginal frailty models overcome the problem of parameter interpretation because they are in fact semi-parametric models in which the marginal laws with survival functions $S_1, S_2, \cdots, S_d$ are modeled separately from the copula. Clayton's copula for the multivariate case may be written as

$$S_k(t_1, \cdots, t_d) = P(X_{k1} \ge t_1, \cdots, X_{kd} \ge t_d) = [\sum_{j=1}^{d} S_{kj}(t_j)^{-\frac{1}{\alpha}} - (d-1)]^{-\alpha} \tag{19.2}$$

for some parameter $\alpha > 0$, where the marginal survival functions $S_1, S_2, \cdots, S_d$ are arbitrary, and often assumed to follow Cox model:

$$\lambda_{kj}(t) = \lambda_0(t) e^{(Z_k'\beta + V_{kj}'\gamma_j)}. \tag{19.3}$$

Here the margin parameters are interpretable in the usual way but the model, dependent as it is on a single association parameter α, allows no possibility of modeling within cluster dependence on the individual level covariate V_{kj}, $j = 1, 2, \ldots, d$. The model may therefore represent a first order approximation to the far more complicated dependence structure likely to be present in real survival data. Our model which breaks down the dependence structure in the data into hierarchical components without resorting to the use of random effects allows a finer representation of dependence. Moreover it allows an interpretation of the parameters as log odds ratios for failure.

19.3 The Model

Let $d = max\{n_k, k \leq K\}$ be the maximal cluster size and T the maximum value of the discrete time. $R_k(t)$ is an element of the set $\{(0,1)^{\bigotimes d}\}$ representing the subset of the elements still at risk just before t, in cluster k, while $Y_k(t)$ is an element of the set $\{(0,1)^{\bigotimes d}\}$ representing the subset, at time t, of those elements of cluster k who jump at time t. Finally, Z_k is an observed covariate that may depend on time t. Following Arjas and Haara (1988), we define the "past" σ-field recurrently by:

$$\begin{aligned} \mathcal{F}_k(0) & \quad \text{is the initial information on cluster } k \\ \mathcal{G}_k(t) &= \mathcal{F}_k(t-1) \bigvee \sigma(R_k(t), Z_k(t)) \\ \mathcal{F}_k(t) &= \mathcal{G}_k(t) \bigvee \sigma(Y_k(t)). \end{aligned} \tag{19.4}$$

Assuming that the censoring law does not depend explicitly on the parameter of interest θ, the likelihood of parameters $\theta = (\theta_1, \cdots, \theta_{l_p})'$ of interest may be written as

$$\begin{aligned} V(\theta) &= \prod_{k=1}^{K} \prod_{t=1}^{T} P^\theta [Y_k(t) = y_k(t) | \mathcal{F}_k(t-1), R_k(t)] \\ & \quad \prod_{k=1}^{K} \prod_{t=1}^{T} P^\theta [R_k(t) = r_k(t) | \mathcal{F}_k(t-1)] \\ &= V_1(\theta) V_2(\theta). \end{aligned} \tag{19.5}$$

Here $y_k(t)$ and $r_k(t)$ denote the observed values of $R_k(t)$ and $Y_k(t)$ for $k = 1, 2, \cdots, K$, and $t = 1, 2, \cdots, T$. For uncensored data $r_k(t) = r_k(t-1) - y_k(t)$ and $V(\theta) = V_1(\theta)$. In case of censoring, we assume that $\theta = (\theta_1, \theta_2)$ where θ_1 is the parameter of interest, so that we can use the partial likelihood V_1 for inference on θ_1. Without further comment, we shall drop the subscript 1 from

θ. The fundamental assumption in our model is

$$\begin{aligned} P(Y_k(t) = y \mid \mathcal{F}_k(t-1), R_k(t) = r) &= P(Y_k(t) = y \mid R_k(t) = r) \\ &= P(Y(t) = y \mid R(t) = r). \quad (19.6) \end{aligned}$$

This is a Markov like assumption that characterizes our model. The likelihood

$$V(\theta) = \prod_{k=1}^{K} P^{\theta}[Y_k(t) = y_k(t) \mid R_k(t) = r_k(t)] \tag{19.7}$$

is a complete likelihood for our model in the uncensored case and a partial likelihood in the right censored case. It may also be viewed, in light of (19.5-19.6) as a partial likelihood for θ for censored or uncensored data when the true model for the complete, possibly unobserved, data, does not satisfy assumption (19.6). Defining now

$$N(r, y, t) = \#(k : R_k(t) = r, Y_k(t) = y)$$

we may write the log-likelihood of (19.7) as

$$L = \sum_{k=1}^{K} \sum_{t=1}^{T} \sum_{r} \sum_{y \le r} N(r, y, t) \ln(P^{\theta}[Y(t) = y \mid R(t) = r]). \tag{19.8}$$

We now write $P^{\theta}[Y(t) = y \mid R(t) = r]$ as $P_{nn}(r, y, t)/c(r, t)$ and parameterize the non normalized probabilities $P_{nn}(r, y, t)$ using a set of parameters $p_{r,y}(t)$ indexed by r and $0 < y \le r$ as follows:

$$P_{nn}(r, y, t) = e^{\sum_{0<r' \le r} \sum_{0<y' \le r' \wedge y} p_{r',y'}(t)} \tag{19.9}$$

and

$$c(r, t) = \sum_{y \le r} e^{\sum_{0<r' \le r} \sum_{0<y' \le r' \wedge y} p_{r',y'}(t)}. \tag{19.10}$$

When no restrictions are placed on the new parameters $p_{r,y}(t)$, our model is saturated and imposes no further restrictions on the model beyond the basic Markov assumption (19.6). For $d = 2$ we have a five p-parameters set, $p_{11,11}(t)$, $p_{11,01}(t)$, $p_{11,10}(t)$, $p_{10,10}(t)$ and $p_{01,01}(t)$.

Theorem 19.3.1 *For clusters of size d, the law defined by (19.7)–(19.10) with $K = 1$ for complete data defines a law for independent $X_1, \cdots, X_d$ if and only if $p_{r,y}(t) \equiv 0$ for all r with $|r| > 1$ and all $y \le r$, and marginal laws defined by*

$$P(X_j = t) = \prod_{1 \le t' \le t} (1 + e^{p_j(t')})^{-1} e^{p_j(t')} \tag{19.11}$$

for $p_j(t) = p_{r,y}(t)$ with $r_{j'} = \delta_{jj'}$ and $y = r$.

We state Theorems 19.3.2 and 19.3.3 for the bivariate case. Extensions to the general d-dimensional case are straightforward but require cumbersome notation.

Theorem 19.3.2 *For the bivariate law defined by (19.7)–(19.10) with $K = 1$, $d = 2$ for the complete data (without censoring), we have:*

$$
\begin{aligned}
& p_{11,10} \\
&= \ln\left(\frac{P[X_1 = t, X_2 > t | X_1 \geq t, X_2 \geq t] / P[X_1 > t, X_2 > t | X_1 \geq t, X_2 \geq t]}{P[X_1 = t, X_2 < t | X_1 \geq t, X_2 < t] / P[X_1 > t, X_2 < t | X_1 \geq t, X_2 < t]}\right)
\end{aligned} \tag{19.12}
$$

and similarly for $p_{11,01}$. Also

$$
p_{11,11} = \ln\left(\frac{P(X_1 = t, X_2 = t) P(X_1 > t, X_2 > t)}{P(X_1 = t, X_2 > t) P(X_1 > t, X_2 \geq t)}\right). \tag{19.13}
$$

PROOF. Immediate. ■

This shows that parameter $p_{11,10}$ is the log of the odds ratio of the failure risks for one subject in the cluster according to the presence or absence at risk of the other one in the cluster.

Theorem 19.3.3 *A general discrete bivariate law on $E_T \bigotimes E_T$ belongs to the family of laws defined by (19.7)–(19.10) for $K = 1$, $d = 2$, as defined for complete data, if and only if:*

$$
\begin{aligned}
& P(X_2 = t | X_2 \geq t, X_1 < t) = P(X_2 = t | X_2 \geq t, X_1 = t_1) \ \forall\, 0 < t_1 < t \\
& P(X_1 = t | X_1 \geq t, X_2 < t) = P(X_1 = t | X_1 \geq t, X_2 = t_2) \ \forall\, 0 < t_2 < t.
\end{aligned} \tag{19.14}
$$

PROOF. We prove first that (19.14) is a sufficient condition for being in our family, as for any bivariate law, we can write for $0 < t_1 < t_2 \leq T$,

$$
\begin{aligned}
P(X_1 = t_1, X_2 = t_2) &= P(X_1 \geq t_1, X_2 \geq t_1) \\
&\quad \times P(X_1 = t_1, X_2 > t_1 | X_1 \geq t_1, X_2 \geq t_1) \\
&\quad \times P(X_2 \geq t_2 | X_1 = t_1, X_2 \geq t_1 + 1) \\
&\quad \times P(X_2 = t_2 | X_2 \geq t_2, X_1 = t_1)
\end{aligned} \tag{19.15}
$$

$$
\begin{aligned}
&= \prod_{t=1}^{t_1 - 1} P(X_1 > t, X_2 > t | X_1 \geq t, X_2 \geq t) \\
&\quad \times P(X_1 = t_1, X_2 > t_1 | X_1 \geq t_1, X_2 \geq t_1) \\
&\quad \times \prod_{t=t_1+1}^{t_2 - 1} P(X_2 > t | X_1 = t_1, X_2 \geq t) \\
&\quad \times P(X_2 = t_2 | X_1 = t_1, X_2 \geq t_2).
\end{aligned} \tag{19.16}
$$

Under assumption (19.14), we may replace $X_1 = t_1$ by $X_1 < t$ and $X_1 < t_2$ respectively inside the two last factors, obtaining thus

$$
\begin{aligned}
P(X_1 = t_1, X_2 = t_2) &= \prod_{t=1}^{t_1-1} P(X_1 > t, X_2 > t | X_1 \geq t, X_2 \geq t) \\
&\times P(X_1 = t_1, X_2 > t_1 | X_1 \geq t_1, X_2 \geq t_1) \\
&\times \prod_{t=t_1+1}^{t_2-1} P(X_2 > t | X_1 < t, X_2 \geq t) \\
&\times P(X_2 = t_2 | X_1 < t_2, X_2 \geq t_2). \qquad (19.17)
\end{aligned}
$$

By symmetry, we have the same result for the case $T \geq t_1 > t_2 > 0$. For $0 \leq t_1 = t_2 = t \leq T$, we can always write

$$
\begin{aligned}
P(X_1 = t, X_2 = t) &= \prod_{t'=1}^{t-1} P(X_1 > t', X_2 > t' | X_1 \geq t', X_2 \geq t') \\
&\times P(X_1 = t, X_2 = t | X_1 \geq t, X_2 \geq t). \qquad (19.18)
\end{aligned}
$$

This proves the if part of the theorem. We note that the formulation of this result in terms of our parameters P_{nn} and $c(r,t)$ is

$$
\begin{aligned}
P(X_1 = t_1, X_2 = t_2) &= \prod_{t=1}^{t_1-1} [\frac{1}{c((1,1),t)}] \frac{P_{nn}((1,1),(1,0),t_1)}{c((1,1),t_1)} \\
& \prod_{t=t_1+1}^{t_2-1} [\frac{1}{c((0,1),t)}] \frac{P_{nn}((0,1),(0,1),t_2)}{c((0,1),t_2)} \\
P(X_1 = t_1, X_2 = t_1) &= \prod_{t=1}^{t_1-1} \frac{1}{c((1,1),t)} \frac{P_{nn}((1,1),(1,1),t_1)}{c((1,1),t_1)} .
\end{aligned}
$$

Conversely, if $\mathcal{L}(X_1, X_2)$ is in our family, (19.14) follows directly from the basic assumption of our model (19.6). Thus, a test criterion for being in our family may be based on equations (19.14). This completes the proof of the theorem. ■

Remarks

1. In case of censoring, the only unusable observations are those for which the first time is a censoring time. Otherwise, one can stratify on the first observed "death" time the number of the second "deaths" at each time t and check for independence. For consistency and asymptotic results for our partial maximum likelihood estimate see Gross and Huber (2000).

2. In applications, when $(X_{k1}, \cdots, X_{kd})$ are not exchangeable, their indices can represent "structural covariates". In some applications X_{k1} may represent survival of the treated and X_{k2} survival of the non treated. In

the skin grafts example below there are up to 4 closely matched grafts and up to 4 poorly matched grafts per patient. One representation of the data would then involve up to $d = 8$ grafts, where $X_{k1}, \cdots, X_{k4}$ represent survival of closely matched grafts, and $X_{k5}, \cdots, X_{k8}$ represent survival of poorly matched grafts. We chose a more parsimonious representation below that reflects the exchangeability within the poorly matched and within closely matched grafts.

19.4 An Application to Skin Grafts Rejection

19.4.1 Description of the data

The data from Bachelor and Hackett (1970), see Table 19.1, is concerned with skin grafts on severely burnt patients, having 30% to 50% of the surface of their skin burnt. There are at most $N = 4$ grafts per patient, and the HLA matching of the donor and the receiver is divided into two classes: 1 for close and 2 for poor. Only $K = 16$ patients (clusters) are available in all. The question is what is the impact of the HLA matching on the duration of the graft, taking into account the dependence of the different grafts on the same patient. Censoring time is death of patient before rejection. The data are as follows. The notation for the risk and jump sets is slightly different, as one has to take into account the fact that some elements in the cluster are exchangeable leading to equality of the corresponding parameters $p's$. Examples of risk sets, jump sets and parameters are given in Table 19.2. Since the data refer to 16 patients in all, with a total of 34 allografts, a completely non parametric model that leaves all main effect parameters $p_{1,1}(t)$ and $p_{2,2}(t)$, for t between 11 and 93 days, is not feasible. We therefore attempted a model with time-linear main effects and constant interaction parameters. That is model 1 (see Table 19.3). We dropped the time dependency from the model using Akaike's criterion and chose model 2. We then proceeded to search for the most parsimonious model with constant p-parameters. In the model chosen for its smallest AIC, model 8, only three parameters remain: $p_{11,1}$, $p_{12,1}$ and $p_{12,1} = p_{12,2}$. Thus the single interaction parameter in the model does not depend on the quality of the match (1=close, 2=poor). Since $p_{1,1}$ is a log odds and so is $p_{2,2}$, we may estimate closeness of match effect by $p_{1,1} - p_{2,2} = -1.42$ with a 95% confidence interval equal to $(-2.21, -.62)$ leading to $e^{(p_{1,1}-p_{2,2})} = .24$, with a 95% confidence interval equal to $(.11, .53)$. The model we chose, model 8, shows a main effect for graft match, but no second order effect since $p_{12,1} = p_{12,2}$ in this model. This should be compared to the results of Nielsen *et al.* (1992). They fit a Cox model with gamma frailty and obtained an estimate of the "treatment effect" $e^{\beta} = 0.31$. The likelihood ratio test of $H_0 : \beta = 0$ was 10.28. No confidence interval is reported. They mention that Kalbfleisch and Prentice (1980) obtained an

Table 19.1: Bachelor and Hackett (1970) skin grafts data on severely burnt patients

Patient	sex	age	% burnt	HLA	time until rejection	censoring
1	m	7	38	p	19	1
2	f	8	40	c	24	1
3	m	2.5	40	p	18	1
				p	18	1
4	f	7	50	p	29	1
				c	37	1
5	m	22	35	c	19	1
				p	13	1
6	f	14	35	p	19	1
				p	19	1
7	f	63	40	c	57	0
				c	57	0
				p	15	1
8	m	47	45	c	93	1
				p	26	1
9	f	60	20	c	16	1
				p	11	1
10	m	57	18	c	22	1
				p	17	1
11	m	17	45	c	20	1
				p	26	1
12	m	16	30	p	21	1
				c	18	1
13	m	32	55	c	77	1
				c	63	1
				p	43	1
				c	29	1
14	f	31	50	p	28	0
				p	28	0
15	f	42	30	c	29	1
				p	15	1
				p	18	1
16	m	62	45	p	41	1
				c	60	0

Table 19.2: Some risk sets R and jump sets S for skin grafts data

Risk sets:	
$R_k(i) = (1122)$	2 close and 2 poor grafts are present at time i^- in patient k.
$R_{k'}(i) = (1120)$	2 close and 1 poor grafts are present at time i^- in patient k'.
$R_{k''}(i) = (1100)$	2 close grafts are present at time i^- in patient k''.
Jump sets:	
$S_k(i) = (1200)$	1 close and 1 poor grafts were rejected at time i in patient k.
$S_{k'}(i) = (2000)$	1 poor graft was rejected at time i in patient k'.
$S_{k''}(i) = (0000)$	No graft was rejected at time i in patient k''.

Table 19.3: Model selection for burn data

Model #	Model description	AIC
1	$p_{1,1}(t) = a_1 + b_1 t$	221.97
	$p_{2,2}(t) = a_2 + b_2 t$	
	$p_{11,1}, p_{22,2}, p_{12,1}, p_{12,2}$ constant	
2	$p_{1,1}, p_{2,2}, p_{11,1}, p_{22,2}, p_{12,1}, p_{12,2}$ constant	220.10
3	$p_{1,1}, p_{2,2}, p_{12,1}$ constant	219.59
4	$p_{1,1}, p_{2,2}, p_{11,1}$ constant	220.52
5	$p_{1,1}, p_{2,2}, p_{12,2}$ constant	216.52
6	$p_{1,1} = p_{2,2}, p_{12,2}$ constant	227.79
7	$p_{1,1} = p_{2,2}, p_{12,1} = p_{12,2}$ constant	226.72
8	$p_{1,1}, p_{2,2}, p_{12,1} = p_{12,2}$ constant	215.21*
9	$p_{1,1}, p_{2,2},$ constant	218.90
10	$p_{1,1}, p_{2,2}, p_{22,2}$ constant	217.28
11	$p_{1,1}, p_{2,2}, p_{12,1}, p_{12,2}$ constant	217.21
12	$p_{1,1}, p_{2,2}, p_{12,1}, p_{22,2}$ constant	217.73

Table 19.4: Parameters estimation in model 8 having the smallest AIC

Parameter	Estimate	95% confidence interval	AIC
$p_{1,1}$	-3.1438226	-3.816822 -2.470823	215.2077
$p_{2,2}$	-1.7249390	-2.177313 -1.272564	
$p_{12,1} = p_{12,2}$	-0.6581479	-1.951248 0.634952	

effect of 0.22 with a 95% confidence interval of (0.048, 1.03) based on a subset of the data. Our result is certainly in rough agreement with theirs. Nielsen *et al.* (1992) could not reject the hypothesis that the variance parameter of the frailty gamma distribution is zero, in other words, the hypothesis that allografts of a given patient are independent. In our fitted model we can reject the null hypothesis that allograft survivals in a single patient are independent (although the 95% confidence interval for $p_{12,1} = p_{12,2}$ contains zero, but just barely). The likelihood ratio test for $H_0 : p_{12,1} = p_{12,2} = 0$ in model 8 is LR $= 5.7$, which is significant with $p = 0.02$. Note that our model contains only one dependence parameter and that it does not depend on graft match.

References

1. Arjas, E. and Haara, P. (1988). A note on the asymptotic normality in the Cox regression model, *The Annals of Statistics*, **16**, 1133–1140.

2. Bachelor, J. R. and Hackett, M. (1970). HL-A Matching in treatment of burned patients with skin allografts, *Lancet*, **19**, 581–583.

3. Clayton, D. G. and Cuzick, J. (1985). Multivariate Generalizations of the Proportional Hazards Model (with discussion), *Journal of the Royal Statistical Society, Series A*, **148**, 82–117.

4. Clegg, L. X., Cai, J., and Sen, P. K. (1999). A marginal mixed baseline hazards model for multivariate failure time data, *Biometrics*, **55**, 805–812.

5. Efron, B. (1988). Logistic regression, survival analysis, and the Kaplan-Meier curve, *Journal of the American Statistical Association*, **83**, 414–425.

6. Gross, S. T. and Huber-Carol C. (2000). Hierarchical dependency models for multivariate survival data with censoring, *Lifetime Data Analysis*, **6**, 299–320.

7. Hanley, J. A. and Parnes, M. N. (1983). Nonparametric estimation of a multivariate distribution in the presence of censoring, *Biometrics*, **39**, 129–139.

8. Huster, W. J., Brookmeyer, R., and Self, S. G. (1989). Modeling paired survival data with covariates, *Biometrics*, **45**, 145–156.

9. Kalbfleisch, J. G. and Prentice (1980). *The Statistical Analysis of Failure Time Data*, New York: John Wiley & Sons.

10. Nielsen, G. G., Gill, R. D., Andersen, P. K., and Sorensen, T. I. A. (1992). *Scandinavian Journal of Statistics*, **19**, 25–43.

11. Parner, E. (1998). Asymptotic theory for the correlated Gamma-frailty model, *Annals of Statistics*, **26**, 183–214.

12. Ross, E. A. and Moore, D. (1999). Modeling clustered discrete or grouped time survival data with covariates, *Biometrics*, **55**, 813–819.

13. Scheike, T. H., Petersen, J. H., and Martinussen, T. (1999). Retrospective ascertainment of recurrent events: An application to time to pregnancy, *Journal of the American Statistical Association*, **94**, 713–725.

14. Vaupel, J. W., Manton, K. G., and Stallard, E. (1979). The impact of heterogeneity in individual frailty and the dynamics of mortality, *Demography*, **16**, 439–447.

20

Discrimination Index, the Area Under the ROC Curve

Byung-Ho Nam and Ralph B. D'Agostino

Boston University, Boston, Massachusetts

Abstract: The accuracy of fit of a mathematical predictive model is the degree to which the predicted values coincide with the observed outcome. When the outcome variable is dichotomous and predictions are stated as probabilities that an event will occur, models can be checked for good discrimination and calibration. In case of the multiple logistic regression model for binary outcomes (event, non-event), the area under the ROC (Receiver Operating Characteristic) curve is the most used measure of model discrimination. The area under the ROC curve is identical to the Mann-Whitney statistic. We consider shift models for the distributions of predicted probabilities for event and non-event. From the interval estimates of the shift parameter, we calculate the confidence intervals for the area under the ROC curve. Also, we present the development of a general description of an overall discrimination index C (*overall* C) which we can extend to a survival time model such as the the Cox regression model. The general theory of rank correlation is applied in developing the *overall* C. The *overall* C is a linear combination of three independent components: event vs. non-event, event vs. event and event vs. censored. By showing that these three components are asymptotically normally distributed, the *overall* C is shown to be asymptotically normally distributed. The expected value and the variance of the *overall* C are presented.

Keywords and phrases: Discrimination, calibration, logistic regression model, receiver operating characteristic, Cox regression model, discrimination index, health risk appraisal function, confidence interval, Mann-Whitney statistic, censoring

20.1 Introduction

Background: Performance measures in mathematical predictive models

Consider a vector of variables $\vec{V}(V_1, V_2, \ldots, V_k)$, independent variables in a regression, or risk factors, and a variable W, the dependent variable, or outcome variable having 1 for positive outcome and 0 for negative outcome. Here, 'positive outcome' indicates occurrence or presence of an event such as coronary heart disease.

Health Risk Appraisal functions (HRAF) are mathematical models that are functions of the data ($\vec{V}$), which relates to the probability of an event (W). Symbolically, for the configuration of $\vec{V}$ of the data

$$f(\vec{V}) = f(V_1, V_2, \ldots, V_k) = P(W = 1) = P$$

where $P(W = 1)$ is the probability of a positive outcome or an event.

Our focus is to evaluate the performance of a HRAF with regard to its ability to predict the outcome variable. We consider two models: logistic regression model, and survival model such as Cox regression model.

First, the logistic regression model is a model that relates $\vec{V}$ to the development of an event over a period of time t, and the following is its mathematical expression:

$$P(W = 1) = \frac{1}{1 + \exp((-\beta'\vec{V}))} = \frac{1}{1 + \exp(-(\beta_0 + \beta_1 V_1 + \beta_2 V_2 + \cdots + \beta_k V_k))}.$$

Second, the Cox regression model is a survival analysis model that relates $\vec{V}$ to the development of an event over a period of time t, but in this model, we take into consideration the time to event and censoring, for example, dropouts, lost to follow-up. The following is its mathematical expression:

$$P(T \leq t) = S_0(T = t/\bar{V})^{\exp(\beta'\vec{V} - \beta'\bar{V})},$$

where $S_0(T = t/\bar{V})$ is the survival probability for those with the mean vector values, $\bar{V}$.

Measures of predictive accuracy

The accuracy of a model is the degree to which the predicted values coincide the observed outcomes. When the outcome variable is dichotomous and predictions are stated as probabilities that an event will occur, models can be checked for two general concepts, discrimination and calibration. Our focus here is to evaluate the performance of a model with regard to its discrimination. Discrimination refers to a model's ability to correctly distinguish the two classes of outcomes. Perfect discrimination would result in two non-overlapping sets of predicted probabilities from the model, one set for the positive outcomes, the other for the negative outcomes.

20.2 Nonparametric Confidence Interval for Area under the ROC Curve

20.2.1 Discrimination in logistic regression

The area under the Receiver Operating Characteristic (ROC) Curve is one the most used measures for a model discrimination. The following is how to construct a ROC curve from the logistic regression:

Suppose we have n subjects. All n subjects have their predicted probabilities $(Y_1, Y_2, \ldots, Y_n)$. Then, select some probability value Y^* and state as the decision rule that all subjects with predicted probabilities equal to or above that value Y^* will be classified as positive, and all of less will be classified as negative. Hence, for each Y^*, a two by two table such as follows can be generated:

Call subject + if $Y_i > Y^*$

		+	−
True State	+	a	b
	−	c	d

From this table, *sensitivity* $= \frac{a}{a+b}$, and *specificity* $= \frac{d}{c+d}$ can be calculated. If one selects all possible values of Y^* for decision cutoff points and plots *sensitivity* on the Y axis and 1−*specificity* on the X axis in a two dimensional graph and connect the plots by a line curve, then the resulting curve is called the Receiver Operating Characteristic (ROC) Curve. The area under this curve is a measure for model discrimination. Interpretation of this area, also called C statistic, is that it is the estimated probability for positive outcome being higher than that for a negative outcome. Thus,

$$\begin{aligned} C \text{ statistic} &= C = \text{ area under the ROC curve} \\ &= \hat{P}(Y_1 > Y_2) \end{aligned}$$

where

$$\begin{aligned} Y_1 &= \text{predicted probabilities for those who had events} \\ Y_2 &= \text{predicted probabilities for those without events.} \end{aligned}$$

The value of C varies from 0.5 with no discrimination ability to 1 with perfect discrimination and is related only to the ranks of the predicted probabilities. Bamber (1975) recognized that the area under the ROC curve is an unbiased estimator of the probability of correctly ranking a (event, no-event) pair and that this probability is closely connected with the Mann-Whitney statistic.

Hanley and McNeil (1982) elaborated the relationship between the area under the ROC curve and the Mann-Whitney statistic and showed that the two are identical, i.e.,

$$C \text{ statistic} = \frac{1}{N_1 \cdot N_2} W_{Y_1 Y_2} \tag{20.1}$$

where

$$\begin{aligned} N_1 &= \text{number of those who had events} \\ N_2 &= \text{number of those without events} \\ W_{Y_1Y_2} &= [\text{number of pairs } (Y_1, Y_2) \text{ with } Y_1 > Y_2] \\ &\quad + \frac{1}{2}[\text{number of pairs } (Y_1, Y_2) \text{ with } Y_1 = Y_2]. \end{aligned}$$

Lehmann (1975) showed that the C statistic is asymptotically normally distributed. See the appendix for detail. Several methods for constructing confidence intervals or confidence bounds for the area under the ROC curve have been proposed: Sen (1967) and Govindarajulu (1991) proposed confidence intervals based upon asymptotic normality of the Mann-Whitney statistic and the consistent estimator of its standard error. Birnbaum and McCarthy (1958) developed a method for constructing confidence bounds that do not depend on asymptotic normality, but which are very conservative. Ury (1972) proposed a method that used Chebychev's inequality in place of asymptotic normality, which yields a very conservative confidence interval. More recently, Hilgers (1991) proposed distribution- free confidence bounds for ROC curves based on a combination of two local confidence bounds calculated separately for the sensitivity and specificity. Schafter (1994) proposed efficient confidence bounds for ROC curves by elaborating the test statistic introduced by Greenhouse and Mantel, and generalized by Linnet (1987).

In this paper, we propose a nonparametric approach for constructing a confidence interval for the area under a ROC curve. In Section 20.2.1, we reviewed the computation of the area under the ROC curve. In Section 20.2.2, we describe how to obtain point and interval estimates for the shift parameter Δ, the "difference" between the distribution of predicted probabilities of the events and non-events. Using the interval estimates for Δ, we then derive a confidence interval for the area under the ROC curve in Section 20.2.3.

20.2.2 Estimation of the shift parameter Δ under the shift model

As above, let the Y_1's be the predicted probabilities for those who had events and Y_2's be the predicted probabilities for those without events. Say the cumulative distribution function of Y_1 and Y_2 are G and F. The shift model assumes that $G(y) = F(y - \Delta)$, for all y, $\Delta > 0$ so that the distribution G is obtained by shifting F by an amount of Δ.

Point estimate of Δ

Say, we have random samples of n_1 events and n_2 non-events, then under the shift model, the observations $Y_{21}, Y_{22}, \ldots, Y_{2n_2}$, and $Y_{11} - \Delta, Y_{12} - \Delta, \ldots, Y_{1n_1} - \Delta$ have the same distribution. Hence, we could estimate Δ by the amount which the Y_1-values must be shifted to give the best possible agreement with Y_2-values. To do this, we define $D_{jk} = Y_{1j} - Y_{2k}$, for $j = 1$ to n_1, and $k = 1$ to n_2. Then following Lehmann (1975), the estimator $\hat{\Delta}$ would be the **median** of $n_1 n_2$ values of D_{jk} which Lehmann (1975) showed is an unbiased estimator of Δ (i.e., $E(\hat{\Delta}) = \Delta$) and also median unbiased (i.e. Δ is the median of the distribution of $\hat{\Delta}$) if one of the following conditions are satisfied:

(1) The distribution of F is symmetrically distributed about some point μ.

(2) The two sample sizes are equal, that is, $n_1 = n_2$.

Confidence interval of Δ

A distribution-free (i.e., independent of F) confidence interval $(\Delta_{\text{low}}, \Delta_{\text{up}})$ for Δ can now be obtained readily using another theorem of Lehmann's (1975, p. 87). From this theorem, we have

$$P_{\Delta}(\Delta \leq D_{(\ell)}) = P_0(W_{Y_1 Y_2} \leq \ell - 1) \tag{20.2}$$

and δ is the coverage probability such that

$$\delta = P_{\delta}(\Delta \leq D_{(\ell)}) \tag{20.3}$$

where $D_{(\ell)}$ is the ℓth ordered value of D_{jk} for $j = 1$ to n_1 and $k = 1$ to n_2 and δ is the coverage probability. For small n_1 and n_2, we can use the table for the Mann-Whitney test to obtain the value of ℓ and the closest value of δ (e.g. 0.5%, 2.5%, 97.5%, 99.5%). If n_1 and n_2 are large, Lehmann (1975) suggested a normal approximation to get the value of ℓ. Applying the approximation with continuity correction to the right-hand side of (20.3), we have

$$\delta \approx \Phi\left[\frac{\ell - \frac{1}{2}(n_1 n_2 + 1)}{\sqrt{\frac{n_1 n_2 (n_1 + n_2 + 1)}{12}}}\right] \tag{20.4}$$

where $\Phi(\cdot)$ is the cumulative distribution of the standard normal distribution. For a 95% confidence interval such that $P(\Delta_{\text{low}} \leq \Delta < \Delta_{\text{up}}) = 0.95$, the lower bound $\Delta_{\text{low}} = D_{(\ell)}$ where

$$\ell = \frac{1}{2}\left[n_1 n_2 + 1 - 1.96\sqrt{\frac{n_1 n_2(n_1 + n_2 + 1)}{3}}\right] \tag{20.5}$$

so, the Δ_{low} will be the ℓth value from the lowest of $n_1 n_2$ values of D_{jk}. In a similar fashion, the upper bound $\Delta_{\text{up}} = D_{(n_1 n_2 - \ell + 1)}$ is the $(n_1 n_2 - \ell + 1)$th value from the lowest of values of D_{jk}.

20.2.3 Confidence interval for the area under the ROC curve

We now construct the lower and the upper confidence bounds for C $(C_{\text{low}}, C_{\text{up}})$ by using the lower and the upper bounds for Δ. Let $Dis(Y)$ denote the distribution of Y. Then, it can be seen that, under the shift model,

$$Dis(Y_1 - \Delta_0) = Dis(Y_2). \tag{20.6}$$

Hence,

$$Dis(Y_1 + \Delta_1 - \Delta_0) = Dis(Y_2 + \Delta_1). \tag{20.7}$$

For the lower bound (C_{low}), let $\Delta_1 = \Delta_{\text{low}}$, $\Delta_0 = \hat{\Delta}$. Then, from (20.6), (20.7), $Dis(Y_1 - \hat{\Delta}) = Dis(Y_2)$ and

$$Dis(Y_1 + \Delta_{\text{low}} - \hat{\Delta}) = Dis(Y_2 + \Delta_{\text{low}}). \tag{20.8}$$

Now, say we have a new pair of (Y_{2k}, V_j) for $k = 1$ to n_2, $j = 1$ to n_1, where $V_j = Y_{1j} + (\Delta_{\text{low}} - \hat{\Delta})$. Hence, the C_{low} would be

$$\begin{aligned}\frac{1}{mn}[\{&\text{number of pairs } (k,j) \text{ with } Y_{2k} < V_j\} \\ &+\{\text{number of pairs } (k,j) \text{ with } Y_{2k} = V_j\}] \\ &\qquad = \frac{1}{n_1 n_2} W_{VY_2}\,.\end{aligned} \tag{20.9}$$

For the upper bound (C_{up}), let $\Delta_1 = \Delta_{\text{up}}$, $\Delta_0 = \hat{\Delta}$. Then, from (20.6), (20.7), $Dis(Y_1 - \hat{\Delta}) = Dis(Y_2)$. So,

$$Dis(Y_1 + \Delta_{\text{up}} - \hat{\Delta}) = Dis(Y_2 + \Delta_{\text{up}}). \tag{20.10}$$

Now, say we have a new pair of (Y_{2k}, U_j) for $k = 1$ to n_2, $j = 1$ to n_1, where $U_j = Y_j + (\Delta_{\text{up}} - \hat{\Delta})$. Hence, the C_{up} would be $\frac{1}{n_1 n_2} W_{UY_2}$. Therefore, the confidence interval for C, the area under the ROC curve would be

$$(C_{\text{low}}, C_{\text{up}}) = \left(\frac{1}{n_1 n_2} W_{VY_2}\,,\ \frac{1}{n_1 n_2} W_{UY_2}\right). \tag{20.11}$$

Since $W_{Y_1Y_2}$ is monotonic non-decreasing, this interval preserves the confidence coverage probability of $(\Delta_{\text{low}}, \Delta_{\text{up}})$.

For example, below we have 20 non-events and 5 events and their predicted probabilities:

$Y1:$ 0.111 0.148 0.189 0.237 0.251

$Y2:$ 0.034 0.067 0.095 0.107 0.114 0.121 0.128 0.133 0.139 0.142
0.147 0.152 0.155 0.164 0.175 0.187 0.193 0.216 0.227 0.243

From (20.1), we compute $W_{Y_1Y_2} = 70$ and $C = \frac{70}{5*20} = 0.7$.

Point estimate of $\Delta(\hat{\Delta})$

$\hat{\Delta}$ is the median of mn values of D_{jk}. Here, we have $mn = 100$, we see $\hat{\Delta}$ is

$$\frac{1}{2}[D_{(50)} + D_{(51)}] = \frac{1}{2}[0.041 + 0.042] = 0.0415.$$

Confidence interval for Δ

From (20.5), the ℓ value for 95% confidence interval is $\ell = 22$. Hence, Δ_{low} is $D_{(22)} = -0.022$. Similarly, Δ_{up} is $D_{(100-22+1)} = D_{(79)} = 0.104$. So, the 95% confidence interval for Δ is $(-0.022, 0.104)$.

Confidence interval for C

For C_{low}, we get $V_j = Y_{1j} + (\Delta_{\text{low}} - \hat{\Delta}) = Y_{1j} + (-0.022 - 0.0415) = Y_{1j} - 0.0635$ for $j = 1$ to 5. Then, we calculate $W_{VY_2} = 38$, hence $C_{\text{low}} = \frac{38}{100} = 0.38$. For C_{up}, we get $U_j = Y_{1j} + (\Delta_{\text{up}} - \hat{\Delta}) = Y_{1j} + 0.104 - 0.0415 = Y_{1j} + 0.0625$ for $j = 1$ to 5. Then we calculate $W_{UY_2} = 91$, hence $C_{\text{up}} = \frac{91}{100} = 0.91$. Therefore, the 95% confidence interval for C is (0.38, 0.91).

20.3 Extension of C Statistic to Survival Analysis

Suppose we have n individuals, among which n_1 developed events in time t (*event*), n_2 did not develop events in time t (*non-event*) and n_3 were censored in time t (*censored*) $(n = n_1 + n_2 + n_3)$.

Define

$$\begin{aligned} T_i &= \text{survival time for } i\text{th individual}, i = 1, 2, \ldots, n \\ Y_i &= \text{predicted probability for developing an event in time } t \text{ for} \\ &\quad\ i\text{th individual}, i = 1, 2, \ldots, n. \end{aligned}$$

Then, we have n pairs of $(T_1, Y_1), (T_2, Y_2), \ldots, (T_n, Y_n)$.

Define

$$\text{Overall } C = \frac{1}{Q} \sum_{i=1}^{n-1} \sum_{j=1}^{n} a_{ij} b_{ij}$$

where,

$$\begin{aligned}
Q &= \text{ the total number of comparisons made} \\
a_{ij} &= 1 \text{ if } T_i < T_j, \text{ and at least one of pair } (T_i, T_j) \text{ is for event,} \\
&\quad\; i, j = 1, 2, \ldots, n \\
&= 0 \text{ otherwise} \\
b_{ij} &= 1 \text{ if } Y_i > Y_j, \text{ and at least one of pair } (Y_i, Y_j) \text{ is for event,} \\
&\quad\; i, j = 1, 2, \ldots, n \\
&= 0 \text{ otherwise.}
\end{aligned}$$

Hence we have

$$\begin{aligned}
T_{1i} &= \text{survival time for event}, i = 1, 2, \ldots, n_1 \\
Y_{1i} &= \text{predicted probability for event}, i = 1, 2, \ldots, n_1 \\
T_{2j} &= \text{survival time for non-event}, j = 1, 2, \ldots, n_2 \\
Y_{2j} &= \text{predicted probability for non-event}, j = 1, 2, \ldots, n_2 \\
T_{3j} &= \text{survival time for censored}, j = 1, 2, \ldots, n_3 \\
Y_{3j} &= \text{predicted probability for censored}, j = 1, 2, \ldots, n_3.
\end{aligned}$$

From the above, we have three sets of comparisons:

1. *event* vs. *non-event*: comparing those who developed events against those who did not

2. *event* vs. *event*: comparing those who developed events against those who also developed events

3. *event* vs. *censored*: comparing those who developed events against those who were censored

Note that these three comparisons are independent of one another.

Now, we examine the first component of overall C (event vs. non-event): Define C statistic, C_1

$$C_1 = \frac{1}{Q_1} \sum_{i=1}^{n_1} \sum_{j=1}^{n} a_{ij} b_{ij} \tag{20.12}$$

where

$$\begin{aligned} Q_1 &= \text{the total number of comparisons made} \\ a_{ij} &= 1 \text{ if } T_{1i} < T_{2j} \\ &= 0 \text{ otherwise} \\ b_{ij} &= 1 \text{ if } Y_{1i} > Y_{2j}. \end{aligned}$$

Here, since all the survival times for those who did not develop events are longer than the maximum value of the event time for those who developed events, it is obvious that a_{ij} is always equal to 1. Hence ,

$$C_1 = \frac{1}{Q_1} \sum_{i=1}^{n_1} \sum_{j=1}^{n} b_{ij}. \tag{20.13}$$

The numerator in C_1 is exactly the same as the Mann-Whitney statistic for continuous data when we compare the predicted probabilities for dichotomous outcomes where $Q_1 = n_1 \cdot n_2$. Thus, C_1 can be expressed as

$$C_1 = \frac{1}{n_1 \cdot n_2} W_{Y_1 Y_2}. \tag{20.14}$$

Therefore, C_1 is asymptotically normally distributed.

Next, we examine the second component of overall C (*event* vs. *event*). One important assumption to make here is that one who developed an event *earlier* in time has *higher* predicted probability for an event. Define

$$C_2 = \frac{1}{Q_2} \sum_{i=1}^{n_1 - 1} \sum_{j=1}^{n_1} a_{ij} b_{ij} \tag{20.15}$$

where

$$\begin{aligned} a_{ij} &= 1 \text{ if } T_{1i} < T_{1j},\ i,j = 1,2,\ldots,n_1,\ i<j \\ &= 0 \text{ otherwise} \\ b_{ij} &= 1 \text{ if } Y_{1i} > Y_{1j},\ i,j = 1,2,\ldots,n_1,\ i<j \\ &= 0 \text{ otherwise.} \end{aligned}$$

C_2 is very closely related to the rank correlation coefficient $\hat{\tau}$. $\hat{\tau}$ has total score in its numerator. Instead, C_2 has the total positive score, PS in its numerator. Both $\hat{\tau}$ and C_2 have the same denominator Q_2, which is equal to $\frac{1}{2}n_1(n_1 - 1)$. It can be shown that C_2 has a linear relationship with $\hat{\tau}$,

$$C_2 = \frac{1}{2}(\hat{\tau} + 1). \tag{20.16}$$

Kendall (1970) showed that it is asymptotically normally distributed. See the appendix for its mean and variance. Hence C_2 is also asymptotically normal.

Now, we examine the third component of overall C (*event* vs. *censored*). Define

$$C_3 = \frac{1}{Q_3} \sum_{i=1}^{n_1} \sum_{j=1}^{n_3} a_{ij} b_{ij} \tag{20.17}$$

where

$$\begin{aligned} a_{ij} &= 1 \text{ if } T_{1i} < T_{3j},\ i = 1, 2, \ldots, n_1,\ j = 1, 2, \ldots, n_3 \\ &= 0 \text{ otherwise} \\ b_{ij} &= 1 \text{ if } Y_{1i} > Y_{3j},\ i = 1, 2, \ldots, n_1,\ j = 1, 2, \ldots, n_3 \\ &= 0 \text{ otherwise.} \end{aligned}$$

In this component, we only take a pair of comparison $[(T_{1i}, Y_{1i}), (T_{3j}, Y_{3j})]$ where the censored time (T_{3j}) is longer than the event time (T_{1i}). Thus, the numerator of C_3 can be expressed as $\sum_{i=1}^{n_1}\sum_{j=1}^{n_3}(b_{ij} \mid a_{ij} = 1)$. Also, Q_3 should be equal to $\sum_{i=1}^{n_1}\sum_{j=1}^{n_3} a_{ij}$. An important assumption that we make here is that the censoring occurs completely at random so that C_3 is independent of C_1 and C_2. The numerator of C_3 is the same as the Mann-Whitney statistic given that we only include the pairs where the censored time is longer than the event time. Hence, C_3 can be described as a conditional Mann-Whitney statistic. We express C_3 as

$$C_3 = \frac{1}{Q_3} W^C_{Y_1 Y_3}. \tag{20.18}$$

Since the Mann-Whitney statistic is asymptotically normally distributed, we can argue that C_3 is also asymptotically normal. See appendix for its mean and variance.

Now, we combine all three components into one overall discrimination index C. From (20.1), (20.2) and (20.3), we have

$$\begin{aligned} C &= \frac{1}{Q_1 + Q_2 + Q_3}(W_{Y_1 Y_1} + P + W^C_{Y_1 Y_3}) \\ &= \frac{Q_1}{Q_1 + Q_2 + Q_3}\,\frac{1}{Q_1} W_{Y_1 Y_2} + \frac{Q_2}{Q_1 + Q_2 + Q_3}\,\frac{1}{Q_2}\,\frac{1}{2}(\hat{\tau} + 1) \\ &\quad \frac{Q_3}{Q_1 + Q_2 + Q_3}\,\frac{1}{Q_3} W^C_{Y_1 Y_3} \\ &= a \cdot C_1 + b \cdot C_2 + (1 - a - b) \cdot C_3 \end{aligned} \tag{20.19}$$

so, C is a linear combination of C_1, C_2 and C_3 where

$$a = \frac{Q_1}{Q_1 + Q_2 + Q_3}\,,\quad b = \frac{Q_2}{Q_1 + Q_2 + Q_3}\,.$$

Now, we can argue that the overall C tends to normality since C_1, C_2 and C_3 are all independent of one another and each of them is asymptotically normal. See appendix for the mean and variance of the overall C.

Appendix

1. Mean and Variance of C statistic, the area under the ROC curve in logistic regression

$$\begin{aligned} E(C) &= P(Y_1 > Y_2) = P_1 \\ \mathrm{Var}(C) &= \frac{1}{n_1 n_2}[P_1(1-P_1) + (n_2-1)(P_{12} - P_1^2) \\ &\qquad + (n_1-1)(P_{13} - P_1^2)] \end{aligned}$$

where,

$$\begin{aligned} P_{12} &= P(Y_2 < Y_1 \text{ and } Y_2 < Y_1') \text{ for } Y_1 \neq Y_1' \\ P_{13} &= P(Y_2 < Y_1 \text{ and } Y_2' < Y_1) \text{ for } Y_2 \neq Y_2'. \end{aligned}$$

2. Mean and Variance of $(\hat{\tau})$

$$\begin{aligned} E(\hat{\tau}) &= \tau \\ \mathrm{Var}(\hat{\tau}) &= \frac{4(n_1-2)}{n_1(n_1-1)}\mathrm{Var}(\tau_1) + \frac{2}{n_1(n_1-2)}(1-\tau^2) \end{aligned}$$

where, τ is the rank correlation coefficient and $\mathrm{Var}(\tau_i)$ is an unknown quantity and

$$\mathrm{Var}(\tau_i) \leq \frac{1}{2}(1-\tau^2).$$

3. C_3 tends to normality with

$$\begin{aligned} E(C_3) &= P(Y_1 > Y_3 | T_1 < T_3) = P_3 \\ \mathrm{Var}(C_3) &= \frac{1}{Q_3^2}[Q_3 \cdot P_3(1-P_3) + A \cdot (P_{32} - P_3^2) + B \cdot (P_{33} - P_3^2)] \end{aligned}$$

where,

$$\begin{aligned} P_{32} &= P(Y_1 > Y_3, Y_1' > Y_3 | T_1 < T_3, T_1' < T_3) \\ P_{33} &= P(Y_1 > Y_3, Y_1 > Y_3' | T_1 < T_3, T_1 < T_3') \end{aligned}$$

and Q_3, A and B are unknown quantities.

4. Mean and Variance of overall C

$$\begin{aligned}
E[C] &= E[aC_1 + bC_2 + (1-a-b)C_3] \\
&= aP_1 + b\frac{1}{2}(\tau+1) + (1-a-b)\cdot P_3 \\
&= \frac{n_1n_2}{n_1n_2 + \frac{1}{2}n_1(n_1-1) + \sum_{i=1}^{n_1}\sum_{j=1}^{n_3} a_{ij}} P_1 \\
&\quad + \frac{\frac{1}{4}n_1(n_1-1)}{n_1n_2 + \frac{1}{2}n_1(n_1-1) + \sum_{i=1}^{n_1}\sum_{j=1}^{n_3} a_{ij}} (\tau+1) \\
&\quad + \frac{\sum_{i=1}^{n_1}\sum_{j=1}^{n_3} a_{ij}}{n_1n_2 + \frac{1}{2}n_1(n_1-1) + \sum_{i=1}^{n_1}\sum_{j=1}^{n_3} a_{ij}} P_3.
\end{aligned}$$

$$\begin{aligned}
&\mathrm{Var}[C] \\
&= \mathrm{Var}[aC_1 + bC_2 + (1-a-b)C_3] \\
&= a^2\mathrm{Var}[C_1] + b^2\mathrm{Var}[C_2] + (1-a-b)^2\mathrm{Var}[C_3] \\
&= a^2\frac{1}{n_1n_2}\{P_1(1-P_1) + (n_1-1)(p_{12}-P_1^2) + (n_2-1)(P_{13}-P_1^2)\} \\
&\quad + b^2\frac{1}{4}\left[\frac{4(n_1-2)}{n_1(n_1-1)}\mathrm{Var}(\tau_i) + \frac{2}{n_1(n_1-1)}(1-\tau^2)\right] \\
&\quad + (1-a-b)^2\frac{1}{Q^2)3}[Q_3P_3(1-P_3) + A(P_{32}-P_3^2) + B(P_{33}-P_3^2)] \\
&= \frac{n_1n_2}{\left\{n_1n_2 + \frac{1}{2}n_1(n_1-1) + \sum_{i=1}^{n_1}\sum_{j=1}^{n_3} a_{ij}\right\}^2} \\
&\qquad \times\{P_1(1-P_1) + (n_1-1)(P_{12}-P_1^2) + (n_2-1)(P_{13}-P_1^2)\} \\
&\quad + \frac{\{\frac{1}{2}n_1(n_1-1)\}^2}{\left\{n_1n_2 + \frac{1}{2}n_1(n_1-1) + \sum_{i=1}^{n_1}\sum_{j=1}^{n_3} a_{ij}\right\}^2} \\
&\qquad \times\frac{1}{4}\left[\frac{4(n_1-2)}{n_1(n_1-1)}\mathrm{Var}(\tau_i) + \frac{2}{n_1(n_1-1)}(1-\tau^2)\right] \\
&\quad + \frac{1}{\left\{n_1n_2 + \frac{1}{2}n_1(n_1-1) + \sum_{i=1}^{n_1}\sum_{j=1}^{n_3} a_{ij}\right\}^2} \\
&\qquad \times\left[\sum_{i=1}^{n_1}\sum_{j=1}^{n_3} a_{ij}P_3(1-P_3) + A(P_{32}-P_3^2) + B(P_{33}-P_3^2)\right].
\end{aligned}$$

References

1. Bamber, D. (1975). The area above the ordinal dominance graph and the area below the receiver operating graph, *Journal of Mathematical Psychology*, **12**, 387–415 .

2. Birnbaum, Z. W. and McCarthy, R. C. (1958). A distribution-free upper bounds for $\Pr\{Y < X\}$, based on independent samples of X and Y, *Annals of Mathematical Statistics*, **29**, 558–562.

3. Govindarajulu, Z. (1991). Distribution-free confidence bounds for $P(X < Y)$, *Methods of Information in Medicine*, **30**, 96–101 .

4. Hanley, J. A. and McNeil, B. J. (1982). The meaning and use of the area under a receiver operating characteristic (ROC) curve, *Radiology*, **143**, 29–36.

5. Hilgers, R. A. (1991). Distribution-free confidence bounds for ROC curves, *Methods of Information in Medicine*, **30**, 96–101.

6. Kendall, M. G. (1970). *Rank Correlation Methods*, London: Griffin.

7. Lehmann, E. L. (1975). *Nonparametrics (Statistical Methods based on Ranks)*, California: Holden-Day.

8. Linnet, K. (1987). Comparison of quantitative diagnostic tests: type I error, power, and sample size, *Statistics in Medicine*, **6**, 147–158.

9. Schafter, H. (1994). Efficient confidence bounds for ROC curves, *Statistics in Medicine*, **13**, 1551–1561.

10. Sen, P. K. (1967). A note on asymptotically distribution-free confidence bounds for $P\{X < Y\}$, based on two independent samples, *Sankhyā, Series A*, **29**, 95–102,

11. Ury, H. K. (1972). On distribution-free confidence bounds for $\Pr\{Y < X\}$, *Technometrics*, **14**, 577–581.

21

Goodness-of-Fit Tests for Accelerated Life Models

Vilijandas Bagdonavičius and Mikhail S. Nikulin

Vilnius University, Lithuania
Université Bordeaux 2, Bordeaux, France

Abstract: Goodness-of-fit test for the generalized Sedyakin's model is proposed when accelerated experiments are done under step-stresses. Various alternatives are considered. Power of the test against approaching alternatives is investigated.

Keywords and phrases: Accelerated life testing, accelerated failure time (AFT), Cox model, Generalized Sedyakin's model (GS), goodness-of-fit, power function, proportional hazards (PH), Sedyakin's model, step-stress

21.1 Introduction

In accelerated life testing (ALT) units are tested at higher-then-usual levels of stress to induce early failures. The results are extrapolated to estimate the lifetime distribution at the design stress using models which relate the lifetime to the stress.

Many models for constant-over-time stresses are known. An important tool for generalization of such models to the case of time-varying stresses is the *physical principle in reliability* formulated by Sedyakin (1966) for simple step-stresses and generalized by Bagdonavičius (1978) for general time-varying stresses.

Some of the well-known accelerated life models [see, for example, Bagdonavičius *et al.* (2000)] for time-varying stresses as, for example, the *accelerated failure time* (AFT) model, verify this principle, some do not. An example is the case of the *proportional hazards* (PH) model when the failure time distribution is not exponential under constant stresses. In this paper a goodness-of-fit test is given for the *generalized Sedyakin's* (GS) model when the data are obtained from accelerated experiments with step-stresses.

Suppose that stress is a deterministic time function:

$$x(\cdot) = (x_1(\cdot), ..., x_s(\cdot))^T : [0, \infty) \to B \subset \mathbf{R}^s, \quad x(\cdot) \in E,$$

where E is a set of possible stresses. If $x(\cdot)$ is constant in time, we shall write x instead of $x(\cdot)$ and we note $E_1 = \{x\}$ a set of constant-in-time stresses. Denote informally by $T_{x(\cdot)}$ the time-to-failure under stress $x(\cdot)$ and by $S_{x(\cdot)}(t)$ the survival function.

The sense of accelerated life models is best seen if they are formulated in terms of the hazard rate function

$$\alpha_{x(\cdot)}(t) = \lim_{h \downarrow 0} \frac{1}{h} \mathbf{P}\{T_{x(\cdot)} \in (t, t+h] \mid T_{x(\cdot)} > t\} = -\frac{S'_{x(\cdot)}(t)}{S_{x(\cdot)}(t)}.$$

Each specified accelerated life model relates the hazard rate (or other function) to stress in some particular way. Denote by $A_{x(\cdot)}(t) = -ln\{S_{x(\cdot)}(t)\}$ the accumulated hazard rate under stress $x(\cdot)$.

21.2 Generalized Sedyakin's Model

The idea of the physical principle in reliability is the following. For two identical populations of units functioning under different stresses x_1 and x_2, two moments t_1 and t_2 are equivalent if the probabilities of survival until these moments are equal:

$$\mathbf{P}\{T_{x_1} > t_1\} = S_{x_1}(t_1) = S_{x_2}(t_2) = \mathbf{P}\{T_{x_2} > t_2\}.$$

If after these equivalent moments the units of both groups are observed under the same stress x_2 , i.e. the first population is observed under step-stress

$$x(\tau) = \begin{cases} x_1, & 0 \le \tau < t_1, \\ x_2, & \tau \ge t_1, \end{cases}$$

and the second always under the constant stress x_2, then for all $s > 0$

$$\alpha_{x(\cdot)}(t_1 + s) = \alpha_{x_2}(t_2 + s).$$

Using the idea of Sedyakin, Bagdonavičius generalized the model to the case of any time-varying stresses by supposing that the hazard rate $\alpha_{x(\cdot)}(t)$ at any moment t is a function of the value of stress at this moment and of the probability of survival until this moment. It is formalized by the following definition.

Definition 21.2.1 The generalized Sedyakin's (GS) model holds on a set of stresses E if there exists a positive on $E \times \mathbf{R}^+$ function g such that for all $x(\cdot) \in E$

$$\alpha_{x(\cdot)}(t) = g\left(x(t), S_{x(\cdot)}(t)\right). \tag{21.1}$$

Equivalently, the model can be written in the form $\alpha_{x(\cdot)}(t) = g_1\left(x(t), A_{x(\cdot)}(t)\right)$, where $g_1(x,s) = g(x, exp\{-s\})$. On sets E_1 of constant stresses the equality (21.1) always holds. It is seen from the following proposition.

Proposition 21.2.1 *If the hazard rates $\alpha_x(t) > 0,\ t > 0$ exist on a set of constant stresses E_1 then the GS model holds on E_1.*

PROOF. For all $x \in E_1$ we have:

$$\alpha_x(t) = \alpha_x(A_x^{-1}(A_x(t))) = g_1\left(x, A_x(t)\right),$$

with $g_1(x,s) = \alpha_x(A_x^{-1}(s))$. ∎

Thus, the GS model does not give any relations between the hazard rates (or survival functions) under different constant stresses. This model only shows the influence of variability of stress in time on survival and gives the rule of the hazard rate (or survival) function construction under any time-varying stress from the hazard rate (or survival) functions under different constant stresses. It is seen from the following proposition.

Proposition 21.2.2 *If the GS model holds on a set $E \supset E_1$ of stresses $x(\cdot)$: $\mathbf{R}^+ \to E_1$, then for all $x(\cdot) \in E$*

$$\alpha_{x(\cdot)}(t) = \alpha_{x_t}\left(A_{x_t}^{-1}(A_{x(\cdot)}(t))\right), \tag{21.2}$$

where x_t is constant stress equal to the value of time-varying stress $x(\cdot)$ at the moment t.

PROOF. If the GS model holds on a set $E \supset E_1$ then for all $x \in E_1$

$$g_1(x,s) = g_1\left\{x, A_x(A_x^{-1}(s))\right\} = \alpha_x(A_x^{-1}(s)).$$

Thus,

$$\alpha_{x(\cdot)}(t) = g_1\left\{x(t), A_{x(\cdot)}(t)\right\} = \alpha_{x_t}\left\{A_{x_t}^{-1}(A_x(\cdot)(t))\right\}.$$

∎

Restrictions of the GS model when not only the rule (21.2) but also some relations between survival under different constant stresses are assumed can be considered. These narrower than GS models can be formulated by using models for constant stresses and the rule (21.2).

Let us consider the meaning of the rule (21.2) for step-stresses of the form:

$$x(\tau) = x_i, \quad \text{if} \quad \tau \in [t_{m-1} \le \tau < t_m) \quad i = 1, \cdots, m, \tag{21.3}$$

where $x_1, \cdots, x_m$ are constant stresses. If $m = 2$, step-stress is called *simple*.

Sets of step-stresses of the form (21.3) will be usually denoted by E_m.

Proposition 21.2.3 *If the GS model holds on E_2 then the survival function under stress $x(\cdot) \in E_1$ verifies the equality*

$$S_{x(\cdot)}(t) = \begin{cases} S_{x_1}(t), & 0 \le \tau < t_1, \\ S_{x_2}(t - t_1 + t_1^*), & \tau \ge t_1; \end{cases} \tag{21.4}$$

the moment t_1^* is determined by the equality $S_{x_1}(t_1) = S_{x_2}(t_1^*)$.

PROOF. Set

$$a = A_{x_1}(t_1) = A_{x(\cdot)}(t_1) = A_{x_2}(t_1^*). \tag{21.5}$$

The accumulated hazard rate verifies the integral equation (we denote by g the function g_1 in what follows)

$$A_{x(\cdot)}(t) = \int_0^t g\left(x(u), A_{x(\cdot)}(u)\right) du. \tag{21.6}$$

The equalities (21.5) and (21.6) imply that for all $t \ge t_1$

$$A_{x(\cdot)}(t) = a + \int_{t_1}^t g\left(x_2, A_{x(\cdot)}(u)\right) du$$

and

$$\begin{aligned} A_{x_2}(t - t_1 + t_1^*) &= a + \int_{t_1^*}^{t-t_1+t_1^*} g\left(x_2, A_{x_2}(u)\right) du \\ &= a + \int_{t_1}^t g\left(x_2, A_{x_2}(u - t_1 + t_1^*)\right) du. \end{aligned}$$

So for all $t \ge t_1$ the functions $A_{x(\cdot)}(t)$ and $A_{x_2}(t - t_1 + t_1^*)$ satisfy the integral equation

$$h(t) = a + \int_{t_1}^t g\left(x_2, h(u)\right) du$$

with the initial condition $h(t_1) = a$. The solution of this equation is unique, therefore we have

$$A_{x(\cdot)}(t) = A_{x_2}(t - t_1 + t_1^*), \quad \text{for all} \quad t \ge t_1.$$

It implies the equality (21.4). ■

Let us consider a set E_m of more general stepwise stresses of the form (21.3). Set $t_0 = 0$. Using Proposition 21.2.3 and by recurrence the following proposition can be shown.

Proposition 21.2.4 *If the GS model holds on E_m then the survival function $S_{x(\cdot)}(t)$ verifies the equalities:*

$$S_{x(\cdot)}(t) = S_{x_i}(t - t_{i-1} + t^*_{i-1}), \quad \text{if} \quad t \in [t_{i-1}, t_i),\ i = 1, 2, \ldots, m, \tag{21.7}$$

where t^*_i verify the equations

$$\begin{aligned} S_{x_1}(t_1) &= S_{x_2}(t^*_1), \ldots, S_{x_i}(t_i - t_{i-1} + t^*_{i-1}) \\ &= S_{x_{i+1}}(t^*_i), \qquad i = 1, \ldots, m-1. \end{aligned} \tag{21.8}$$

In the literature on ALT [see Nelson (1990)] the model (21.7) is also called the *basic cumulative exposure model.*

N. M. Sedyakin called his model the physical principle in reliability meaning that this model is very wide. Nevertheless, this model and it's generalization can be not appropriate in situations of periodic and quick change of stress level or when switch-on's or switch-off's of stress from one level to another can imply failures or shorten the lifetime.

Let us consider an example which shows how the GS model can be used for generalization of models for constant stresses to the case of time-varying stresses. Suppose that under different constant stresses the survival functions differ only in scale: for any $x \in E_1$

$$S_x(t) = G\{r(x)\, t\}. \tag{21.9}$$

Let us generalize the model (21.9) to the case of time-varying stresses by supposing that the GS model also holds, i.e. the hazard rates under time-varying stresses are obtained from the hazard rates under constant stresses by the rule (21.2).

Proposition 21.2.5 *The GS model with the survival functions (21.9) on E_1 holds on $E \supset E_1$ iff there exist a positive on E function r and a positive on $[0, \infty)$ function q such that for all $x(\cdot) \in E$*

$$\alpha_{x(\cdot)}(t) = r\{x(t)\}\, q\{S_{x(\cdot)}(t))\}. \tag{21.10}$$

Let us find the expression of the survival function under time-varying stresses for the model (21.10).

Proposition 21.2.6 *The model (21.10) holds on a set of stresses E iff there exists a survival function G such that for all $x(\cdot) \in E$*

$$S_{x(\cdot)}(t) = G\left(\int_0^t r\{x(\tau)\}\, d\tau\right). \tag{21.11}$$

The model (21.11) is *the accelerated failure time model.*

21.3 Alternatives to the GS Model

21.3.1 Proportional hazards model

In survival analysis the most used model describing the influence of covariates on the lifetime distribution is the *proportional hazards* (PH) or *Cox model.* In terms of stresses it is formulated as follows.

Suppose that under different constant stresses $x \in E_1$ the hazard rates are proportional to a baseline hazard rate:

$$\alpha_x(t) = r(x)\, \alpha_0(t). \tag{21.12}$$

In the statistical literature the following formal generalization of the PH model to time-varying stresses is used.

Definition 21.3.1 The proportional hazards (PH) model holds on a set of stresses E if for all $x(\cdot) \in E$

$$\alpha_{x(\cdot)}(t) = r\{x(t)\}\, \alpha_0(t). \tag{21.13}$$

The model (21.13) is not natural when items are aging under usual constant stress. Indeed, denote by x_t constant-in-time stress equal to the value of time-varying stress $x(\cdot)$ at the moment t. Then the PH implies that

$$\alpha_{x(\cdot)}(t) = \alpha_{x_t}(t).$$

For *any* t the intensity under time-varying stress $x(\cdot)$ at the moment t *does not depend on the values of stress* $x(\cdot)$ *before the moment* t but only on the value of stress at this moment. It is not natural when the hazard rates are not constant under constant stresses, i.e. when times-to-failure under constant stresses are not exponential under constant stresses. When is the PH model also the GS model? It is given in the following proposition.

Proposition 21.3.1 [Bagdonavičius and Nikulin (2000)] *Suppose that the PH model holds on the set E including a set of constant stresses E_1 and all stresses of the form*

$$x_s(\tau) = \begin{cases} x_1, & 0 \le \tau < s, \\ x_2, & \tau \ge s, \end{cases}$$

with $s < \delta$, where δ is any positive number and $x_1, x_2 \in E_1$ are fixed stresses.

The GS model also holds on E iff the time to-failure is exponential for all $x \in E_0$.

The proposition implies that on sets of time-varying stresses the PH model is an alternative to the GS model if the time-to-failure distribution under constant stresses is not exponential.

Similarly as for the AFT model the PH model for constant stresses can be generalized to the case of time-varying stresses using the rule (21.2). The obtained model is different from the model (21.13) and is much more natural.

21.3.2 Model including influence of switch-up's of stresses on reliability

Suppose that an item is observed under step-stress (21.3) and after the switch-up at the moment t_i from constant stress x_i to constant stress x_{i+1} the survival function has a jump:

$$S_{x(\cdot)}(t_i) = S_{x(\cdot)}(t_i-)\,\delta_i;$$

here δ_i is the probability for an item not to fail because of the switch-up at the moment t_i. In this case the GS model for step-stresses can be modified as follows:

$$S_{x(\cdot)}(t) = S_{x_i}(t - t_{i-1} + t^{**}_{i-1}), \tag{21.14}$$

where

$$t^{**}_1 = S^{-1}_{x_2}\{S_{x_1}(t_1)\,\delta_1\}, \quad t^{**}_i = S^{-1}_{x_{i+1}}\{S_{x_i}(t_i - t_{i-1} + t^{**}_{i-1})\,\delta_i\}. \tag{21.15}$$

21.4 Test Statistic for the GS Model

Suppose that a group of n_0 units is tested under the step-stress (21.3) and m groups of $n_1, \ldots, n_m$ items are tested under constant stresses $x_1, \ldots, x_m$ ($x_1 < \cdots < x_m$), respectively. The units are observed time t_m given for the experiment. We write $x(\cdot) < y(\cdot)$ if $S_{x(\cdot)}(t) > S_{y(\cdot)}(t)$ for all $t > 0$. The idea of goodness-of-fit is based on comparing two estimators $\hat{A}^{(1)}_{x(\cdot)}$ and $\hat{A}^{(2)}_{x(\cdot)}$ of the accumulated hazard rate $A_{x(\cdot)}$. One estimator can be obtained from the experiment under the step-stress (21.3) and another from the experiments under stresses $x_1, \cdots, x_m$ by using the equalities (21.7) and (21.8).

Denote by $N_i(t)$ and $Y_i(t)$ the numbers of observed failures in the interval $[0, t]$ and the number of items at risk just prior to the moment t, respectively, for the group of items tested under stress x_i and $N(t)$, $Y(t)$ the analogous numbers for the group of items tested under stress $x(\cdot)$.

Set $\alpha_i = \alpha_{x_i}$, $\alpha = \alpha_{x(\cdot)}$, $A_i = A_{x_i}$, $A = A_{x(\cdot)}$ $i = 1, \ldots, m$. The first estimator $\hat{A}^{(1)}$ of the accumulated hazard A is the well-known *Nelson-Aalen estimator* obtained from the experiment under the step-stress (21.3):

$$\hat{A}^{(1)}(t) = \int_0^t \frac{dN(v)}{Y(v)}.$$

The second is suggested by the GS model [formulae (21.7) and (21.8)] and is obtained from the experiments under constant stresses:

$$\hat{A}^{(2)}(t) = \hat{A}_{i+1}(t - t_i + \hat{t}_i^*),\ t \in [t_i, t_{i+1}),\ i = 0, ..., m-1,$$

where

$$\begin{aligned} \hat{t}_0^* &= 0, \\ \hat{t}_1^* &= \hat{A}_2^{-1}(\hat{A}_1(t_1)),\ \dots,\ \hat{t}_{i+1}^* = \hat{A}_{i+2}^{-1}(\hat{A}_{i+1}(t_{i+1} - t_i + \hat{t}_i^*)),\ i = 0, ..., m-2, \end{aligned}$$

$$\hat{A}_i^{-1}(s) = \inf\{u : \hat{A}_i(u) \geq s\}, \quad \hat{A}_i(t) = \int_0^t \frac{dN_i(v)}{Y_i(v)} \quad i = 1, ..., m.$$

The test is based on the statistic

$$T_n = \int_0^{t_m} K(v)\, d\{\hat{A}^{(1)}(t) - \hat{A}^{(2)}(t)\}, \tag{21.16}$$

where K is the weight function.

We shall consider the weight functions of the type: for $v \in [t_i, t_i + \Delta t_i)$

$$K(v) = \frac{1}{\sqrt{n}} \frac{Y(v) Y_{i+1}(v - t_i + \hat{t}_i^*)}{Y(v) + Y_{i+1}(v - t_i + \hat{t}_i^*)}\, g\left(\frac{Y(v) + Y_{i+1}(v - t_i + \hat{t}_i^*)}{n}\right),$$

where g is a nonnegative bounded continuous function with bounded variation on $[0, 1]$ and $n = \sum_{i=0}^m n_i$. The condition $x_1 < \cdots < x_m$ implies that

$$\mathbf{P}\{T_n \text{ is defined}\} \to 1 \quad \text{as} \quad n_i \to \infty.$$

21.5 Asymptotic Distribution of the Test Statistic

Let us consider at first the asymptotic distribution of the estimators $\hat{t}_i^*$. Denote by $\xrightarrow{\mathcal{D}}$ the convergence in distribution. **Assumptions A.**

(a) The hazard rates α_i are positive and continuous on $(0, \infty)$;

(b) $A_i(t) < \infty$ for all $t > 0$;

(c) $n \to \infty, \quad n_i/n \to l_i, \quad l_i \in (0, 1)$.

Lemma 21.5.1 *Suppose that Assumptions A hold. Then*

$$\sqrt{n}(\hat{t_j^*} - t_j^*) \xrightarrow{\mathcal{D}} a_j \sum_{l=1}^{j} d_{jl}\{U_l(t_{l-1}^* + \Delta t_{l-1}) - U_{l+1}(t_l^*)\}, \tag{21.17}$$

where

$$d_{jl} = \prod_{s=l}^{j-1} c_s, \quad l = 1, ..., j-1; \quad d_{jj} = 1,$$

$$a_j = \frac{1}{\alpha_{j+1}(t_j^*)}, \quad c_s = \frac{\alpha_{s+1}(t_s^* + \Delta t_s)}{\alpha_{s+1}(t_s^*)},$$

$U_1, \cdots, U_m$ *and* U *are independent Gaussian martingales with* $U_i(0) = U(0) = 0$ *and*

$$\mathbf{Cov}\,(U_i(s_1), U_i(s_2)) = \frac{1}{l_i}\,\frac{1 - S_i(s_1 \wedge s_2)}{S_i(s_1 \wedge s_2)} := \sigma_i^2(s_1 \wedge s_2),$$

$$\mathbf{Cov}\,(U(s_1), U(s_2)) = \frac{1}{l_0}\,\frac{1 - S(s_1 \wedge s_2)}{S(s_1 \wedge s_2)} := \sigma^2(s_1 \wedge s_2)$$

with $S_i = \exp\{-A_i\}$, $S = \exp\{-A\}$.

PROOF. Under Assumptions A for any $t \in (0, t_m)$ the estimators $\hat{A}_i$ and $\hat{A}^{(1)}$ are uniformly consistent on $[0, t]$, and

$$\sqrt{n}(\hat{A}_i - A_i) \xrightarrow{\mathcal{D}} U_i, \quad \sqrt{n}(\hat{A}^{(1)} - A) \xrightarrow{\mathcal{D}} U \tag{21.18}$$

on $D[0, t]$, the space of cadlag functions on $[0, t]$ with Skorokhod metric.

We prove (21.17) by recurrence. If $i = 1$ then

$$\begin{aligned}\sqrt{n}(\hat{t_1^*} - t_1^*) &= \sqrt{n}(\hat{A}_2^{-1}(\hat{A}_1(t_1)) - A_2^{-1}(\hat{A}_1(t_1))) \\ &\quad + \sqrt{n}(A_2^{-1}(\hat{A}_1(t_1)) - A_2^{-1}(A_1(t_1))).\end{aligned} \tag{21.19}$$

For any $0 < s_1 < s_2 < \infty$ [see Andersen *et al.* (1993)]

$$\sqrt{n}(\hat{A}_2^{-1} - A_2^{-1}) \xrightarrow{\mathcal{D}} U_2^* \tag{21.20}$$

on $D[s_1, s_2]$, where

$$U_2^*(s) = -\frac{e^{-s}U_2(A_2^{-1}(s))}{p_2(A_2^{-1}(s))}$$

and p_i is the density of T_{x_i}. Consistence of the estimator $\hat{A}_1(t_1)$ and the convergence (21.20) imply that

$$\sqrt{n}\{\hat{A}_2^{-1}(\hat{A}_1(t_1)) - A_2^{-1}(\hat{A}_1(t_1))\} \xrightarrow{\mathcal{D}} -\frac{U_2(t_1^*)}{\alpha_2(t_1^*)} = -a_1 U_2(t_1^*). \tag{21.21}$$

Using the delta method and the convergence (21.18), we obtain

$$\sqrt{n}\{A_2^{-1}(\hat{A}_1(t_1)) - A_2^{-1}(A_1(t_1))\} \xrightarrow{\mathcal{D}} \frac{1}{\alpha_2(t_1^*)} U_1(t_1) = a_1 U_1(t_0^* + \Delta t_0). \tag{21.22}$$

Thus (21.19), (21.21) and (21.22) imply that

$$\sqrt{n}(\hat{t}_1^* - t_1^*) \xrightarrow{\mathcal{D}} a_1 d_{11}\{U_1(t_0^* + \Delta t_0) - U_2(t_1^*)\}.$$

Suppose that (21.17) holds for $i = j$. Then similarly as in the case $i = 1$ we have

$$\begin{aligned}
&\sqrt{n}(\hat{t}_{j+1}^* - t_{j+1}^*) \\
&\quad = \sqrt{n}\{\hat{A}_{j+2}^{-1}(\hat{A}_{j+1}(\hat{t}_j^* + \Delta t_j)) - A_{j+2}^{-1}(A_{j+1}(t_j^* + \Delta t_j))\} \\
&\quad = a_{j+1}\{U_{j+1}(t_j^* + \Delta t_j) - U_{j+2}(t_{j+1}^*)\} + a_{j+1}\frac{c_j}{a_j}\sqrt{n}(\hat{t}_j^* - t_j^*) + \Delta_n,
\end{aligned}$$

where $\Delta_n \xrightarrow{P} 0$ as $n \to 0$. The last formula and the assumption of recurrency imply that

$$\begin{aligned}
&\sqrt{n}(\hat{t}_{j+1}^* - t_{j+1}^*) \xrightarrow{\mathcal{D}} \\
&\quad a_{j+1}\Big\{U_{j+1}(t_j^* + \Delta t_j) - U_{j+2}(t_{j+1}^*) \\
&\qquad + \sum_{l=1}^{j}\prod_{s=l}^{j} c_s\{U_l(t_{l-1}^* + \Delta t_{l-1}) - U_{l+1}(t_l^*)\}\Big\} \\
&\quad = a_{j+1}\sum_{l=1}^{j+1} d_{j+1,l}\{U_l(t_{l-1}^* + \Delta t_{l-1}) - U_{l+1}(t_l^*)\}.
\end{aligned}$$

Let us consider the limit distribution of the statistic T_n. Note that

$$\frac{K(v)}{\sqrt{n}} \xrightarrow{\mathbf{P}} k(v) = \frac{l_0 l_{i+1}}{l_0 + l_{i+1}} S(v)\, g\left((l_0 + l_{i+1})S(v)\right),\ v \in [t_i, t_{i+1}].$$

Set

$$\begin{aligned}
e_j &= a_j\{k(t_j)\,\alpha_{j+1}(t_j^*) - k(t_j + \Delta t_j)\,\alpha_{j+1}(t_j^* + \Delta t_j) \\
&\quad + \int_{t_j^*}^{t_j^* + \Delta t_j} \alpha_{i+1}(t_i^* + \Delta t_i)(v) d\, k(v + t_j - t_j^*) \quad j = 1, ..., m-1,
\end{aligned}$$

and $f_0 = f_m = 0$, $f_i = \sum_{j=i}^{m-1} e_j d_{ji}$, $i = 1, ..., m-1$.

Proposition 21.5.1 *Under Assumptions A*

$$T_n \xrightarrow{\mathcal{D}} \int_0^{t_m} k(v)d\,U(v) + \sum_{i=0}^{m-1}\{f_{i+1}U_{i+1}(t_i^* + \Delta t_i) - f_i U_{i+1}(t_i^*) - \int_{t_i^*}^{t_i^*+\Delta t_i} k(v + t_i - t_i^*)d\,U_{i+1}(v)\}. \tag{21.23}$$

PROOF. The statistic (21.16) can be written in the form

$$\begin{aligned}
T_n &= \int_0^{t_m} K(v)\, d\{\hat{A}^{(1)}(t) - A(t)\} \\
&\quad + \sum_{i=1}^{m-1}\int_{t_i^*}^{\hat{t}_i^*} K(v + t_i - \hat{t}_i^*)\, d\hat{A}_{i+1}(v) \\
&\quad - \sum_{i=1}^{m-1}\int_{t_i^*+\Delta t_i}^{\hat{t}_i^*+\Delta t_i} K(v + t_i - \hat{t}_i^*)\, d\hat{A}_{i+1}(v) \\
&\quad - \sum_{i=1}^{m-1}\int_{t_i^*}^{t_i^*+\Delta t_i} \{K(v + t_i - \hat{t}_i^*) - K(v + t_i - t_i^*)\}\, d\hat{A}_{i+1}(v) \\
&\quad - \sum_{i=0}^{m-1}\int_{t_i^*}^{t_i^*+\Delta t_i} K(v + t_i - t_i^*)\, d\{\hat{A}_{i+1}(v) - A_{i+1}(v)\} \\
&= \int_0^{t_m} k(v)dU(v) + \sum_{i=1}^{m-1} k(t_i)\alpha_{i+1}(t_i^*)\sqrt{n}(\hat{t}_i^* - t_i^*) \\
&\quad - \sum_{i=1}^{m-1} k(t_i + \Delta t_i)\, \alpha_{i+1}(t_i^* + \Delta t_i)\, \sqrt{n}(\hat{t}_i^* - t_i^*) \\
&\quad + \sum_{i=1}^{m-1}\int_{t_i^*}^{t_i^*+\Delta t_i} \alpha_{i+1}(v)\, d\,k(v + t_i - t_i^*)\sqrt{n}(\hat{t}_i^* - t_i^*) \\
&\quad - \sum_{i=0}^{m-1}\int_{t_i^*}^{t_i^*+\Delta t_i} k(v + t_i - t_i^*)\, dU_{i+1}(v) + o_p(1).
\end{aligned} \tag{21.24}$$

where $o_p(1) \xrightarrow{\mathbf{P}} 0$ as $n \to \infty$. The lemma implies that

$$\begin{aligned}
&\sum_{i=1}^{m-1}\{k(t_i)\alpha_{i+1}(t_i^*) - k(t_i + \Delta t_i)\alpha_{i+1}(t_i^* + \Delta t_i) \\
&\qquad + \int_{t_i^*}^{t_i^*+\Delta t_i} \alpha_{i+1}(v)d\,k(v + t_i - t_i^*)\}\sqrt{n}(\hat{t}_i^* - t_i^*) \\
&\qquad = \sum_{i=0}^{m-2} f_{i+1}U_{i+1}(t_i^* + \Delta t_i) - \sum_{i=1}^{m-1} f_i U_{i+1}(t_i^*) + o_p(1).
\end{aligned} \tag{21.25}$$

The formulae (21.24) and (21.23) imply the result of the proposition. ■

Corollary 21.5.1 *Under the assumptions of the theorem* $T_n \xrightarrow{\mathcal{D}} N(0, \sigma_T^2)$, *where*

$$\begin{aligned}\sigma_T^2 &= \int_0^{t_m} k^2(v)\, d\sigma^2(v) + \sum_{i=0}^{m-1} \{f_{i+1}^2 \sigma_{i+1}^2(t_i^* + \Delta t_i) \\ &+ f_i(f_i - 2f_{i+1})\sigma_{i+1}^2(t_i^*) - 2f_{i+1} \int_{t_i^*}^{t_i^*+\Delta t_i} k(v + t_i - t_i^*)\, d\sigma_{i+1}^2(v) \\ &+ \int_{t_i^*}^{t_i^*+\Delta t_i} k^2(v + t_i - t_i^*)\, d\sigma_{i+1}^2(v)\}. \end{aligned} \tag{21.26}$$

Remark 21.5.1 The variance σ_T^2 can be consistently estimated by the statistic

$$\begin{aligned}\hat{\sigma}_T^2 &= \int_0^{t_m} \hat{k}^2(v) d\hat{\sigma}^2(v) + \sum_{i=0}^{m-1} \{\hat{f}_{i+1}^2 \hat{\sigma}_i^2(\hat{t}_i^* + \Delta t_i) \\ &+ \hat{f}_i(\hat{f}_i - 2\hat{f}_{i+1})\hat{\sigma}_i^2(\hat{t}_i^*) - 2\hat{f}_{i+1} \int_{\hat{t}_i^*}^{\hat{t}_i^*+\Delta t_i} \hat{k}(v + t_i - \hat{t}_i^*) d\,\hat{\sigma}_i^2(v) \\ &+ \int_{\hat{t}_i^*}^{\hat{t}_i^*+\Delta t_i} \hat{k}^2(v + t_i - \hat{t}_i^*)\, d\,\hat{\sigma}_{i+1}^2(v)\}, \end{aligned}$$

where

$$\hat{k}(v) = K(v)/\sqrt{n}, \quad \hat{\sigma}^2(v) = \frac{n}{n_0}\left(\frac{1}{\hat{S}(v)} - 1\right), \quad \hat{\sigma}_i^2(v) = \frac{n}{n_i}\left(\frac{1}{\hat{S}_i(v)} - 1\right),$$

$\hat{S}$ and $\hat{S}_i$ are the empirical survival functions,

$$\begin{aligned}\hat{f}_0 &= \hat{f}_m = 0, \\ \hat{f}_i &= \sum_{s=i}^{2m-1} \hat{e}_s \hat{d}_{si} \quad i = 1, \ldots, m-1, \\ \hat{e}_s &= \hat{a}_s\Big(\hat{k}(t_s) - \hat{k}(t_s + \Delta t_s) - \hat{k}(t_s)\,\hat{\alpha}_{s+1}(\hat{t}_s^*) + \hat{k}(t_s + \Delta t_s)\,\hat{\alpha}_{s+1}(\hat{t}_s^* + \Delta t_s) \\ &- \int_{\hat{t}_s^*}^{\hat{t}_s^*+\Delta t_s} \hat{k}(v + t_s - \hat{t}_s^*) d\,\hat{A}_{s+1}(v)\Big), \\ \hat{d}_{si} &= \prod_{l=i}^{s-1} \hat{c}_l, \quad i = 1, \ldots, s-1, \quad \hat{d}_{ss} = 1, \\ \hat{c}_l &= \frac{\hat{\alpha}_{l+1}(\hat{t}_l^* + \Delta t_l)}{\hat{\alpha}_{l+1}(\hat{t}_l^*)}, \quad \hat{a}_s = \frac{1}{\hat{\alpha}_{s+1}(\hat{t}_s^*)}, \end{aligned}$$

and $\hat{\alpha}_{s+1}(\hat{t}_s^*)$, $\hat{\alpha}_{s+1}(\hat{t}_s^* + \Delta t_s)$ are the kernel estimators:

$$\begin{aligned}\hat{\alpha}_{s+1}(\hat{t}_s^*) &= \frac{1}{b}\int_0^{t_m} Ker\left(\frac{\hat{t}_i^* - u}{b}\right) d\hat{A}_{i+1}(u),\\ \hat{\alpha}_{s+1}(\hat{t}_s^* + \Delta t_s) &= \frac{1}{b}\int_0^{t_m} Ker\left(\frac{\hat{t}_i^* + \Delta t_s - u}{b}\right) d\hat{A}_{i+1}(u);\end{aligned}$$

here Ker is some kernel function.

21.6 The Test

The hypothesis

$$H_0 : GS\ model\ holds\ on\ E = \{x_1, \cdots, x_m, x(\cdot)\}$$

is rejected with the approximative significance level α, if

$$\left(\frac{T_n}{\hat{\sigma}_T}\right)^2 > \chi^2_{1-\alpha}(1),$$

where $\chi^2_{1-\alpha}(1)$ is the $(1-\alpha)$-quantile of the chi-square distribution with one degree of freedom.

21.7 Consistency and the Power of the Test Against Approaching Alternatives

Let us find the power of the test against the following alternatives:

H^*: PH model with specified non-exponential time-to-failure distributions under constant stresses.

Under H^*:

$$\begin{aligned}\hat{A}^{(1)}(v) &\xrightarrow{\mathbf{P}} A_*^{(1)}(v) = A_{i+1}(v) - A_{i+1}(t_i) + \mathbf{1}\{i>0\}\sum_{l=1}^{i}\{A_i(t_l) - A_i(t_{l-1})\},\\ \hat{A}^{(2)}(v) &\xrightarrow{\mathbf{P}} A^{(2)}(v) = A_i(v - t_i + t_i^*),\ v \in [t_i, t_{i+1}),\ i = 0, \cdots, m-1,\\ \frac{1}{\sqrt{n}}K(v) &\xrightarrow{\mathbf{P}} k^*(v), \quad v \in [t_i, t_{i+1}),\end{aligned}$$

where

$$k^*(v) = \frac{l_0 l_{i+1} S_*^{(1)}(v) S_i(v - t_i + t_i^*)}{l_0 S_*^{(1)}(v) + l_{i+1} S_i(v - t_i + t_i^*)} \times g\left(l_0 S_*^{(1)}(v) + l_{i+1} S_i(v - t_i + t_i^*)\right), \tag{21.27}$$

and $S_*^{(1)}(v) = exp\{-A_*^{(1)}(v)\}$. Convergence is uniform on $[0, t_m]$.

Proposition 21.7.1 *Assume that Assumptions A hold and*

$$\Delta^* = \int_0^{t_m} k^*(v)\, d\{A_*^{(1)}(v) - A^{(2)}(v)\} \neq 0.$$

Then the test is consistent against H^*.

PROOF. Write the test statistic in the form

$$T_n = \int_0^\infty K(v)\, d\{\hat{A}^{(1)}(v) - A_*^{(1)}(v)\} - \int_0^\infty K(v)\, d\{\hat{A}^{(2)}(v) - A^{(2)}(v)\} + \int_0^\infty K(v)\, d\{A_*^{(1)}(v) - A^{(2)}(v)\} = T_{1n} + T_{2n} + T_{3n}. \tag{21.28}$$

Under H^*

$$T_{1n} + T_{2n} \xrightarrow{\mathcal{D}} N(0, {\sigma_T^*}^2),$$

where ${\sigma_T^*}^2$ has the same form (21.26) with the only difference that k is replaced by k^* and $\sigma^2(t)$ is replaced by

$$(\sigma^{(1)})^2(t) = \frac{1}{l_0}\left(\frac{1}{S^{(1)}(t)} - 1\right).$$

Under H^* we have

$$\hat{\sigma}_T^2 \xrightarrow{\mathbf{P}} {\sigma_T^*}^2 \tag{21.29}$$

and

$$\frac{T_{1n} + T_{2n}}{\hat{\sigma}_T} \xrightarrow{\mathcal{D}} N(0, 1). \tag{21.30}$$

The third member in (21.28) can be written in the form

$$T_{3n} = \sum_{i=1}^{m-1} \int_{t_i}^{t_{i+1}} K(v)\, \{\alpha_{i+1}(v) - \alpha_{i+1}(v - t_i + t_i^*)\}\, dv. \tag{21.31}$$

The assumptions of the proposition and the equalities (21.28)–(21.31) imply that under H^*

$$\frac{1}{\sqrt{n}} T_{3n} \xrightarrow{\mathbf{P}} \Delta^*, \quad \frac{T_n}{\hat{\sigma}_T} \xrightarrow{\mathbf{P}} \infty.$$

Thus under H^*

$$\mathbf{P}\left\{\left(\frac{T_n}{\hat{\sigma}_T}\right)^2 > \chi^2_{1-\alpha}(1)\right\} \to 1.$$

■

Remark 21.7.1 If α_i are increasing (decreasing) then the test is consistent against H^*.

PROOF. The inequalities $x_1 < \cdots < x_m$ imply that $t_i > t_i^*$ for all i. If α_i are increasing (decreasing) then $\Delta^* > 0$ ($\Delta^* < 0$) under H^*. Proposition 21.7.1 implies the consistency of the test. ■

Let us consider the sequence of approaching alternatives

$$H_n^* : \ PH \ with \ \alpha_i(t) = \left(\frac{t}{\theta_i}\right)^{\frac{\varepsilon}{\sqrt{n}}}$$

with fixed $\varepsilon > 0$ $(i = 1, \cdots, m)$. Then

$$T_{3n} \xrightarrow{\mathbf{P}} \mu = -\varepsilon \sum_{i=1}^{m-1} \int_{t_i}^{t_{i+1}} k^*(v) \ln(1 + \frac{t_i^* - t_i}{v}) dv > 0,$$

and

$$\frac{T_n}{\hat{\sigma}_T} \xrightarrow{\mathcal{D}} N(a, 1), \quad \left(\frac{T_n}{\hat{\sigma}_T}\right)^2 \xrightarrow{\mathcal{D}} \chi^2(1, a),$$

where $a = \mu/\sigma_T^*$ and $\chi^2(1, a)$ denotes the chi-square distribution with one degree of freedom and the non-centrality parameter a (or a random variable having such distribution).

The power function of the test is approximated by the function

$$\beta = \lim_{n\to\infty} \mathbf{P}\left\{\left(\frac{T_n}{\hat{\sigma}_T}\right)^2 > \chi^2_{1-\alpha}(1) \mid H_n\right\} = \mathbf{P}\left\{\chi^2(1, a) > \chi^2_{1-\alpha}(1)\right\}. \quad (21.32)$$

Let us find the power of the test against the following alternatives:

H^{**}: *the model (21.14) with specified time-to-failure distributions under constant stresses*

Under H^{**}

$$\hat{A}^{(1)}(v) \xrightarrow{\mathbf{P}} A_{**}^{(1)}(v) = A_i(v - t_i + t_i^{**}), \ v \in [t_i, t_{i+1}), \ i = 0, \cdots, m-1,$$

$$\hat{A}^{(2)}(v) \xrightarrow{\mathbf{P}} A^{(2)}(v) = A_i(v - t_i + t_i^*), \ v \in [t_i, t_{i+1}), \ i = 0, \cdots, m-1,$$

$$\frac{1}{\sqrt{n}} K(v) \xrightarrow{\mathbf{P}} k^{**}(v),$$

where $k^{**}(v)$ has the same form as $k^*(v)$ with the only difference that $S_*^{(1)}$ is replaced by $S_{**}^{(1)}(v) = \exp\{-A_{**}^{(1)}(v)\}$. Convergence is uniform on $[0, t_m]$.

Proposition 21.7.2 *Assume that Assumptions A hold and*

$$\Delta^{**} = \int_0^\infty k^{**}(v)\, d\{A_{**}^{(1)}(v) - A^{(2)}(v)\} \neq 0.$$

Then the test is consistent against H^{**}.

Remark 21.7.2 If α_i are increasing (decreasing) then the test is consistent against H^{**}.

Let us consider the sequence of approaching alternatives

H_n^{**}: *the model (21.14) with specified time-to-failure distributions under constant stresses* and $\delta_i = 1 - \frac{\varepsilon_i}{\sqrt{n}}$.

Similarly as in the case of the alternatives H_n^* it can be shown that

$$\frac{T_n}{\hat{\sigma}_T} \xrightarrow{\mathcal{D}} N(a,1), \quad \left(\frac{T_n}{\hat{\sigma}_T}\right)^2 \xrightarrow{\mathcal{D}} \chi^2(1, \mid a \mid),$$

where $a = \mu/\sigma_T^{**}$ and

$$\mu = -\sum_{i=1}^{m-1} a_i \sum_{j=1}^{i} d_{ij}\, \varepsilon_j \int_{t_i}^{t_{i+1}} k(v)\, d\,\alpha_{i+1}(v - t_i + t_i^*).$$

The parameter μ is positive (negative) if the functions α_i are convex (concave).

The power function of the test is approximated by the function (21.32) with $a = \mu/\sigma_T^{**}$.

Acknowledgement. This research was supported by the Conseil Regional d'Aquitane, Grant 20000204009.

References

1. Andersen, P. K., Borgan, O., Gill, R. D., and Keiding, N. (1993). *Statistical Models Based on Counting Processes*, New York: Springer-Verlag.

2. Bagdonavičius, V. (1978). Testing the hypothesis of the additive accumulation of damages, *Probability Theory and its Applications*, **23**, 403–408.

3. Bagdonavičius, V. and Nikulin, M. (2000). Semiparametric estimation in accelerated life testing, In: *Recent Advances in Reliability Theory. Methodology, Practice and Inference* (Eds., N. Limnios and M. Nikulin), pp. 405–418, Boston: Birkhäuser.

4. Bagdonavičius, V., Gerville-Réache, L., Nikoulina, V., and Nikulin, M. (2000). Expériences accélérées: analyse statistique du modèle standard de vie acclérée, *Revue de Statistique Appliquée*, **XLVIII**, 5–38.

5. Nelson, W. (1990). *Accelerated Testing: Statistical Models, Test Plans, and Data Analyses*, New York: John Wiley & Sons.

6. Sedyakin, N. M. (1966). On one physical principle in reliability theory, *Technical Cybernetics*, **3**, 80–87.

PART VI

Graphical Methods and General Goodness-of-Fit Tests

22

Two Nonstandard Examples of the Classical Stratification Approach to Graphically Assessing Proportionality of Hazards

Niels Keiding

University of Copenhagen, Copenhagen, Denmark

Abstract: Goodness-of-fit assessment of the crucial proportionality and log-linearity assumptions of the Cox (1972a,b) proportional hazards regression models for survival data and repeated events has necessitated several new developments. This contribution presents two concrete examples of nonstandard application of these ideas: in discrete-time regression for the retro-hazard of the reporting delay time in a multiple sclerosis registry, and in analysing repeated insurance claims in a fixed time window.

Keywords and phrases: Complementary log-log model, Cox regression model, goodness-of-fit, multiple sclerosis, non-life insurance, nonparametric maximum likelihood, renewal process, retro-hazard, semi-Markov process

22.1 Introduction

Survival analysis has contributed a number of specialised approaches to the general methodology of goodness-of-fit; for recent textbook surveys see e.g. Andersen *et al.* (1993, Section VII.3), Hill *et al.* (1996, Chapter 7), Klein and Moeschberger (1997, Chapter 11), Hosmer and Lemeshow (1999, Chapter 6) or Therneau and Grambsch (2000, Chapter 6).

Beyond asserting the role of the specific approaches to goodness-of-fit in survival analysis and more general event history analysis, the purpose of this presentation is to report on two recent examples from my own experience, where the classical stratification approach to graphically assessing proportionality of hazards was used in nonstandard contexts.

22.2 Some Approaches to Testing Proportionality of Hazards

Cox (1972a) proposed the regression model

$$\lambda(t|z) = \lambda_0(t)e^{\beta z}$$

for the hazard $\lambda(t|z)$ at time t for given covariates z. Here $\lambda_0(t)$ is a freely varying "underlying" hazard which together with the log-linear dependence on covariates makes the model *semi-parametric*.

The proportionality assumption is restrictive, but nevertheless often taken for granted. The most common specific graphical checks and numerical tests focus attention on one covariate component at a time, i.e. asks whether proportionality holds "with respect to w" in the model

$$\lambda(t|z, w) = \lambda_0(t)e^{\beta z + \gamma w}.$$

Cox (1972a) pointed out that his model and the estimation techniques allowed *time-dependent covariates* $z(t)$, and he was there primarily motivated by wishing to test numerically for proportional hazards. As an example in our situation, this could be done by defining an additional covariate $w_1(t) = w \cdot \log t$ and testing the hypothesis $\delta = 0$ in the extended model

$$\lambda(t|z, w) = \lambda_0(t)e^{\beta z + \gamma w + \delta w \log t}.$$

In this presentation we link to a different tradition, most obvious for discrete covariates w. Assume for simplicity w dichotomous, assuming values 0 or 1. Then define the model *stratified on* w

$$\lambda(t|z, w) = \lambda_w(t)e^{\beta z}$$

and check (usually graphically) proportionality of $\lambda_0(t)$ and $\lambda_1(t)$. Note that β does not depend on w. Under proportionality $\lambda_1(t) = e^{\gamma}\lambda_0(t)$, i.e. $\Lambda_1(t) = e^{\gamma}\Lambda_0(t)$ and $\log \Lambda_1(t) = \gamma + \log \Lambda_0(t)$ with $\Lambda_w(t) = \int_0^t \lambda_w(u)du$.

Thus, the curves $(t, \log \Lambda_w(t), w = 0, 1,$ are parallel, and $(\Lambda_0(t), \Lambda_1(t))$ is the line through (0,0) with slope e^{γ}. The two most common plots are therefore $(t, \log \hat{\Lambda}_w(t))$ and $(\hat{\Lambda}_0(t), \hat{\Lambda}_1(t))$, see Andersen *et al.* (1993, Section VII.3).

We shall here give two non-standard examples of this basic *stratification* approach to testing proportional hazards.

22.3 "Proportionality" in Discrete-Time Regression for Retro-Hazard

Esbjerg, Keiding, and Koch-Henriksen (1999) studied models for reporting delay to the Danish Multiple Sclerosis Registry. As is well known particularly from the large literature on AIDS epidemiology, reporting delays are by nature *right-truncated*, and as discussed in detail by Kalbfleisch and Lawless (1989, 1991), Keiding and Gill (1990), Gross and Huber-Carol (1992) and Keiding (1992), the *retro-hazard* is then the most convenient parameterization, as seen by the following argument.

Esbjerg, Keiding, and Koch-Henriksen (1999) worked in discrete time, with X denoting the calendar time of onset and T the time between onset of multiple sclerosis (MS) and diagnosis. T will be called the reporting delay. In the analysis we consider X and T independent discrete random variables. They are only observed if $X+T \leq x^*$, where x^* is some fixed calendar time when the collection of data stops. Now condition on $X+T \leq x^*$ and $X = x$, and define $\tau = x^* - x$. Because of the independence of X and T

$$\begin{aligned} f_T(t|\tau, x) &= P(T = t|X = x, X + T \leq x^*) \\ &= f(t)/F(\tau) \\ &= f_T(t|\tau) \quad 0 \leq t \leq \tau \end{aligned}$$

which is independent of x. It follows that the *retro-hazard* or the reverse time hazard defined by

$$\rho(t) = P(T = t|T \leq t) = f(t)/F(t)$$

satisfies

$$\rho(t) = \frac{f_T(t|\tau)}{F_T(t|\tau)};$$

that is, the retro-hazard is the same in the marginal distribution of T and the distribution right-truncated by τ.

Esbjerg, Keiding, and Koch-Henriksen (1999) assumed the so-called c log log link model for the distribution of the reporting delay T for a patient with covariate vector z given by the retro-hazard

$$P(T = u|T \leq u) = \rho(u|z) = 1 - \exp(-\exp(\gamma_u + z'\beta)) \ .$$

The two crucial assumptions in this model for the retro-hazard are the assumption of proportionality of the logarithm of one minus the hazard function and linearity in the covariates of the link-function-transformed distribution function. In the following the two assumptions will be denoted the assumptions of proportionality and linearity.

Under the assumed model the ratio between the logarithms of one minus the hazard for persons i and j is

$$\frac{\log\left(1-\rho(u|\mathbf{z}_i)\right)}{\log\left(1-\rho(u|\mathbf{z}_j)\right)} = \exp\left((\mathbf{z}_i-\mathbf{z}_j)'\beta\right)$$

which is independent of time u. It is then clear that the ratio

$$\frac{\log(F(u|\mathbf{z}_i))}{\log(F(u|\mathbf{z}_j))}$$

is also independent of time. This property resembles the proportional hazards assumption in the Cox-model. One way to check the assumption for a categorical time-independent covariate z^m with k levels is as follows: Represent z^m by the vector $\mathbf{v}^m = (v^{m_1}, \cdots, v^{m_k})$ where $v^{m_1} = I\left(z^m = l\right), l = 1, \cdots, k$. Let $\mathbf{z} = \left(z^1, \cdots, z^{m-1}, \mathbf{v}^m\right) = \left(\mathbf{z}^0, \mathbf{v}^m\right)$, and consider the model where all other covariates are included and stratify by the levels of z^m. The baseline distribution functions from the different strata are $F_{0m_1}, \cdots, F_{0m_k}$ and the distribution function in stratum j for person i is given as

$$F_{m_j}(u|\mathbf{z}_i) = \left(F_{0m_j}\right)^{\exp\left(\mathbf{z}_i^{0t}\beta_0\right)}$$

while in the original model it is given as

$$F(u|\mathbf{z}_i) = (F_0(u))^{\exp\left(v_i^{m_j}\beta_{m_j}\exp\mathbf{z}_i^{0t}\beta_0\right)}, j = 1, \cdots, k$$

with $\beta_{m_k} \equiv 0$. Plotting $\log(-\log(F_{0m_1})), \cdots, \log(-\log(F_{0m_k}))$ against time u_1 the distance between the lines $\log(-\log(F_{0m_1}))$ and $\log(-\log(F_{0m_k}))$ should be constant and approximately equal to β_{m_j}.

Using these ideas Esbjerg, Keiding, and Koch-Henriksen (1999) showed that the cloglog-link model was not satisfied for the covariate age at onset, so that young and old patients had to be analysed separately.

22.4 The Renewal Assumption in Modulated Renewal Processes

Keiding, Anderson, and Fledelius (1998) studied insurance claims data collected in the time window from 1 January 1988 to 31 December 1991.

Observing a simple renewal process in an observation window $[t_1, t_2]$ involves four different elementary observations:

1. Times x_i from one renewal to the next, contributing the density $f(x_i)$ to the likelihood.

2. Times from one renewal to t_2, right-censored observations of F, contributing factors of the form $1 - F(t_2 - T_j)$ to the likelihood.

3. Times from t_1 to the first renewal (forward recurrence times), contributing, in the stationary case, factors of the form $(1 - F(T_j - t_1))/\mu$ to the likelihood.

4. Knowledge that no renewal happened in $[t_1, t_2]$, being right-censored observations of the forward recurrence time, contributing in the stationary case a factor

$$\int_{t_2-t_1}^{\infty} (1 - F(u))\, du/\mu \ .$$

In the stationary case the resulting nonparametric maximum likelihood estimation problem can be solved by an EM-type algorithm [Vardi (1982) and Soon and Woodroofe (1996)]. Keiding, Anderson, and Fledelius (1998) wanted to generalize the estimation method from iid variables to the modulated renewal process proposed by Cox (1972b), and also preferred to avoid the stationarity condition required for inclusion of the (uncensored and censored) forward recurrence times of types 3 and 4.

This is possible by restricting attention to (uncensored and censored) times since a renewal, that is, observations of types 1 and 2. As discussed repeatedly by Gill (1980, 1983), see also Aalen and Husebye (1991) and Andersen *et al.* (1993, Example X.1.8), the likelihood based on observations of type 1 and 2 is identical to one based on independent uncensored and censored life times from the renewal distribution F. Therefore the standard estimators (Kaplan-Meier, Nelson-Aalen) from survival analysis are applicable, and their usual large sample properties may be shown (albeit with new proofs) to hold.

The above analysis is sensitive to departures from the assumption of homogeneity between the iid replications of the renewal process. Restricting attention to time since first renewal will be biased (in the direction of short renewal times) if there is unaccounted heterogeneity, as will the re-use of second, third, ... renewals within the time window. As always, incorporation of observed covariates may reduce the unaccounted heterogeneity, but the question is whether this will suffice.

These ideas were implemented on the above mentioned data on insurance claims over a four-year interval. For property claims Figure 22.1 shows the Kaplan-Meier estimate (with pointwise 95% confidence intervals) based on observations of types 1 and 2, (that is, durations after a claim), while the curve marked "RT-algorithm" is the nonparametric maximum likelihood estimate based on all four types of observations. As envisioned, the durations after a claim are shorter than they should be under the stationary renewal hypothesis, indicating heterogeneity between the insurance takers.

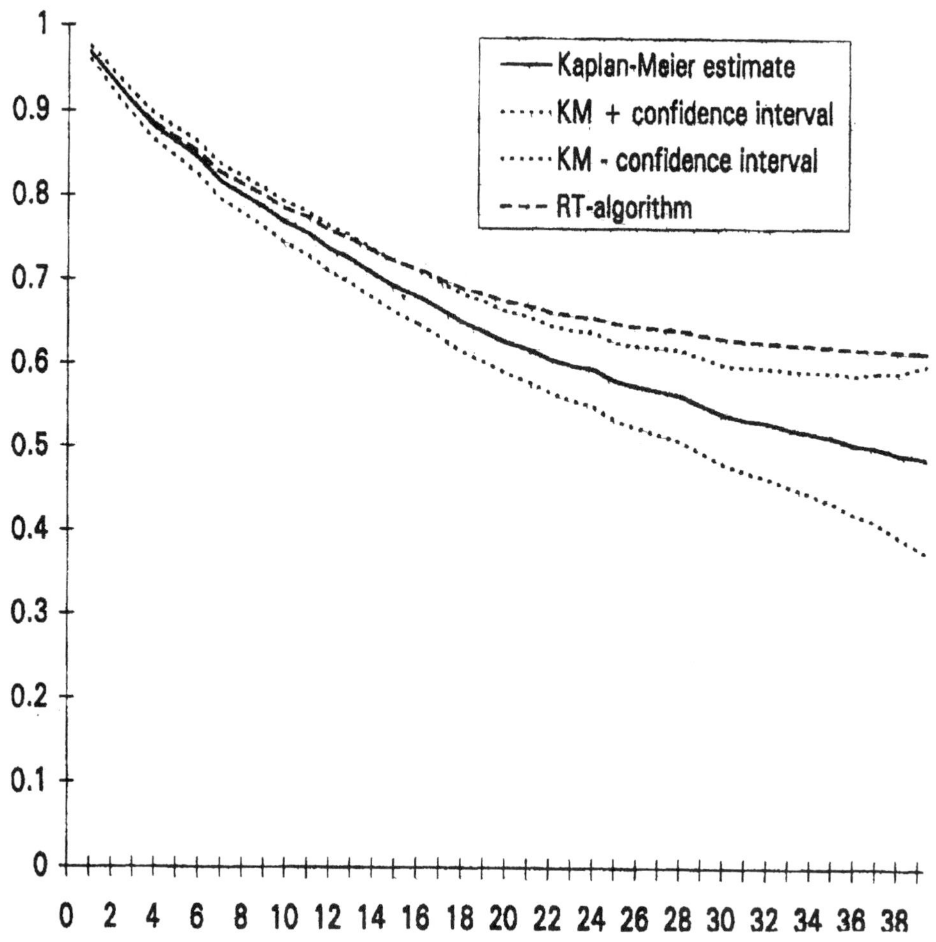

Figure 22.1: The probability of remaining property-claim free calculated by the Kaplan-Meier estimate based on durations since an observed claim (with pointwise 95% confidence limits) and by the nonparametric maximum likelihood estimate based on all observations in the assumed stationary renewal process. It is seen that the durations after an observed claim are generally *shorter*. [From Andersen and Fledelius (1996)].

Keiding, Anderson, and Fledelius (1998) went on to develop a Cox regression model for a modulated renewal process that could account for some of the heterogeneity. Here the occurrence of claims of type h for policy holder i at duration t since last claim of the type is governed by a Cox regression model with intensity

$$\lambda_{hi}(t) = \alpha_{0h}(t) \exp\left[\beta_h' Z_{hi}(t)\right] Y_{hi}(t)$$

where $\alpha_{0h}(t)$ is a freely varying so-called underlying intensity function common to all policy holders i but specific to insurance type h. The indicator $Y_{hi}(t)$ is 1 if policy holder i is at risk to make a claim of type h at time t, 0 otherwise. The covariate process $Z_{hi}(t)$ indicates fixed exogenous as well as time-dependent endogenous covariates. Finally the vector β_h contains the regression coefficients on the covariates $Z_{hi}(t)$. For this model Dabrowska (1995) proved asymptotic results for the "usual" profile likelihood based inference, under the crucial assumption that the covariates $Z_{hi}(t)$ depend on time only through (the backwards recurrence time) t.

The claim occurrences are viewed through a fixed time window, but under the model valid inference may be based on the likelihood composed of the product of contributions from the distribution of time from first to second claim, second to third claim, and so on, the last being right-censored. The expected deviation from the model is that time from claim $j = 1$ is longer than times from claims $j = 2, 3, \ldots$ Keiding, Anderson, and Fledelius (1998) therefore extended the model to the Cox regression model

$$\lambda_{hij}(t) = \alpha_{0hj}(t) \exp\left[\beta_{hj}' Z_{hi}(t)\right] Y_{hi}(t) \ .$$

In practice the regression coefficients β_{hj} and the underlying intensities $\alpha_{0hj}(t)$ after claim j are assumed identical for $j = 2, 3, \ldots$. A good evaluation of the fit of the Cox model can be based on first assessing identity of regression coefficients ($\beta_{h1} = \beta_{h2}$) and then, refitting in a so-called *stratified* Cox regression model with identical β_{hj} but freely varying $\alpha_{h0j}(t)$ over j, comparing the underlying intensities ($\alpha_{0h1}(t) = \alpha_{0h2}(t)$) after first and after later claims. For the first hypothesis a standard log partial likelihood ratio test may be performed, for the second Keiding, Anderson, and Fledelius (1998) documented a series of graphical checks as surveyed by Andersen *et al.* (1993, Section VII.3). Further development of this goodness-of-fit approach might follow the lines of Andersen *et al.* (1983).

Based on generalizations of the standard graphs mentioned above, Keiding, Anderson, and Fledelius (1998) concluded that property and auto claims could not be described by the postulated modulated renewal processes, while the model for household claims was not rejected by this approach.

References

1. Aalen, O. O. and Husebye, E. (1991). Statistical analysis of repeated events forming renewal processes, *Statistics in Medicine*, **10**, 1227–1240.

2. Andersen, C. and Fledelius, P. (1996). Individuel tarifering - En anvendelse af Cox' regressionsmodel i skadesforsikring, *M.Sc. thesis in actuarial mathematics, University of Copenhagen.*

3. Andersen, P. K., Borgan, Ø., Gill, R. D., and Keiding, N. (1993). *Statistical Models Based on Counting Processes*, New York: Springer Verlag.

4. Andersen, P. K., Christensen, E., Fauerholdt, L., and Schlichting, P. (1983). Evaluating prognoses based on the proportional hazards model, *Scandinavian Journal of Statistics*, **10**, 141–144.

5. Cox, D. R. (1972a). Regression models and life tables (with discussion), *Journal of the Royal Statistical Society, B*, **34**, 187–220.

6. Cox, D. R. (1972b). The statistical analysis of dependencies in point processes, In *Stochastic Point Processes* (Ed., P.A.W. Lewis), pp. 55–66, New York: John Wiley.

7. Dabrowska, D. M. (1995). Estimation of transition probabilities and bootstrap in a semiparametric Markov renewal model, *Nonparametric Statistics*, **5**, 237–259.

8. Esbjerg, S., Keiding, N., and Koch-Henriksen, N. (1999). Reporting delay and corrected incidence of multiple sclerosis, *Statistics in Medicine*, **18**, 1691–1706.

9. Gill, R. D. (1980). Nonparametric estimation based on censored observations of a Markov renewal process, *Zeitschrift für Wahrscheinlichkeitstheorie und verwandte Gebiete*, **53**, 97–116.

10. Gill, R. D. (1983). Discussion of the papers by Helland and Kurtz, *Bulletin of the International Statistical Institute*, **50**, 239–243.

11. Gross, S. T. and Huber-Carol, C. (1992). Regression models for truncated survival data, *Scandinavian Journal of Statistics*, **19**, 193–213.

12. Hill, C., Com-Nougué, C., Kramar, A., Moreau, T., O'Quigley, J., Senoussi, R., and Chastang, C. (1996). *Analyse statistique des donnes de survie*, 2e éd., Paris: Flammarion.

13. Hosmer, D. W. and Lemeshow, S. (1999). *Applied Survival Analysis. Regression Modeling of Time to Event Data*, New York: John Wiley & Sons.

14. Kalbfleisch, J. D. and Lawless, J. F. (1989). Inference based on retrospective ascertainment: an analysis of the data on transfusion-related AIDS, *Journal of the American Statistical Association*, **84**, 360–372.

15. Kalbfleisch, J. D. and Lawless, J. F. (1991). Regression models for right truncated data with applications to AIDS incubation times and reporting lags, *Statistica Sinica*, **1**, 19–32.

16. Keiding, N. (1992). Independent delayed entry, In *Survival Analysis: State of the Art* (Eds., J. P. Klein and P. K. Goel), pp. 309–326, Dordrecht: Kluwer.

17. Keiding, N., Andersen, C., and Fledelius, P. (1998). The Cox regression model for claims data in non-life insurance, *ASTIN Bulletin*, **28**, 95–118.

18. Keiding, N. and Gill, R. D. (1990). Random truncation models and Markov processes, *The Annals of Statistics*, **18**, 582–602.

19. Klein, J. P. and Moeschberger, M. L. (1997). *Survival Analysis. Techniques for Censored and Truncated Data*, New York: Springer Verlag.

20. Soon, G. and Woodroofe, M. (1996). Nonparametric estimation and consistency for renewal processes, *Journal of Statistical Planning and Inference*, **53**, 171–195.

21. Therneau, T. M. and Grambsch, P. M. (2000). *Modeling Survival Data. Extending the Cox Model*, New York: Springer.

22. Vardi, Y. (1982). Nonparametric estimation in renewal processes, *Annals of Statistics*, **10**, 772–785.

23

Association in Contingency Tables, Correspondence Analysis, and (Modified) Andrews Plots

Ravindra Khattree and Dayanand N. Naik

Oakland University, Rochester, Missouri
Old Dominion University, Norfolk, Virginia

Abstract: Andrews plots [Andrews (1972)], as a tool to graphically interpret multivariate data, have recently gained considerable recognition. We present the use of Andrews plots and Modified Andrews plots which are recently introduced by the authors [Khattree and Naik (2001)], in graphically exhibiting the association in the cross classified data and in the context of correspondence analysis. A new alternative to traditional correspondence analysis introduced by C. R. Rao (1995) and implementation of our approach in this case are also discussed.

Keywords and phrases: Andrews plots, correspondence analysis, Hellinger distance, modified Andrews plots

23.1 Introduction

Andrews plots [Andrews (1972)] as a tool for the graphical representation of multivariate data have recently gained considerable popularity due to the simplicity of their plotting and many desirable and attractive mathematical properties. Khattree and Naik (2001) provide a review of Andrews and other related plots along with their merits and demerits.

For a p-dimensional multivariate observation $\mathbf{y} = (y_1, \ldots, y_p)'$, the Andrews curve in argument t is defined by

$$f_{\mathbf{y}}(t) = \frac{y_1}{\sqrt{2}} + y_2 \sin(t) + y_3 \cos(t) + y_4 \sin(2t) + y_5 \cos(2t) + \ldots \qquad -\pi \leq t \leq \pi. \tag{23.1}$$

Clearly the p-dimensional observation is then represented by a curve and hence can be plotted in two-dimensions, without any loss of information, which usually occurs in other graphical representation such as the scatter plots or the dimensionality reduction based techniques such as Bi-plots.

A plot containing Andrews curves corresponding to all the multivariate observations is called an Andrews plot. Khattree and Naik (2001) recently suggested a modified Andrews plot as a plot consisting of curves

$$\begin{aligned} g_{\mathbf{y}}(t) &= \frac{1}{\sqrt{2}}\Big\{y_1 + y_2[\sin(t)+\cos(t)] \\ &\quad + y_3[\sin(t)-\cos(t)] + y_4[\sin(2t)+\cos(2t)] \\ &\quad + y_5[\sin(2t)-\cos(2t)] + \dots\Big\}, \\ &\qquad -\pi \le t \le \pi, \end{aligned} \tag{23.2}$$

which has all the properties of the Andrews plots yet avoids some of the shortcomings of Andrews plots. Specifically, the odd numbered terms in the Andrews plot given in (23.1) simultaneously vanish at $t = 0$. This is a major disadvantage since human eyes tend to concentrate on and notice changes or similarities more easily and more immediately in the central regions of the graph (that is, around $t = 0$); yet, for Andrews plot around $t = 0$, these similarities or dissimilarities are mostly due to even numbered variables $y_2, y_4, \dots$. This is not so, for the modified Andrews plots given by (23.2). It can be seen from the fact that $\frac{1}{\sqrt{2}}[\sin(kt) \pm \cos(kt)] = \sin\left(kt \pm \frac{\pi}{4}\right) = \sin\left(k\left(t \pm \frac{\pi}{4k}\right)\right)$, which has the phase angle $\pm\frac{\pi}{4k}$. Thus, none of the terms in (23.2) has the common phase angle and hence will not vanish at the same point. Therefore, for any t, the value of $f_{\mathbf{y}}(t)$ depends on at least $(p-1)$, y_i values.

The objective of this article is to use the Andrews plots to graphically represent the associations in the contingency tables. Often such associations are depicted through correspondence analysis where the rows and columns of the contingency table are represented by points in a two-or three-dimensional scatter plot. As mentioned earlier, this may not be completely satisfactory due to loss of information caused by the dimensionality reduction. Andrews and modified Andrews plots are clearly the ideal alternatives to avoid such loss. We describe the approach in the next section.

23.2 (Modified) Andrews Plots in Correspondence Analysis

For an $a \times b$ contingency table $\mathbf{N} = ((n_{ij}))$ the levels of two discrete variables A and B are represented by rows and columns, respectively. The correspondence analysis is a mathematical approach to represent these as the points in an m-dimensional space, where $m = \min(a, b) - 1$. This is done by using the singular value decomposition of the correspondence matrix $\mathbf{Q} = ((q_{ij}))$, where $q_{ij} = \frac{n_{ij}}{n}$, $n = \sum_{i=1}^{a}\sum_{j=1}^{b} n_{ij}$. Clearly the correspondence matrix $\mathbf{Q}$ is the matrix of relative frequencies. The above singular value decomposition provides the representation of each row and column of $\mathbf{N}$ as a point in m-dimensional space. Projections of these points in two or three dimensions provide a way to present the information graphically. To present things more formally yet briefly, from the correspondence matrix $\mathbf{Q}$, we define

$$\mathbf{r} = \mathbf{Q1}$$

and

$$\mathbf{c} = \mathbf{Q'1},$$

where $\mathbf{1} = (1, \ldots, 1)'$ is a column vector of appropriate dimension, containing 1 as the entry at each place. The vectors $\mathbf{r}$ and $\mathbf{c}$ are called, respectively, the row and column masses. We also define

$$\mathbf{D_r} = \text{diag}\,(\mathbf{r}),$$
$$\mathbf{D_c} = \text{diag}\,(\mathbf{c}),$$

and let

$$\mathbf{F} = \mathbf{D_r^{-1}}(\mathbf{Q} - \mathbf{rc'})\mathbf{D_c^{-1}}(\mathbf{Q} - \mathbf{rc'})'.$$

One can show with little manipulation that

$$n\text{tr}\,(\mathbf{F}) = \sum_{i=1}^{a}\sum_{j=1}^{b} \frac{\left(n_{ij} - \frac{n_{i.}n_{.j}}{n}\right)^2}{\frac{n_{i.}n_{.j}}{n}}$$

which is the Pearson's chi-squared test statistic for testing independence between A and B. The quantity $T_I = \text{tr}(\mathbf{F})$ is referred to as the total inertia. It is just a multiple of Pearson's chi-squared and can be interpreted as a measure of the magnitude of the total row (or column) squared deviations. This interpretation avails us an opportunity to attempt to partition T_I into several components and allocate these components (which would essentially be the eigenvalues of $\mathbf{F}$) to various dimensions. This is done using the generalized singular value decomposition [see, Jobson (1992)] of the matrix $\mathbf{Q}-\mathbf{rc}'$. In fact, one can directly take the generalized singular value decomposition of the correspondence matrix $\mathbf{Q}$ rather than $\mathbf{Q}-\mathbf{rc}'$. The two decompositions are essentially equivalent except for one trivial (and in fact, redundant for our purpose) singular value which is always equal to 1.

In the traditional correspondence analysis only the first two or three dimensions are retained and using these, one explores any associations or other striking features in the plots. However, the graphical aspect at this stage can be handled more effectively through (modified) Andrews plots. Instead of working with projections of all $m \times 1$ row and column points, we instead transform these points to Andrews or (modified) Andrews curves. Consequently, the information available in the data is used to its fullest. Row (or column) points closer to each other transform to curves nearer to each other. Any exceptional or influential categories can also be readily identified by these curves. Further, assuming a reasonable degree of symmetry in the correspondence matrix $\mathbf{Q}$, association of any row category to any column category will be reflected in the similarities between the corresponding (modified) Andrews curves. Alternatively, the asymmetric version of (modified) Andrews plots can be used [Greenacre (1984)]. Independence in the contingency table should result in these plots showing no specific patterns.

It is more instructive to illustrate the techniques, approach and ways of interpretation through some examples. The next section provides these illustrations using certain real data sets.

23.3 Some Examples

Example 23.3.1 Fisher *et al.* (1982) present a study of reproducibility of coronary arteriography diagnostic procedure which plays an important role in deciding if a bypass surgery on a patient may be appropriate. It is thus important to monitor the quality of arteriography and ascertain the amount of agreement between the quality control site readings and the clinical site readings. The problem is essentially that of interlaboratory testing in a categorical data context. Table 23.1 presents the results for 870 randomly selected patients with two readings. For these readings the degree or intensity of disease was classified as none, zero vessel disease but some disease and one-, two-, and three-vessel disease. For identification in plots, these groups for the quality control site are denoted by 1, 2, 3, 4 and 5, respectively, and for the clinical site, the notation used are A, B, C, D and E, respectively.

Table 23.1: Agreement with respect to number of diseased vessels

Quality Control Site Reading	Clinical Site Reading				
	Normal (A)	Some (B)	One (C)	Two (D)	Three (E)
Normal (1)	13	8	1	0	0
Some (2)	6	43	19	4	5
One (3)	1	9	155	54	24
Two (4)	0	2	18	162	68
Three (5)	0	0	11	27	240

The row and column points $\mathbf{y} = (y_1, y_2, y_3, y_4)'$ for the data are obtained by performing the generalized singular value decomposition and these are reported in Table 23.2. Using these, we obtain the functions $f_{\mathbf{y}}(t)$ and $g_{\mathbf{y}}(t)$ for each of the five rows and columns. These plots are given in Figures 23.1 and 23.2. In the case of perfect agreement between the clinical site and quality control site readings, all row points should be the same as the corresponding column points, and thus the corresponding Andrews curves should be identical also. As seen in the plots, the agreement between the two readings appears to be quite good for each of the five classification groups and the (modified) Andrews curves of the corresponding points follow each other very closely in case of $f_{\mathbf{y}}(t)$ as well as $g_{\mathbf{y}}(t)$.

The same argument can be made for the plots in Figures 23.3, 23.4 and 23.5, where the components of vector $\mathbf{y}$ have been permuted so that different y_i are assigned to the constant $\frac{1}{\sqrt{2}}$ in the expressions of $g_{\mathbf{y}}(t)$ (similar plots corresponding to $f_{\mathbf{y}}(t)$ are not shown here). It is not surprising, since in the present case, none of the four dimensions can be deemed redundant in that each explains a significant percent of total inertia. Specifically, these percentages are 43.80, 28.59, 16.31 and 11.30%. However, in many cases, the major portion (up to 85 or 90%) of the total inertia can be attributed to the first two dimensions. Whenever this does not happen, as in the present case, the use of only the first two dimensions to obtain a plot of points will not be very effective compared to the use of Andrews plots which utilize all dimensions.

Table 23.2: Clinical and QC site evaluations: Row and column points

Points		Coordinates			
		y_1	y_2	y_3	y_4
Row	Normal	3.14513	1.72789	0.91817	1.52739
(QC Site)	Some	1.67179	0.25096	-0.32230	-0.97903
	One	0.24290	-0.84605	-0.32512	0.27160
	Two	-0.34541	-0.16330	0.72270	-0.15729
	Three	-0.61364	0.68014	-0.34912	0.05434
Column	Normal	3.12841	1.74992	0.96954	1.71600
(Clinical)	Some	1.94382	0.40705	-0.25836	-1.07768
	One	0.35725	-0.89110	-0.45605	0.26900
	Two	-0.26104	-0.32340	0.72042	-0.12930
	Three	-0.56821	0.59772	-0.26196	0.02836

Example 23.3.2 Data in Table 23.3 are taken from Srole *et al.* (1978) and are reported in Agresti (1990) as well. The objective of the study was to examine the relationship, if any, between the mental impairment and parents' socioeconomic status. Six levels of socioeconomic status, 1 (high) to 6 (low) and four levels of mental health status, A (well), B (mild symptom formation), C (moderate symptom formation) and D (impairment) are taken. Data are obtained on a sample of 1660 residents of Manhattan. It may be pointed out that both variables are ordinal in nature.

As per the structure of the data for plotting, we have six row points and four column points in the three-dimensional space. However, the first dimension alone is able to explain 93.95% of the total inertia, and the first two dimensions combined explain almost 99% of that. Thus, the third dimension contains no significantly useful information for all practical purposes. The ten row and column points are listed in Table 23.4. The modified Andrews plots are given in Figures 23.6 and 23.7. The following observations are made from Figure 23.6.

(a) The two variables are ordered categorical in nature and this fact clearly shows up in Figure 23.6. Also, it can be observed that there is possibly a further subgrouping within the socioeconomic status. Status 1 and 2 are similar to each other; 3 and 4 are similar and to some extent, there

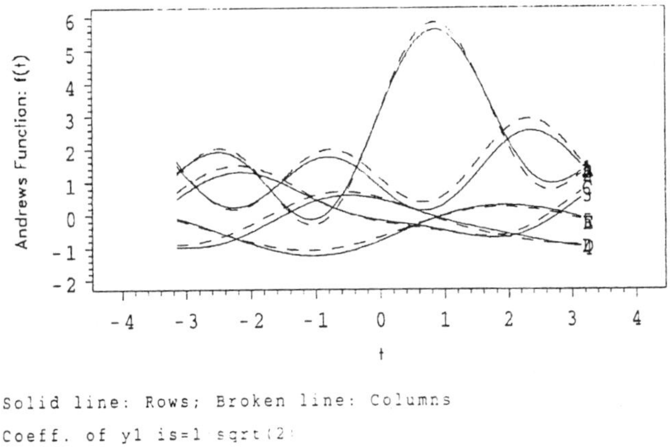

Figure 23.1: Agreement w.r.t. no. of diseased vessels

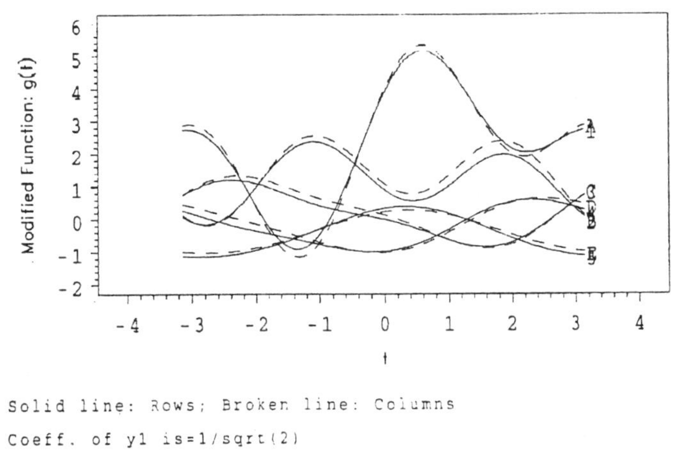

Figure 23.2: Agreement w.r.t. no. of diseased vessels

is some similarity between status 5 and 6. However, as subgroups, these three subgroups clearly separate themselves out in these plots.

(b) With respect to mental health status, the health status B (mild symptom formation) and C (moderate symptom formation) appear to be similar, but are clearly different from A (well) and D (impaired). Further, the ordinal nature of each health status is self evident in these plots.

(c) There appears to be positive association between the mental health status of children and parents' socioeconomic status, with mental health of children to be generally better at the higher levels of parents' socioeconomic status. Further, the upper two levels of parents' socioeconomic status groups generally correspond to a "well" mental health status; the next

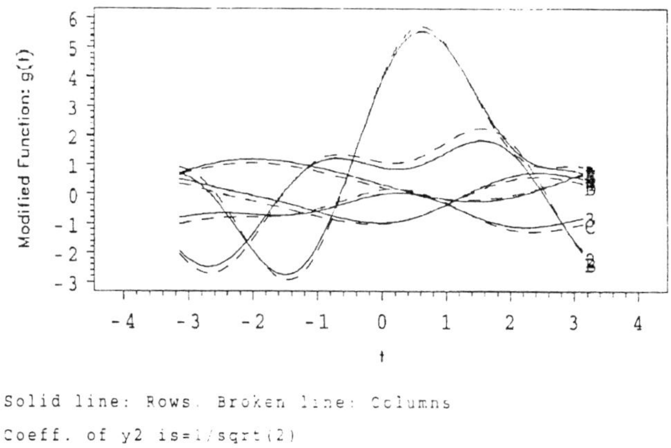

Figure 23.3: Agreement w.r.t. no. of diseased vessels

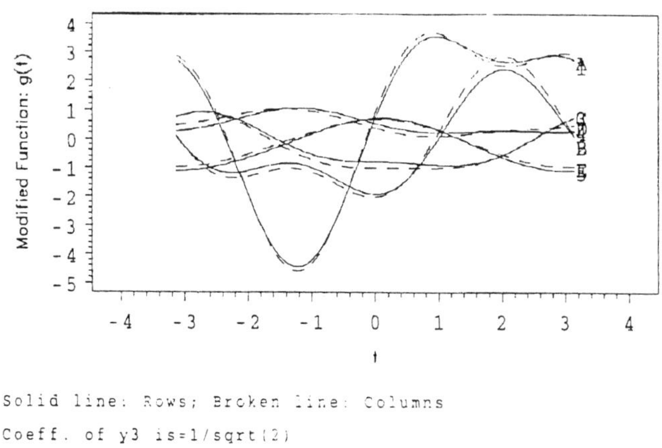

Figure 23.4: Agreement w.r.t. no. of diseased vessels

two levels are closely related to "mild" or "moderate symptom formation" and the lowest two levels of parents' socioeconomic status generally severely affect the mental health status of children towards impairment. It may be remarked that ideally one should use the *asymmetric plots* to assess the closeness of row and column points. However, in the present case our regular plot provides the interpretable results.

The plot Figure 23.7 corresponds to the case when y_2 has been assigned to the constant $\frac{1}{\sqrt{2}}$ in the function $g_{\mathbf{y}}(t)$. This plot reveals a few additional interesting observations.

(a) With respect to parents' socioeconomic status, curves corresponding to groups 1 and 2 show the patterns just opposite to those corresponding to

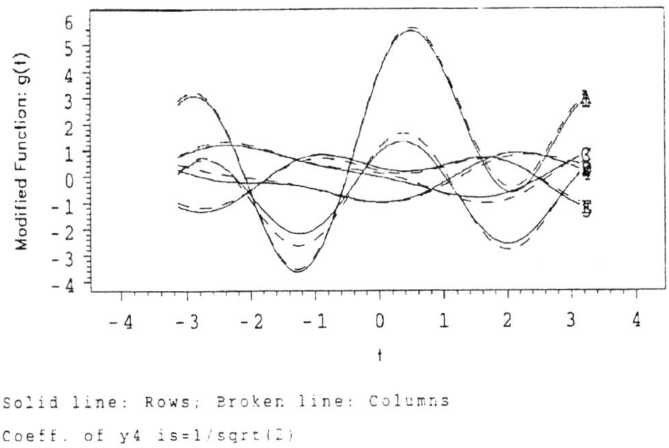

Figure 23.5: Agreement w.r.t. no. of diseased vessels

groups 5 and 6.

(b) With respect to mental health status, groups A (well) and D (impaired) follow the patterns opposite to each other.

(c) There, possibly, is very little difference between groups 3 and 4 of parents' socioeconomic status and between groups B and C with respect to mental health status. Conceivably, in each case, the corresponding two groups can be combined. If this is done, then there is a positive association between the combined groups (3, 4) and (B, C). Plots for this situation are not shown.

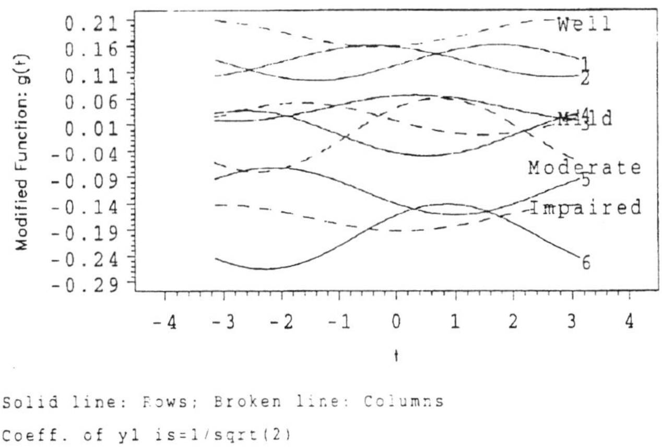

Figure 23.6: Agreement w.r.t. no. of diseased vessels

Table 23.3: Cross-classification of mental health status and parents' socioeconomic status

Parents' Socioeconomic Status	Mental Health Status (Well) A	(Mild Symptom Formation) B	(Moderate Symptom Formation) C	(Impaired) D
1 (high)	64	94	58	46
2	57	94	54	40
3	57	105	65	60
4	72	141	77	94
5	36	97	54	78
6 (low)	21	71	54	71

It may be emphasized that so much information could not have been extracted from these data if traditional two-dimensional scatter plots of correspondence analysis were constructed or if the usual Pearson's chi-squared test for independence was applied. Further, we note that the suggested modification of Andrews plots provides curves which are easier to interpret and are more informative.

23.4 Modified Andrews Plots and Rao's Correspondence Analysis

Rao (1995) suggested the use of Hellinger distance instead of chi-squared distance for measuring the distance between two row (or column) profiles in correspondence analysis. This was done for the two main reasons (a) the chi-squared distance is not just the function of individual profiles under consideration but it also depends on the average of all profiles and (b) for the reason given in (a) and while measuring the affinities between profiles, undue emphasis may be given to categories with low frequencies. However, one important point to be emphasized here is that regardless of the choice of distance, (modified) Andrews plots can still be used in the context of correspondence analysis. To discuss the

Table 23.4: Mental health and parents' socioeconomic status: row & column points

Points		Coordinates		
Row		y_1	y_2	y_3
(Socioeconomic)	1	0.180932	0.019248	0.027525
	2	0.184996	0.011625	-.027386
	3	0.059031	0.022197	-.010575
	4	-.008887	-.042080	0.011025
	5	-.165392	-.043606	-.010368
	6	-.287690	0.061994	0.004824
Column	Well	0.259536	-.012102	0.022589
(Mental Health)	Mild	0.029588	-.023651	-.019818
	Moderate	-.014210	0.069901	-.003230
	Impaired	-.237392	-.018897	0.015848

approach briefly, let $\mathbf{R} = ((r_{ij}))$ be the $a \times b$ matrix of row profiles defined as

$$\mathbf{R} = \mathbf{D_r^{-1}Q} = \begin{bmatrix} \frac{p_{11}}{p_{1.}} & \frac{p_{12}}{p_{1.}} & \cdots & \frac{p_{1b}}{p_{1.}} \\ \frac{p_{21}}{p_{2.}} & \frac{p_{22}}{p_{2.}} & \cdots & \frac{p_{2b}}{p_{2.}} \\ \vdots & & & \\ \frac{p_{a1}}{p_{a.}} & \frac{p_{a2}}{p_{a.}} & \cdots & \frac{p_{ab}}{p_{a.}} \end{bmatrix}.$$

Define

$$\mathbf{H} = ((h_{ij})) = (\sqrt{r_{ij}}) = \begin{bmatrix} \mathbf{h}_1' \\ \mathbf{h}_2' \\ \vdots \\ \mathbf{h}_a' \end{bmatrix},$$

where the vector $\mathbf{h}_i$ is defined as

$$\mathbf{h}_i = (\sqrt{r_{i1}}, \sqrt{r_{i2}}, \ldots, \sqrt{r_{ib}})', i = 1, 2, \ldots, a.$$

The Hellinger distance between the ith and j^{th} profile is defined as

$$\delta_{ij}^2 = (\mathbf{h}_i - \mathbf{h}_j)'(\mathbf{h}_i - \mathbf{h}_j),$$

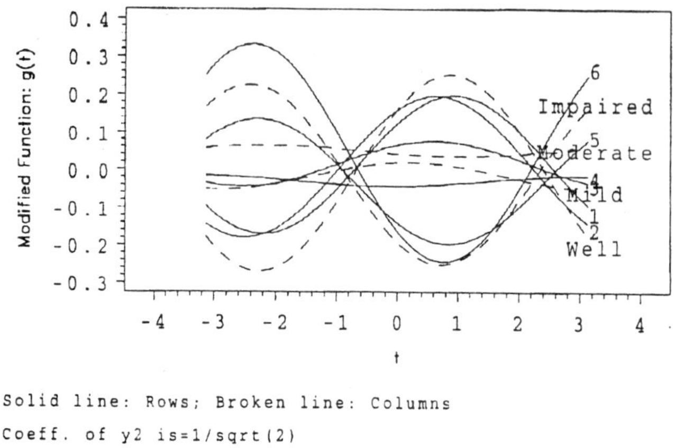

Figure 23.7: Agreement w.r.t. no. of diseased vessels

and as stated earlier, it depends only on the ith and j^{th} profiles.

The canonical coordinates for representing row profiles are determined from the singular value decomposition of the $a \times b$ matrix $\mathbf{D}_{\mathbf{r}}^{\frac{1}{2}}(\mathbf{H} - \mathbf{1}'\boldsymbol{\eta})$, where Rao suggests the choice of vector $\boldsymbol{\eta}$ as

$$\boldsymbol{\eta} = \left(\sqrt{\frac{n_{.1}}{n}}, \sqrt{\frac{n_{.2}}{n}}, \ldots, \sqrt{\frac{n_{.b}}{n}}\right)'$$

or

$$\boldsymbol{\eta} = \mathbf{H}'\mathbf{r}.$$

It must be noted that the dimension of the space m, in this case, is the rank of $\mathbf{D}_{\mathbf{r}}^{\frac{1}{2}}(\mathbf{H} - \mathbf{1}'\boldsymbol{\eta})$ and is not necessarily equal to $\min(a, b) - 1$. If the singular value decomposition of $\mathbf{D}_{\mathbf{r}}^{\frac{1}{2}}(\mathbf{H} - \mathbf{1}'\boldsymbol{\eta})$ is

$$\mathbf{D}_{\mathbf{r}}^{\frac{1}{2}}(\mathbf{H} - \mathbf{1}'\boldsymbol{\eta}) = \lambda_1 \mathbf{u}_1 \mathbf{v}_1' + \lambda_2 \mathbf{u}_2 \mathbf{v}_2' + \cdots + \lambda_m \mathbf{u}_m \mathbf{v}_m',$$

then canonical coordinates for row profiles are given by

$$\lambda_1 \mathbf{D}_{\mathbf{r}}^{-\frac{1}{2}} \mathbf{u}_1,\ \lambda_2 \mathbf{D}_{\mathbf{r}}^{-\frac{1}{2}} \mathbf{u}_2, \ldots, \lambda_m \mathbf{D}_{\mathbf{r}}^{-\frac{1}{2}} \mathbf{u}_m,$$

and the column profiles are given by standard coordinates

$$\lambda_1 \boldsymbol{\Delta}_{\mathbf{c}}^{-1} \mathbf{v}_1,\ \lambda_2 \boldsymbol{\Delta}_{\mathbf{c}} \mathbf{v}_2, \ldots, \lambda_m \boldsymbol{\Delta}_{\mathbf{c}}^{-1} \mathbf{v}_m,$$

where

$$\boldsymbol{\Delta}_{\mathbf{c}} = \left[\operatorname{diag}[(\mathbf{H} - \mathbf{1}\boldsymbol{\eta}')'(\mathbf{H} - \mathbf{1}\boldsymbol{\eta}')]\right]^{\frac{1}{2}}.$$

These can be plotted in the same plot. If one wishes to have the canonical coordinates for column profiles and standard coordinates for row profiles, above formulas can be suitably modified.

We will illustrate the modified Andrews plots based on Rao's correspondence analysis using a drug efficacy data of Calimlin *et al.* (1982).

Example 23.4.1 The data are reported in Table 23.5. The objective of the study was to examine the ratings assigned by the hospital patients to four drugs, with an objective to determine whether a particular drug or a group of drugs are favored by the patients. The four drugs are named as Z100, EC4, C60 and C15 and the five ratings on ordinal scale are from poor to excellent. We compute here the canonical coordinates for row profiles representing various drugs and standard coordinates for the column profiles representing the ratings. The coordinates are not shown here to save space. Also, we present here only the modified Andrews plots as Figures 23.8 and 23.9.

Table 23.5: Results of a survey on analgesic efficacy of drugs

	Ratings of Drug's Efficacy					
Drug	Poor	Fair	Good	Very Good	Excellent	Total
Z100	5	1	10	8	6	30
EC4	5	3	3	8	12	31
C60	10	6	12	3	0	31
C15	7	12	8	1	1	29
Total	27	22	33	20	19	121

Figure 23.8 clearly illustrates the closeness of the modified Andrews curves corresponding to drugs C15 and C60 and that of curves corresponding to EC4 and Z100. In the same plot, one also observes the anticipated clustering of the efficacy ratings, namely, {Excellent and Very Good} forming one cluster and the remaining three ratings forming the other cluster. The proximity of EC4 and Z100 to the higher two efficacy ratings and proximity of C15 and C60 to the lowest three ratings is also noted. This indicates that the EC5 and Z100 are perceived as the superior choices by patients.

Figure 23.9 represents modified Andrews plot when the roles of y_1 and y_2 (that is, the scores corresponding to the first and second dimensions) in (23.2) have been interchanged. This obviously changes the curves. However, in this plot the similarities and dissimilarities between various drugs and ratings and associations between the particular drugs and ratings are depicted not by the relative closeness of corresponding curves but by their particular patterns (in terms of ups and downs). The curves corresponding to EC4 and Z100 have

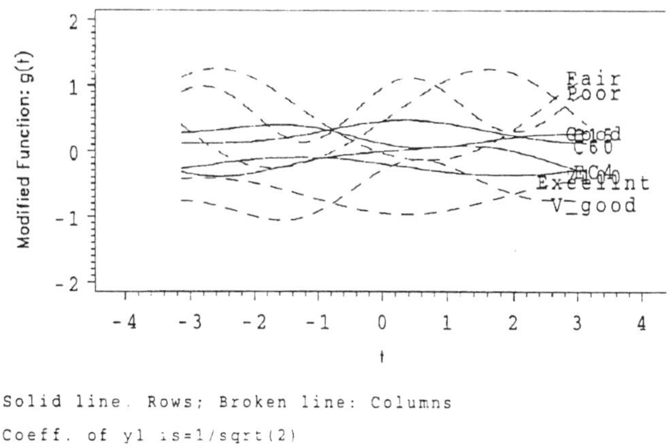

Figure 23.8: Drug vs. efficacy rating

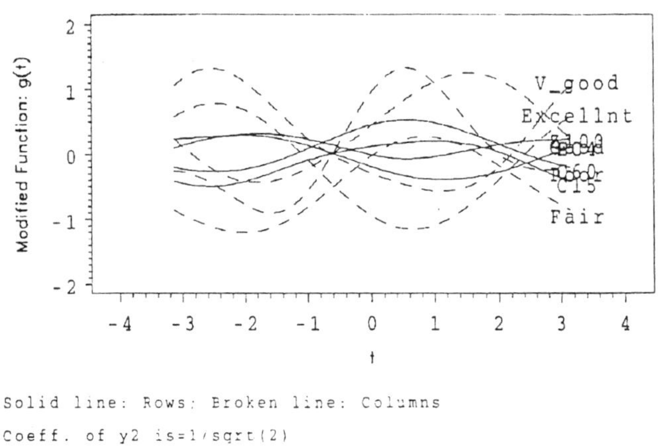

Figure 23.9: Drug vs. efficacy rating

larger amplitudes compared to the other two drugs. A similar observation is made about the efficacy ratings. At the same time, it is also observed that the curves for EC4 and Z100 and those corresponding to C15 and C60 follow the opposite directions (when the former ones go up, the latter ones tend to go down). At the same time such changes for {EC4, Z100} and efficacy ratings {Excellent, Very Good} are in the same direction, thereby indicating the strong association in this case. A similar statement can be made for drugs {C15, C60} and efficacy ratings {Poor, Fair, Good}. It must be emphasized that all these features are not as distinct in the two-dimensional plots (not shown here) obtained by appropriate dimensionality reduction.

23.5 Conclusions

Andrews plots as a powerful graphical technique for multivariate data are not only useful in traditional clustering or outlier detection problems, but they can also be very important tools in analyzing experimental data in other discrete or descriptive multivariate analysis problems, such as contingency tables and correspondence analysis. The fact that there is no loss of dimension, and hence no approximations before the graphical displays of the data, allows these plots to be much more informative than traditional scatter plots or biplots. Further, the modified Andrews plots are even more informative and useful in the graphical displays of these data.

References

1. Agresti, A. (1990). *Categorical Data Analysis*, New York: John Wiley & Sons.

2. Andrews, D. F. (1972). Plots of high dimensional data, *Biometrics*, **28**, 125–136.

3. Calimlin, J. F., Wardell, W. M, Davis, H. T., Lasagna, L., and Gillies, A. J. (1982). Analgesic efficacy of an orally administered combination of Pentazocine and Aspirin: With observations on the use of and statistical efficacy of global subjective efficacy ratings, *Clinical Pharmacology and Therapeutics*, **21**, 34–43.

4. Fisher, L. D., Judkins, M. P. Lesperance, J., Cameron, A., Swaye, P., Ryan, T. J., Maynard, C., Bourassa, M., Kennnedy, J. W., Gosselin, A., Kemp, H., Faxon, D., Wexler, L., and Davis, K. (1982). Reproducibility of coronary arteriographic reading in the coronary artery surgery study (CASS), *Catheterization and Cardiovascular Diagnosis*, **8**, 565–575.

5. Greenacre, M. J. (1984). *Theory and Applications of Correspondence Analysis*, London: Academic Press.

6. Jobson, J. D. (1992). *Applied Multivariate Data Analysis*, Vol. II, New York: Springer-Verlag.

7. Khattree, R. and Naik, D. N (2001). Andrews plots for multivariate data: Some new suggestions and applications, *Journal of Statistical Planning and Inference*, to appear.

8. Rao, C. R. (1995). A review of canonical coordinates and an alternative to correspondence analysis using Hellinger distance, *Quwstiio*, **19**, 23–63.

9. Srole, L. , Langner, T. S., Michael, S. T., Kirkpatrick, P., Opler, M. K., and Rennie, T. A. C. (1978). *Mental Health in the Metropolis: The Midtown Manhattan Study*, Revised edition, New York: NYU Press.

24

Orthogonal Expansions and Distinction Between Logistic and Normal

Carles M. Cuadras and Daniel Cuadras

University of Barcelona, Barcelona, Spain

Abstract: We propose a graphical goodness-of-fit test which consists in representing a sample along principal dimensions of a random variable. This method is used to distinguish the logistic from the normal distribution. Some simulations are given.

Keywords and phrases: Maximum correlation, Wasserstein distance, principal components, continuous scaling

24.1 Introduction

Given an ordered sample $\mathcal{X} = \{x_1 \leq ... \leq x_n\}$, where the $x_i's$ are iid observations of a r.v. X with continuous cdf F, the classical statistics for testing the hypothesis $H_0 : F = F_0$ are based on

$$D_n = \sup_{-\infty<x<\infty} |F_n(x) - F_0(x)|$$

and

$$W_n^2(\Psi) = n \int_{-\infty}^{\infty} \Psi(F_0(x))(F_n(x) - F_0(x))^2 dF_0(x), \tag{24.1}$$

where F_n is the empirical cdf and Ψ is a suitable function. Recently, some goodness-of-fit tests have been proposed by studying the Wasserstein distance between F_n and F_0

$$\mathcal{W}_n^2 = \int_0^1 [F_n^{-1}(u) - F_0^{-1}(u)]^2 du$$

[del Barrio, Cuesta-Albertos, and Matrán (2000)]. Another related test is based on the maximum correlation between the sample and X

$$r_n^+ = (E_{H_n}(\mathcal{X}\cdot X) - \overline{x}\mu)/s\sigma,$$

where $\overline{x}, \mu$ are the means, s^2, σ^2 the variances and

$$E_{H_n}(\mathcal{X}\cdot X) = \int_0^1 F_n^{-1}(u)F_0^{-1}(u)du. \tag{24.2}$$

[Cuadras and Fortiana (1993)]. These statistics can be justified as follows. Let X, Y be two r.v.'s with cdf's F, G and finite means and variances $\mu_1, \mu_2, \sigma_1^2, \sigma_2^2$, respectively. The Wasserstein distance between F, G is

$$\mathcal{W}^2 = \inf_H E(X-Y)^2 = \int_0^1 [F^{-1}(u) - G^{-1}(u)]^2 du,$$

and the maximum correlation is

$$\rho^+ = \sup_H \rho(X,Y) = (\int_0^1 F^{-1}(u)G^{-1}(u)du - \mu_1\mu_2)/\sigma_1\sigma_2,$$

where H is a bivariate distribution with marginals F, G. The minimum for $E(X-Y)^2$ and the maximum for the correlation $\rho(X,Y)$ is reached for the distribution $H^+(x,y) = \min\{F(x), G(y)\}$, and then $F(X) = G(Y)$ (a.s.), i.e., we obtain $\mathcal{W}^2$ and ρ^+ when there is a functional relation between X, Y. Thus $\mathcal{W}_n^2$ and r_n^+ may be used as agreement measures between F_n and F_0. However, $\mathcal{W}^2$ and ρ^+ are related by

$$\mathcal{W}^2 = \sigma_1^2 + \sigma_2^2 - 2\sigma_1\sigma_2\rho^+ + (\mu_1 - \mu_2)^2,$$

and the goodness-of-fit tests based on $\mathcal{W}_n^2$ and r_n^+ are essentially equivalent.

In this paper we relate the orthogonal expansion of a r.v. in principal components to r_n^+ and use this methodology to distinguish two very similar distributions such as logistic and normal.

24.2 Orthogonal Expansions in Principal Components

Let X be a continuous random variable with range $[a,b]$ and cdf F. Let (λ_j, ψ_j) be the countable orthonormal set of eigenvalues and eigenfunctions of the integral operator $\mathcal{K}$ with kernel $K(s,t) = \min\{F(s), F(t)\} - F(s)F(t)$

$$(\mathcal{K}\varphi)(t) = \int_a^b K(s,t)\,\varphi(s)ds,$$

i.e., (λ_j, ψ_j) satisfies $\mathcal{K}\psi_j = \lambda_j\psi_j$. Cuadras and Fortiana (1995, 2000) proved that defining

$$h_j(x) = \int_a^x \psi_j(s)\,ds,$$

then the sequence $X_j = h_j(X), j \geq 1$, is a countable set of uncorrelated random variables, providing several orthogonal expansions for X, e.g.,

$$\begin{aligned} X &= x_0 + \textstyle\sum_{j=1}^{\infty} h_j(b)(X_j - h_j(x_0)), \\ X &= x_0 + \textstyle\sum_{j=1}^{\infty}(X_j^2 - h_j(b)h_j(x_0)), \\ X &= \mu_0 + \textstyle\sum_{j=1}^{\infty} h_j(b)(X_j - \mu_j), \end{aligned} \tag{24.3}$$

where

$$\mu_0 = E(X), \mu_j = E(X_j).$$

The sequence (X_j) can be obtained by Karhunen-Loève expansion on the stochastic process $\mathbf{X} = \{X_t,\ t \in [a,b]\}$, where X_t for each $t \in [a,b]$ is the indicator of $[X > t] = X^{-1}(t, \infty)$, or by continuous scaling on the distance function $\delta(x, x') = (\mid x - x' \mid)^{1/2}$. It can be proved that

$$\operatorname{tr}(\mathcal{K}) = \frac{1}{2}E[\mid X - X' \mid] = \sum_{j=1}^{\infty} \lambda_j,$$

where X, X' are iid and $V = \frac{1}{2}E[\mid X - X' \mid]$ is the so-called geometric variability of X with respect to distance δ. Moreover $\lambda_j = \mathrm{Var}(X_j)$ and each eigenvalue accounts for the geometric variability V, which is a dispersion measure for X. Thus

$$X_1 = h_1(X) = \int_a^b X_t\psi_1(t)dt,$$

with variance λ_1, is the first principal component of $\mathbf{X}$. Examples of principal components $h_j(X)$ and the corresponding variances λ_j are [Cuadras and Fortiana (1995) and Cuadras and Lahlou (2000)]:

1. $\sqrt{2}(1 - \cos j\pi X)/(j\pi)$, $\lambda_j = 1/(j\pi)^2$, if X is $(0,1)$ uniform and

$$\operatorname{tr}(\mathcal{K}) = \sum_{j=1}^{\infty} \frac{1}{(j\pi)^2} = \frac{1}{6}.$$

2. $[2J_0(\xi_j \exp(-X/2)) - 2J_0(\xi_j)]\,/\xi_j J_0(\xi_j)$, $\lambda_j = 4/\xi_j^2$, if X is exponential with mean 1, where $J_1(\xi_j) = 0$ and J_0, J_1 are the Bessel functions of the first kind of order 0 and 1, and

$$\operatorname{tr}(\mathcal{K}) = \sum_{j=1}^{\infty} \frac{4}{\xi_j^2} = \frac{1}{2}.$$

3. $\sqrt{(2j+1)/j(j+1)}L_j(F(X))$, $\lambda_j = 1/j(j+1)$, if X is standard logistic with $F(x) = (1+e^{-x})^{-1}$ and

$$\mathrm{tr}(\mathcal{K}) = \sum_{j=1}^{\infty} \frac{1}{j\,(j+1)} = 1,$$

where $L_1(x) = \sqrt{3}(2x-1), L_2(x) = \sqrt{5}(6x^2-6x+1), \ldots$ are the shifted Legendre polynomials on $[0,1]$.

Motivation for the use of these expansions is as follows:

1. If X is $(0,1)$ uniform and writing the second expansion in (24.3) as

$$X = \sum_{j=1}^{\infty} \frac{U_j^2}{j^2\pi^2},$$

where $U_j = \sqrt{2}(1 - \cos j\pi X), j \geq 1$, is a countable set of uncorrelated equally distributed random variables, we have a formal analogy with the expansion

$$W^2 = \sum_{j=1}^{\infty} \frac{Y_j^2}{j^2\pi^2},$$

where W^2 is the limit distribution of the Cramér-von Mises statistic W_n^2, and $Y_1, Y_2, \ldots$ are iid $N(0,1)$.

2. If X is logistic, we also obtain a formal analogy with the Anderson-Darling statistic A_n^2 [see Shorack and Wellner (1986, p. 225)]. Such an analogy may be due to the function Ψ in defining (24.1), which is $\Psi^{-1} = 1$, the uniform density, for the Cramér-von Mises statistics, and $\Psi^{-1} = t(1-t)$, giving the logistic density $f = F(1-F)$ for the Anderson-Darling statistics. This suggests that by setting Ψ^{-1} = a probability density function, we obtain a general form for $W_n^2(\Psi)$.

3. Let us define the "maximum correlations" $r_1, r_2, \ldots$ between the sample and the principal dimensions $X_1, X_2, \ldots$ of a r.v. X as the correlations obtained by considering the bivariate cdf $H_n(x,y) = \min\{F(x), F_n(y)\}$. Assuming $\mu_0 = 0$ and $\mathrm{var}(X) = 1$, writing the last expansion in (24.3) as

$$X = \sum_{j=1}^{\infty} \rho_j Y_j,$$

where each Y_j now has variance 1, it is clear that $\rho_j = \rho(X, X_j)$ is the correlation coefficient between X and the principal dimension X_j. The agreement between F_n and F may then be seen by comparing $r_1, r_2, \ldots$ to

$\rho_1, \rho_2, ...$ Taking the expectation $E(\mathcal{X}\cdot X)$, where $\mathcal{X}$ is the sample, with respect to H_n, we obtain the following relation between these correlations and the overall maximum correlation:

$$r_n^+ = \sum_{j=1}^{\infty} \rho_j r_j.$$

Thus r_n^+ has an expansion like A_n^2 and W_n^2, where $\rho_1, \rho_2, ...$ are constant coefficients and $r_1, r_2, ...$ are random but not independent.

24.3 Maximum Correlation for the Logistic Distribution

In this section we obtain the maximum correlations for the logistic distribution and use this goodness-of-fit approach to distinguish this distribution from the normal distribution.

A. If X follows the standard logistic distribution $F(x) = (1 + e^{-x})^{-1}$, then $F^{-1}(u) = -\log\left(u^{-1} - 1\right)$, and

$$E_{H_n}(\mathcal{X}\cdot X) = \int_0^1 F_n^{-1}(u) G^{-1}(u) du = \sum_{i=1}^{n} x_i \int_{\frac{i-1}{n}}^{\frac{i}{n}} \left(-\log\left(u^{-1} - 1\right)\right) du \quad (24.4)$$

and since $\int \left(\log\left(u^{-1} - 1\right)\right) du = -(1-u)\log(1-u) - u\log u$, after several simplifications (24.4) can be written as

$$E_{H_n}(\mathcal{X}\cdot X) = \frac{1}{n}\sum_{i=1}^{n} x_i \left[A_i - B_i\right],$$

where

$$\begin{aligned} A_i &= (n-i)\log(n-i) + i\log(i), \\ B_i &= (n-i+1)\log(n-i+1) + (i-1)\log(i-1), \end{aligned}$$

with $0\log 0 = 0$. Thus the maximal Hoeffding correlation between the sample and the logistic variable X, with mean 0 and variance $\pi^2/3$, is given by

$$r_n^+ = \frac{\sqrt{3}}{n}\sum_{i=1}^{n} x_i \left[A_i - B_i\right]/(s\pi).$$

B. Correlations between X and the principal dimensions.

From the expansion for the logistic variable, as $h_j(+\infty) = 0$ if j is even, we have

$$X/\sigma = \sum_{j=1}^{\infty} 2c_{2j-1}^2 L_{2j-1}(F(X))/\sigma = \sum_{j=1}^{\infty} \rho_j Y_j,$$

where $\sigma = \pi/\sqrt{3}, c_j = \sqrt{(2j+1)/j(j+1)}, Y_j = \sqrt{2j+1}L_j(X)$ and

$$\rho_j = \begin{cases} 0 & \text{if j is even,} \\ 2\sqrt{3(2j+1)}/(\pi j(j+1)) & \text{if j is odd.} \end{cases}$$

The first four correlations are

$$\rho_1 = 0.997, \rho_2 = 0, \rho_3 = 0.1103, \rho_4 = 0.$$

C. Correlations between the sample $\mathcal{X}$ and the principal dimensions.

A tedious computation shows that the covariances $m_k = E_{H_n}(\mathcal{X} \cdot F(X))$ are given by

$$m_k = \sum_{i=1}^{n} x_i(i^{k+1} - (i-1)^k).$$

Thus, the first four correlations between the sample $\mathcal{X}$ and the principal dimensions are:

$$\begin{aligned} r_1 &= \sqrt{12}\,(m_1 - \overline{x}/2)\,/s, \\ r_2 &= \sqrt{180}\,(m_2 - m_1 + \overline{x}/6)\,/s, \\ r_3 &= \sqrt{28}\,(10m_3 - 15m_2 + 6m_1 - \overline{x}/2)\,/s, \\ r_4 &= \sqrt{90}\,(7m_4 - 14m_3 + 9m_2 - 2m_1 + \overline{x}/10)\,/s, \end{aligned}$$

where $\overline{x}, s$ are the mean and standard deviation of the sample. The expansion of r_n^+ is

$$r_n^+ = \frac{2}{\pi} \sum_{j\,\mathrm{odd}} \frac{\sqrt{3(2j+1)}}{j(j+1)} r_j.$$

D. Plotting the principal dimensions.

Figures 24.1–24.4 give the principal dimensions $h_i(X), i = 1, \dots, 4$, where X is standard logistic. By using the following approximation [Abramowitz and Stegun (1972)]:

$$y(p) = \sqrt{-2\log p}$$
$$- \frac{2.515517 + 0.802853\sqrt{\log(1/p^2)} + 0.010328\log(1/p^2)}{1 + 1.432788\sqrt{\log(1/p^2)} + 0.189269\log(1/p^2) + .001308(\log(1/p^2))^{1.5}},$$

where p is $(0,1)$ uniform, we generate the $N(0,1)$ r.v. $Z = y(1-p)$. Thus $Y = (\pi/\sqrt{3})Z$ is a normal r.v. with the same mean and variance as X with standard logistic cdf F. Then we also plot $h_i(Y), i = 1, \ldots, 4$, where Y is normal, see Figures 24.1–24.4.

These plots have been obtained as follows. We write $h_i(X)$ in terms of $F(X) = p$, and $h_i(Y)$ in terms of $F(Y)$, i.e., $F(G^{-1}(p))$, where G is the cdf of Z. As it is described below, in Figures 24.5–24.8 we perform a similar plot, but replacing X, Y by a logistic and a normal sample, respectively.

24.4 Distinction Between Logistic and Normal

Suppose that X is standard logistic and Y is normal with the same mean 0 and variance $\pi^2/3$. Both distributions are very similar in shape, as noted by Johnson, Kotz, and Balakrishnan (1995, p. 119), although the value of the kurtosis β_2 is 4.2 for logistic, quite different from the normal $\beta_2 = 3$. But this has a small relative effect on the cdf. Consequently, an ommnibus test based on the Kolmogorov D_n or the maximum correlation r_n^+ may not be conclusive for deciding whether a given sample comes from a logistic or from a normal distribution. As a procedure, we propose to examine the correlations between the sample and the principal dimensions and represent the sample along the principal axes.

As an illustration, we generated 10 normal and 10 logistic samples of size $n = 50$, computed the maximum correlations and represented the curves. The results for the 20 samples are:

Logistic samples							Normal samples						
r_L^+	r_N^+	D_L	D_N	L/N	Li	Gr	r_L^+	r_N^+	D_L	D_N	L/N	Li	Gr
0.983	0.985	0.07	0.10	1.7	L	L	0.976	0.983	0.06	0.08	1.3	L	N
0.986	0.985	0.23	0.11	1.3	L	L	0.982	0.988	0.09	0.07	0.7	N	N
0.976	0.974	0.06	0.10	8.1	L	L	0.958	0.978	0.08	0.08	0.2	N	N
0.975	0.984	0.12	0.10	1.5	L	L	0.987	0.988	0.08	0.07	2.1	L	N
0.972	0.973	0.08	0.10	4.1	L	L	0.975	0.985	0.05	0.07	0.6	N	N
0.975	0.980	0.11	0.09	0.5	N	L	0.979	0.987	0.07	0.08	1.2	L	N
0.961	0.967	0.13	0.13	2.6	L	L	0.976	0.986	0.11	0.10	0.3	N	N
0.971	0.970	0.07	0.13	5.5	L	L	0.983	0.986	0.11	0.09	1.7	L	L
0.962	0.958	0.14	0.09	4.8	L	L	0.976	0.991	0.08	0.06	0.2	N	N
0.982	0.991	0.06	0.07	0.7	N	L	0.983	0.992	0.11	0.08	0.4	N	N

where:

1. The ten left and the ten right samples are generated as logistic and normal, respectively.

2. r_L^+, r_N^+ are the maximum correlations and D_L, D_N are the Kolmogorov statistics obtained assuming logistic and normal distribution [Stephens

(1979)]. L/N is the likelihood ratio. If $L/N > 1$ the sample may be logistic (L), otherwise normal (N).

3. Li indicates the decision (L=logistic, or N=normal) according to the likelihood ratio L/N.

4. Gr indicates which decision is taken by looking at the plot along the principal dimensions. The graphical decision is: L (logistic distribution), N (normal distribution or no logistic).

Thus, for the first logistic sample $r_L^+ = 0.983 < r_N^+ = 0.985$ and the fit to the normal appears better. However, the sample correlations with the first four dimensions are:

$$r_1 = 0.9507, r_2 = -0.0574, r_3 = 0.2606, r_4 = 0.0155,$$

and the agreement with the theoretical correlations is quite good. For the first normal sample with the same size we have $r_L^+ = 0.976 < r_N^+ = 0.983$.

Figures 24.5–24.8 give a plot of the first to the fourth dimensions. The continuous line contains the theoretical values, the $\triangle$-line is the first logistic sample and the ■-line corresponds to the first normal sample. The fit to the theoretical line is better for the logistic sample, whereas the normal sample trend is similar to its theoretical curve, see Figures 24.5–24.8. So both samples can be identified correctly.

Some conclusions are:

1. r^+ may not distinguish between logistic and normal when the sample is logistic, but provides a correct distinction when the sample is normal.

2. However, D_L, D_N and L/N are quantities which cannot discriminate between logistic and normal.

3. The decision from the graphical display of the principal dimensions is correct in all cases of logistic sample and in nine cases of normal sample.

4. The graphical decision could not be 100% conclusive due to the proximity between logistic and normal curves, but may help the user when the other tests are unable to make a clear distinction.

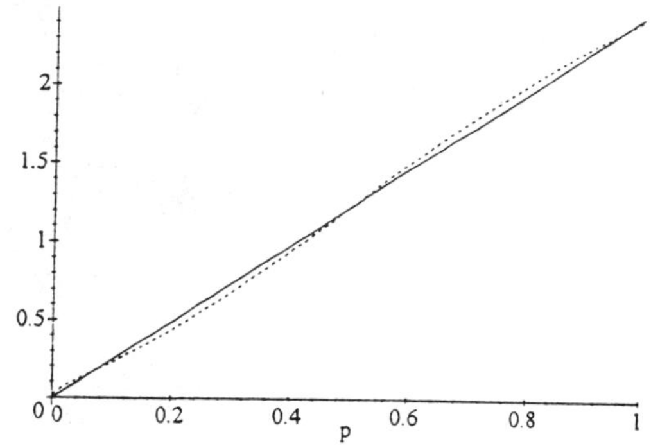

Figure 24.1: Plot of the theoretical principal dimensions $h_1(X)$, $h_1(Y)$, where X, Y follow the logistic (solid line) and normal (dashed line) distribution respectively

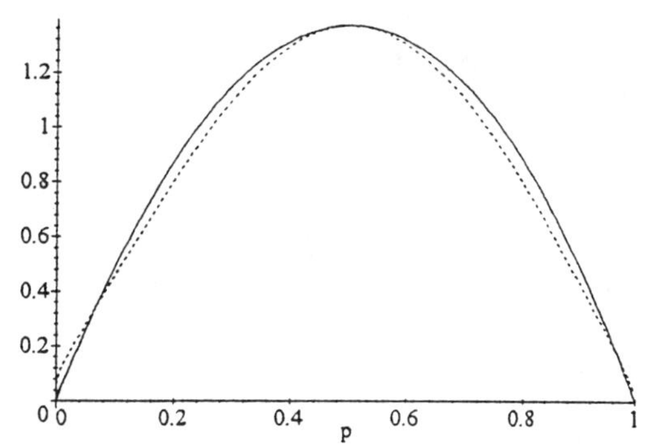

Figure 24.2: Plot of the theoretical principal dimensions $h_2(X)$, $h_2(Y)$, where X, Y follow the logistic (solid line) and normal (dashed line) distribution respectively

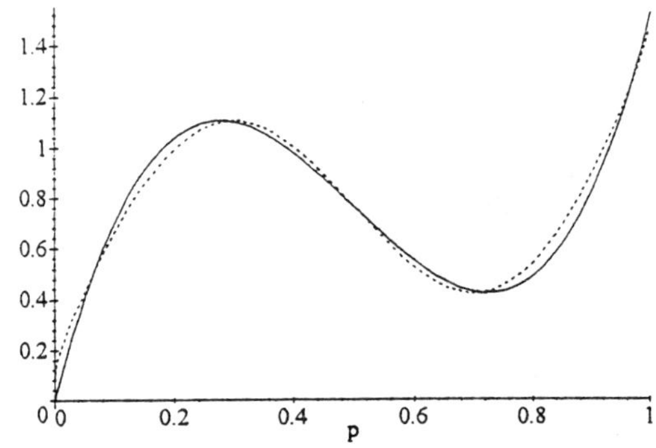

Figure 24.3: Plot of the theoretical principal dimensions $h_3(X)$, $h_3(Y)$, where X, Y follow the logistic (solid line) and normal (dashed line) distribution respectively

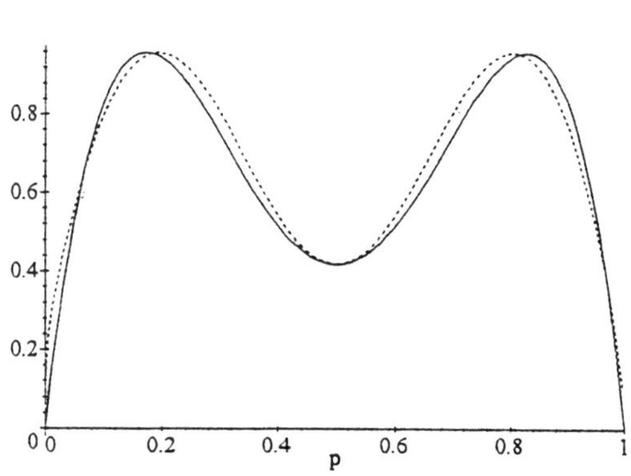

Figure 24.4: Plot of the theoretical principal dimensions $h_4(X)$, $h_4(Y)$, where X, Y follow the logistic (solid line) and normal (dashed line) distribution respectively

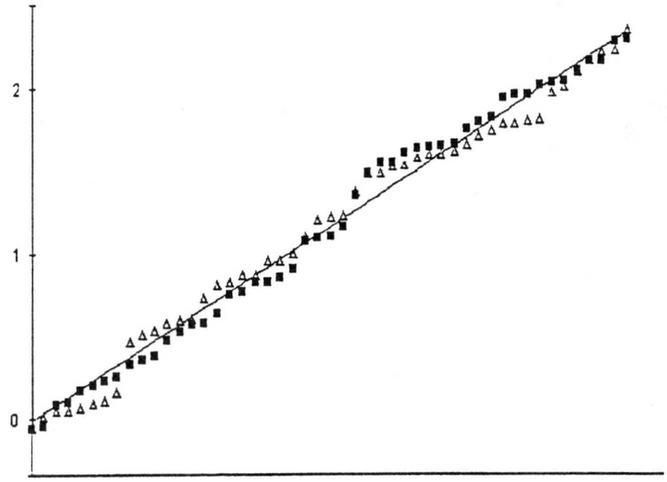

Figure 24.5: First logistic dimension: continuous line. Logistic sample: Δ-line. Normal sample: ■-line. Compare to Figure 24.1

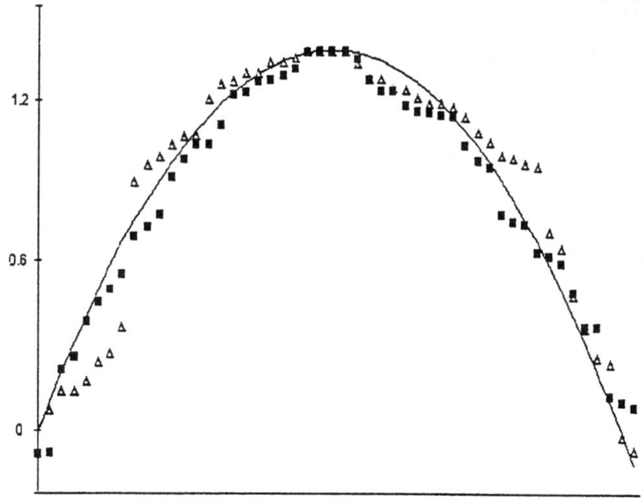

Figure 24.6: Second logistic dimension: continuous line. Logistic sample: Δ-line. Normal sample: ■-line. Compare to Figure 24.2

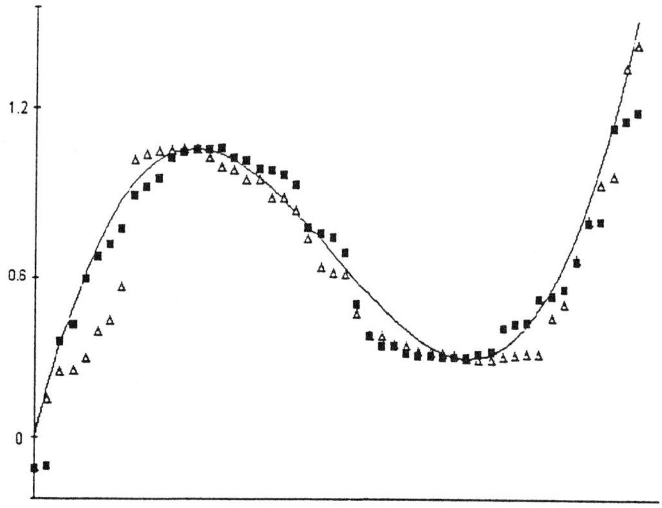

Figure 24.7: Third logistic dimension: continuous line. Logistic sample: Δ-line. Normal sample: ■-line. Compare to Figure 24.3

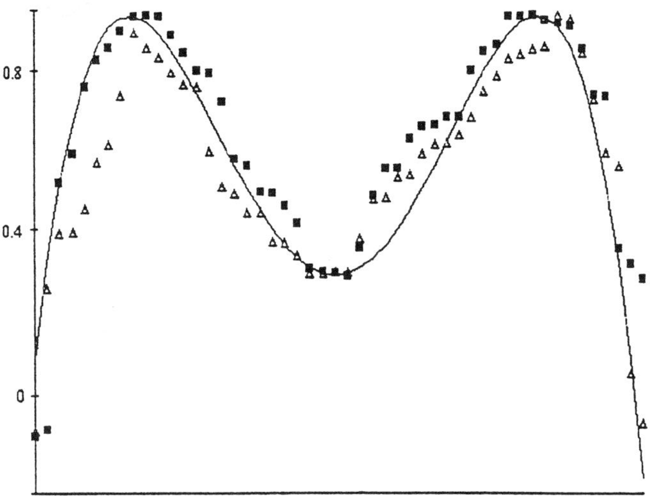

Figure 24.8: Fourth logistic dimension: continuous line. Logistic sample: Δ-line. Normal sample: ■-line. Compare to Figure 24.4

References

1. Abramowitz, M. and Stegun, I. A. (1972). *Handbook of Mathematical Functions*, New York: Dover Publications.

2. Cuadras, C. M. and Fortiana, J. (1993). Continuous metric scaling and prediction, In *Multivariate Analysis, Future Directions 2* (Eds., C. M. Cuadras and C. R. Rao), pp. 47–66, Amsterdam:Elsevier Science Publishers B. V. (North–Holland).

3. Cuadras, C. M. and Fortiana, J. (1995). A continuous metric scaling solution for a random variable, *Journal of Multivariate Analysis*, **52**, 1–14.

4. Cuadras, C. M. and Fortiana, J. (1996). Weighted continuous metric scaling, In *Multidimensional Statistical Analysis and Theory of Random Matrices* (Eds., A. K. Gupta and V. L. Girko), pp. 27–40, The Netherlands: VSP, Zeist.

5. Cuadras, C. M. and Fortiana, J. (2000). The Importance of Geometry in Multivariate Analysis and some Applications, In *Statistics for the 21st Century* (Eds., C. R. Rao and G. Szekely), pp. 93–108, New York: Marcel Dekker.

6. Cuadras, C. M. and Lahlou, Y. (2000). Some orthogonal expansions for the logistic distribution, *Communications in Statistics—Theory and Methods*, **29** (in press).

7. del Barrio, E., Cuesta-Albertos, J. A., and Matrán, C. (2000). Contributions of empirical and quantile processes to the asymptotic theory of goodness-of-fit tests, *TEST*, **9**, 1–96.

8. Johnson, N. L., Kotz, S., and Balakrishnan, N. (1995). *Continuous Univariate Distributions*, Second Edition, New York: John Wiley & Sons.

9. Shorack, G. R. and Wellner, J. A. (1986). *Empirical Processes with Applications to Statistics*, New York: John Wiley & Sons.

10. Stephens, M. A. (1979). Tests of fit for the logistic distribution based on the empirical distribution function, *Biometrika*, **66**, 591–595.

25

Functional Tests of Fit

Denis Bosq

Université Paris VI, Paris, France

Abstract: In this paper we define and study a class of functional tests of fit. These smooth tests are associated with projection density estimators. The Pearson's X^2-test belongs to this class since it is associated with the histogram.

Using a multidimensional Berry-Esseen type inequality we obtain the asymptotic behaviour in distribution of the test's statistics. As a consequence we obtain consistency with rates. Exponential rate holds in a special case.

We also study asymptotic efficiency under adjacent hypothesis and calculate the Bahadur slope.

Finally we give criteria for choosing a test in the class and present some numerical simulations.

Keywords and phrases: Test of fit, X^2-test , adjacent hypothesis, contigüity, projection density estimator, Berry-Esseen inequality, asymptotic efficiency, Bahadur efficiency

25.1 Introduction

In this paper we study a large class of goodness-of-fit tests in a general framework. This class contains the smooth test introduced by Neyman (1937) and the Pearson's X^2-test (1900).

The functional tests of fit (FTF) are based on the deviation of a density estimator with respect to the true density. They have been considered by several authors, we may quote Bickel and Rosenblatt (1973), Nadaraja (1976), Henze (1997), Hart (1997), Gregory (1980), among others.

We now describe the class of tests that is studied in this paper: Let $X_1, \dots,$ X_n be i.i.d. random variables with values in a measurable space $(E, \mathcal{B})$. We want to test

$$H_0 : X_1 \text{ has the distribution } \mu(X_1 \sim \mu).$$

For this purpose we consider an orthonormal system $(e_{jn},\ j \geq 0)$ in $L^2(\mu)$ with $e_{on} = 1$ and a sequence $(\lambda_{jn},\ j \geq 0)$ of real numbers such that $\lambda_{on} = 1$ and $\sum_{j=0}^{n} \lambda_{jn}^2 < \infty$. We always suppose that $L^2(\mu)$ is an infinite dimensional separable Hilbert space.

Now we construct the kernel

$$K_n(x, y) = \sum_{j=0}^{\infty} \lambda_{jn} e_{jn}(x) e_{jn}(y), \qquad x, y \in E \tag{25.1}$$

and the associated density estimator defined as

$$f_n(x) = \frac{1}{n} \sum_{i=1}^{n} K_n(x, X_i), \qquad x \in E. \tag{25.2}$$

The associated test is based on the statistic

$$T_n = \sqrt{n}(f_n - 1), \tag{25.3}$$

it rejects H_0 for large values of $\|T_n\|$ where $\|\cdot\|$ denotes the $L^2(\mu)$-norm. Note that the X^2-test is the FTF test based on the kernel

$$K_{n,o}(x, y) = \sum_{j=0}^{k_n} \mu(A_{jn})^{-1} 1_{A_{jn}}(x) 1_{A_{jn}}(y) \tag{25.4}$$

where (A_{jn}) is a partition of $(E, \mathcal{B})$.

$K_{n,o}$ has clearly the form (25.1) since it appears as the reproducing kernel of $\mathrm{sp}(1_{A_{on}}, \ldots, 1_{A_{k_n n}})$, a space that contains the constant 1. It is then possible to rewrite $K_{n,o}$ under the form (25.1).

The organization of this paper is as follows: Section 25.2 describes the behaviour of $\|T_n\|$ in distribution under H_0 and under the alternative hypothesis H_1. Section 25.3 deals with consistency and exponential rate of convergence. Section 25.4 provides limits in distribution under adjacent hypothesis while the choice of K_n is discussed in Section 25.5. Section 25.6 is devoted to local efficiency of the test and efficiency associated with the Bahadur slope. Indications concerning the proofs are given in Section 25.7. Finally numerical applications appear in Section 25.8.

25.2 Behaviour of $\|T_n\|$ in Distribution

First we give an inequality that provides approximation of T_n's distribution for fixed n. For this purpose we need some notation and assumptions. We set

$$\begin{aligned} \ell_n &= \frac{1}{\pi}\left(1 + \frac{1}{(\lambda_{3n})^3\sqrt{2}}\right) \quad (\lambda_{3n} \neq 0), \\ M_n &= \sup_{j \geq 0} \|e_{jn}\|_\infty < \infty, \end{aligned}$$

where $\|\cdot\|_\infty$ denotes essential supremum,

$$\Lambda_{r,n} = \sum_{j>r} \lambda_{jn}^2, \qquad r \geq 0,$$

$$U_n = \sum_{j=1}^{\infty} \lambda_{jn} \xi_j e_{jn}$$

where $(\xi_j,\ j \geq 1)$ is an i.i.d. auxiliary sequence of $\mathcal{N}(0,1)$ random variables. Now we have the following bound

Theorem 25.2.1 *If $X_1 \sim \mu$ then*

$$\sup_{j\geq 0} |P(\|T_n\|^2 \leq a) - P(\|U_n\|^2 \leq a)| \leq C_n, \quad n \geq 1,\ r \geq 4, \tag{25.5}$$

with

$$C_n = (6 + 2\ell_n + M^2)\Lambda_{r,n}^{2/3} + 3c_0 M_n \frac{r^4}{\sqrt{n}}\ , \tag{25.6}$$

where c_0 is a universal constant.

This explicit bound holds in particular for the X^2-test: if $r = k_n$ and $\mu(A_{jn}) = (k_n+1)^{-1}$, $0 \leq j \leq k_n$, one obtains the bound $3c_0 \frac{(k_n+1)^{9/2}}{\sqrt{n}}$.

Now from (25.5) it is easy to derive limits in distribution for $\|T_n\|^2$. We consider two cases:

(A) $\lambda_{jn} \xrightarrow[n\to\infty]{} \lambda_j$, $j \geq 1$ with $|\lambda_{jn}| \leq \lambda_j'$, $j \geq 1$ where $\sum_j \lambda_j'^2 < \infty$.

(B) There exists (r_n) such that

$$\frac{r_n^4}{\sqrt{n}} \to 0,\ \sum_{j>r_n} \lambda_{jn}^2 \to 0,\ \sum_{j=1}^{\infty} \lambda_{jn}^4 \to \infty.$$

Corollary 25.2.1 *If A holds, then*

$$\|T_n\|^2 \xrightarrow{\mathcal{D}} \|U\|^2 = \sum_{j=1}^{\infty} \lambda_j^2 \xi_j^2. \tag{25.7}$$

If B holds, then

$$\frac{\|T_n\|^2 - \sum_1^\infty \lambda_{jn}^2}{\sqrt{2}\left(\sum_1^\infty \lambda_{jn}^4\right)^{1/2}} \xrightarrow{\mathcal{D}} N \sim \mathcal{N}(0,1). \tag{25.8}$$

We now give an approximation if $X_1 \sim \nu \neq \mu$ for the special kernel.

$$K_{n,1}(x,y) = 1 + \sum_{j=1}^{k_n} e_j(x)e_j(y), \qquad x,y \in E \tag{25.9}$$

where $e_0 = 1$, $e_1, e_2, \ldots$ is a fixed orthonormal system in $L^2(\mu)$ such that $m_n = \max_{0 \leq j \leq k_n} \|e_j\|_\infty < \infty$.

Concerning ν we suppose that the covariance matrix

$$\Gamma_{\nu,n} \left(\int e_j e_\ell d\nu - \int d_j d\nu \int e_\ell d\nu \right)_{1 \leq j,\ \ell \leq k_n}$$

is regular with eigenvalues $\gamma_1 \geq \cdots \geq \gamma_{k_n} > 0$.

Then we have the following bound

Theorem 25.2.2 *If $X_1 \sim \nu$ with $\Gamma_{\nu,n}$ regular and if T_n is constructed with the kernel $K_{n,1}$ then*

$$\sup_{a \geq 0} |P(\|T_n\|^2 \leq a) - G_{k_n}(a)| \leq 8c_0 m_n \sup_{1 \leq \ell \leq k_n} \gamma_\ell^{-3/2} \frac{k_n^{1\frac{1}{2}}}{\sqrt{n}} \tag{25.10}$$

where G_{kn} is the distribution function associated with the characteristic function $[\det(I_{k_n} - 2it\Gamma\nu, n)]^{-1/2}$.

In particular if $k_n = k$, $\|T_n\|^2 \xrightarrow{\mathcal{D}} T \sim \eta_k$ where η_k is the distribution associated with G_k.

25.3 Consistency of FTF Tests and Rate of Convergence

In order to study consistency we specify H_1, in the general case, by setting

$$H_1 = \left\{ \nu : \exists\ j_n(\nu),\ n \geq 1 :\ \underline{\lim} \left| \int \lambda_{j_n(\nu),n} \int e_{j_n(\nu),n} d\nu \right| > 0 \right\}.$$

If $e_{j_n} = e_j$, $n \geq 1$ and $\underline{\lim} \lambda_{j_n} > 0$, $j \geq 1$, this is equivalent to

$$H_1 = \left\{ \nu : \exists\ j(\nu) : \int e_{j(\nu)} d\nu \neq 0 \right\}.$$

Thus, if (e_j) is a complete orthonormal system in $L^2(\mu)$, $H_0 + H_1$ contains all the probabilities of the form

$$\nu = f \cdot \mu + \gamma$$

where $\emptyset \neq f \in L^2(\mu)$ and γ is orthogonal to μ.

Now as usual we will say that the test $\|T_n\| > W_n$ is *consistent* if

- $\alpha_n = P\mu(\|T_n\| > w_n) \underset{n\to\infty}{\longrightarrow} 0$, and
- $\beta_n(\nu) = P_\nu(\|T_n\| > w_n) \underset{n\to\infty}{\longrightarrow} 1$, $\nu \in H_1$.

The following statement gives necessary and sufficient conditions for consistency.

Theorem 25.3.1 *If B holds the FTF test is consistent if and only if*

$$\frac{w_n}{n} \to 0, \tag{25.11}$$

and

$$\frac{w_n - \sum_1^\infty \lambda_{jn}^2}{\left(\sum_1^\infty \lambda_{jn}^4\right)^{1/2}} \to \infty. \tag{25.12}$$

Now it is possible to obtain bounds for α_n and β_n. In the next statement $K_n' = K_n - 1$ and $w_n = c\sqrt{n}$.

Theorem 25.3.2 *(1) We have*

$$\alpha_n \leq 2\exp\left(-n\frac{c^2}{2a_n + (3/2)b_n c}\right) \tag{25.13}$$

with

$$a_n = \int \|K_n'(x,\cdot)\|d\mu(x),$$

and

$$b_n = \|K_n(X_1,\cdot)\|_\infty.$$

(2) If $\gamma_n := \|\int K_n(x,\cdot)d\nu(x)\| - c > 0$, *then*

$$\beta_n(\nu) \geq 1 - 2\exp\left(-n\frac{\gamma_n^2}{2a_n' + (2/3)b_n'\gamma_n}\right) \tag{25.14}$$

where

$$a_n' = E_\nu\|K_n'(X_1,\cdot) - E_\nu K_n'(X_1,\cdot)\|^2,$$

and

$$b_n' = \|K_n'(X_1,\cdot) - E_\nu K_n'(X_1,\cdot)\|_\infty.$$

If $K_n = \sum_{j=0}^k e_j \otimes e_j$ where k is fixed and $\|e_j\|_\infty < \infty$, $j = 1,\ldots,k$, Theorem 25.3.2 shows that the level and the power of the test tend to zero at an *exponential rate*. More generally the same property holds if

$$K_n = \sum_{j=0}^\infty \lambda_j e_j \otimes e_j$$

is fixed and bounded.

25.4 Adjacent Hypothesis

In this section we study the local behaviour of FTF tests. We first consider the simple case where the kernel is

$$K(x,y) = 1 + \sum_{j=1}^{k} e_j(x)e_j(y); \qquad x, y \in E. \tag{25.15}$$

Thus K is the reproducing kernel of a finite dimensional subspace of $L^2(\mu)$, say $\mathcal{E}$, which contains the constants and such that $\|\varphi\|_\infty < \infty$ for each φ in $\mathcal{E}$.

The local situation is described by observation of independent r.v.'s $X_{1n}, \dots, X_{nn}$ with common distribution ν_n $(\nu \geq 1)$.

We set $K' = K - 1$ and $\mathcal{E}' = \text{sp}\{e_1, \dots, e_k\}$ and consider the following conditions:

(C1) $\sqrt{n} \int K'(x, \cdot) d\nu_n(x) \underset{n\to\infty}{\longrightarrow} g$, with $\|g\|^2 = \lambda^2 \neq 0$,
where convergence takes place in $\mathcal{E}'$.

(C2) $\Gamma_{n_\nu}(j,\ell) = \int e_j e_\ell d\nu_n - \int e_j d\nu_n \int e_\ell d\nu_n \underset{n\to\infty}{\longrightarrow} \delta_{j,\ell}$, $1 \leq j,\ \ell \leq k$,
where $\delta_{j,\ell} = 1$ if $j = \ell$; 0 if $j \neq \ell$.

If (C1) and (C2) are satisfied we will say that the sequence (ν_n) is *adjacent* to μ. Note that if

$$\nu_n = (1 + h_n) \cdot \nu \tag{25.16}$$

where $h_n \in \mathcal{E}'$, $n \geq 1$ then (C1) and (C2) may be replaced by

(C) $\sqrt{n} \left\| 1 - \frac{d\nu_n}{d\mu} \right\|_{L^2(\mu)} \to \lambda \neq 0$

a condition that implies *contiguity* of (ν_n) with respect to μ [see Roussas (1978) or Van der Vaart (1998)].

Under adjacency we have the following asymptotic result.

Theorem 25.4.1 *If (C1) and (C2) [resp. (C)] hold then*

$$\|T_n\|^2 \xrightarrow{\mathcal{D}} Q_k^{(\lambda)} \sim X^2(k, \lambda) \tag{25.17}$$

where $X^2(k,\lambda)$ is a X^2 with k degrees of freedom and parameter λ.

Moreover, if (C) holds with $h_n = \frac{g}{\sqrt{n}}$ then

$$\sup_{a \geq 0} |P(\|T_n\|^2 \leq a) - P(Q_k^{(\lambda)} \leq a)| \leq \frac{c}{\sqrt{n}} \tag{25.18}$$

where $c > 0$ is constant.

We now turn to the more general case where

$$K_n(x,y) = 1 + \sum_{j=1}^{k_n} e_{j_n}(x)e_{j_n}(y); \quad x, y \in E, \ n \geq 1 \text{ with } (k_n) \to \infty. \quad (25.19)$$

Here we replace (C1) , (C2) by

(C_1') $\frac{n}{\sqrt{k_n}} \sum_{j=1}^{k_n} \left(\int e_{jn} d\nu_n\right)^2 \to \ell \neq 0.$

(C_2') $\overline{\lim} \sum_{j,\ell=1}^{k_n} \left(\int e_{jn} e_{\ell n} d\nu_n - \int e_{jn} d\nu_n \int e_{\ell' n} d\nu_n - \delta_{j\ell}\right)^2 < 1.$

We then have

Theorem 25.4.2 *If* $\sup_{j,n} \|e_{jn}\|_\infty < \infty$, $k_n \to \infty$, $\frac{k_n^{11}}{\sqrt{n}} \to 0$ *then*

$$\frac{\|T_n\|^2 - k_n}{\sqrt{2k_n}} \xrightarrow{\mathcal{P}} N \sim \mathcal{N}\left(\frac{\ell}{\sqrt{2}}, 1\right). \quad (25.20)$$

Theorems 25.4.1 and 25.4.2 allow us to obtain asymptotic power of the FTF test $\|T_n\| > w_n$. For the kernel defined by (25.15) one may use (25.6) in Corollary 25.2.1 to obtain a test of asymptotic level α by setting $w_n^2 = X_{k,\alpha}^2$ where $P(Q_h(0) > X_{k,\alpha}^2) = \alpha$. The asymptotic power β is given by Theorem 25.4.1, we have

$$\beta = P(Q_k(\lambda) > X_{k,\alpha}^2)$$

thus

$$\beta = P(Q_k(0) > X_{k,\alpha}^2 - \lambda^2). \quad (25.21)$$

Now for the kernel defined by (25.19) one may use (25.8) in Corollary 25.2.1. Set $w_n^2 = k_n + \sqrt{2k_n} N_\alpha$ then $\|T_n\| > w_n$ has asymptotic level α and Theorem 25.4.2 provides the asymptotic power

$$\beta = P\left(N > N_\alpha - \frac{\ell}{\sqrt{2}}\right). \quad (25.22)$$

25.5 Choosing a Kernel

One of the major interests of FTF tests is possibility of choosing a reasonable kernel that takes into account the alternative hypothesis H_1. We discuss such a choice in this section.

(a) Testing μ against a mixture

Consider the case where

$$H_1 = \left\{ \nu : \nu = \left(\sum_{i=0}^{k} a_i f_i \right) \cdot \mu;\ a_i > 0,\ \sum_{i=0}^{k} a_i = 1 \right\},$$

where $1 = f_0, f_1, \ldots, f_k$ are given densities.

In that case a natural kernel should be the reproducing kernel of $\mathrm{sp}\{1, f_1, \ldots, f_k\}$ since Theorem 25.3.2 shows that the associated test converges at an exponential rate.

A typical example is the Gaussian case where $\mu = \mathcal{N}(m_0, \sigma_0^2)$ and $\nu = \sum_{i=0}^{k} a_i \mathcal{N}(m_i, \sigma_i^2)$. Note that it is eventually possible to replace (m_i, σ_i^2), $0 \leq i \leq k$ by suitable estimators without changing asymptotic properties of the test.

(b) Choosing a kernel under adjacent hypothesis

Suppose that $\nu_n = \left(1 + \frac{g_n}{\sqrt{n}}\right) \cdot \mu$ where $g_n \to g \neq 0$ weakly in $L^2(\mu)$, with $g \perp 1$ and take a kernel of the form

$$K(x,y) = 1 + \frac{g(x)}{\|g\|} \cdot \frac{g(y)}{\|g\|} + \sum_{j=2}^{k} e_j(x) e_j(y) \tag{25.23}$$

where $1, \frac{g}{\|g\|}, e_2, \ldots, e_k$ is an orthonormal system.

Then the associated test has a *maximal asymptotic power among the class of FTF tests*. One may extend this property by considering various orthogonal adjacent hypothesis, namely

$$\nu_{n-i} = \left(1 + \frac{g_{n,i}}{\sqrt{n}}\right) \cdot \mu,$$

$g_{n,i} \to g_i \neq 0$, weakly, $1 \leq i \leq k$; with $1, g_1, \ldots, g_k$ orthogonal. This situation leads to the choice

$$K(x,y) = 1 + \sum_{j=1}^{k} \frac{g_i(x)}{\|g_i\|} \frac{g_i(y)}{\|g_i\|} \ . \tag{25.24}$$

Concerning the λ_j's they can be useful to measure a weight for each "part" of H_1.

25.6 Local Efficiency of FTF Tests

In order to determine local efficiency of FTF tests we consider adjacent hypothesis (see Section 25.5) and use the optimal kernels defined in the previous subsection. In the second subsection we study efficiency associated with Bahadur slope.

(a) Local efficiency

Since the yardstick is the Neyman–Pearson (NP) test we first study the asymptotic behaviour of the NP test under adjacent hypothesis. Assumptions in the next lemma are a little more general than (C).

Lemma 25.6.1 *If* $\nu_n = \left(1 + \frac{c_n g_n}{\sqrt{n}}\right)\mu$ *where* $\|g_n\|^2 \to \lambda^2$ *and* (c_n) *is a sequence of real numbers such that*

$$\frac{\max(1, c_n^2)\max(1, \|g_n\|_\infty^3)}{\sqrt{n}} \to 0,$$

then

(1) If $X_{1n} \sim \mu$, $n \geq 1$

$$c_n^{-1}\sum^{n} -j = 1 \log\left(1 + \frac{c_n}{\sqrt{n}} g_n(X_{jn})\right) + \frac{c_n}{2}\|g_n\|^2 \xrightarrow{\mathcal{P}} N \sim \mathcal{N}(0, \gamma^2). \tag{25.25}$$

(2) If $X_{1n} \sim \nu_n$, $n \geq 1$

$$c_n^{-1}\sum^{n} -j = 1 \log\left(1 + \frac{c_n}{\sqrt{n}} g_n(X_{jn})\right) - \frac{c_n}{2}\|g_n\|^2 \xrightarrow{\mathcal{P}} N \sim \mathcal{N}(0, \gamma^2). \tag{25.26}$$

This lemma is, in some sense, more general than those generally given since it does not suppose that $(c_n\|g_n\|^2)$ has a limit.

Here we only use Lemma 25.6.1 in the particular case where $c_n \to 1$. It is then easy to see that, if the N.P. test has asymptotic level $\alpha \in]0, 1[$, its asymptotic power is given by

$$\beta_0 = P(N > N_\alpha - \lambda). \tag{25.27}$$

Now if (C) holds one may use (25.21) and (25.27) to obtain asymptotic efficiency of the optimal FTF test. We have

$$E_k = \frac{P(Q_k(0) > X_{k,\alpha}^2 - \lambda^2)}{P(N > N_\alpha - \lambda)}\ . \tag{25.28}$$

If $k = 1$ this efficiency may be written under the form

$$E_1 = \frac{P(|N - \lambda| > N_{\alpha/2})}{P(N > N_\alpha - \lambda)}\ , \tag{25.29}$$

which shows that the FTF test has a good asymptotic behaviour.

(b) Bahadur efficiency

In order to calculate the *Bahadur slope* of ETF tests we consider a kernel of the form $K = 1 + K'$ where

$$K' = \sum_{j=1}^{\infty} \lambda_j e_j \otimes e_j \tag{25.30}$$

with $\sup_j \|e_j\|_\infty < \infty$, $|\lambda_j| \downarrow 0$ and $\sum_j \lambda_j^2 < \infty$. On the other hand we set

$$\Delta(\nu) = \|K(x,\cdot)d\nu(x)\|, \qquad \nu \in H_1.$$

Then we have:

Theorem 25.6.1 *The Bahadur slope of the test defined by* $\|T_n\| \geq w_n$ *is*

$$c_T^{(K')}(\nu) = \frac{\Delta^2(\nu)}{\lambda_1^2}(1 + o(1)) \text{ as } \Delta(\nu) \to 0. \tag{25.31}$$

Note that a similar result appears in Gregory (1980). Now if $\nu = (1+h)\mu$ with $|h| < 1$, $\int h d\mu = 0$, $\int h^2 d\mu > 0$ are obtained

$$c_T^{(K')}(\nu) = \sum_{j\geq 1} \left(\frac{\lambda_j}{\lambda_1}\right)^2 \left(\int h e_j d\mu\right)^2.$$

Hence the best kernel of the form (25.30) is

$$K'_h = \frac{h}{\|h\|} \otimes \frac{h}{\|h\|} \tag{25.32}$$

and

$$c_T^{(K'h)} = \|h\|^2. \tag{25.33}$$

Now the Bahadur slope of the N.P. test is given by

$$C_T^{(\text{N.P.})} = 2\int (1+h)\log(1+h)d\mu$$

[see Van der Vaart (1998, p. 308)]. Thus the Bahadur relative efficiency of the FTE test based on K'_h is

$$E_B(h) = \frac{\|h\|^2}{2\int(1+h)\log(1+h)d\mu}. \tag{25.34}$$

Note that

$$\lim_{\|h\|_\infty \to 0} E_B(h) = 1. \tag{25.35}$$

25.7 Indications Concerning the Proofs

Proof of Theorem 25.2.1 uses the decomposition

$$T_n = \sum_{j=1}^{r_n} \lambda_{jn}\hat{a}_{jn}e_j + \sum_{j>r_n} \lambda_{jn}\hat{a}_{jn}e_j := Z_{n_1} + Z_{n_2}$$

where

$$\hat{a}_{jn} = \frac{1}{2}\sum_{i=1}^{n} e_j(X_j), \qquad j \geq 1.$$

Concerning Z_{n_1} one uses Sazonov (1968) inequality that gives the bound $3c_0 M_n \frac{r^4}{\sqrt{n}}$. The bound $(6 + 2\ell_n + M^2)\Lambda_{r,n}^{2/3}$ is obtained by using Tchebychev inequality at the order 4. Details which are rather intricate, appear in Bosq (1980).

Corollary 25.2.1 is an easy consequence of Theorem 25.2.1. Proof of Theorem 25.2.2 is similar to that of Theorem 25.2.1. Theorem 25.3.1 is a consequence of Theorems 25.2.1 and 25.2.2. Theorem 25.3.2 is easily established by using exponential type inequalities in Hilbert space [see Pinelis-Sakhanenko (1985) and Bosq (2000)].

Proofs of Theorems 25.4.1, 25.4.2 and Lemma 25.6.1 are given in Bosq (1983).

PROOF OF THEOREM 25.6.1. (sketch) The FTF test is based on the statistics $U_n = \left\| \frac{1}{n}\sum_{i=1}^{n} K'(X_i, \cdot) \right\|$.

The strong law of large numbers in a Hilbert space entails

$$U_n \to \left\| \int K'(x, \cdot) d\nu(x) \right\|, \qquad \nu \text{ a.s.}; \ \nu \in H_1. \tag{25.36}$$

On the other hand the Sethuraman theorem [see Nikitin (1995, p. 23)] implies

$$\frac{1}{n}\log P_\mu(U_n > \varepsilon) \underset{n\to\infty}{\longrightarrow} \ell(\varepsilon), \qquad \varepsilon > 0 \tag{25.37}$$

where

$$\ell(\varepsilon) = -\frac{\varepsilon^2}{2\sigma^2}(1 + o(1)) \text{ as } \varepsilon \to 0 \tag{25.38}$$

with

$$\sigma^2 = \sup_{\|x^*\|=1} \operatorname{Var} x^*[K'(X_1, \cdot)].$$

It is easy to see that

$$\sigma^2 = lambda_1^2. \tag{25.39}$$

Now we are in a position to use the Bahadur Theorem [see Nikitin (1995, pp. 6–7)]. We obtain

$$C_t(\nu) = \frac{1}{\lambda_1^2} \left\| \int K'(x, \cdot) d\nu(x) \right\|^2 (1 + o(1)) \quad \text{hence (25.31).}$$

25.8 Simulations

The simulations presented below have been performed by Izraelewitch *et al.* (1988). The kernel has the form (25.15) where $e_1, \ldots, e_k$ are Legendre polynomials over $[-1, +1]$.

Here m is the uniform distribution on $[-1, +1]$. The goal of these simulations is comparison between the power of X^2 test and the power of the FTF test based on the Legendre polynomials under various alternatives.

For each alternative the problem is transported over $[-1, +1]$ by putting

$$Y_i = 2F(X_i) - 1; \qquad 1 \leq i \leq n$$

where F is the distribution function of ν.

The power of each test appears in ordinates while the number of cells in the X^2 test and the degree of the Legendre polynomial used appears in abscissas.

In general the smooth test is better than the X^2-test especially if the variance is greater under the alternative than under H_0.

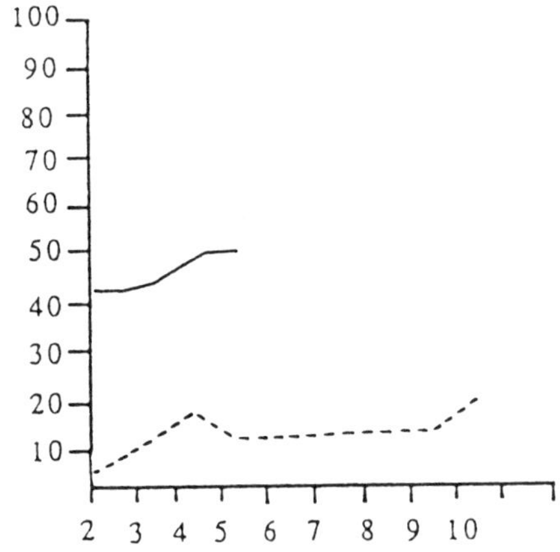

Figure 25.1: $H : N(0,1)$; $H_a : N(0, 25/16)$; $n = 50$

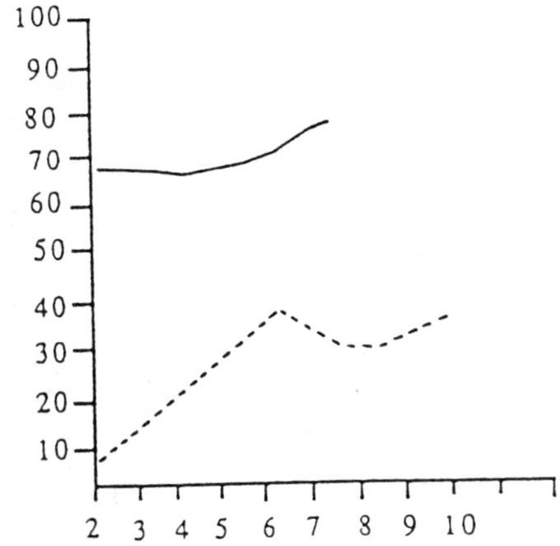

Figure 25.2: $H : N(0,1)$; $H_a : N(0, 25/16)$; $n = 100$

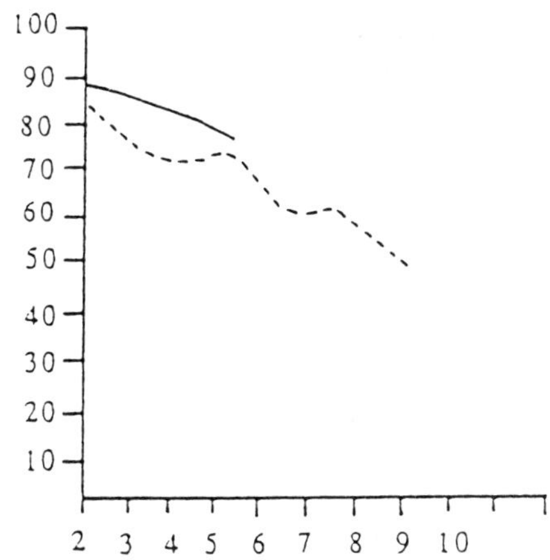

Figure 25.3: $H : N(0,1)$; $H_a : N(0,5,1)$; $n = 50$

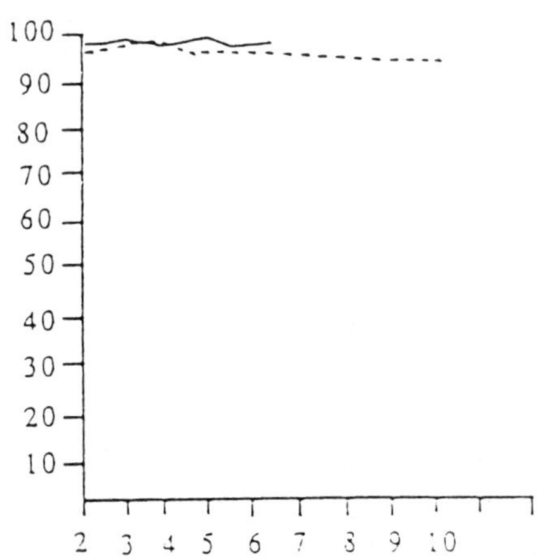

Figure 25.4: $H : N(0,1)$; $H_a : N(0,5,1)$; $n = 100$

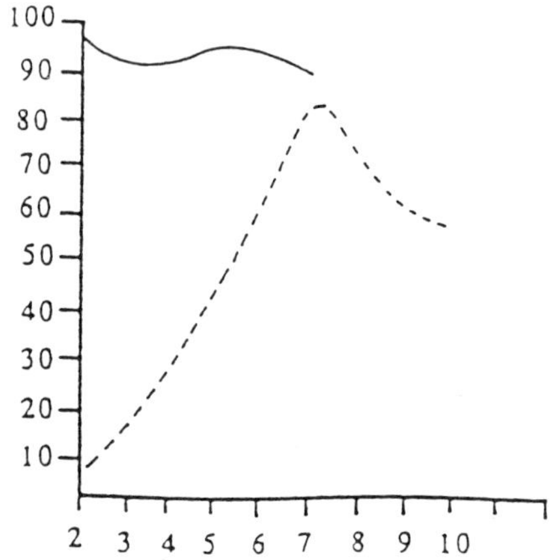

Figure 25.5: H : CAUCHY $(0,1)$; H_a : STUDENT (25); $n = 50$

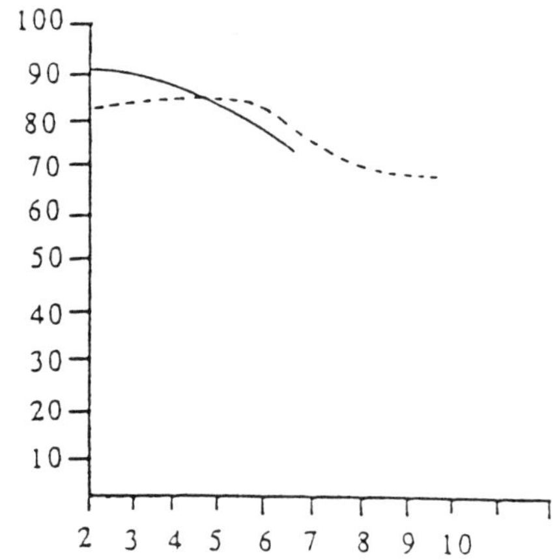

Figure 25.6: $H : U(0,1)$; H_a : BETA$(3/2, 1)$

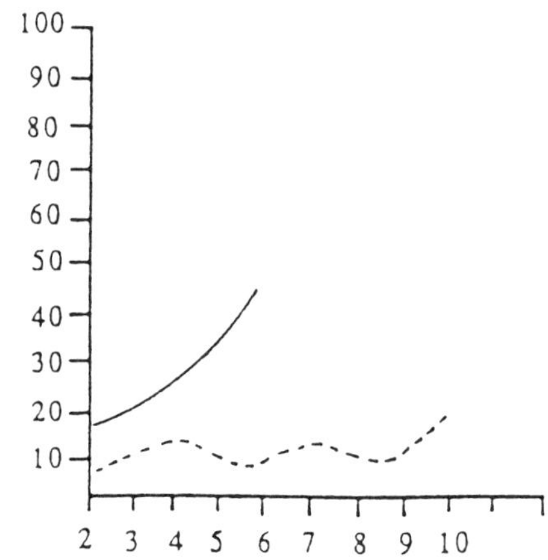

Figure 25.7: $H : N(0,1)$; $H_a : 0,9N(0,1) + 0,1N(0,25)$; $n = 50$

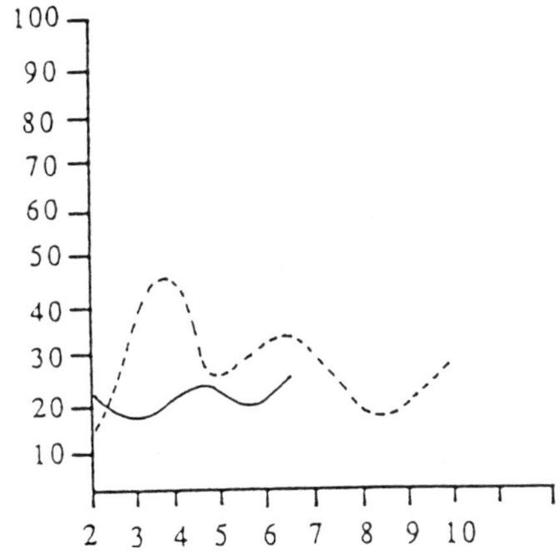

Figure 25.8: $H : N(0,1)$; $H_a : 0,8N(0,1) + 0,2N(0,0,04)$

References

1. Bickel, P. J. and Rosenblatt, M. (1973). On some global measures of the deviations of density function estimates, *Annals of Statistics*, **1**, 1071–1095.

2. Bosq, D. (1980). Sur une classe de tests qui contient le test du X^2, *Publ. ISUP* fasc. 1-2 p. 1–16.

3. Bosq, D. (1983). Lois limites et efficacité asymptotique des tests Hilbertiens, *Stat. et Anal. des données*, **8**, 1–40.

4. Bosq, D. (1989). Tests du X^2 généralisés. Comparaison avec le test du X^2 classique, *Rev. Statist. Appliquée*, XXXVII (1), 43–52.

5. Bosq, D. (2000). *Linear Processes in Function Spaces - Lecture Notes in Statistics*, New York: Springer-Verlag.

6. Gregory, G. G. (1980). On efficiency and optimality of quadratic tests, *Annals of Statistics*, **8**, 116–131.

7. Hart, J. D. (1997). *Nonparametric Smoothing and its Applications in Lack-of-Fit Testing*, New York: Springer-Verlag.

8. Henze, N. (1997). Do component of smooth tests of fit have diagnostic properties? *Metrika*, **45**, 121–130.

9. Izraelewitch, E., Lafitte, I., Lavault, Z., and Roubert, B. (1988). Le test du X^2 et le test de Legendre, *Projet ISUP - Paris.*

10. Nadaraja, E. A. (1976). A quadratic measure of the deviation of a density estimator, *Theory of Probability and its Applications*, **21**, 843–850.

11. Neyman, J. (1937). Smooth test for goodness of fit, *Skand. Aktuar*, **20**, 119–128.

12. Nikitin, Y. (1995). *Asymptotic Efficiency of Nonparametric Tests*, Cambridge University Press.

13. Pearson, K. (1900). On the criterion that a given system of deviations from the probable in the case of correlated system of variables is such that it can be reasonably supposed to have arisen from random sampling, *Philosophical Magazine*, **50**, 157–175.

14. Pinelis, I. F. and Sakhanenko, I. (1985). Remarks on inequalities for large deviation probabilities, *Theory of Probability and Analysis, I*, 157–214.

15. Roussas, G. (1978). *Contiguity of Probability Measures*, Cambridge University Press.

16. Sazanov, V. V. (1968). On the multidimensional central limit theorem, *Sankhyā, Series A*, **30**, 191–204.

17. Sazanov, V. V. (1968). On ω^2 criterion, *Sankhyā, Series A*, **30**, 205–210.

18. Tenreiro, C. (2000). On a class of integral statistics for testing goodness of fit, *Preprint*, University Coimbra, Portugal.

19. Van der Vaart, A. W. (1998). *Asymptotic Statistics*, Cambridge University Press.

26

Quasi Most Powerful Invariant Tests of Goodness-of-Fit

Gilles R. Ducharme and Benoît Frichot

Université Montpellier II, Montpellier, France

Abstract: In this chapter, we consider the problem of testing the goodness-of-fit of either one of two location-scale families of density when these parameters are unknown. We derive an $O(n^{-1})$ approximation to the densities of the maximal invariant on which the most powerful invariant test is based. The resulting test, which we call quasi most powerful invariant, can be applied to many situations. The power of the new procedure is studied for some particular cases.

Keywords and phrases: Most powerful invariant test, Laplace approximation

26.1 Introduction

Let $x_1, ..., x_n$ be i.i.d. observations from a real random variable X with density $f(.)$ on $\mathbb{R}$. We consider the problem of testing

$$\mathcal{H}_0 : f(.) = \frac{1}{\sigma} f_0 \left(\frac{. - \mu}{\sigma} \right) \quad \text{against} \quad \mathcal{H}_1 : f(.) = \frac{1}{\sigma} f_1 \left(\frac{. - \mu}{\sigma} \right) \tag{26.1}$$

where f_0 and f_1 are two densities with known form and $(\mu, \sigma) \in \mathbb{R} \times \mathbb{R}_*^+$ are the location-scale parameters. If (μ, σ) are given under both hypotheses, the Neyman-Pearson Lemma gives the most powerful test for (26.1). Otherwise, it is natural to restrict attention to the class of tests invariant to the group of affine-linear transformations

$$\mathcal{G} = \{x_1, ..., x_n \to cx_1 + b, ..., cx_n + b; (c, b) \in \mathbb{R}_*^+ \times \mathbb{R}\}.$$

Lehmann (1959) gives in this context the most powerful invariant (MPI) test for (26.1), who rejects $\mathcal{H}_0$ for large values of $\log(q_1/q_0)$ where, for $j = 0, 1$ and

ℓ_j the log-likelihood of the data

$$q_j(x_1,...,x_n) = \int_0^\infty \int_{-\infty}^\infty \prod_{i=1}^n f_j(cx_i+b)\; c^{n-2}\; db\; dc \qquad (26.2)$$
$$= \int_0^\infty \int_{-\infty}^\infty e^{\ell_j(\mu,\sigma)} \frac{1}{\sigma}\; d\mu\; d\sigma \qquad (26.3)$$

which is proportional to the density of the maximal invariant linked to $\mathcal{G}$. The quantity (26.2) has been obtained explicitly for few distributions only : normal – Hájek and Šidák (1967), uniform and exponential – Uthoff (1970), double exponential – Uthoff (1973), Cauchy – Franck (1981). Otherwise, it is complicated to calculate (26.3) and this impairs the use of this approach. In this work, we propose an approximation of (26.2) which allows us to circumvent this difficulty.

26.2 Laplace Approximation

The approach developed here is based on the Laplace approximation for integrals, which can be found in for example Barndorff-Nielsen and Cox (1989). Let $D = \mathbb{R} \times \mathbb{R}_*^+$, $(\hat{\mu}_j, \hat{\sigma}_j) = \arg\max_{(\mu,\sigma)\in D} \ell_j(\mu,\sigma)$ the maximum likelihood estimators (MLE) and $\ddot{\ell}_j(\mu,\sigma)$, the Hessian matrix of ℓ_j, supposed non-singular. It's existence is assured by regularity conditions on f_j.

Proposition 26.2.1 *If the derivatives of order 2 of ℓ_j are continuous in the neighborhood of $(\hat{\mu},\hat{\sigma})$ and if $(\hat{\mu},\hat{\sigma})$ are in the interior of D, then*

$$q_j = \frac{e^{\ell_j(\hat{\mu}_j,\hat{\sigma}_j)}(2\pi)}{n\hat{\sigma}_j|\det\ \ddot{\ell}_j(\hat{\mu}_j,\hat{\sigma}_j)|^{1/2}}(1+O(n^{-1})) = \hat{q}_j(1+O(n^{-1})).$$

If $\hat{\mu}$ is located at a border of D, we can adapt the approximation. For example, if $D = (-\infty, \hat{\mu}_j] \times \mathbb{R}_*^+$ and if $\partial\ell_j(\mu,\sigma)/\partial\mu < 0$ on D, we find the following corollary :

Corollary 26.2.1 *If $\partial\ell_j/\partial\mu$ and $\partial^2\ell_j/\partial\sigma^2$ are continuous in the neighborhood of $(\hat{\mu}_j,\hat{\sigma}_j)$ in D, (26.2) can be written*

$$q_j = \frac{e^{\ell_j(\hat{\mu}_j,\hat{\sigma}_j)}\sqrt{2\pi}}{n^{3/2}\hat{\sigma}_j\ (-\partial\ell_j(\hat{\mu}_j,\hat{\sigma}_j)/\partial\mu)\ |\partial^2\ell_j(\hat{\mu}_j,\hat{\sigma}_j)/\partial\sigma^2|^{1/2}}(1+O(n^{-1})) \qquad (26.4)$$
$$= \hat{q}_j(1+O(n^{-1})). \qquad (26.5)$$

26.3 Quasi Most Powerful Invariant Test

The preceding results suggest to approach the MPI test by the test that rejects $\mathcal{H}_0$ for large values of $\log(\hat{q}_1/\hat{q}_0)$. We call this test quasi most powerful invariant (QMPI) because the statistic of the test is the approximation to the order n^{-1} of the statistic of the MPI test and is invariant to the action of $\mathcal{G}$.

• **Normal distribution:** the exact value of (26.2) is obtained by Hájek and Šidák (1967). They give $q_N = \dfrac{\Gamma(\frac{n-1}{2})}{2n^{n/2}(\pi\hat{\sigma}^2)^{(n-1)/2}}$ with $\hat{\sigma}^2 = \dfrac{1}{n}\sum_{i=1}^{n}(x_i - \hat{\mu})^2$ and $\hat{\mu} = \bar{x}$. Proposition (26.2.1) gives $\hat{q}_N = \dfrac{\exp(-\frac{n}{2})}{\sqrt{2}n(2\pi)^{(n-2)/2}\hat{\sigma}^{(n-1)}}$. Going from q_N to $\hat{q}_N$ is made by Stirling's formula.

• **Logistic distribution:** the logistic case is interesting because the density of the maximal invariant has not been calculated to our knowledge. We have

$$q_L = \int_0^\infty \int_{-\infty}^\infty e^{\ell(\mu,\sigma)}\,\frac{1}{\sigma}d\mu\; d\sigma \tag{26.6}$$

$$= \int_0^\infty \int_{-\infty}^\infty \prod_{i=1}^{n} \frac{\exp\left(\frac{x_i-\mu}{\sigma}\right)}{\sigma\left(1+\exp\left(\frac{x_i-\mu}{\sigma}\right)\right)^2}\,\frac{1}{\sigma}d\mu\; d\sigma. \tag{26.7}$$

Proposition (26.2.1) gives the approximation

$$\hat{q}_L = \frac{e^{\ell(\hat{\mu},\hat{\sigma})}(2\pi)}{n\hat{\sigma}\sqrt{|AB-C^2|}}$$

with $(\hat{\mu},\hat{\sigma})$ the maximum likelihood estimators of (μ,σ) and

$$A = \frac{1}{\hat{\sigma}^2} - \frac{2}{n\hat{\sigma}^2}\sum_{i=1}^{n}\frac{\exp\left(2\frac{x_i-\hat{\mu}}{\hat{\sigma}}\right)}{(1+\exp\left(\frac{x_i-\hat{\mu}}{\hat{\sigma}}\right))^2} \tag{26.8}$$

$$B = \frac{1}{\hat{\sigma}^2} + \frac{2}{n\sigma^2}\sum_{i=1}^{n}\frac{\left(\frac{x_i-\hat{\mu}}{\hat{\sigma}}\right)^2\exp\left(\frac{x_i-\hat{\mu}}{\hat{\sigma}}\right)}{1+\exp\left(\frac{x_i-\hat{\mu}}{\hat{\sigma}}\right)} - \frac{2}{n\hat{\sigma}^2}\sum_{i=1}^{n}\left[\frac{\left(\frac{x_i-\hat{\mu}}{\hat{\sigma}}\right)\exp\left(\frac{x_i-\hat{\mu}}{\hat{\sigma}}\right)}{1+\exp\left(\frac{x_i-\hat{\mu}}{\hat{\sigma}}\right)}\right]^2 \tag{26.9}$$

$$C = \frac{\bar{x}-\mu}{\hat{\sigma}^3} + \frac{1}{\hat{\sigma}^2} - \frac{2}{n\hat{\sigma}^2}\sum_{i=1}^{n}\frac{\left(\frac{x_i-\hat{\mu}}{\hat{\sigma}}\right)\exp\left(2\frac{x_i-\hat{\mu}}{\hat{\sigma}}\right)}{(1+\exp\left(\frac{x_i-\hat{\mu}}{\hat{\sigma}}\right))^2} \tag{26.10}$$

The QMPI test of the hypothesis $\mathcal{H}_0$ of normal distribution against a logistic distribution under $\mathcal{H}_1$ rejects $\mathcal{H}_0$ for large values of $\log(\hat{q}_L/\hat{q}_N)$. The following Monte Carlo simulations show the gain of power of the QMPI test as compared to some classical tests for normality ($\sqrt{b_1}$: kurtosis, b_2: skewness, SW: Shapiro-Wilks, KS: Kolmogorov-Smirnov, CvM: Cramér-von Mises) described in D'Agostino and Stephens (1986).

Table 26.1: Empirical power of tests of normality based on 10,000 samples of size $n = 50$ from a logistic distribution

level α	$\sqrt{b_1}$	b_2	SW	KS	CvM	$QPPI$
0.10	24.7	27.3	24.1	12.4	14.1	38.6
0.05	17.8	21.3	17.0	6.4	5.1	31.4

• **Exponential distribution:** Uthoff (1970) has obtained $q_E = \dfrac{\Gamma(n-1)}{n^n \tilde{\sigma}^{n-1}}$ with $\tilde{\sigma} = \overline{x} - \tilde{\mu}$ and $\tilde{\mu} = \min\{x_1, ..., x_n\}$. Corollary (26.2.1) gives $\hat{q}_E = \dfrac{e^{-n}\sqrt{2\pi}}{n^{3/2}\tilde{\sigma}^{n-1}}$. Here again, Stirling's formula shows the closeness of the approximation. Suppose now that we want to test an exponential distribution against an alternative of normality. The QMPI test rejects $\mathcal{H}_0$ for large values of $\log\tilde{\sigma} - \log\hat{\sigma}$. One find here the test of Shapiro and Wilk (1972), which explains its good properties.

We can remark here that we have $\log(q_N/q_E) = \log(\hat{q}_N/\hat{q}_E) + a_n$ where a_n doesn't depend on the data but only on the size n of the sample. This shows that the QMPI test gives exactly the same power than the MPI test in this case.

References

1. Barndorff-Nielsen, O. E. and Cox, D. R. (1989). *Asymptotic Techniques for Use in Statistics*, New York: Chapman & Hall.

2. D'Agostino, R. B. and Stephens, M. A. (1986). *Goodness-of-fit Techniques*, New York and Basel: Marcel Dekker Inc.

3. Franck, W. E. (1981). The most powerful invariant test of normal versus Cauchy with applications to stable alternatives, *Journal of the American Statistical Association*, **76**, 1002–1005.

4. Hájek, J. and Šidák, F. (1967). *Theory of Rank Tests*, New York: Academic Press.

5. Lehmann, E. L. (1959). *Testing Statistical Hypotheses*, New York: John Wiley & Sons.

6. Shapiro, S. S. and Wilk, M. B. (1972). An analysis of variance test for the exponential distribution (complete samples), *Technometrics*, **14**, 355–370.

7. Uthoff, V. A. (1970). An optimum test property of two well-known statistics. *Journal of the American Statistical Association*, **65**, 1597-1600.

8. Uthoff, V. A. (1973). The most powerful scale and location invariant test of the normal versus the double exponential, *Annals of Statistics*, **1**, 170–174.

PART VII

Model Validity in Quality of Life

27

Test of Monotonicity for the Rasch Model

Jean Bretagnolle

University of Paris-Sud, Orsay, France

Abstract: Sums of independent non-identically distributed indicators have been studied by numerous authors. We give an application of them to some questionnaire models.

Keywords and phrases: Concentration, monotonicity test

27.1 Results in the Literature

Let A_i, $i = 1, 2, \ldots . n$ be n independent events with respective probabilities p_i. Let $S = \Sigma 1_{A_i}$ be the number of success, with mean $np = \Sigma p_i$, where p is the mean probability of success. Let T, with probability law $Bin(n, p)$, have a binomial distribution with parameters n et p. For these two variables, we define:

their equal mean, $E(S) = E(T) = np$,

their median $\mu(S)$, $\mu(T)$ (definition $\mu(X) = Inf(k|P(X \leq k) > 1/2)$),

their modes $m(S)$, $m(T)$ (definition $m(X) = Argmax P(X = k)$).

Let k, k' be two integers bounding np : $k = \sup(k|k \leq np)$, $k' = \inf(k|k \geq np)$. When $X = S$ and when $X = T$, $\mu(X)$ equal k or k', $m(X)$ equal k or k'.

This result is known for the binomial distribution, and its extension to S is due to Samuels (1965) for the mode and Hamza (1995) for the median, who give the more precise following result $|\mu(S) - E(S)| \leq \log 2$.

In another respect, Hoeffding (1955) shows that:

$$\forall x < k,\ P(S \leq x) \leq P(T \leq x)\ ,\ \forall z > k',\ P(T \geq z) \leq P(S \geq z). \tag{27.1}$$

This last result is about concentration around the mean. A consequence (more easy to prove directly) is:

$$\text{if } \varphi \text{ is convex, } E\varphi(S) \leq E\varphi(T). \tag{27.2}$$

27.2 Extension of Hoeffding Result

Now let A_{ij}, with probability p_{ij}, be independent, $i \in I_j,\ j \in J$; for convenience, we suppose by now that $I_j = I$ does not depend upon j, and that I, J are the cardinals of those sets. Let S_j be the number of successes in the column j and T_j the corresponding $Bi(I, p_j)$ where $Ip_j = \Sigma_i p_{ij}$; finally S is the sum of S_j, and T that of T_j. Let p the probability "mean mean" defined by $I \times J \times p = \Sigma_j p_j = \Sigma_{ij} p_{i,j}$ and T' the binomial $Bi(I \times J, p)$.

The previous result gives immediately that:

S is more concentrated around it's mean than T'. But $Var(T')$ is in general greater than $Var(T)$.

Combining these three results of 27.1 we can extend, with a loss of only a lag of one unit, the concentration result.

Let, K, and K' be two equal or consecutive integers bounding $ET = I \times J \times p = I \times \Sigma p_i = \Sigma_{i,j} p_{i,j}$. Then, we have:

$$\begin{aligned} \forall x &< K - 1,\ P(S \leq x) \leq P(T \leq x)\ , \\ \forall z &> K' + 1,\ P(T \geq z) \leq P(S \geq z)(1) \end{aligned} \tag{27.3}$$

and, of course, for every φ convex of J variables

$$E(\varphi(S_1, \ldots, S_J)) \leq E(\varphi(T_1, \ldots, T_J)). \tag{27.4}$$

Then, to evaluate the level of a test about the partial sums,we can increase it under the additional hypothesis, that in every column,the p_{ij} does not depend on i:

- for the tests on the total sum
- for the tests on a convex function of the partial sums, conditionally to replacing its exact level by its convex majoring (particularly, for the Large Deviation approximation).

27.3 A Questionnaire Model

I persons must answer to J questions (independent answers).The model assumption is that exact response probability, the $p_{i,j}$,could be stochastic, but conditionally to their *abilities* θ_i, must verify: For a permutation σ not depending on i, $p_{i,\sigma(j)}$ is monotone on j which imply (with previous notations) the weaker hypothesis: For a permutation σ, $p_{\sigma(j)}$ is monotone on j.

(In the Rasch Model, we suppose that the odds ratio $\lambda = p/(1-p)$ can be written $\lambda_{ij} = \exp(\theta_i - \beta_j)$.)

The *good ordering* of the questions correspond to the case with $\sigma = id$.

We have **one** observation, $(\xi_{i,j})$ (independent conditionally to the $\lambda_{i,j}$), with law $B(1, p_{i,j})$ (in the following, we consider the λ as deterministic). Let S_j be the score of the question j. We want to test the hypothesis $\sigma = id$ against $\sigma \neq id$. (Actually, as there is an equal number of parameters and observations,we cannot estimate them.) Let then

$$H_o: \text{ The } p_j \text{ are increasing on } j. \tag{27.5}$$

Let us consider the *associated problem*, where the independent T_j follow respective laws $B(I, p_j)$. Note HA the additional hypothesis:

$$HA: p_{i,j} \text{ does not depend on } i. \tag{27.6}$$

The natural estimates are the $\widehat{p}_j = S_j/I$. But under H_o and HA, we can use another estimate called *isotonic*:

Let F_j be the cumulative suit: $F_o = 0, \ldots, F_j = F_{j-1} + S_j, \ldots$.

The maximum likelihood estimate, under monotonicity constraint, is constructed as follows: Let G_j be the bigger convex function smaller than F_j. The estimate is defined as

$$\widetilde{p}_j = (G_j - G_{j-1})/I. \tag{27.7}$$

By *nodes suit* we mean the suit of integers where$G$ and F coincide, and by *structure* of the observation, the statistic

$$\Sigma = (U,\ \nu_o = 0, F_0 = 0,\ \nu_1, F_{\nu_1}, \cdots, \nu_U = J, F_J) \tag{27.8}$$

where U is the number of nodes, ν the positions of these nodes, the $F_\nu = G_\nu$ the values in these nodes.

Among conceivable tests, we particularly study two, the first of type L_1, the second of type Kolmogorov-Smirnov, based on the scores.

$$IFL = \Sigma_j(F_j - G_j),\ FL = Sup_j(F_j - G_j) \tag{27.9}$$

(note that $F_j - G_j \geq 0$ by construction) with rejection set $IFL \geq C,\ FL \geq C'$.

(A) Let $f_j = EF_j$, $\Delta = Sup(F_j - f_j)$, $\delta = Sup(f_j - F_j)$. Under H_o, f is convex, and then $FL \leq \delta + \Delta$. Using the Paul Levy inequalities and the control of the medians, it follows that

$$P(\delta + \Delta > x) \leq 4P(F_J - f_J > x - 2) + 4P(f_J - F_J > x - 2)$$

[this upper bound is asymptotically precise, see formula 2 on Page 146 of the Borodin and Saalminen (1996)]. Finally, using the concentration result, it follows, in general, for $x \geq 3$ that

$$P(FL > x) \leq 4P(T \geq x - 3) + 4P(T < 3 - x) \tag{27.10}$$

where the right-hand depend only on the p_j. We will then get a good approximation by substituting the $\widehat{p}_j$ or the $\widetilde{p}_j$.

This upper bounding-evaluation is valid underH_o, but unfortunately, it is very conservative because the inequality$FL \leq \delta + \Delta$ is very bad (in a similar problem of isotonic estimation of a density or a regression, the order of size is not the same).

(B) The statistic FL is convex on the scores. Note $P_{(p_j)}$ any probability with mean (p_j), and $P_{(p_j),A}$ the binomial probability with the same mean. Then, for any φ increasing convex, we have

$$P_{(p_j)}(IF \geq x) \leq \ inf_{\varphi \in \Phi} E_{(p_j),A} \varphi(IF)/\varphi(x). \tag{27.11}$$

We can check by simulation that the right-hand satisfies an inequality of the form

$$\inf_{\varphi \in \Phi} E_{(p_j),A} \varphi(IF)/\varphi(x) \ \leq \ M \times P_{(p_j),AD}(IF \geq x) \tag{27.12}$$

where M is of order 2. Moreover, using the crude approximation described in the last section which gives the order of the queue, we can justify that the factor M must be lower than e. But the loss of this factor is, again, too large.

(C) Applying the bootstrap, we replace the distribution by the distribution $\widetilde{P}$ (which depends only on Σ), *and not on the distribution* $\widehat{P}$.

We simulate under HA, (that is, the binomial simulation) keeping only those observations which verify $\Sigma(Z) = \Sigma(X)$. Note that m the number of such observations.

Let $\widetilde{P}_{\Sigma,m}$ be the empirical obtained. $\widetilde{P}_{\Sigma,m}(IF(Z) \geq C) = \alpha$ gives the threshold $C = C(\Sigma, \alpha)$ of a test with level α (as for the test about IFL). (We do not try to construct the test when the structure satisfies $U = J{+}1$, i.e., when the X_j, and then the two estimates, are increasing). We repeat the operation ℓ times, under the same probability p_{ij}, of course.

27.4 Simulations about the Level in the Conditional Test Case

Our simulation shows that, with equal p_j, the level is largely dominated by the level under the additional hypothesis, as the concentration inequality try to involve it.

Under HA, the bootstrap level is very good, but slightly larger than the theoretical level when the p_j are constants, which is not surprising for specialists who well know that the isotonic estimation is most problematic when the estimate is not strictly monotonic. This phenomenon will correct itself when I increases (else, the test is conservative). Let a the theoretical level, α the bootstrap level.

For $J = 4, 5, 6$, $\ell = 400$, $m = 200$, by $I = 10$, for the two tests, with p_j constant

$$\begin{aligned} a &= 5\%, & \alpha - a &< 1 \cdot 9\% \\ a &= 10\%, & \alpha - a &< 3\% \\ a &= 20\%, & \alpha - a &< 4\%. \end{aligned}$$

27.5 Simulations about the Power under HA

We wanted to compare our two tests to Gaussian tests based on the following well-known asymptotic property: if we let $A_j = 2Arcsin\sqrt{X_j/I}$, then $A_j \approx p_j + \mathcal{N}_j$, where the $\mathcal{N}_j$ are independent $\mathcal{N}(0,1)$, which allows us to test the monotonicity of the means. But, even for $I = 40$ or 80, this approximation turned out to be very bad (for instance the distribution of the $Sup_j(\ (A_j - A_{j+1})^+\ )$ or of their sums are roughly false).

27.6 Conclusion

Our present applications are promising, but do not include the expected result which could be:

> for tests about the gap between the observation and its isotonic, the maximum level is obtained under the binomial case, and that will allow us to use a test function only of the scores.

References

1. Borodin, A. and Salminen, P. (1996). *Handbook of Brownian Motion Facts and Formulae*, Basel: Birkhäuser Verlag.

2. Hamza, K. (1995). The smallest uniform upperbound on the distance between the mean and the median of the binomial and Poisson distributions, *Statistics and Probability Letters*, **23**, 21–25.

3. Hoeffding, W. (1955). On the distribution of the number of successes in independent trials, *Annals of Mathematical Statistics*, **27**, 713–721.

4. Samuels, S. (1965). On the number of successes in independent trials,*Annals of Mathematical Statistics*, **36**, 1272–1278.

28

Validation of Model Assumptions in Quality of Life Measurements

A. Hamon, J. F. Dupuy, and M. Mesbah

University of South Brittany, Vannes, France

Abstract: The measurement of Quality of Life (QoL) has become an increasingly used outcome measure in clinical trials over the past few years. QoL can be assessed by self-rated questionnaires, i.e. a collection of items intended to measure several characteristics of the QoL. When new tests are developed, the goal is to produce a valid measure. The measuring process needs to construct a statistical model and then try to approach it in the best way. Different statistical "methods," in fact models, are used during the validation process of QoL questionnaires. In this paper, we will present the hypothesis underlying each model and which tests are used in practice.

Keywords and phrases: Measurement model, reliability, Cronbach alpha coefficient, Rasch model, goodness-of-fit tests

28.1 Introduction

For the past ten years, almost all major clinical trials have included programs designed to evaluate Quality of Life (QoL). Now that the data from these trials are being analyzed as part of the results, we are faced with the validity of the instrument which produced such measures.

"Increasing and sometimes indiscriminate use of QoL measures has provoked concern about these method in the (health) context, especially when important consequences, such as treatment decisions or resource allocation, depend on them" [Cox *et al.* (1992)]. Generally we can assume that QoL is a complex notion which can be divided in multiple components. Each component is related to a specific domain, such as sociability, communication or mobility, and is evaluated using a set of well-chosen items. Hence a QoL instrument is composed of different subsets of items referred to as dimensions.

The validation process needs to construct a statistical model and then to test its goodness-of-fit. In practice, the validation of the QoL instrument is mainly based on a survey where the QoL questionnaires are assessed by people chosen randomly from the population of interest. The statistical validation consists generally of the following steps: 1- assumption of an underlying stochastic model generating the individual responses to the form; 2- checking the validity of the model or some of its specific properties by examining the real data obtained. This step could be done by an adequate goodness-of-fit test. In practice, the use of goodness-of-fit test is unfortunately limited. Graphical exploratory methods and computing of some well-known measure of association mostly correlation, are more popular. Classical methods are based on mixed linear models. We present in Section 28.2, the most popular measurement model used in this field: the parallel model. Modern methods use a mixed generalized linear model, and the Rasch model is since 1960 the most frequently used measurement model. After presenting briefly this model, various goodness-of-fit tests are presented.

28.2 Classical Theory

We denote by X_{ij} the response of person i $(i = 1, .., n)$ to item j $(j = 1, .., k)$. One of the most popular model is a mixed one way models with the item as random factor:

$$X_{ij} = \mu_j + a_i + e_{ij}. \tag{28.1}$$

In this equation, the random variables a_i and e_{ij} are uncorrelated. They have an expectation equal to 0 and their variances are respectively denoted by σ_a^2 and σ_e^2. The variable $\tau_{ij} = \mu_j + a_i$ is called the true measure so that the observed score is the sum of the true measure and the measurement error e_{ij}.

The reliability ρ of the instrument is defined as the ratio of the true measure variance and the observed measure variance. In the model defined by (28.1) we can show that $\rho = \frac{\sigma_a^2}{\sigma_a^2+\sigma_e^2}$, which is also the constant correlation between any two items. The reliability coefficient ρ can be easily interpreted as a correlation coefficient between the true and the observed measure. The k straight regression lines (X_{ij}, τ_{ij}), corresponding to the items $(j = 1, .., k)$ are parallel and $Corr\,(X_{ij}, \tau_{ij}) = \rho^{1/2}$. The reliability $\tilde{\rho}$ of the sum of k items is equal to

$$\tilde{\rho} = \frac{k\rho}{(k-1)\,\rho + 1} \tag{28.2}$$

This formula is known as the Spearman-Brown formula, it shows that when the number of items increases, the reliability tends to 1. Its maximum likelihood estimator, under the assumption of a normal distribution of the error and the true score, is known as the Cronbach Alpha Coefficient (CAC) [Kristof (1963)]:

$$\alpha = \frac{k}{k-1}\left[1 - \frac{\sum_{j=1}^{k} S_j^2}{S_{tot}^2}\right] \tag{28.3}$$

where $S_j^2 = \frac{1}{n-1}\sum_{i=1}^{n}(x_{ij} - \bar{x}_j)^2$ and $S_{tot}^2 = \frac{1}{nk-1}\sum_{i=1}^{n}\sum_{j=1}^{k}(x_{ij} - \bar{x})^2$. CAC can be computed to find the most reliable subset of items [Moret *et al.* (1993)]. As a first step, all items are used to compute the CAC. Then at every step, one item is removed from the scale. The removed item is the one which gives for the scale without the item the maximum CAC. This procedure is repeated until only two items remain. If the parallel model is true, it can be shown, using the Spearman-Brown formula that increasing the number of items increases the reliability of the total score which is estimated by Cronbach alpha. Thus, a decrease of such a curve when adding an item could strongly bring us to suspect that the given item is a bad one (in term of goodness-of-fit of the model). A supplementary and popular way to assess the influence of the item on the goodness-of-fit of the parallel model is by examining the empirical correlations item to total (or to total minus the given item). Assuming the parallel model, these correlations must be equal. A low correlation indicates a bad item. We now present a real data example.

28.3 SIP Mobility Data (I)

Classical validation of a QoL questionnaire will now be illustrated using data from a clinical trial in the depression field. 483 persons answer to the SIP (Sickness Impact Profile) questionnaire as reported by Bergner *et al.*. We use here a French translation. The SIP consists on 136 dichotomous items distributed in 12 dimensions. As an example, we apply the described method to the *Mobility* scale. This dimension contains 10 items. Individuals are asked to respond "Yes" to those items that described them on the present day. Items of the Mobility dimension and the numbers of persons who answered positively are presented in Table 28.1. In Table 28.2, we give the distribution of the individuals scores. The step by step procedure with the Alpha coefficient explained in the last section, is given in Figure 28.1.

If an instrument is already validated, the curve is monotonously increasing. Figure 28.1 shows a non-increasing curve. We can choose to increase the reliability of the instrument by deleting 1 item: the number 3.

Table 28.1: Items of the Mobility dimension; $n = 466$

Item	Contents	frequency	proportion
1	I am getting around only with one building	97	0.21
2	I stay within a room	59	0.13
3	I am staying in bed more	235	0.50
4	I am staying in bed most of the time	88	0.19
5	I am not now using public transportation	133	0.26
6	I stay home most of the time	296	0.63
7	I am only going to places with restrooms nearby	20	0.04
8	I am not going into town	167	0.36
9	I stay away from home only for brief periods of time	275	0.59
10	I do not get around in the dark or in unlit places without someone's help	50	0.11

Table 28.2: Distribution of the individuals scores for the Mobility dimension; $n = 466$

Score	0	1	2	3	4	5	6	7	8	9	10
Frequency	69	56	62	93	74	54	34	19	7	2	0

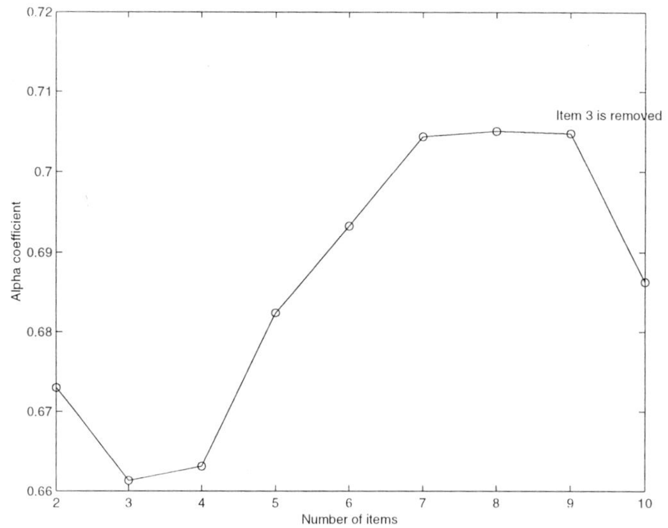

Figure 28.1: Step by step procedure with the CAC for the Mobility dimension

28.4 The Rasch Model

The Rasch model assumes that each of n persons labelled i $(i = 1, \ldots, n)$ answered to k items labelled j $(j = 1, \ldots, k)$. For each individual and each item, a binary response X_{ij} equal to 1 or 0 (yes or no) is recorded. Rasch (1960) proposed the model

$$P\left(X_{ij} = x_{ij} | \theta_i, \beta_j\right) = \frac{\exp\left[x_{ij}\left(\theta_i - \beta_j\right)\right]}{1 + \exp\left[\theta_i - \beta_j\right]} \tag{28.4}$$

where θ_i $(\theta_i \in \mathbb{R})$ is an individual parameter representing an individual's level of quality of life and β_j $(\beta_j \in \mathbb{R})$ is known as the difficulty parameter, or, in the context of quality of life, it can be interpreted as the quality of life level an individual must possess to achieve a 0.5 probability to answer positively to the item. Individuals are assumed to be independent. The variables X_{ij} $(j = 1, ..., k)$ are locally independent, that is if θ_i is held fixed the answers are independent.

One of the possible estimation methods is based on the particular property of the Rasch model, namely sufficiency of the total score. This is the conditional maximum likelihood (CML) estimation method, first proposed by Andersen (1972). This method is important because many goodness-of-fit tests are based on CML estimations. We will now explain how it is possible to test the fit of a set of items to the Rasch model in order to evaluate the validity of the questionnaire.

28.4.1 Goodness-of-fit tests

First, we must point out some tests that are constructed by remarking that the Rasch model can also be formulated as a log linear model. In this context, Mellenbergh and Vijn (1981) have presented some goodness-of-fit tests. Relying on another point of view, Tjur (1982) introduced a random raw effect Rasch model and related it to a conditional multiplicative Poisson model. This connection leads also to some goodness-of-fit tests.

There exists an extensive literature on goodness-of-fit tests in the Rasch model [see Glas and Verhelst (1995) for a good review]. Since statistical properties of tests (asymptotic distribution, power) are not always well stated, we will present only some of the well-established tests. For all these goodness-of-fit tests, the null hypothesis is that the data are realizations of random variables distributed according to the Rasch model.

The first test we report was proposed by Andersen (1973) and is based on the following properties of the Rasch model. Firstly, the raw score $S_i = \sum_j X_{iJ}$ is a sufficient statistic for the individual parameter θ_i. Secondly, the CML estimators are consistent and asymptotically normally distributed. Finally the CML estimators based on the sub-sample of individuals who achieved a score s ($s = 1, ..., k-1$), are consistent and asymptotically normally distributed when $n_s \to \infty$ [Andersen (1973)]. The conditional likelihood denoted by $L_C(x, \beta)$ can be written

$$L_C(X_1, ..., X_n, \beta_1, ..., \beta_k) = -\sum_{i=1}^{n} \left\{ \sum_{j=1}^{k} x_{ij}\beta_j + \log(\gamma_k(s_i)) \right\} \tag{28.5}$$

where the coefficients γ_k are defined

$$\gamma_k(s) = \sum_{u/\sum_j u_j = s} \left(\prod_{j/u_j=1} \exp(-\beta_j) \right). \tag{28.6}$$

The CML estimators will be denoted by $\hat{\beta}_C$. The conditional likelihood based on the sub-sample of persons with score s will be denoted by $L_C^{(s)}(x, \beta)$ and the associated maximum likelihood estimators by $\hat{\beta}_C^{(s)}$. Andersen (1973) proposed a test based on the ratio of the estimated likelihood $L_C(x, \hat{\beta}_C)$ and the sum of the sub-samples likelihoods $L_C^{(s)}(x, \hat{\beta}_C^{(s)})$. The test statistic is :

$$Z = 2\sum_{s=1}^{k-1} L_C^{(s)}(x, \hat{\beta}_C^{(s)}) - 2L_C(x, \hat{\beta}_C). \tag{28.7}$$

Theorem 28.4.1 [Andersen (1973)] *If $n_s \to \infty$ for $s = 1, ..., k-1$ and if the assumptions of the Rasch model hold then*

$$Z \longrightarrow \chi^2_{(k-1)(k-2)}. \tag{28.8}$$

This theorem implies that, only when all the size (n_s) of the sub samples of persons with a score s is sufficiently large, can we consider that the distribution of Z can be approximated by a χ^2-distribution. These requirements are not generally achieved in practice. Most often, we find that the number of persons with a low or large score is too small. To take this problem into account, Andersen (1973) proposed to group the values s to obtain larger and more homogeneous sub-samples. Let $s_0, ..., s_r$, be r integers ($r > 1$) so that we have $s_0 = 0 < s_1 < ... < s_{r-1} < s_r = k-1$. Then we group in a sub-sample all the persons who achieved a score between 1 and s_1, in another sub-sample we group the individuals with scores between $s_1 + 1$ and s_2, and so on.... In each subsample, we can obtain CML estimators $\hat{\beta}_C^{(l)}$. Andersen proves that with slight modifications, the previous theorem is always true. Now the test statistic is Z_1

$$Z_1 = 2\sum_{l=1}^{r} \sum_{s=s_{l-1}+1}^{s_l} L_C^{(s)}(x, \hat{\beta}_C^{(l)}) - 2L_C(x, \hat{\beta}_C) \tag{28.9}$$

where n_l is the number of individuals with a score between $s_{l-1}+1$ and s_l. If $n_l \to \infty$ for $l = 1, ..., r$ then the statistic Z_1 is asymptotically χ^2-distributed with $(r-1)(k-1)$ degrees of freedom. Andersen (1973) considers the power of this statistic against the specific alternative of the logistic two parameters model. In this model the probability of a positive answer now depends on two parameters

$$P(X_{ij} = 1|\theta_i, \beta_j) = \frac{\exp\{\alpha_j(\theta - \beta_j)\}}{1 + \exp\{\alpha_j(\theta - \beta_j)\}}, \tag{28.10}$$

α_j is called the discrimination parameter; this allows the items' characteristic curves to have different slopes. A special case is the Rasch model in which all the discrimination parameters are equal.

Glas (1988) proposed two χ^2 goodness-of-fit tests. The first test, denoted by R_1 and called first order test, is based on the differences between the observed and expected number of individuals with a score s and who achieved item j. The second test (R_2, second order test) is constructed with the number of persons who positively answer to two questions j and l ($j \neq l$).

Let N_{sj} be the number of persons with score s and who respond 1 to item j. Instead of computing the expectation of N_{sj}, Glas (1988) proposed to compute the conditional expectation $\mathbb{E}(N_{sj}, \beta/N_s = n_s)$. After some computations we can show that

$$\mathbb{E}(N_{sj}, \beta/N_s = n_s) = \frac{\gamma_{k-1}(s-1)\exp(-\beta_j)}{\gamma_k(s)} \sum_{i=1}^{n} p(S_i = s/N_s = n_s). \tag{28.11}$$

This expectation will be estimated

$$\mathbb{E}(N_{sj}, \hat{\beta}_C/N_s = n_s) = n_s \frac{\hat{\gamma}_{k-1}(s-1)\exp(-\hat{\beta}_j^C)}{\hat{\gamma}_k(s)}, \tag{28.12}$$

Let $d_s^C = (d_{sj}^C)_{j=1,\dots,k}$ be the vector of the differences $n_{sj} - \mathbb{E}(N_{sj}, \hat{\beta}_C/N_s = n_s)$ and $\hat{V}_s^C$ its estimated variance-covariance matrix.

Theorem 28.4.2 [Glas (1988)] *Let R_1 be defined by*

$$R_1 = \sum_{s=1}^{k-1} \frac{{}^t d_s^C \hat{V}_s^{C(-1)} d_s^C}{n_s}. \tag{28.13}$$

If for all s ($1 \leq s \leq k-1$) the size of the sub sample of persons with score s tends to infinity then the statistic R_1 is asymptotically χ^2 distributed with $(k-1)(k-2)$ degrees of freedom.

The same argument with the number of persons who positively answer to two items l and j leads to the construction of the R_2 statistic. This statistic has an asymptotic χ^2 distribution with $\frac{k(k+1)}{2} - 2(k-1)$ degrees of freedom.

In a validating process, we aim at detecting bad items (in the sense that they do not fit well), and sometimes also subsets of items that fit the model. Molenaar (1983) proposes a graphical method to select homogeneous subsets of items. We will briefly present this interesting method in the next section.

28.4.2 A graphical method

The graphical method suggested by Molenaar (1983) serves to identify items weakly related to the latent trait and subscales of items measuring the same latent trait. We briefly describe this method and then apply it in the following section.

This graphical method is based on the following consideration: the CML estimations of the item difficulties being independent, the same difficulty estimates should be obtained apart from random fluctuations, for a division of the respondents into two subgroups, according to their response to a particular item.The principle of the method is as follows:

(a) for each particular item i in its turn, separate the respondents into two subgroups (respectively denoted by G_0 and G_1) according to their response (respectively *yes* and *no*) to the chosen item i;

(b) estimate the difficulties of the other items j ($j \neq i$) in the two groups G_0 and G_1;

(c) plot the estimates of the difficulties in G_1 versus the estimates of the difficulties in G_0.

If the items i and j are locally independent, the position of the item j should lie close to the first diagonal. If there exists some dependence between these two items, item j should lie below or above the diagonal.

28.5 SIP Mobility Data (II)

One popular graphical method to assess the monotonicity of the item response function is to plot the item trace (Figure 28.2). The item trace is the representative graph of the proportion of success against the corresponding raw score. If the Rasch model is valid and if the raw score is replaced by the true latent trait θ, these curves are logistics and "parallels" or shifted.

To compute the goodness-of-fit tests presented before, we use the software RSP [Glas and Ellis (1993)]. For all these tests, the split of the sample based on raw scores is given in Table 28.3. The likelihood ratio test provides a significant result ($Z_1 = 95$ with 27 degrees of freedom and the p-value is less than 0.0000), indicating that the assumptions of the Rasch model do not hold for the studied

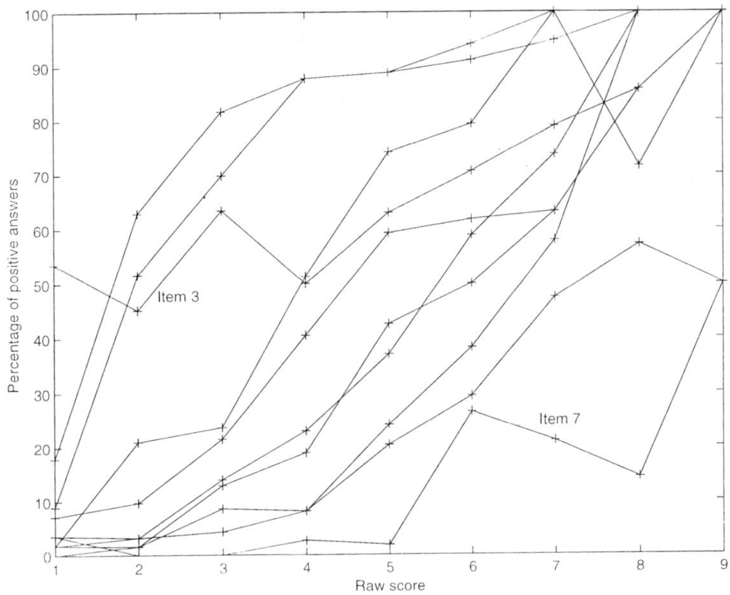

Figure 28.2: Traces of the Mobility dimension items

Table 28.3: Division into 4 subgroups; $n = 466$

Group	from score	size
1	1 to 2	118
2	3 to 3	93
3	4 to 4	70
4	5 to 9	116

data. The R_1 test is also significant. This test is based on the observed and expected frequency of the positive answers to each item in each subgroup. In Table 28.4, we present these numbers for the item 3 because it is the worst item of all, in the sense of a large difference between observed and expected frequency. If this item is removed, the statistic R_1 is no longer significant. It value is 20 and the asymptotic distribution is a χ^2 with 24 degrees of freedom (p-value $= 0.68$). With this set of 9 items, the R_2 test is highly significant, indicating that at least two items of the questionnaire are not locally independent. Applying the described graphical method results in the following conclusions. Separation of respondents is made on item 2 of the scale.

Table 28.4: Expected and observed frequency of positive answers to item 3 in each subgroup

Group	expected	observed
1	31.7	58
2	52.3	59
3	50	37
4	101	81

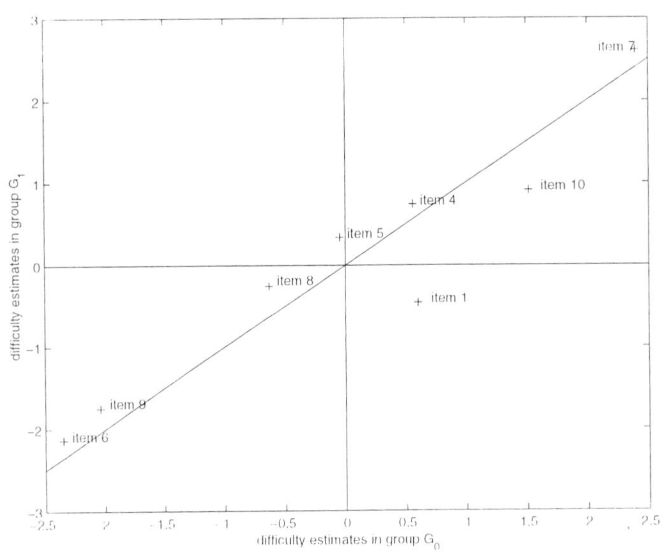

Figure 28.3: Difficulty estimates in each group formed by the individuals who positively answer to item 2 (G_1) and negatively answer to item 2 (G_0)

Figure 28.3 shows that item 1 appears to be dependent on item 2 ; it appears to be more difficult for people scoring 0 to item 2 than for people scoring 1 to item 2. To a less extent, item 10 seems to be correlated with item 2. Other items seem to lie close to the diagonal, indicating that the response to each of them is independent of the response to item 2. This method is repeated for each item of the scale. We present one more example. It involves separation on item 10, in order to check the conclusions drawn from the previous graph. Item

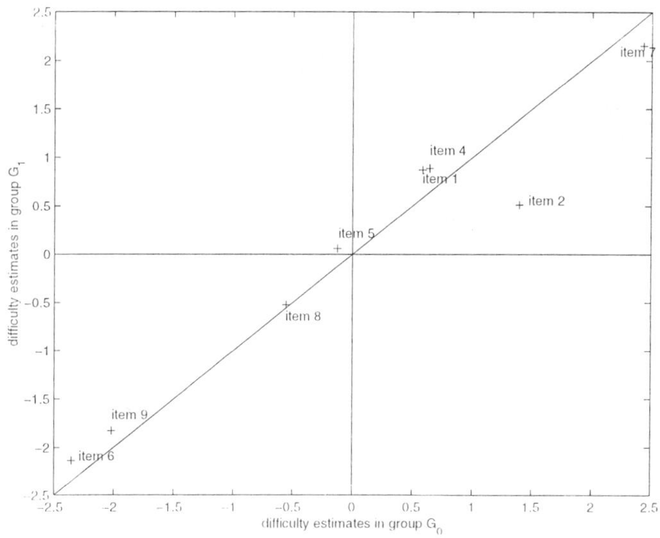

Figure 28.4: Difficulty estimates in each group formed by the individuals who positively answer to item 10 (G_1) and negatively answer to item 10 (G_0)

10 is used to split and gives the graph 28.4. Item 2 appears to be correlated with item 10 whereas there is no indication that the other items depend on the response to item 10.

All the items appear to be locally independent of item 8 (the figure is not presented here). Similar plots are obtained when using items $1, 4, 5, 6, 7$ and 9 to separate the respondents. Hence the scale Mobility contains a group of three items correlated with each other $\{1, 2, 10\}$. In order to obtain a Rasch homogeneous instrument, we choose one item from $\{1, 2, 10\}$ and form a scale by gathering this item and the items $\{4, 5, 6, 7, 8, 9\}$.

For making a choice, we calculate the likelihood ratio test. We check that it is not significant for each of the three possible sets of items. As a scale for Mobility we choose the set for which the level of significance of the test is the largest.

28.6 Conclusion

A century after the famous Karl Pearson paper, fields of application where goodness-of-fit tests are useful, are increasing. The validation of quality of life questionnaire is undoubtedly one of those. Nevertheless, checking underlying properties of those ideal measurement models is in practice, often done by simple graphics or by way of simple statistics.

References

1. Andersen, E. B. (1973). Asymptotic properties of conditional maximum likelihood estimators, *Journal of the Royal Statistical Society, Series B*, **32**, 283–301.

2. Cox, D., Fitzpatrick, R., Fletcher, A. E., Gore, S. M., Spiegelhalter, D. J., and Jones, D. R. (1992). Quality-of-life Assessment: Can we Keep It Simple?, *Journal of the Royal Statistical Society, Series A*, **155**, 353–393.

3. Glas, C. A. W. (1988). The derivation of some tests for the Rasch model from the multinomial distribution, *Psychometrika*, **53**, 525–546.

4. Glas, C. A. W. and Ellis, J. (1993). *Rasch Scaling Program: User's Manual Guide*, iec ProGAMMA, Groningen, The Netherlands.

5. Glas, C. A. W. and Verhelst, N. D. (1995). Testing the Rasch model, In *Rasch Models Foundations, Recent Developments and Applications* (Eds., G. Fischer and I. Molenaar), New York: Springer-Verlag.

6. Kristof, W. (1963). The statistical theory of stepped-up reliability coefficients when a test has been divided into several equivalent parts, *Psychometrika*, 28, 221–238.

7. Mellenbergh, G. J. and Vijn, P. (1981). The Rasch model as a loglinear model, *Applied Psychological Measurement*, **5**, 369–376.

8. Molenaar, I. W. (1983). Some improved diagnostics for failure of the Rasch model, *Psychometrika*, **48**, 49–72.

9. Moret, L., Mesbah, M., Chwalow, J., and Lellouch, J. (1993). Validation statistique interne d'une chelle de mesure : relation entre analyse en composantes principales, coefficient α de Cronbach et coefficient de corrélation intra-classe, *Revue d'Épidmiologie et de Santé Publique*, 41, 179–186.

10. Rasch, G. (1960). *Probabilistic models for some intelligence and attainment tests*, Danmarks Paedagogiske Institut, Copenhagen.

11. Tjur, T. (1982). A connection between Rasch's item analysis model and a multiplicative Poisson model, *Scandinavian Journal of Statistics*, **9**, 23–30.

PART VIII

Tests of Hypotheses and Estimation with Applications

29

One-Sided Hypotheses in a Multinomial Model

Richard M. Dudley and Dominique M. Haughton

Massachusetts Institute of Technology, Cambridge, Massachusetts
Bentley College, Waltham, Massachusetts

Abstract: Several independent data sets are represented as coordinates of one i.i.d. data set. We consider model selection for several 2×2 contingency tables including models with and without a common odds ratio and one-sided models such as models where a treatment is beneficial. We use Jeffreys priors on each model. For studies of long-term treatment with aspirin after a heart attack, we find that the treatment appeared to be beneficial if it began within six months after the heart attack; if begun later it had no effect on mortality.

Keywords and phrases: Information criteria, Jeffreys prior, aspirin

29.1 Introduction

In this paper we continue [Dudley and Haughton (1997)] to study model selection based on an extended Schwarz (1978) BIC method for multiple data sets and where individual models may be half-spaces or half-lines.

Multiple clinical trials of the same treatment against placebo give multiple data sets, where the true or pseudo-true parameter values may differ between studies. We also consider models where one parameter, a common odds ratio, is the same for all data sets, while others vary. Some models of interest are one-sided; specifically, models where the treatment is beneficial or harmful. For one treatment, the long-term use of aspirin after a heart attack (myocardial infarction, or MI), of seven individual studies, six did not show a significant effect of aspirin on overall mortality at the 5% level, and one showed a mildly significant benefit. The AMIS (1980) study had the largest sample size and found a (non-significant) negative effect for aspirin. Several meta-analyses of the mortality data have been done, e.g. Canner (1987), DerSimonian and Laird (1986), and Gaver *et al.* (1992), which included 6 of the 7 studies, but not Vogel, Fischer, and Huyke (1979), the one study showing a significant benefit of aspirin. In summary, the meta-analyses found a point estimate showing an overall benefit for aspirin, for example with 10.7% mortality in the placebo groups and 9.9% mortality in the aspirin groups [Canner (1987)] but with a 95% confidence interval for the common odds ratio which contains 1, or not, by a narrow margin, depending on the specifics of each analysis. We will do an apparently new kind of analysis, based on a mixture model device and Jeffreys priors, and considering not only the full data but also data separated according to whether treatment began within 6 months after the MI [cf. Canner (1987)]. We will need to have in mind possible "multiple comparisons" issues.

29.2 Putting Multiple Data Sets Into an i.i.d. Form

Let P_θ, $\theta \in \Theta$, be laws on a sample space X, where each P_θ has a density $f(\theta, x)$, $\theta \in \Theta$, $x \in X$, with respect to a σ-finite measure μ on X. For $k = 1, ..., K$ let $x_{k1}, ..., x_{kn(k)}$ be i.i.d. $P_{\theta(k)}$ for some $\theta(k) \in \Theta$. For different k let the data sets be independent. Let $N := n(1) + \cdots n(K)$. Let's see how the x_{ki} can be taken to be coordinates of N i.i.d. observations. For $r = 1, ..., K$ let e_r be the rth standard unit vector in $\mathbf{R}^K$ so that $(e_r)_k := \delta_{rk} := 1_{\{r=k\}}$. Let Z be the set of K points e_r. Let ν be counting measure on Z. Consider the family of laws on $X \times Z$ having densities with respect to $\mu \times \nu$ given by

$$\Pi_{k=1}^{K} [v_k f(\theta(k), x)]^{z_k} \tag{29.1}$$

where $0^0 := 1$, $\{v_k\}_{k=1}^K \in \mathcal{V}_K := \{\{v_i\}_{i=1}^K : v_i \geq 0, \sum_{i=1}^K v_i = 1\}$, and $z = (z_1, ..., z_K) \in Z$. If we take N i.i.d. observations $(x_i, z(i))$ from such a distribution, then the numbers $n_k := n(k)$ of observations having $z(i) = e_k$ for $k = 1, ..., K$ have a multinomial $(N, v_1, ..., v_K)$ distribution and the observations x_i with $z(i) = e_r$ will be i.i.d. $P_{\theta(r)}$ for each r. The likelihood function (29.1) is the special case where all observations are "categorized" of the mixture model given e.g. by Titterington, Smith, and Makov (1985, p. 3, (1.2.3)).

29.3 Model Selection Criteria

Let $\mathcal{P} = \{P_\theta, \theta \in \Theta\}$, where Θ is an open set in a Euclidean space. Let $m_1, ..., m_J$ be a finite set of models included in Θ, where $m_1 = \Theta$. For $k = 1, ..., K$, the kth independent data set $X^{(k)} := (X_{k1}, ..., X_{kn_k})$ will consist of n_k observations i.i.d. with a law $P^{(k)}$ which may not be in $\mathcal{P}$. We assume that the maximum likelihood estimate $\hat{\theta}_k$ of $\theta \in \Theta$ based on $X^{(k)}$ as $n_k \to \infty$ eventually exists almost surely and converges to some θ_{0k}, the *pseudo-true* value of θ for $P^{(k)}$ [Berk (1966), Huber (1967) and Poskitt (1987)]. We will need some regularity conditions on models m_j [Dudley and Haughton (1997, 2000)].

One is that each m_j is either a manifold, or a manifold-with-boundary included in some m_i where in a suitable parameterization, m_i is an open set in a Euclidean space and m_j is a half-space intersected with m_i.

We call a model m_j the *best* model if it is the unique smallest model (in the sense of inclusion) containing θ_{0k} for $k = 1, ..., K$. A best model will always exist if for any i and $j = 1, ..., J$, $m_i \cap m_j = m_r$ for some r, as will hold for the models we consider. Our object is to choose the best model m_j, under loss functions which may differ for different wrong choices of models.

Well-known model selection criteria are based on penalized maximum likelihood. Let $ML_k(m_j)$ be the supremum of the likelihood for the kth data set over $\theta \in m_j$. Let $MLL_{kj} := \log ML_k(m_j)$. Suppose first that there is only one data set ($K = 1$) and all the models are manifolds. One chooses the model for which $MLL_{1j} - \Delta_j$ is largest, where the penalty Δ_j increases with the dimension d_j of m_j. In the BIC of G. Schwarz (1978), $\Delta_j = d_j(\log n)/2$ where $n = n_1$ is the sample size. It has been shown under some conditions [Haughton (1988) and Poskitt (1987)] that as $n_1 \to \infty$, the BIC is consistent, i.e. the probability of choosing the best model converges to 1. Moreover, the BIC is equivalent to choosing the model having the highest value of the leading terms in the logarithm of its posterior probability of being the best model, under any prior probabilities having strictly positive, continuous densities. The leading terms do not involve the particular choices of priors.

The BIC has not been found to work optimally, so one may look beyond the leading terms. In our cases the well-known Jeffreys (1946) prior measure J_j exists on each m_j. The density of J_j is the square root of the determinant of the Fisher information matrix, independently of the parameterization, see also Kass (1989). If $0 < J_j(m_j) < \infty$, as is true in our cases, then the Jeffreys prior probability on m_j will be called $\mu_j := J_j/J_j(m_j)$.

In the hierarchical Bayes method of meta-analysis, e.g. DerSimonian and Laird (1986) and Gaver *et al.* (1992), a parameter θ_i gives an effect size, e.g. a difference in mortality rates, for the treatment vs. placebo in the ith study, where θ_i are i.i.d. with a hyper-parametric prior. Such priors may be more flexible and realistic, but we use Jeffreys priors to avoid subjective choices.

To approximate posterior probabilities we will need to approximate integrals $\int lik_{(n)} d\mu_j$ for $k = 1, ..., K$, $j = 1, ..., J$, where $lik_{(n)}$ is a likelihood function. Here n may be n_k for some k, when $lik_{(n)}$ is the likelihood function for the kth data set, or n may be the total sample size $N = n_1 + \cdots + n_K$ in (29.1). If m_j is a manifold, and $ML(m_j) := \sup_{m_j} lik_{(n)}$, then we use the approximation

$$\int lik_{(n)} d\mu_j \sim (2\pi/n)^{d_j/2} ML(m_j) J_j(m_j)^{-1} \tag{29.2}$$

given by a sharpened Laplace method, which holds with a relative error of $O_p(n^{-1})$ [Poskitt (1987)], if the pseudo-true parameter θ_0 is in m_j and under some further hypotheses. If $\theta_0 \notin m_j$, both the actual integral and its approximant become exponentially small in n relative to the integral over any m_i containing θ_0 [Haughton (1988, 1989), Poskitt (1987) and Dudley and Haughton (2000)]. The simple and accurate approximation (29.2) applies if and only if μ_j is the Jeffreys prior. The factors $n^{-d_j/2}ML(m_j)$ correspond to the BIC; the remaining constant factors provide the sharpening.

If m_j is a half-space intersected with a parameterized manifold m_i, and m_r is the boundary of m_j, so that $d_j = d_i = d_r + 1$, then J_j is J_i restricted to m_j. Let $MLL_u := \sup_{m_u} \log lik_{(n)}$. We use the approximation

$$\int_{m_j} lik_{(n)} d\mu_j / \int_{m_i} lik_{(n)} d\mu_i \sim J_i(m_i) J_i(m_j)^{-1} \Phi(s_{ij} x_{ir}) \tag{29.3}$$

where Φ is the standard normal distribution function, $s_{ij} = 1$ if $MLL_j = MLL_i$ and $s_{ij} = -1$ if $MLL_j < MLL_i$, and $x_{ir} = [2(MLL_i - MLL_r)]^{1/2}$. According to Dudley and Haughton (2000), either (29.3) is valid with a relative error approaching 0 as $n \to \infty$, or $\int_{m_j} lik_{(n)} d\mu_j / \int_{m_u} lik_{(n)} d\mu_u$ approaches 0 exponentially for any m_u containing the pseudo-true θ_0.

29.4 Application to 2×2 Contingency Tables

We consider seven models $M_j(t)$, $j = 1, ..., 7$, for K independent 2×2 contingency tables, $k = 1, ..., K$. In each model $t = (t_1, ..., t_K)$ where $0 < t_k < 1$ for each k. For our application, the probabilities in the kth table will be written as

	Died	Survived	Total
Treatment	$r_k t_k$	$(1 - r_k) t_k$	t_k
Control	$p_k(1 - t_k)$	$(1 - p_k)(1 - t_k)$	$1 - t_k$

The treatment probability t_k was set by design as $t_k = 1/2$ in six of the seven studies we consider; in the PARIS study, $t_k = 2/3$. In $M_1(t)$, p_k and r_k are unrestricted in $[0, 1]$. The other models $M_j(t)$ will be subsets of $M_1(t)$. In $M_2(t)$, $0 \le p_k \le r_k \le 1$ for each k: the treatment is not beneficial. In $M_3(t)$, $0 \le r_k \le p_k$ for each k: the treatment is not harmful. In $M_j(t)$, $j = 4, 5, 6, 7$, $0 < p_k < 1$, $0 < r_k < 1$ for each k and there is a common odds ratio ψ, i.e.

$$[r_k(1 - p_k)]/[(1 - r_k)p_k] = \psi, \quad k = 1, ..., K. \tag{29.4}$$

$M_7(t)$ will be the full common odds ratio model where $0 < \psi < \infty$. $M_4(t)$, $M_5(t)$ and $M_6(t)$ will be the submodels of $M_7(t)$ where $\psi = 1$, $\psi \ge 1$ and $\psi \le 1$ respectively. Thus $M_4(t)$ is the null hypothesis that the treatment has no effect. Since $M_7(t) = M_5(t) \cup M_6(t)$, $M_7(t)$ can never be the best model.

The models $M_j(t)$ for $K = 1$ will be called $m_j(t)$, where $0 < t = t_1 < 1$ in this case. Then $m_1(t) = m_7(t)$, $m_2(t) = m_5(t)$, and $m_3(t) = m_6(t)$. The total Jeffreys prior measures of these models are, for $0 < t < 1$,

$$J_4(m_4(t)) = \pi, \quad J_1(m_1(t)) = \pi^2[t(1-t)]^{1/2}, \tag{29.5}$$

$$J_2(m_2(t)) = J_5(m_5(t)) = J_3(m_3(t)) = J_6(m_6(t)) = J_1(m_1(t))/2. \tag{29.6}$$

Each model $M_4(t)$ is a flat hyperplane if ψ is a coordinate parameter for $M_7(t)$. Thus $M_5(t)$ and $M_6(t)$ are half-spaces in $M_7(t)$.

29.5 Common Odds Ratio Profile Likelihoods

Suppose there is a common odds ratio ψ as in (29.4). First let $K = 1, p := p_1$, and $y := p/(1-p)$. In terms of y and ψ we find that the likelihood function for a 2×2 table with entries $n_{ij} > 0$, $i, j = 1, 2$, with a dot representing summation over an index, is $\psi^{n_{11}} y^{n_{\cdot 1}} [(1-t)/(1+y)]^{n_{2\cdot}} [t/(1+\psi y)]^{n_{1\cdot}}$ in the $m_1(t)$ model. For fixed ψ and t, the likelihood function clearly is maximized where its derivative equals 0, giving a quadratic equation in y [Yu (1992, eq. (2.4))]. We take the unique positive root, which is the MLE of y for fixed ψ. We will get a $\hat{y}_k$ for each data set, not depending on $t = t_k$. Inserting $\hat{y}_k$ as a function of ψ given the data n_{ij} for each k into the likelihood function and multiplying over k, we get a function $p\ell(\cdot)$ of one variable, ψ, the *profile likelihood*, that has to be maximized to find the MLE $\hat{\psi}$ for our model. Here $\hat{\psi}$ also does not depend on $t_1, ..., t_K$.

Suppose that $p\ell$ is unimodal, as it was in our cases. Then maximizing it on a computer is straightforward. To get, approximately, a 95% credible interval for ψ, i.e. an interval whose posterior probability conditional on $M_7(t) \setminus M_4(t)$ is 0.95, we seek ψ_0 and ψ_1 such that the posterior probabilities of $\psi \leq \psi_0$ and $\psi \geq \psi_1$ are each 0.025. We can do this, with good asymptotic accuracy [Dudley and Haughton 2000)], by finding $\psi_0 < \hat{\psi} < \psi_1$ such that $\log p\ell$ at ψ_0 and at ψ_1 is smaller than at $\hat{\psi}$ by $1.92 \doteq 1.96^2/2$. We do not know a proof that $p\ell$ is unimodal under general conditions. In our cases it appeared to be logarithmically concave on an interval starting at 0 and containing $\hat{\psi}$, but logarithmically convex for larger ψ.

29.6 Jeffreys Priors for Mixture Models

Let $\mathcal{M}_1(t)$ be the model given by (29.1) where the parameters $\theta(k)$, $k = 1, ..., K$, are those of $M_1(t)$, namely $\theta(k) = (p_k, r_k)$, $k = 1, ..., K$. Likewise for $j = 2, ..., 7$ let $\mathcal{M}_j(t)$ be the model included in $\mathcal{M}_1(t)$ whose parameters $\theta(k)$ are those of $M_j(t)$. For each model $\mathcal{M}_j(t)$, the Fisher information matrix has a block structure $F_j = [\begin{smallmatrix} V & 0 \\ 0 & C_j \end{smallmatrix}]$ where V is the $(K-1)\times(K-1)$ Fisher information matrix for the multinomial family $(1, v_1, ..., v_K)$ with the parameters $v_1, ..., v_{K-1}$. Thus $v_K \equiv 1-v_1-\cdots v_{K-1}$, $V_{kk} = v_k^{-1}+v_K^{-1}$ for $k = 1, ..., K-1$ and $V_{kr} = V_{rk} = v_K^{-1}$, $1 \le k < r \le K-1$. We have $\det V = 1/\Pi_{k=1}^K v_k$, as can be seen by way of the parameters $w_k = v_k^{1/2}$: the Jeffreys prior measure for the multinomial $(1, w_1^2, ..., w_K^2)$ family is $2^{K-1}w_K^{-1}dw_1\cdots dw_{K-1} = (\Pi_{k=1}^K v_k)^{-1/2}dv_1\cdots dv_{K-1}$, which is the surface area measure in the positive orthant of the sphere $w_1^2 + \cdots + w_K^2 = 1$, e.g. Kass (1989).

The matrix $C_1 = C_2 = C_3$ is a $2K \times 2K$ diagonal matrix with diagonal entries $v_k t_k/[r_k(1-r_k)]$ and $v_k(1-t_k)/[p_k(1-p_k)]$ for $k = 1, ..., K$. So

$$(\det C_1)^{1/2} = \Pi_{k=1}^K v_k[t_k(1-t_k)]^{1/2}/\{p_k(1-p_k)r_k(1-r_k)\}^{1/2}.$$

Thus the Jeffreys prior measure on $\mathcal{M}_j(t)$ for $j = 1, 2$, or 3 is a product measure on a product of $K+1$ spaces, one being $\mathcal{V}_K$ and the others being squares ($j = 1$) or triangles ($j = 2, 3$) with coordinates (p_k, r_k), $k = 1, ..., K$. Since the likelihood function has the same product form, so does any posterior distribution. Each triangle is the intersection of a half-plane with the corresponding square. Thus the results of Dudley and Haughton (2000) apply for each k. We have

$$\int_0^1\int_0^1\cdots\int_0^1 \sqrt{\det C_1}\,dp_1dr_1\cdots dp_Kdr_K = \pi^{2K}\Pi_{k=1}^K v_k[t_k(1-t_k)]^{1/2}.$$

By symmetry, if the integral is taken only over the region where $r_k \le p_k$ for each k, for $M_3(t)$, it is multiplied by 2^{-K}, and likewise for the $M_2(t)$ region. By normalization of Dirichlet distributions, e.g. Johnson and Kotz (1972, Chap. 40, Sec. 5), the total Jeffreys prior measures of $\mathcal{M}_j(t)$ for $j = 1, 2, 3$ are

$$\mathcal{J}_1(\mathcal{M}_1(t)) = \pi^{2K}\Gamma(3/2)^K\Gamma(3K/2)^{-1}\left[\Pi_{k=1}^K t_k(1-t_k)\right]^{1/2}, \tag{29.7}$$

$$\mathcal{J}_2(\mathcal{M}_2(t)) = \mathcal{J}_3(\mathcal{M}_3(t)) = \mathcal{J}_1(\mathcal{M}_1(t))/2^K. \tag{29.8}$$

The matrix C_4 for $\mathcal{M}_4(t)$ is a $K\times K$ diagonal matrix with entries $v_k/[p_k(1-p_k)]$, $k = 1, ..., K$, so the total Jeffreys prior measure of $\mathcal{M}_4(t)$ is

$$\mathcal{J}_4(\mathcal{M}_4(t)) = \pi^K/(K-1)!. \tag{29.9}$$

For the common odds ratio model $\mathcal{M}_7(t)$ and its submodels $\mathcal{M}_5(t)$ and $\mathcal{M}_6(t)$, the matrix $C_5 = C_6 = C_7$ is a $(K+1)\times(K+1)$ non-diagonal matrix C with

$$C_{11} = \frac{1}{\psi}\sum_{k=1}^K \frac{t_kv_ky_k}{(1+\psi y_k)^2}, \quad C_{r+1,r+1} = \frac{v_r}{y_r}\left[\frac{1-t_r}{(1+y_r)^2} + \frac{\psi t_r}{(1+\psi y_r)^2}\right]$$

and $C_{1,r+1} = C_{r+1,1} = t_r v_r/(1+\psi y_r)^2$ for $r = 1, ..., K$, and $C_{uv} = C_{vu} = 0$ for $2 \le u < v \le K+1$. By row reduction we get a matrix B with

$$B_{11} = C_{11} - \sum_{r=2}^{K+1} \frac{C_{1r}^2}{C_{rr}} = \frac{1}{\psi} \sum_{k=1}^{K} \frac{v_k y_k t_k (1-t_k)}{(1-t_k)(1+\psi y_k)^2 + \psi t_k (1+y_k)^2}, \quad (29.10)$$

$B_{1r} = 0$ for $r = 2, ..., K+1$, $B_{uv} = C_{uv}$ for $2 \le u \le K+1$, $1 \le v \le K+1$, so

$$\det C = \det B = B_{11} \Pi_{r=1}^{K} \frac{v_r[(1-t_r)(1+\psi y_r)^2 + \psi t_r (1+y_r)^2]}{y_r (1+y_r)^2 (1+\psi y_r)^2}. \quad (29.11)$$

We have $\mathcal{J}_7(\mathcal{M}_j(t)) = \mathcal{J}_j(\mathcal{M}_j(t))$ for $j = 5, 6$. By symmetry, $\mathcal{J}_7(\mathcal{M}_5(t)) = \mathcal{J}_7(\mathcal{M}_6(t)) = \mathcal{J}_7(\mathcal{M}_7(t))/2$. Perhaps surprisingly, from the form of B_{11}, the y_k are not conditionally independent given ψ in the Jeffreys prior, even if $t_k = 1/2$ for all k. We integrated $(\det C \det V)^{1/2}$ over its $2K$-dimensional domain, where all the parameters are nonnegative and $v_1 + \cdots + v_{K-1} \le 1$, and $t_1, ..., t_K$ are fixed, to estimate the total Jeffreys prior measure $\mathcal{J}_7(\mathcal{M}_7(t))$ by Monte Carlo calculations for $K \ge 2$. Writing $\mathcal{J}_j(\mathcal{M}_j, K)$ to indicate dependence on K, where in our cases $K = 3, 6$ or 7, while $t_j = 1/2$ except for one value of j where $t_j = 2/3$, our simulations (to 50,000 iterations) gave $\mathcal{J}_5(\mathcal{M}_5, 3) \doteq 16.7$, $\mathcal{J}_5(\mathcal{M}_5, 6) \doteq 14.5$, $\mathcal{J}_5(\mathcal{M}_5, 7) \doteq 9.0$.

29.7 Posterior Probabilities that Models are Best

Our overall prior probability will be $\frac{1}{6}\sum_{j=1}^{6} \nu_j$ where $\nu_j = \mathcal{J}_j/\mathcal{J}_j(\mathcal{M}_j(t))$ is the Jeffreys prior probability on $\mathcal{M}_j(t)$ for $j = 1, ..., 6$. Recall that $\mathcal{M}_7(t)$ cannot be the best model, so ν_7 is omitted from the prior. Let $\mathcal{M}_j := \mathcal{M}_j(t)$.

Let $lik_{(N)}$ be the likelihood function for all K data sets together, given by (29.1). The posterior distribution is then $\pi_N = (\sum_{j=1}^{6} lik_{(N)}\nu_j)/D$, where D is the total mass of $\sum_{j=1}^{6} lik_{(N)}\nu_j$ on $\mathcal{M}_1$. The posterior probability that $\mathcal{M}_j$ is the best model is then $\pi_N(\tilde{\mathcal{M}}_j)$ where $\tilde{\mathcal{M}}_j = \mathcal{M}_j \setminus \bigcup\{\mathcal{M}_r; \mathcal{M}_r \subset \mathcal{M}_j, \mathcal{M}_r \ne \mathcal{M}_j\}$ is the set where $\mathcal{M}_j$ is the best model. The maximum likelihood estimate of v_k is as usual n_k/N where n_k is the sample size for the kth data set and N is the grand total sample size.

Let d_j be (now) the dimension of $\mathcal{M}_j$ for each j. The models $\mathcal{M}_j$ are manifolds for $j = 1, 4$ and 7. For $K \ge 2$ these three models are of different dimensions, with $\mathcal{M}_4 \subset \mathcal{M}_7 \subset \mathcal{M}_1$. Let $ML_{(j)}$ be the maximum of the likelihood over $\mathcal{M}_j$ for N observations. Then by (29.2) we can approximate the ratio of integrals $\int lik_{(N)} d\nu_i / \int lik_{(N)} d\nu_j$ by A_i/A_j where $A_r = ML_{(r)}(2\pi/N)^{d_r/2}/\mathcal{J}_r(\mathcal{M}_r)$ for i and $j = 1, 4$ or 7. Here the maximum likelihoods with respect to the v_k divide out, so we can omit them. Let $\lambda_1 := \int lik_{(N)} d\nu_1$.

Define s_{ijk} and x_{irk} as s_{ij} and x_{ir} respectively for the kth data set in (29.3) with $n = n_k$. We use (29.3) for each data set separately and (29.8) to get for $j = 2, 3$, $\pi_N(\tilde{\mathcal{M}}_j) \sim \lambda_1(2^K + 1)\Pi_{k=1}^K \Phi(s_{1jk}x_{14k})/D$, as $N \to +\infty$, where $v_k > 0$ for all k and $\sim$ denotes asymptotic equivalence except possibly for exponentially small probabilities (Dudley and Haughton, 2000). We also get: $\pi_N(\tilde{\mathcal{M}}_1) \sim \lambda_1[1 - \Pi_{k=1}^K \Phi(s_{12k}x_{14k}) - \Pi_{k=1}^K \Phi(s_{13k}x_{14k})]/D$. By (29.2) and (29.9), $\pi_N(\tilde{\mathcal{M}}_4) = \pi_N(\mathcal{M}_4) \sim (2/N)^{(2K-1)/2} ML_{(4)}(K-1)!/(D\sqrt{\pi})$. Applying (29.2) to $\mathcal{M}_1$ gives $\lambda_1 \sim (2\pi/N)^{(3K-1)/2} ML_{(1)}/\mathcal{J}_1(\mathcal{M}_1)$ where $\mathcal{J}_1(\mathcal{M}_1)$ is given by (29.7). For $\mathcal{M}_7$ we get $\int lik_{(N)} d\nu_7 \sim (2\pi/N)^K ML_{(7)}/\mathcal{J}_7(\mathcal{M}_7)$. Let s_{ij} and x_{ir} be as in (29.3) for the full data set with $n = N$ and m_u replaced by $\mathcal{M}_u$ for each u. Then for $j = 5$ or 6, since $\mathcal{M}_j$ is a half-space in $\mathcal{M}_7$, $\int lik_{(N)} d\nu_j / \int lik_{(N)} d\nu_7 \sim 2\Phi(s_{7j}x_{74})$.

29.8 Data on Long-Term Aspirin Therapy after an MI

The following contingency tables give survival data after long-term treatment (averaging a year or more) following an MI, for clinical trials in which patients were given aspirin or a placebo, for the full data from 7 different studies.

FULL DATA FROM SEVEN STUDIES

	Died	Survived	Total
Aspirin[a]	49	566	615
Placebo	65	559	624
Total	114	1125	1239

	Died	Survived	Total
Aspirin[b]	85	725	810
Placebo	52	354	406
Total	137	1079	1216

	Died	Survived	Total
Aspirin[c]	49	623	672
Placebo	71	597	668
Total	120	1220	1340

	Died	Survived	Total
Aspirin[d]	45	713	758
Placebo	65	706	771
Total	110	1419	1529

	Died	Survived	Total
Aspirin[e]	27	290	317
Placebo	32	277	309
Total	59	567	626

	Died	Survived	Total
Aspirin[f]	103	744	847
Placebo	127	751	878
Total	230	1495	1725

	Died	Survived	Total
Aspirin[g]	246	2021	2267
Placebo	219	2038	2257
Total	465	4059	4524

a: Elwood *et al.* (1974); b: PARIS (1980) study; c: Vogel et al. (1979); d: CDPRG (1980) study; e: Breddin et al. (1980); f: Elwood and Sweetnam (1980); g: AMIS (1980) study.

The numbers in the above seven tables are as in Appendix I of the ATC (1994) survey, which included some updates from the original publications, except for the first row of the table from the PARIS study, not given in ATC (1994).

Patients entered the CDPRG study on average 7 years after their last heart attack. For two other studies, it was also a long time between last MI and entry into the study for some patients. The next two tables give data only on those patients who began treatment within 6 months after their heart attack in those two studies. In the other four studies, all or nearly all patients entered the studies within 6 months of their last MI.

SUBSAMPLES WHO BEGAN TREATMENT WITHIN 6 MONTHS

	Died	Survived	Total
Aspirin[b]	16	157	173
Placebo	18	77	95
Total	34	234	268

	Died	Survived	Total
Aspirin[g]	35	249	284
Placebo	29	234	263
Total	64	483	547

b: PARIS (1980) study (for two surviving patients taking aspirin, the time since last MI was unknown); g: AMIS (1980) study (estimated).

We will also analyze separately the study (CPDRG) and subsamples of two other studies, tabulated as follows, where patients entered treatment later.

SUBSAMPLES WHO BEGAN TREATMENT AFTER 6 MONTHS

	Died	Survived	Total
Aspirin[b]	69	566	635
Placebo	34	277	311
Total	103	843	946

	Died	Survived	
Aspirin[g]	211	1772	1983
Placebo	190	1804	1994
	401	3576	3977

b: PARIS (1980) study; g: AMIS (1980) study (estimated).

29.9 Numerical Results

For the full data sets or either of the subdivisions according to when treatment began, the estimated posterior probabilities of being best, $\hat{\pi}_N(\tilde{\mathcal{M}}_j)$, never rose above 0.0004 for $j = 1, 2, 3$, apparently because of the higher dimension of these models. Thus our analysis selects common odds ratio models.

For the full data from all seven studies, $\hat{\pi}_N(\mathcal{M}_4) = 0.916$ for the null hypothesis, so it is preferred, and $\hat{\pi}_N(\tilde{\mathcal{M}}_6) = 0.083$ for a benefit of aspirin. Exact frequentist confidence intervals, in a sense, for common odds ratios can be computed [Mehta and Walsh (1992)] by the package StatXact. The approximate 95% credible interval for ψ conditional on $\mathcal{M}_7 \setminus \mathcal{M}_4$, and the frequentist approximate 95% confidence interval, is $[0.783, 0.993]$, not quite containing 1; the interval also equals the "mid-p corrected exact" confidence interval for ψ given by StatXact. Thus in a frequentist sense the null hypothesis would be rejected, but by an uncomfortably narrow margin. The contrast between this and the Bayesian outcome is analogous to the well-known "Lindley's paradox," Lindley (1957), see also e.g. Bernardo and Smith (1994, pp. 394, 406-7, 415-6, 422). The unconditional MLE ψ_{MLE} of ψ is 0.882, as are the conditional MLE ψ_{CML}, e.g. Nam (1993), and the Mantel-Haenszel (1959) estimator ψ_{MH}.

For patients who entered the studies within 6 months of their last MI (six data sets), $\hat{\pi}_N(\tilde{\mathcal{M}}_6) = 0.660$ for a benefit of treatment and $\hat{\pi}_N(\mathcal{M}_4) = 0.339$ for the null hypothesis. We have $\psi_{MLE} = \psi_{CML} = 0.770$ and $\psi_{MH} = 0.771$. The 95% credible interval for ψ in $\mathcal{M}_7 \setminus \mathcal{M}_4$ is $[0.651, 0.911]$, which again equals the mid-p corrected exact confidence interval. The null hypothesis $\mathcal{M}_4$ is rejected in a frequentist sense with a 2-sided p-value by the likelihood ratio test equal to 0.0023, or by more precise StatXact computations, 0.0024 or 0.0026. If with probability 2/3 aspirin reduces mortality (by an estimated 23%) and with probability 1/3 is equivalent to a placebo, then consideration of plausible loss functions suggests that aspirin should be recommended.

For patients entering studies more than 6 months after their last MI (three data sets), $\hat{\pi}_N(\mathcal{M}_4) = 0.971$, strongly supporting the null hypothesis.

29.10 Discussion and Conclusions

Since the separation of the data according to the time treatment began was done retrospectively it does not give a statistically clear outcome. The number of possible multiple comparisons is not well-defined. It appears however that medical researchers would consider it unethical to do any further prospective studies of aspirin vs. placebo in the situation of the seven studies we considered.

Aspirin appears to be beneficial only if treatment begins within 6 months after a heart attack. Canner's survey (1987) found benefits of aspirin only during the first year or two of treatment. ATC (1994, p. 96) suggest to the contrary that aspirin (or other antiplatelet) treatment be continued indefinitely after a heart attack. It may be that lower doses of aspirin avoid enough of its negative consequences to provide a benefit beyond two years. We would suggest further clinical trials of that and related questions.

Acknowledgment. We thank Michael Woodroofe for telling us about the paper of Berk (1966).

References

1. AMIS (1980). The Aspirin Myocardial Infarction Study Research Group. The Aspirin Myocardial Infarction Study: Final results, *Circulation*, **62** (suppl. V), V79-V84.

2. ATC (1994). Antiplatelet Trialists' Collaboration. Collaborative overview of randomised trials of antiplatelet therapy–I: Prevention of death, myocardial infarction, and stroke by prolonged antiplatelet therapy in various categories of patients, *British Medical Journal*, **308**, 81–106.

3. Berk, R. H. (1966). Limiting behavior of posterior distributions when the model is incorrect, *Annals of Mathematical Statistics*, **37**, 51–58.

4. Bernardo, J. M. and Smith, A. F. M. (1994). *Bayesian Theory*, Chichester: John Wiley & Sons.

5. Breddin, K., Loew, D., Lechner, K., Oberla, K., and Walter, E. (1980). The German-Austrian aspirin trial: A comparison of acetylsalicylic acid, placebo and phenprocoumon in secondary prevention of myocardial infarction, *Circulation*, **62** (suppl. V), V63–V72.

6. Canner, P. L. (1987). An overview of six clinical trials of aspirin in coronary heart disease, *Statistics in Medicine*, **6**, 255–263.

7. CDPRG (1980). The Coronary Drug Project Research Group. Aspirin in coronary heart disease, *Circulation*, **62** (suppl. V), V59-V62.

8. DerSimonian, R. and Laird, N. (1986). Meta-analysis in clinical trials, *Controlled Clinical Trials*, **7**, 177–188.

9. Dudley, R. M. and Haughton, D. (1997). Information criteria for multiple data sets and restricted parameters, *Statistica Sinica*, **7**, 265–284.

10. Dudley, R. M. and Haughton, D. (2000). Asymptotic normality with small relative errors of posterior probabilities of half-spaces, Preprint.

11. Elwood, P. C., Cochrane, A. L., Burr, M. L., Sweetnam, P. M., Williams, G., Welsby, E., Hughes, S. J., and Renton, R. (1974). A randomized controlled trial of acetyl salicylic acid in the secondary prevention of mortality from myocardial infarction, *British Medical Journal*, **1**, 436–440.

12. Elwood, P. C. and Sweetnam, P. M. (1980). Aspirin and secondary mortality after myocardial infarction, *Circulation*, **62** (suppl. V), V53-V58.

13. Gaver, D. P., Draper, D., Goel, P. K., Greenhouse, J. B., Hedges, L. V., Morris, C. N., and Waternaux, C. (1992). In *Combining Information: Statistical Issues and Opportunities for Research*, Washington D.C.: National Academy Press.

14. Haughton, D. M. A. (1988). On the choice of a model to fit data from an exponential family, *Annals of Statistics,* **16**, 342–355.

15. Haughton, D. (1989). Size of the error in the choice of a model to fit data from an exponential family, *Sankhyā, Series A*, **51** , 45–58.

16. Huber, P. J. (1967). The behavior of maximum likelihood estimates under nonstandard conditions, *Proceedings of the Fifth Berkeley Symposium on Mathematical Statistics and Probability* [1965], vol. **1**, 221–233. Berkeley and Los Angeles: University of California Press.

17. Jeffreys, H. (1946). An invariant form for the prior probability in estimation problems, *Proceedings of the Royal Society of London, Series A*, **186**, 453–461.

18. Johnson, N. L. and Kotz, S. (1972). *Distributions in Statistics* **4**; *Continuous Multivariate Distributions*, New York: John Wiley & Sons.

19. Kass, R. E. (1989). The geometry of asymptotic inference, *Statistical Science,* **4**, 188–219 (with discussion).

20. Lindley, D. V. (1957). A statistical paradox, *Biometrika*, **44**, 187–192.

21. Mantel, N. and Haenszel, W. (1959). Statistical aspects of the analysis of data from retrospective studies of disease, *Journal of the National Cancer Institute*, **22**, 719–748.

22. Mehta, C. R. and Walsh, S. J. (1992). Comparison of exact, mid-p, and Mantel-Haenszel confidence intervals for the common odds ratio across several 2×2 contingency tables, *The American Statistician*, **46**, 146–150.

23. Nam, J.-M. (1993). Bias-corrected maximum likelihood estimator of a log common odds ratio, *Biometrika*, **80**, 688–694.

24. PARIS (1980). The Persantine-Aspirin Reinfarction Study Research Group. Persantine and aspirin in coronary heart disease, *Circulation*, **62**, 449–461.

25. Poskitt, D. S. (1987). Precision, complexity and Bayesian model determination, *Journal of the Royal Statistical Society, Series B*, **49**, 199–208.

26. Schwarz, G. (1978). Estimating the dimension of a model, *Annals of Statistics*, **6**, 461–464.

27. Titterington, D. M., Smith, A. F. M., and Makov, U. E. (1985). *Statistical Analysis of Finite Mixture Distributions*, Chichester: John Wiley & Sons.

28. Vogel, G., Fischer, C., and Huyke, R. (1979). Reinfarktprophylaxe mit Azetylsalizylsäure, *Folia Haematologica*, **106**, 797–803.

29. Yu, K. F. (1992). On estimating standardized risk differences from odds ratios, *Biometrics*, **48**, 961–964.

30

A Depth Test for Symmetry

Peter J. Rousseeuw and Anja Struyf

University of Antwerp, Antwerp, Belgium
FWO, Belgium

Abstract: It was recently shown for arbitrary multivariate probability distributions that angular symmetry is completely characterized by location depth. We use this mathematical result to construct a statistical test of the null hypothesis that the data were generated by a symmetric distribution, and illustrate the test by several real examples.

Keywords and phrases: Angular symmetry, characterization, hypothesis testing, location depth

30.1 Introduction

It is natural to expect of a multivariate location estimator that in the case of a symmetric distribution the population estimate corresponds to the center of symmetry. Rousseeuw and Struyf (2000) prove that for any *angularly symmetric* multivariate distribution the point with maximal location depth [Tukey (1975)] corresponds to the center of angular symmetry, and they give an expression for this maximal depth. Moreover, they show the converse: whenever the maximal depth equals this expression, the distribution has to be angularly symmetric. Based on this characterization we will now construct a test for angular symmetry of a particular distribution, which also gives us more insight in some existing tests for centrosymmetry and uniformity of a spherical distribution.

30.2 Location Depth and Angular Symmetry

Let P be an arbitrary probability distribution on $\mathbb{R}^p$ (with its usual Borel sets) that need not have a density or any moments. We say that P is *angularly symmetric* about a point $\mathbf{c}$ if for any Borel cone A in $\mathbb{R}^p$ (i.e., a Borel set A such that $sA = A$ for any $0 < s < \infty$) it holds that

$$\mathrm{P}(\mathbf{c} + A) = \mathrm{P}(\mathbf{c} - A).$$

It can easily be seen that P is angularly symmetric about $\mathbf{c}$ if and only if $\mathrm{P}|_{\mathbb{R}^p\setminus\{\mathbf{c}\}}$ is angularly symmetric about $\mathbf{c}$. When $\mathrm{P}(\{\mathbf{c}\}) < 1$ we can consider the conditional probability distribution on $\mathbb{R}^p \setminus \{\mathbf{c}\}$ defined as

$$\mathrm{P}'(B) := \frac{\mathrm{P}(B \setminus \{\mathbf{c}\})}{1 - \mathrm{P}(\{\mathbf{c}\})},$$

and then the angular symmetry of P is equivalent to that of P′. Let us now define the mapping $h : \mathbb{R}^p \setminus \{\mathbf{c}\} \to S = S(\mathbf{0}, 1)$ as the radial projection onto the unit sphere, i.e. $h(\mathbf{x}) = (\mathbf{x} - \mathbf{c})/\|\mathbf{x} - \mathbf{c}\|$. Moreover, let

$$\mathrm{P}_h := (\mathrm{P}')_h \tag{30.1}$$

be the law of h. Then P is angularly symmetric about $\mathbf{c}$ if and only if

$$\mathrm{P}_h(B) = \mathrm{P}_h(-B)$$

for any Borel set $B \subset S$, i.e. if P_h is *centrosymmetric* about $\mathbf{0}$. Figure 30.1 illustrates the transformation h and the difference between angularly symmetric and centrosymmetric distributions.

The halfspace location depth was introduced by Tukey (1975) as a tool for analyzing finite data sets. The location depth of any point $\boldsymbol{\theta} \in \mathbb{R}^p$ relative to the data set $X_n = \{\mathbf{x}_1, \ldots, \mathbf{x}_n\} \subset \mathbb{R}^p$ is defined as the smallest number of data points in any closed halfspace with boundary through $\boldsymbol{\theta}$, i.e.

$$ldepth(\boldsymbol{\theta}; X_n) = \min_{\|\mathbf{u}\|=1} \#(H_{\boldsymbol{\theta},\mathbf{u}} \cap X_n)/n \tag{30.2}$$

where $H_{\boldsymbol{\theta},\mathbf{u}} = \{\mathbf{x} \in \mathbb{R}^p; \mathbf{u}'(\mathbf{x} - \boldsymbol{\theta}) \geq 0\}$. This definition can easily be generalized to any probability distribution P on $\mathbb{R}^p$. The location depth of a point $\boldsymbol{\theta}$ relative to P then becomes

$$ldepth(\boldsymbol{\theta}; \mathrm{P}) = \inf_{\|\mathbf{u}\|=1} \mathrm{P}(H_{\boldsymbol{\theta},\mathbf{u}}).$$

Since (30.2) equals zero for $\boldsymbol{\theta}$ lying outside the convex hull of the data, and increases when $\boldsymbol{\theta}$ moves closer to the center of the data, it is often referred to

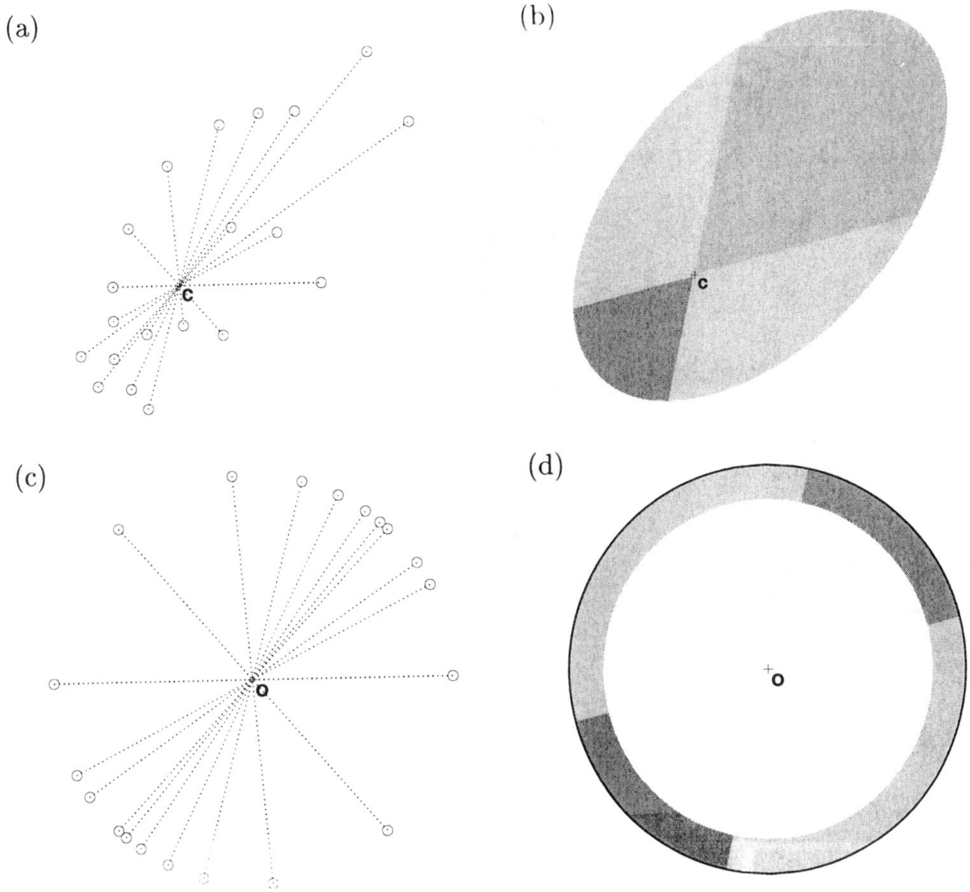

Figure 30.1: Examples of (a) a discrete and (b) a continuous angularly symmetric distribution around $\mathbf{c}$. Transforming (a) and (b) through the mapping $h(\mathbf{x}) = (\mathbf{x}-\mathbf{c})/\|\mathbf{x}-\mathbf{c}\|$ yields the centrosymmetric distributions in (c) and (d)

as *multivariate ranking* [Eddy (1985) and Green (1981)]. This can be visualized by means of the *ldepth regions* D_α given by

$$D_\alpha = \{\boldsymbol{\theta} \in I\!R^p;\ ldepth(\boldsymbol{\theta}; X_n) \geq \alpha\}.$$

These regions are convex sets, with $D_\alpha \subseteq D_{\alpha'}$ for each $\alpha' < \alpha$. The center of gravity of the innermost ldepth region is a point with maximal ldepth, called the *deepest location* or the *Tukey median* of the data set. This multivariate location $T_l^*(\mathrm{P})$ is a robust generalization of the univariate median. Donoho and Gasko (1992) explored the properties of the location depth and of the deepest location for finite data sets. Massé and Theodorescu (1994) and Rousseeuw and Ruts (1999) gave several properties of the location depth for general probability distributions P, that need not have a density. The asymptotic behavior of the depth function was studied by He and Wang (1997) and Massé (1999), and that of the deepest location by Bai and He (1999). Many statistical applications of location depth have been developed. A survey is given in Liu, Parelius, and Singh (1999).

Recently, Rousseeuw, Ruts, and Tukey (1999) proposed the *bagplot*, a bivariate generalization of the univariate boxplot based on location depth. Figure 30.2a depicts the weight of the spleen versus the heart weight of 73 hamsters [Cleveland (1993)]. The Tukey median is given by the black-on-white cross in the center of the data cloud. The dark gray area around the deepest location is called the bag. It is an interpolation of two subsequent depth regions, and contains 50% of the data. The bag corresponds to the box in the classical boxplot. An outlier (plotted as a star) is a point lying outside of the fence, which is obtained by inflating the bag by a factor 3 relative to the Tukey median. The light gray loop ("bolster") is the convex hull of all nonoutlying data points. In this example, 4 hamsters seem to have an extraordinary large spleen and/or heart. The shape of the loop reveals skewness of the data cloud, and suggests a logarithmic transformation of both variables. In Figure 30.2b the bagplot of the transformed data set is given. In this plot, only one outlier remains.

Rousseeuw and Struyf (2000) prove that the location depth can be used to characterize angular symmetry :

Theorem 30.2.1 *A distribution* P *on* $I\!R^p$ *is angularly symmetric about some* $\boldsymbol{\theta}_0$ *if and only if*

$$\mathit{ldepth}(\boldsymbol{\theta}_0;\mathrm{P}) = \frac{1}{2} + \frac{1}{2}\mathrm{P}(\{\boldsymbol{\theta}_0\}).$$

In that case, $\mathit{ldepth}(\boldsymbol{\theta}_0;\mathrm{P})$ *is the maximal value of* $\mathit{ldepth}(\boldsymbol{\theta};\mathrm{P})$ *over all* $\boldsymbol{\theta} \in I\!R^p$.

From Theorem 30.2.1 it follows that any P which is angularly symmetric about some $\boldsymbol{\theta}_0$ with $\mathrm{P}(\{\boldsymbol{\theta}_0\}) > 0$ has a unique center of angular symmetry. Otherwise, there can only be two different centers $\boldsymbol{\theta}_1 \neq \boldsymbol{\theta}_2$ of angular symmetry if P has all its mass on the straight line through $\boldsymbol{\theta}_1$ and $\boldsymbol{\theta}_2$. These corollaries have been proved in another way by Liu (1990) and Zuo and Serfling (2000). A similar property holds for the L^1-median

$$T_{L^1}(\mathrm{P}) = \underset{\mathbf{x}_0}{\operatorname{argmin}}\, \mathrm{E}(\|\mathbf{x} - \mathbf{x}_0\|)$$

since Zuo and Serfling (2000) have proved that a distribution P that is angularly symmetric about a *unique* point $\boldsymbol{\theta}_0$ has $\boldsymbol{\theta}_0$ as an L^1-median.

Remark. Since the condition of the theorem is that $\mathit{ldepth}(\boldsymbol{\theta}_0;\mathrm{P}) = \frac{1}{2} + \frac{1}{2}\mathrm{P}(\{\boldsymbol{\theta}_0\})$ which is at least 1/2, one might think that it would be sufficient to require that $\mathit{ldepth}(\boldsymbol{\theta}_0;\mathrm{P}) \geq \frac{1}{2}$ [this property is called *halfspace symmetry* by

Zuo and Serfling (2000)]. This is not sufficient, however. For instance, take a distribution P_1 which is not angularly symmetric and with $ldepth(0) = \frac{1}{3}$. Then put $P_2 = \Delta_0$ and $P := \frac{1}{2}P_1 + \frac{1}{2}P_2$. For this probability measure P we find $ldepth(0) = \frac{1}{2}(\frac{1}{3}) + \frac{1}{2}(1) = \frac{2}{3} \geq \frac{1}{2}$ although P is not angularly symmetric. As a consequence, Theorem 30.2.1 is stronger than a similar property given by Zuo and Serfling (2000) which is based on halfspace symmetry and requires stricter conditions on P.

For the special case of probability measures with a density it always holds that $\max_{\boldsymbol{\theta}} ldepth(\boldsymbol{\theta}; \mathrm{P}) \leq \frac{1}{2}$, which yields the following corollary of Theorem 30.2.1.

Corollary 30.2.1 *Assume that* P *has a density. Then* P *is angularly symmetric about some* $\boldsymbol{\theta}_0$ *if and only if*

$$ldepth(\boldsymbol{\theta}_0; \mathrm{P}) = \frac{1}{2}.$$

In that case, $ldepth(\boldsymbol{\theta}_0; \mathrm{P}) = \max_{\boldsymbol{\theta}} ldepth(\boldsymbol{\theta}; \mathrm{P})$.

The 'only if' part of this property was previously proved by Rousseeuw and Ruts (1999) in a different way, whereas the 'if' part follows from Theorem 30.2.1.

30.3 A Test for Angular Symmetry

Given $X_n = \{\mathbf{x}_1, \mathbf{x}_2, \ldots, \mathbf{x}_n\}$ and $\boldsymbol{\theta}_0$, Corollary 30.2.1 allows us to use $ldepth(\boldsymbol{\theta}_0; \mathrm{P})$ as a test statistic for the null hypothesis

H_0 : the data come from a continuous distribution P which is angularly symmetric about $\boldsymbol{\theta}_0$.

In the bivariate case, Daniels (1954) gave an expression for the cumulative distribution function

$$F_n(k) = \mathrm{P}(ldepth(\boldsymbol{\theta}_0, X_n) \leq k)$$

under the null hypothesis H_0, namely

$$F_n(k) = \begin{cases} 2(n-2k)\displaystyle\sum_{j=0}^{j'} B\left(n, \frac{1}{2}\right)(n-k+j(n-2k)) & \text{if } k \leq [(n-1)/2], \\ 1 & \text{otherwise.} \end{cases} \tag{30.3}$$

Here $j' = [k/(n-2k)]$ and each term is a probability of the binomial distribution $B(n, \frac{1}{2})$.

The same test statistic has been used by other people to test for different null hypotheses. In two dimensions, the location depth $ldepth(\boldsymbol{\theta}_0, X_n)$ reduces to the bivariate sign test statistic of Hodges (1955) where the null hypothesis H_0 was that P is centrosymmetric about $\boldsymbol{\theta}_0$. By Theorem 30.2.1 we can now see that the real null hypothesis of this test is larger than the original H_0. It actually tests for angular symmetry instead of centrosymmetry, which is a special case. Ajne (1968) uses essentially the same test statistic to test for another null hypothesis, that a distribution on the circle is uniform. Bhattacharyya and Johnson (1969) first noted that both tests use the same test statistic. By the construction in (30.1) and Property 30.2.1 it follows that Ajne's test has a much larger null hypothesis, namely centrosymmetry of the circular distribution. The latter is an illustration of the fact that the masses of all hemispheres of a sphere S in $I\!R^p$ do not suffice to characterize the distribution P on S. Indeed, for *any* centrosymmetric distribution P on S (such as the one in Figure 30.1d) it is true that the mass of each hemisphere equals $\frac{1}{2}$, and hence we cannot distinguish between such distributions on the basis of the masses of hemispheres alone. On the other hand, the masses of all caps of S would be sufficient to characterize P on S by the theorem of Cramér and Wold (1936), since any nontrivial intersection of a halfspace $H \subset I\!R^p$ and S determines a cap of S and vice versa.

Example 1. Let us consider the exchange rates of the German Mark relative to the US Dollar (DEM/USD) and of the Japanese Yen (JPY/USD) from July to December 1998. Every weekday (except on holidays), the exchange rates were recorded at 8PM GMT. Figure 30.3 shows the evolution of the exchange rates over this time period, measured in units of 0.0001 DEM/USD and 0.01 JPY/USD. The data set in Figure 30.4 consists of the 129 differences $(\Delta x, \Delta y)$ between the exchange rates on consecutive days, for both currencies.

From the time series plot in Figure 30.3 as well as from the scatter plot in Figure 30.4 it is clear that Δx and Δy are correlated. We want to test whether these pairs of exchange rate movements come from a bivariate distribution which is angularly symmetric around the origin. Intuitively, we want to test if a movement $(\Delta x, \Delta y)$ of the rates of DEM/USD and JPY/USD with $\Delta y/\Delta x = a$ and $\Delta x > 0$ is equally likely to occur as a movement $(\Delta x, \Delta y)$ with $\Delta y/\Delta x = a$ and $\Delta x < 0$. The location depth of the point $\boldsymbol{\theta}_0 = (0,0)$ can be calculated with the program of Rousseeuw and Ruts (1996). Here, $ldepth(\boldsymbol{\theta}_0, X) = 57$. The p-value equals $F_{129}(57) = 0.88435$, hence we accept the null hypothesis that the data are angularly symmetric around $\boldsymbol{\theta}_0$. Note that large distances or long tails have no effect on this result.

Example 2. The azimuth data [Till (1974, p. 39) and Hand *et al.* (1994)] consist of 18 measurements of paleocurrent azimuths from the Jura Quartzite, Islay. The original measurements (in degrees east of north) are projected onto the circle in Figure 30.5. The location depth of the point $(0, 0)$ relative to this data set equals 1. The p-value is $F_{18}(1) = 0.002197$, so we conclude that the distribution of these data points deviates significantly from angular symmetry.

Example 3. Ferguson *et al.* (1967) described an experiment in which 14 frogs were captured, transported to another place, and then released to see if they would find their way back. The directions in which the 14 frogs started their journey are given by the angles 104°, 110°, 117°, 121°, 127°, 130°, 136°, 145°, 152°, 178°, 184°, 192°, 200°, and 316°. The depth of the origin relative to these data equals 1, which leads to a p-value of 0.020508. Therefore, we reject the null hypothesis that the distribution of the frogs' movements is angularly symmetric around the origin.

30.4 Regression Depth and Linearity of the Conditional Median

Van Aelst *et al.* (2000) present a test similar to the one in Section 30.3 for testing the linearity of the conditional median, using the regression depth [Rousseeuw and Hubert (1999)] as a test statistic. In the simple regression case, the null distribution of that test statistic coincides with the distribution (30.3).

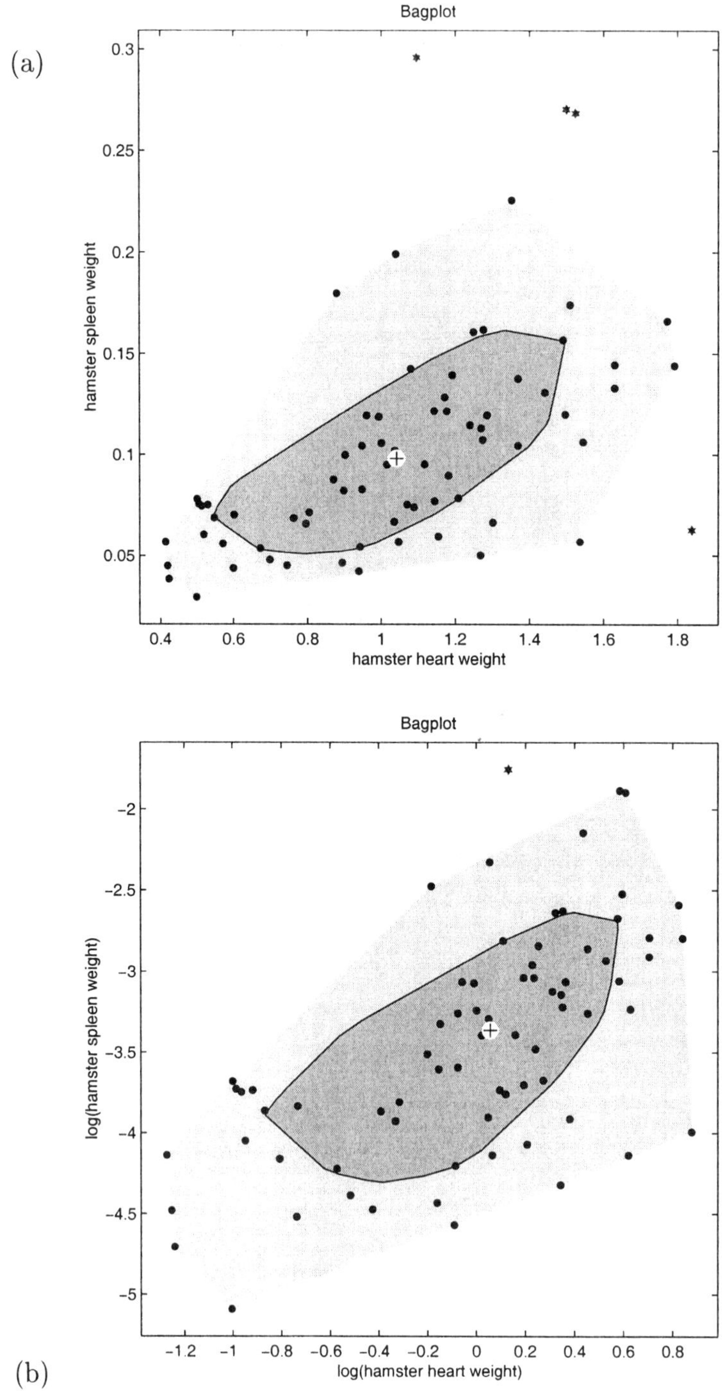

Figure 30.2: (a) Bagplot of the spleen weight versus heart weight of 73 hamsters. (b) Bagplot of the log-transformed data set

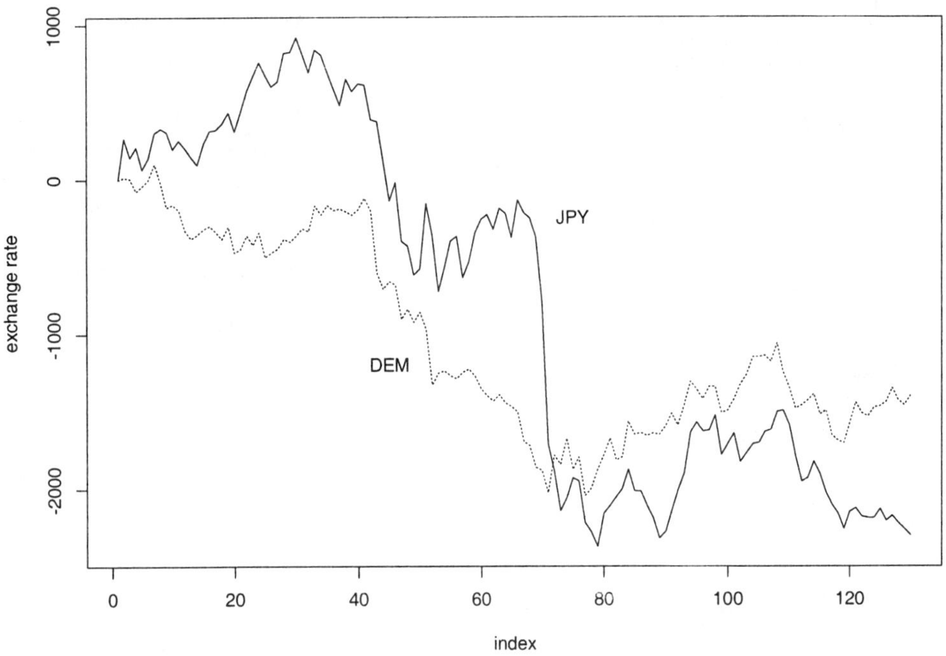

Figure 30.3: Evolution of the exchange rates of DEM/USD (dashed line) and JPY/USD (full line) from July to December 1998

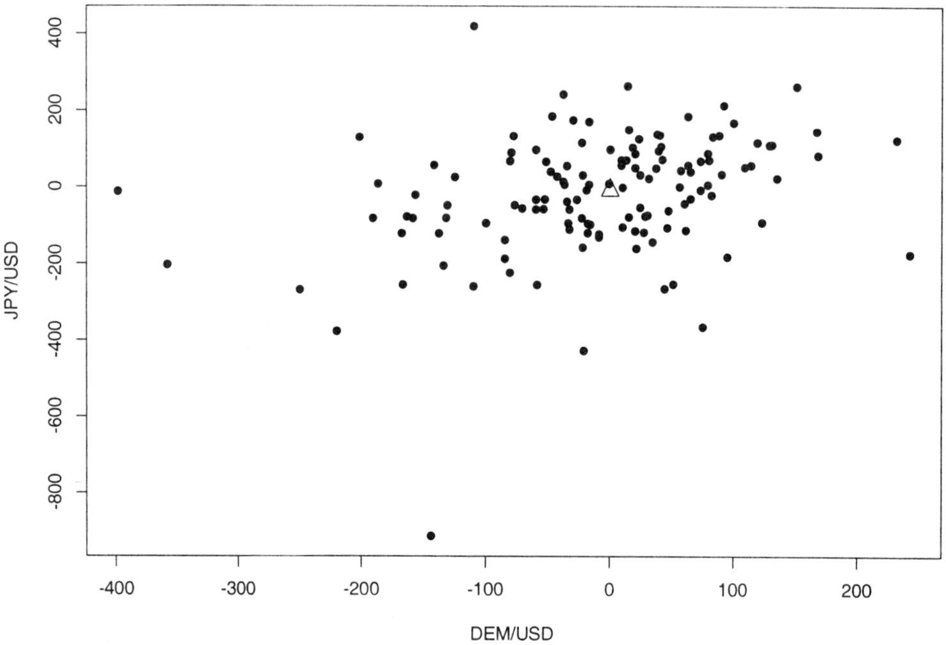

Figure 30.4: Differences between exchange rates on consecutive days for DEM/USD and JPY/USD in the second half of 1998. The origin is depicted as a triangle

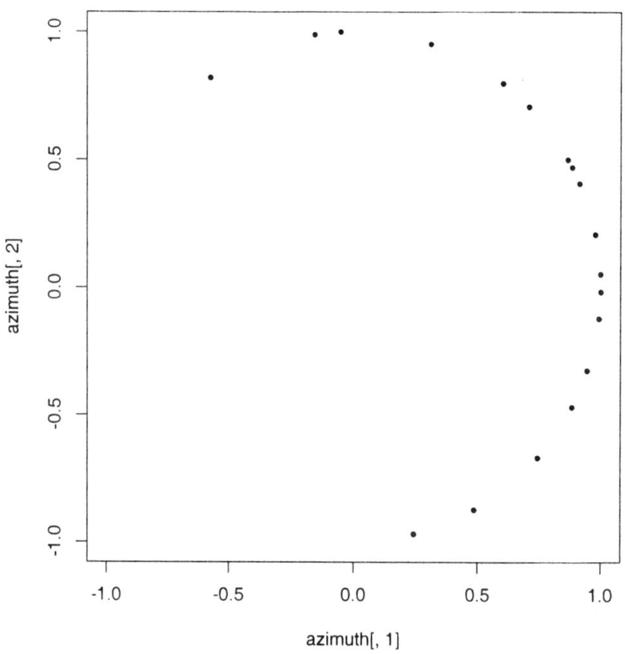

Figure 30.5: The azimuth data

References

1. Ajne, B. (1968). A simple test for uniformity of a circular distribution, *Biometrika*, **55**, 343–354.

2. Bai, Z. and He, X. (1999). Asymptotic distributions of the maximal depth estimators for regression and multivariate location, *Annals of Statistics*, **27**, 1616–1637.

3. Bhattacharyya, G. K. and Johnson, R. A. (1969). On Hodges' bivariate sign test and a test for uniformity of a circular distribution, *Biometrika*, **56**, 446–449.

4. Cleveland, W. S. (1993). *Visualizing Data*, New Jersey: Hobart Press, Summit.

5. Cramér, H. and Wold, H. (1936). Some theorems on distribution functions, *Journal of the London Mathematical Society*, **11**, 290–294.

6. Daniels, H. E. (1954). A distribution-free test for regression parameters, *Annals of Statistics*, **25**, 499–513.

7. Donoho, D. L. and Gasko, M. (1992). Breakdown properties of location estimates based on halfspace depth and projected outlyingness, *The Annals of Statistics*, **20**, 1803–1827.

8. Eddy, W. F. (1985). Ordering of multivariate data, In *Computer Science and Statistics: Proceedings of the 16th Symposium on the Interface* (Ed., L. Billard), pp. 25–30, Amsterdam: North-Holland.

9. Ferguson, D. E., Landreth, H. F., and McKeown, J. P. (1967). Sun compass orientation of the northern cricket frog, *Animal Behaviour*, **15**, 43–53.

10. Green, P. J. (1981). Peeling bivariate data, In *Interpreting Multivariate Data* (Ed., V. Barnett), pp. 3–19, New York: John Wiley & Sons.

11. Hand, D. J., Daly, F., Lunn, A. D., McConway, K. J., and Ostrowski, E. (1994). *A Handbook of Small Data Sets*, London: Chapman & Hall.

12. He, X. and Wang, G. (1997). Convergence of depth contours for multivariate datasets, *Annals of Statistics*, **25**, 495–504.

13. Hodges, J. L. (1955). A bivariate sign test, *Annals of Mathematical Statistics*, **26**, 523–527.

14. Liu, R. Y. (1990). On a notion of data depth based on random simplices, *Annals of Statistics*, **18**, 405–414.

15. Liu, R. Y., Parelius, J., and Singh, K. (1999). Multivariate analysis by data depth: descriptive statistics, graphics and inference, *The Annals of Statistics*, **27**, 783–840.

16. Massé, J. C. (1999). Asymptotics for the Tukey depth, *Technical Report*, Université Laval, Québec, Canada.

17. Massé, J. C. and Theodorescu, R. (1994). Halfplane trimming for bivariate distributions, *Journal of Multivariate Analysis*, **48**, 188–202.

18. Rousseeuw, P. J. and Hubert, M. (1999). Regression depth, *Journal of the American Statistical Association*, **94**, 388–402.

19. Rousseeuw, P. J. and Ruts, I. (1996). Algorithm AS 307: Bivariate location depth, *Applied Statistics (JRSS-C)*, **45**, 516–526.

20. Rousseeuw, P. J. and Ruts, I. (1999). The depth function of a population distribution, *Metrika*, **49**, 213–244.

21. Rousseeuw, P. J., Ruts, I., and Tukey, J. W. (1999). The bagplot: A bivariate boxplot, *The American Statistician*, **53**, 382–387.

22. Rousseeuw, P. J. and Struyf, A. (2000). Characterizing Angular Symmetry and Regression Symmetry, *Technical Report*, University of Antwerp, submitted.

23. Till, R. (1974). *Statistical Methods for the Earth Scientist*, London: MacMillan.

24. Tukey, J. W. (1975). Mathematics and the picturing of data, *Proceedings of the International Congress of Mathematicians*, Vancouver, **2**, 523–531.

25. Van Aelst, S., Rousseeuw, P. J., Hubert, M., and Struyf, A. (2000). The deepest regression method, *Technical Report*, University of Antwerp.

26. Zuo, Y. and Serfling, R. (2000). On the performance of some robust nonparametric location measures relative to a general notion of multivariate symmetry, *Journal of Statistical Planning and Inference*, **84**, 55–79.

31

Adaptive Combination of Tests

Yadolah Dodge and Jana Jurečková

University of Neuchâtel, Neuchâtel, Switzerland
Charles University, Czech Republic

Abstract: In this paper we present a combination of two tests of the linear hypothesis in the linear regression model. The adaptive decision rule which selects the optimal combination of the tests is quite analogous to that which led to the optimal combinations of estimators proposed by the authors.

Keywords and phrases: Regression, adaptive regression, testing hypothesis, adaptive estimation

31.1 Introduction

Consider the linear regression model

$$\mathbf{Y} = \mathbf{X}\beta + \mathbf{z} \tag{31.1}$$

where $\boldsymbol{Y}$ is a $(n \times 1)$ vector of observations with the design matrix $\boldsymbol{X}$ of order $n \times p$ such that $x_{i1} = 1$, $i = 1..., n$, $\boldsymbol{\beta}$ is a $(p \times 1)$ vector parameter and $\boldsymbol{z}$ is an $(n \times 1)$ vector of independent errors, identically distributed according to a distribution function (d.f.) F, which is generally considered as unknown; we only assume that F belongs to some family $\mathcal{F}$ of distribution functions. The problem is that of estimating the parameter $\boldsymbol{\beta}$. Notice that the first component of β is an intercept. Depending on the estimation procedures involved, we shall have to put some more restrictions on the class of underlying distributions, like existence of some moments, positivity of the density in some interval or at least at some points, etc. The regularity conditions are usually mild. Even the condition of symmetry is not necessary and we impose it rather to avoid the problem of eventual nonidentifiability of the intercept.

There are different methods of estimating the unknown parameters. Three of such methods are minimization of (1) sum of squares errors; (2) sum of absolute errors; and (3) maximum of absolute errors. These three methods are members of the class called L_p-estimators which are obtained by minimizing what is known as Minkowsky metric or L_p-norm (criteria) defined as $[\sum | z_i |^p]^{1/p}$ with $p \geq 1$. If we set $p = 1$, we obtain what is known as an absolute or city block metric or L_1-norm. The minimization of this criteria is called L_1-norm or Least Absolute Deviations (LAD) method. If $p = 2$, we have what is known as a Euclidian metric or L_2-norm. The minimization of this distance is known as the least squares (LS) method. The classical approach to the regression problem uses this method. If we minimize the L_p-norm for $p = \infty$ we have the minimax method. There are, however, many other methods for estimating β. For a complete treatment of some of the major methods the reader is referred to Birkes and Dodge (1993) and to Jurečková and Sen (1996).

While the method of least squares enjoys well-known properties within Gaussian parametric models, it is recognized that outliers, which arise from heavy-tailed distributions, have an unusually large influence on the resulting estimates. Outlier diagnostics statistics based on least squares have been developed to detect observations with a large influence on the least squares estimation. For documents related to such diagnostics the reader is referred to Cook and Weisberg (1982, 1994).

While many location estimators were extended in a straightforward way to the linear regression model, this, until recently, was not the case of L-estimators (linear combinations of order statistics). The attempts which were made either were computationally difficult or did not keep the convenient properties of location estimators. In 1978, Koenker and Bassett introduced the concept of regression quantile which provided a basis for L-procedures in the linear model. The trimmed least squares estimator, suggested by the same authors, is an extension of the trimmed mean to the linear model.

At present, there exists a big variety of robust estimators in the linear model. Besides distributional robustness, estimators resistant to the leverage points in the design matrix and possessing a high breakdown point [introduced originally by Hampel (1968); the finite sample version is studied in Donoho and Huber (1983)] were developed and studied.

Summarizing, the last 40 years brought a host of statistical procedures, many of them enjoying excellent properties and being equipped with a computational software. On the other hand, this progress has put an applied statistician into a difficult situation: if he needs to fit his data with a regression hyperplane, he is hesitating which procedure should he use. His decision is then sometimes based on peculiar reasons. If he had some more information on the model, he could choose the estimation procedure accordingly. If his data are automatically collected by a computer and he is not able to make any diagnostics then he might use one of the high breakdown-point estimators but he usually would not, either due to the difficult computation or perhaps due to his scepticism. Then, finally, he might prefer the simplicity to the optimality and good asymptotics and use the classical least squares, LAD-method or one of other reasonably simple methods.

An idea of what to advise for such a situation, instead of to concentrate on one method, is to combine two convenient methods and in such a way diminish the eventual shortages of both. This idea, simple as it is, was surprisingly not very much elaborated until recently. Arthanari and Dodge (1981) introduced an estimation method based on a direct convex combination of LAD- and LS-methods. Dodge (1984) extended this method to a convex combination of LAD and Huber's M-estimation methods and supplemented that by a numerical study based on simulated data.

Later on Dodge and Jurečkovà (1988) observed that the convex combination of two methods could be adapted in the sense that the optimal value of the convex combination methods coefficient, minimizing the resulting asymptotic variance, could be estimated from the observations. The resulting estimator attains a minimum asymptotic variance over all estimators of this kind and for any general distribution with a nuisance scale. Dodge and Jurečková (1988, 1991) then extended the adaptive procedure to the combinations of LAD-method with M-estimation and with trimmed least squares estimation methods. An analogous idea can be used to develop for the combination of two tests of the linear hypothesis in the linear regression model. In Section 31.3 we develop and discuss optimal combination of tests.

In what follows we briefly describe the general idea, leading to a construction of an adaptive convex combination of two estimation methods.

31.2 Adaptive Combination of Estimators

We shall consider a family of symmetric densities indexed by an appropriate measure of scale:

$$\mathcal{F} = \left\{ f : f(z) = s^{-1} f_0\left(z/s\right), \;\; s > 0 \right\} . \tag{31.2}$$

The shape of f_0 is generally unknown; it only satisfies some regularity conditions given later and the unit element $f_0 \in \mathcal{F}$ has $s_0 = 1$. When we would like to combine an L_1-estimator with another class of estimators, then we take $s = 1/f(0)$.

Generally, the scale characteristic s should have a reasonably consistent estimator s_n based on $Y_1, \ldots, Y_n$. Moreover, it would be natural to assume that the estimator s_n is regression-invariant and scale-equivariant, i.e.

$$\begin{aligned}
&\text{(a)} \quad s_n(\boldsymbol{Y}) \xrightarrow{p} s \;\text{ as }\; n \to \infty && (31.3)\\
&\text{(b)} \quad s_n(\boldsymbol{Y} + \boldsymbol{X}\boldsymbol{b}) = s_n(\boldsymbol{Y}) \;\text{ for any }\; \boldsymbol{b} \in R^p \;\text{ (regression-invariance)}\\
&\text{(c)} \quad s_n(c\boldsymbol{Y}) = c s_n(\boldsymbol{Y}) \;\text{ for }\; c > 0 \;\text{ (scale-equivariance).}
\end{aligned}$$

The idea of the adaptative estimator as introduced in Dodge and Jurečkovà (2000) is as follows:

Let $\boldsymbol{T}_n(\delta)$ be a solution of the minimization problem

$$\sum_{i=1}^{n} \rho\left(\frac{Y_i - \boldsymbol{x}_i' \boldsymbol{t}}{s_n}\right) := \min \tag{31.4}$$

with respect to $\boldsymbol{t} \in R^p$, where s_n is a consistent estimator of s and $\rho(z) = \delta\rho_1(z) + (1-\delta)\rho_2(z)$, $0 \le \delta \le 1$, where $\rho_1(z)$ and $\rho_2(z)$ are symmetric (convex) discrepancy functions defining the respective estimators. For instance, $\rho_1(z) = \mid z \mid$ and $\rho_2(z) = z^2$ if we want to combine LAD and LS estimators. Then $\sqrt{n}(\boldsymbol{T}_n(\delta) - \beta)$ has an asymptotically normal distribution $N_p(\mathbf{0}, \boldsymbol{Q}^{-1}\sigma^2(\delta, \rho, f))$, where $\mathbf{Q} = (\mathbf{X}'\mathbf{X})$. Using $\delta = \delta_0$ which minimizes $\sigma^2(\delta, \rho, f)$ with respect to δ, $0 \le \delta \le 1$, we get an estimator $\boldsymbol{T}_n(\delta_0)$ minimizing the asymptotic variance for a fixed distribution shape. Typically, $\sigma^2(\delta, \rho, f)$ depends on f only through two moments of f_0.

That is, in the case of a least squares estimator

$$\sigma_0^2 = \int x^2 f_0(x) dx$$

and

$$E_1^0 = \int \mid x \mid f_0(x) dx,$$

and in the M-estimation case

$$\sigma_0^2 = \int (\psi(x))^2 f_0(x) dx$$

and

$$E_1^0 = \int \mid \psi(x) \mid f_0(x) dx,$$

where $\psi = \rho_2'$.

It is, in fact, a product of s^2 and of an expression containing δ, σ_0^2,and E_1^0 which, being convex in $0 \le \delta \le 1$, could be well minimized with respect to δ. Instead of $\sigma^2(\delta, \rho, f)$, we then minimize its estimation with σ_0^2 and E_1^0 being replaced by their estimators $\hat{\sigma}_0^2$ and $\hat{E}_1^0$ based on the data; denote $\hat{\delta}_0$ the minimizing value.

Then we shall consider the function

$$\rho(z) = \hat{\delta}_0 \rho_1(z) + (1 - \hat{\delta}_0)\rho_2(z)$$

and the minimization (31.4) leads to the estimator $\boldsymbol{T}_n(\hat{\delta}_0)$ such that

$$\sqrt{n}(\boldsymbol{T}_n(\hat{\delta}_0) - \beta) \xrightarrow{\mathcal{D}} N_p(\mathbf{0}, \boldsymbol{Q}^{-1}\sigma_0^2(\rho, f)) \tag{31.5}$$

where

$$\sigma_0^2(\rho, f) = \min_{0 \le \delta \le 1} \sigma^2(\delta, \rho, f). \tag{31.6}$$

Hence, $\boldsymbol{T}_n(\hat{\delta}_0)$ attains the minimum possible asymptotic variance among all solutions of (31.4) corresponding to the pertaining distribution shape. If $\hat{\delta}_0 = 1$ or 0, then $\boldsymbol{T}_n(\hat{\delta}_0)$ coincides with the estimator generated by ρ_1 or ρ_2, respectively.

If we combine L_1-estimation with another estimation procedure, we take $s = 1/f(0)$ due to the fact that the asymptotic covariance matrix of the L_1-estimator is proportional to $1/4(f(0))^2$. The characteristic $s = 1/f(0)$ is simple and robust but not easily estimable, similarly as the density itself.

Recently Dodge and Jurečková (1995) proposed two estimators of s, based on regression quantiles, satisfying (a)–(c) in (31.3). These estimators, histogram and kernel type, respectively, do not need any initial estimator $\hat{\beta}$ of β, and seem to be very convenient for the adaptive convex combinations of estimators of β.

Various procedures based on a combination of two or several estimations, mostly adaptive in some well defined sense are described in Dodge and Jurečkovà (2000). The advantage of these procedures is their simplicity and also that the ideas leading to their construction are acceptable for applied statisticians.

31.3 Adaptive Combination of Tests

We have considered adaptive convex combinations of two kinds of estimators. An analogous idea can be exploited for the combinations of two tests of the linear hypothesis in the linear regression model. It turns out that the same adaptive decision rules, that led to the optimal combinations of estimators also lead to the optimal combinations of two tests. The efficiencies of estimating and testing procedures are also closely related; it is well known that the Pitman efficacy of the test coincides with the reciprocal standard deviation of the asymptotic distribution of the corresponding estimator. Noting that, the adaptive combination of two tests can also be considered as the test corresponding to the adaptive combination of two pertaining estimators.

In this section, we shall briefly illustrate how to use the adaptive procedures developed in Section 31.2 for an adaptive combination of tests. We shall start with some general remarks, and describe in more detail the important special cases of convex combinations of the F-test with the median-type test and the M-test with the median-type test, respectively.

Consider the linear regression model

$$\boldsymbol{Y} = \boldsymbol{X}\beta + \boldsymbol{z} \tag{31.7}$$

where $\boldsymbol{Y}$ is an $(n \times 1)$ vector of observations, $\boldsymbol{X} = \boldsymbol{X_n}$ is an $(n \times p)$ design matrix, $\boldsymbol{\beta}$ is a $(p \times 1)$ vector of unknown parameters and $\boldsymbol{z}$ is an $(n \times 1)$ vector of independent errors, identically distributed with the density $f(z)$ satisfying

$$f(z) = f(-z), \quad z \in R^1 \tag{31.8}$$

$$0 < f(0) < \infty \text{ and } f \text{ has a bounded derivative in a neighborhood of } 0$$

$$0 < \sigma^2 = \int z^2 f(z) dz < \infty$$

$$\text{and } f(z) = (1/s) \ f_0\left(z/s\right), \ s > 0 \tag{31.9}$$

where f_0 is a fixed (but generally unknown) symmetric density such that $f_0(0) = 1$ and the scale statistic s is $s = 1/f(0)$. Denote $\mathcal{F} = \{f : f(z) = (1/s) f_0(z/s), s > 0\}$ the family of densities, satisfying (31.8) to (31.9), indexed by s.

We shall consider the hypothesis

$$H_0: \ \boldsymbol{\beta} = \mathbf{0}; \tag{31.10}$$

but obviously the procedure could also be applied to more general hypotheses of the type $H: \ \mathbf{A}\boldsymbol{\beta} = \mathbf{b}$.

We can generally consider three types of tests of the linear hypothesis:

(i) the Wald type tests,

(ii) the likelihood ratio type tests,

(iii) the score type tests.

(i) The Wald type test of H_0 is based on the quadratic form of an appropriate estimator $\hat{\beta}$ of β,

$$\hat{\beta}'\mathbf{V}^{-1}\hat{\beta}, \tag{31.11}$$

where $\mathbf{V}$ is the covariance matrix of $\hat{\beta}$ or its approximation. Typically, (31.11) has asymptotically χ^2 distribution under H_0 and a noncentral χ^2 distribution with the noncentrality parameter $\hat{\beta}_0'\mathbf{V}^{-1}\hat{\beta}_0$ under the local (Pitman) alternative

$$H_n: \ \hat{\beta} = \hat{\beta}_n = n^{-1/2}\beta_0 \text{ as } n \to \infty \tag{31.12}$$

with a fixed $\beta_0 \in I\!R^p$. The problem may be that of estimating the covariance matrix $\mathbf{V}$.

(ii) and (iii): The likelihood ratio tests and the score type tests are closely related. The latter has a simpler linear form: for instance, for the model $f(x, \theta)$ with the scalar parameter θ, and hypothesis $H^*: \ \theta = \theta_0$, the parametric score test is based on the statistic

$$\sum_{i=1}^{n} \frac{\partial}{\partial \theta} \log \ f(X_i, \theta)|_{\theta_0}. \tag{31.13}$$

The score tests can be performed with less or no estimation of unknown parameters and matrices, compared with the two other tests; moreover, the sign-rank tests, which asymptotically have forms of score tests, need even less estimation due to their invariance. For this reason, we recommend using the ranks rather than the regression quantiles or LAD estimation for testing various hypotheses in the linear model.

The score tests belong to the class of *M-tests* of H_0, which are closely connected with the M-estimation of β. The M-test of H_0 is characterized by the test criterion

$$\frac{1}{\hat{\sigma}_n^2}\mathbf{M}_n'\mathbf{M}_n, \tag{31.14}$$

where

$$\mathbf{M}_n = (n\mathbf{Q}_n)^{-1/2}\sum_{i=1}^{n}\mathbf{x}_i\psi(Y_i), \ \ \mathbf{Q}_n = n^{-1}\mathbf{X}_n'\mathbf{X}_n, \tag{31.15}$$

$\mathbf{x}_i'$ is the ith row of $\mathbf{X}_n$, $i = 1,\ldots,n$, and $\hat{\sigma}_n^2$ is an estimator of the functional $\int_{-\infty}^{\infty}\psi^2(x)dF(x)$; ψ is the (skew-symmetric) score function generating the M-estimator. Then the criterion (31.14) has asymptotically the χ_p^2 distribution under H_0, and the noncentral χ_p^2 distribution under H_n with the noncentrality parameter

$$\frac{\gamma^2(\psi,f)}{\int_{-\infty}^{\infty}\psi^2(x)dF(x)}\beta_0'\mathbf{Q}^{-1}\beta_0, \qquad \gamma(\psi,f) = \int_{-\infty}^{\infty} f(x)d\psi(x). \tag{31.16}$$

The noncentrality parameter (31.16) is equal to the reciprocal square of the *Pitman efficacy* of the test (31.14). Also notice that (31.16) is reciprocal to the asymptotic variance of the M-estimator.

The *sign-rank test* criterion for H_0 has the form

$$(\mathbf{S}_n^+)'\mathbf{S}_n^+, \tag{31.17}$$

where

$$\mathbf{S}_n^+ = (n\mathbf{Q}_n)^{-1/2}\sum_{i=1}^{n}\mathbf{x}_i\varphi^+\Big(\frac{R_i^+}{n+1}\Big), \tag{31.18}$$

where R_i^+ is the rank of $|Y_i|$ among $|Y_1|,\ldots,|Y_n|$, and $\varphi^+ : [0,1) \mapsto \mathbb{R}_+^1$ is a nondecreasing score function, square-integrable on $(0,1)$, and such that $\varphi^+(0) = 0$. Denote also

$$\varphi(u) = \begin{cases} \varphi^+(2u-1) & \text{if} \quad \frac{1}{2} \le u < 1 \\ -\varphi(1-u) & \text{if} \quad 0 < u \le \frac{1}{2}. \end{cases} \tag{31.19}$$

The test criterion (31.17) has asymptotically the χ_p^2 distribution under H_0 and the noncentral χ_p^2 distribution under H_n with noncentrality parameter

$$\frac{\gamma^2(\varphi,f)}{\int_0^1\varphi^2(u)du}\beta_0'\mathbf{Q}^{-1}\beta_0, \qquad \gamma(\varphi,f) = \int_{-\infty}^{\infty} -f'(x)\varphi(F(x))dx. \tag{31.20}$$

Moreover, under H_0, the sign-rank statistic (31.18) admits the asymptotic representation

$$\mathbf{S}_n^+ = (n\mathbf{Q}_n)^{-1/2} \sum_{i=1}^{n} \mathbf{x}_i \varphi(F(Z_i)) + o_p(1) \text{ as } n \to \infty, \tag{31.21}$$

and hence it asymptotically has the form of the M-test.

If we have no knowledge about the shape of the distribution, we could recommend using the pure sign-rank test, which is distribution-free under H_0. However, if we believe that our distribution is close to normal, we can try to use a combination of the F-test of H_0 with the median-type test, which we shall describe in the next subsection. Another possibility would be to combine the F-test with another simple sign-rank test, such as the Wilcoxon. On the other hand, to combine two sign-rank tests (say the median and Wilcoxon ones) makes no sense because the resulting test is distribution-free, and hence would not adapt itself to the underlying distribution.

31.3.1 Adaptive combination of F-test and median-type test

The classical F-test and the median-type test are the counterparts of the LS and LAD estimators, respectively. The classical F-test of H_0 can be described by the criterion (31.14)–(31.15) with $\psi(z) = z, \ z \in I\!R^1$. The F-test of H_0 is based on the criterion

$$\frac{\mathbf{T}_n'\mathbf{T}_n}{\hat{\sigma}_n^2}, \text{ where } \mathbf{T}_n = (n\mathbf{Q}_n)^{-1/2} \sum_{i=1}^{n} \mathbf{x}_i Y_i \tag{31.22}$$

and $\hat{\sigma}_n^2$ is an estimator of σ^2. On the other hand, the median test of H_0 is also of type (31.14)–(31.15) with

$$\mathbf{M}_n = \mathbf{S}_n^+ = (n\mathbf{Q}_n)^{-1/2} \sum_{i=1}^{n} \mathbf{x}_i \, \frac{1}{2} \text{sign } Y_i. \tag{31.23}$$

While the F-test is optimal for f normal, the median test is the locally most powerful signed-rank test of H_0 for the double exponential distribution of errors. We are looking for the optimal convex combination

$$\mathbf{W}_n = (1-\delta)\frac{1}{s}\mathbf{T}_n + \delta \mathbf{S}_n^+, \ 0 \le \delta \le 1. \tag{31.24}$$

The test is considered optimal when it has the maximum Pitman efficacy over $\delta \in [0, 1]$. Notice that the test (31.24) is of the type (31.14) with the ψ function

$$\psi(z) = \frac{1}{s}(1-\delta)z + \delta \, \frac{1}{2}\text{sign } z, \ z \in I\!R^1. \tag{31.25}$$

As it follows from Section 31.2, $\mathbf{W}_n$ is asymptotically normally distributed under the hypothesis H_0; more precisely,

$$\mathbf{W}_n \xrightarrow{\mathcal{D}} N_p(\mathbf{0}, \kappa^2 \mathbf{I}_p), \tag{31.26}$$

where

$$\kappa^2 = (1-\delta)^2 \frac{\sigma^2}{s^2} + \frac{\delta^2}{4} + \delta(1-\delta)\frac{E_1}{s} = \frac{1}{s^2}\sigma^2(\psi, F, s) \tag{31.27}$$

with

$$\sigma^2(\psi, F, s) = \frac{s^2}{4}\left\{4\,(1-\delta)^2\,\sigma_0^2 + 4\delta\,(1-\delta)\,E_1^0 + \delta^2\right\}. \tag{31.28}$$

The asymptotic distribution of $\mathbf{W}_n$ under the Pitman alternative H_n (31.12) will be normal with a nonzero expectation,

$$\mathbf{W}_n \xrightarrow{\mathcal{D}} N_p(\mu, \kappa^2 \mathbf{I}_p), \quad \mu = \mathbf{Q}^{-1/2}\beta_0. \tag{31.29}$$

If we knew the parameters E_1, σ^2, s, we would use the test criterion

$$\mathcal{W}_n = \kappa^{-2}\mathbf{W}_n'\mathbf{W}_n = s^2\sigma^{-2}(\psi, F, s)\mathbf{W}_n'\mathbf{W}_n, \tag{31.30}$$

which is asymptotically χ_p^2 distributed under H_0 and asymptotically noncentral χ_p^2 distributed under H_n with the noncentrality parameter

$$s^2\sigma^{-2}(\psi, F, s)\beta_0'\mathbf{Q}^{-1}\beta_0.$$

The test would have the maximal efficacy for δ minimizing $\sigma^2(\psi, F, s)$ given in (31.28). Thus, we conclude that $\hat{\delta}_n$, leading to the optimal adaptive combination of the least squares and L_1 estimators leads to the optimal adaptive combination of the F-test and the median-type test.

31.3.2 Adaptive combination of M-test and median-type test

Let us consider the combination of the M-test (31.14) with the Huber score function and the median-type test (31.23). We shall deal with the mixed criterion

$$\mathbf{W}_n = (n\mathbf{Q}_n)^{-1/2}\sum_{i=1}^{n}\mathbf{x}_i\left\{(1-\delta)\frac{2}{\gamma_0}\psi\left(\frac{Y_i}{\hat{s}_n}\right) + \delta \text{ sign } Y_i\right\}. \tag{31.31}$$

Equation (31.31) implies that, under H_0, $\mathbf{W}_n$ has an asymptotically normal distribution $N(\mathbf{0}, \tilde{\kappa}^2\mathbf{I}_p)$, where

$$\tilde{\kappa}^2 = (1-\delta)^2\frac{4\sigma_0^2}{\gamma_0^2} + \delta^2 + \delta(1-\delta)\frac{4E_1^0}{\gamma_0} = \frac{4}{s^2}\sigma^2(F, \delta, k) \tag{31.32}$$

with

$$\begin{aligned}\sigma^2(F,\delta,k) &= \frac{s^2}{4}\int \psi^2\left(\frac{x}{s}\right) f(x)dx \qquad (31.33)\\ &= \frac{s^2}{4}\left\{(1-\delta)^2\frac{4\sigma_0^2}{\gamma_0^2}+\delta^2+\frac{4}{\gamma_0}\delta\,(1-\delta)\,E_1^0\right\}.\end{aligned}$$

The asymptotic distribution of $\mathbf{W}_n$ under the Pitman alternative H_n (31.12) will be normal with a non zero expectation,

$$\mathbf{W}_n \xrightarrow{\mathcal{D}} N_p(\mu, \tilde{\kappa}^2\mathbf{I}_p), \quad \mu = \mathbf{Q}^{-1/2}\beta_0. \qquad (31.34)$$

If we knew the parameters E_1^0, σ_2^0, s, we would use the test criterion

$$\mathcal{W}_n = \tilde{\kappa}^{-2}\mathbf{W}_n'\mathbf{W}_n = \frac{s^2}{4}\sigma^{-2}(\psi,\delta,k)\mathbf{W}_n'\mathbf{W}_n, \qquad (31.35)$$

which is asymptotically χ_p^2 distributed under H_0, and asymptotically noncentral χ_p^2 distributed under H_n with the noncentrality parameter $\tilde{\kappa}^{-2}\beta_0'\mathbf{Q}^{-1}\beta_0$. The test would have the maximal efficacy for δ minimizing $\tilde{\kappa}^2$. Hence, the optimal δ would minimize $\sigma^2(F,\delta,k)$ given in (31.33) and would coincide with δ_0 in (31.36). The unknown E_1^0, σ_0^2, and γ_0 we estimate by $\hat{E}_1^0$, $\hat{\sigma}_0^2$, and $\hat{\gamma}_0$ defined in (31.37)–(31.39).

$$\delta_0 = \begin{cases} 0 & \text{if } 2\sigma_0^2 \le E_1^0\gamma_0 \left(\le \frac{\gamma_0^2}{2}\right) \\ \dfrac{4\sigma_0^2 - 2E_1^0\gamma_0}{4\sigma_0^2 - 4E_1^0\gamma_0 + \gamma_0^2} & \text{if } E_1^0 \le \frac{\gamma_0^2}{2} \text{ and } E_1^0\gamma_0 < 2\sigma_0^2 \\ 1 & \text{if } \frac{\gamma_0^2}{2} < E_1^0\gamma_0 < 2\sigma_0^2 \end{cases} \qquad (31.36)$$

$$\begin{aligned}\widehat{E}_1^0 &= \frac{1}{ns_n}\sum_{i=1}^n \left\{\left(Y_i - \boldsymbol{x}_i'\hat{\boldsymbol{\beta}}\left(\tfrac12\right)\right) I\left[\left|Y_i - \boldsymbol{x}_i'\hat{\boldsymbol{\beta}}\left(\tfrac12\right)\right| \le ks_m\right]\right\} \qquad (31.37)\\ &\quad + \frac{k}{n}\sum_{i=1}^n\left\{1 - I\left[\left|Y_i - \boldsymbol{x}_i'\hat{\boldsymbol{\beta}}\left(\tfrac12\right)\right| \le ks_m\right]\right\}\end{aligned}$$

$$\begin{aligned}\widehat{\sigma}_0^2 &= \frac{1}{ns_n^2}\sum_{i=1}^n \left\{\left(Y_i - \boldsymbol{x}_i'\hat{\boldsymbol{\beta}}\left(\tfrac12\right)\right)^2 I\left[\left|Y_i - \boldsymbol{x}_i'\hat{\boldsymbol{\beta}}\left(\tfrac12\right)\right| \le ks_m\right]\right\} \qquad (31.38)\\ &\quad + \frac{k^2}{n}\sum_{i=1}^n\left\{1 - I\left[\left|Y_i - \boldsymbol{x}_i'\hat{\boldsymbol{\beta}}\left(\tfrac12\right)\right| \le ks_m\right]\right\}\end{aligned}$$

$$\widehat{\gamma}_0 = \frac{1}{n}\sum_{i=1}^n I\left[\left|Y_i - \boldsymbol{x}_i'\hat{\boldsymbol{\beta}}\left(\tfrac12\right)\right| \le ks_m\right]. \qquad (31.39)$$

Thus we conclude that $\hat{\delta}_n$, leading to the optimal adaptive combination of the M-estimator and L_1 estimators leads to the optimal adaptive combination of the M-test and the median-type test.

References

1. Arthanari, T. S. and Dodge, Y. (1981). *Mathematical Programming in Statistics*, New York: John Wiley & Sons.

2. Birkes, D. and Dodge, Y. (1993). *Alternative Methods of Regression*, New York: John Wiley & Sons.

3. Cook, R. D. and Weisberg, S. (1982). *Residuals and Inference in Regression*, London: Chapman & Hall.

4. Cook, R. D. and Weisberg, S. (1994). *An Introduction to Regression Graphics*, New York: John Wiley & Sons.

5. Dodge, Y. (1984). Robust estimation of regression coefficient by minimizing a convex combination of least squares and least absolute deviations, *Computational Statistics Quarterly*, **1**, 139–153.

6. Dodge, Y. and Jurečková, J. (1988). Adaptive combination of M-estimator and L_1-estimator in the linear model, In *Optimal Design and Analysis of Experiments* (Eds.,Y. Dodge, V. V. Fedorov, and H. P. Wynn), pp. 167–176, Amsterdam: North-Holland.

7. Dodge, Y. and Jurečková, J. (1991). Flexible L-estimation in the linear model, *Computational Statistics and Data Analysis*, **12**, 211–220.

8. Dodge, Y. and Jurečková, J. (1995). Estimation of quantile density function based on regression quantiles, *Statistics & Probability Letters*, **23**, 73–78.

9. Dodge, Y. and Jurečková, J. (2000). *Adaptive Linear Regression*, New York: Springer-Verlag.

10. Donoho, D. L. and Huber, P. J. (1983). The notion of breakdown point, In *A Festschrift for Erich Lehmann* (Eds., P. J. Bickel, K. A. Doksum, and J. L. Hodges), pp. 157–184, Belmont, California: Wadsworth.

11. Hampel, F. R. (1968). Contributions to the theory of robust estimation, *Ph.D. Thesis*, University of California, Berkeley, CA.

12. Jurečková, J. and Sen, P. K. (1996). *Robust Statistical Inference: Asymptotic and Interrelations*, New York: John Wiley & Sons.

13. Koenker, R. and Bassett, G. (1978). Regression quantiles, *Econometrica*, **46**, 33–50.

32

Partially Parametric Testing

J. C. W. Rayner

University of Wollongong, Wollongong, Australia

Abstract: If a smooth test of goodness-of-fit is applied and the null hypothesis is rejected, the hypothesized probability density function is replaced by a *k parameter alternative* to the original model. Three examples are given of inference based on such a model:

- S-sample smooth tests for goodness-of-fit
- partially parametric alternative tests to the t-test
- tests for the location of modes.

Keywords and phrases: k parameter alternative, modes, orthonormal functions, score test, S-sample goodness-of-fit, Wald test, Wilcoxon test

32.1 Partially Parametric Inference

The smooth tests of goodness-of-fit, as described in Rayner and Best (1989), are based on an idea of Neyman (1937), and a tool of Pierce [see Kopecky and Pierce (1979) and Thomas and Pierce (1979)]. If we wish to test for a distribution with probability density function $f(x;\beta)$, we imbed it in a *k-parameter alternative*

$$g_k(x;\theta,\beta) = C(\theta,\beta)\exp\left\{\sum_{i=q+1}^{q+k}\theta_i h_i(x;\beta)\right\} f(x;\beta), -\infty < x < \infty, \quad (32.1)$$

that involves $k\theta s$. We test $H_0 : \theta = 0$ against $K : \theta \neq 0$. Here $f(x;\beta)$ involves a $q\times 1$ vector of nuisance parameters β (for example, composed of the mean μ and standard deviation σ when testing for normality with unspecified parameters). The $\{h_i(x;\beta)\}$ may be taken to be a set of functions orthonormal on $f(x;\beta)$ and $C(\theta,\beta)$ is a normalizing constant that ensures $g_k(x;\theta,\beta)$ is a proper probability density function. Care must be taken because $C(\theta,\beta)$ may not exist or may only exist over restricted domains. See Figures 32.1 and 32.2 for examples of smooth probability density functions.

Neyman (1937) used this model with $f(x;\beta)$ the continuous uniform distribution and $\{h_i(x;\beta)\}$ the Legendre polynomials. He assumed there were no nuisance parameters, so the probability integral transform could be used to convert completely specified distributions to uniformity. If testing for normality the orthonormal system is usually taken to be the normalized Hermite-Chebyshev polynomials. Kopecky and Pierce (1979) and Thomas and Pierce (1979) both permitted nuisance parameters but took $h_r(x;\beta) = x^r$, with a consequent lack of convenience.

Here by *Partially Parametric Inference* I mean inference based on the more complicated probability density function of the form $g_k(x;\theta,\beta)$ rather than $f(x;\beta)$. One way this may arise is if a smooth goodness-of-fit test for $f(x;\beta)$ has been applied and rejected. Often goodness-of-fit testing is preliminary to some other use, or is a retrospective verification that use was valid. So normality may be assessed if the analysis of variance is contemplated. If normality is rejected, the alternative specified by a smooth test of goodness-of-fit is of the form $g_k(x;\theta,\beta)$. This leads to generalizations of the tests recommended in Section 32.3. In general the term *Partially Parametric Inference* could be used for more general forms. For example, the family $\{h_i(x;\beta)\}$ may not be the orthonormal polynomials; a different form of nesting family other than that specified in (32.1) may be used. An example of the latter occurred in Rayner and Best (1996), where a smooth alternative to independence was proposed and used to find new tests of independence.

If the summation in (32.1) is "full" (k maximal), then virtually all distributions are possible. Pearson's X_P^2 is an example. If the summation is "empty", the probability density function is $f(x;\beta)$ and the inference is parametric. The more terms involved in the summation, the richer the family $g_k(x;\theta,\beta)$ and the more "distribution-free" the inference. The focus here is on having only one or two terms in the summation, towards the parametric end of what might be called the *partially parametric continuum*. *Semiparametric* generally has a different meaning. Three examples will be outlined.

32.2 S-Sample Smooth Tests for Goodness-of-Fit

If we wish to assess if S random samples have been drawn from the same population, we may assume each population probability density function has the form (32.1), but with possibly different θ, and test to see if the corresponding θ in different populations are consistent. See Rayner and Rayner (1997, 1998). One problem with this approach is the choice of $f(x;\beta)$, that we call a *target* probability density function.

As an example of the testing procedure, consider the following example from Rayner and Rayner (1998). Teaching assessments from five class surveys are given in Table 32.1. These results relate to one question in the survey, *"Because of the lecturer, in this subject my enthusiasm has ... "*, which was answered in five categories, labeled from "decreased greatly" (here coded as 0) to "increased greatly" (here coded as 4).

Table 32.1: Class survey results

	0	1	2	3	4	Total	Mean
S1	0	1	8	6	1	16	2.438
S2	0	0	4	11	6	21	3.095
S3	0	0	4	6	0	10	2.600
S4	1	14	42	19	3	79	2.114
S5	4	6	19	16	12	56	2.456

Table 32.2: Components V_{rs} using a discrete uniform target and normalized Chebyshev polynomials

Survey		Order (r)	
(s)	1	2	3
1	1.237	-3.137	-1.591
2	3.549	-0.913	-2.467
3	1.342	-2.646	-2.683
4	0.716	-7.323	-0.636
5	2.435	-2.216	-1.124

A convenient "target" distribution is the discrete uniform on five points. The Chebyshev orthogonal polynomials were used. The testing procedure involves calculating *standardized sums* V_{rs} to assess rth *order* effects in the sth survey. The V_{rs} are asymptotically standard normal and asymptotically mutually independent, and are the same as the components used in one sample testing for goodness-of-fit.

Roughly, the order r statistic reflects rth moment differences between the data and the target distribution. Of the five surveys, survey four has smallest mean and survey two has greatest mean. All the means are above that of the target distribution, while all the dispersions are below that of the target distribution. This suggests the target may not have been well chosen. If only the first two orders are thought relevant, then the third order terms in Table 32.2

may be neglected, with the caveat that the conclusions only apply in respect of, roughly, location and dispersion effects.

Rayner and Rayner (1997) derive score tests for the situation when hierarchic testing is appropriate; Rayner and Rayner (1998) construct the same test statistics nonhierarchically as contrasts in the V_{rs}. From this approach an LSD assessment can be made. Ultimately we find a first order difference between samples two and four, and a second order difference between samples four and five. The location differences suggest the relatively small class responded better than the larger class, a well-known effect. The second order difference between the larger classes reflect greater polarization of one class relative to the other. In fact this was due to using different teaching methods that some students in the polarized class responded well to, while others responded poorly. This effect is not usually accessed by current methods of analysis of such surveys.

Using different targets may affect the conclusion, but the example in Rayner and Rayner (1997) shows remarkable robustness to the targets considered.

I suspect the sort of modifications made by the data-driven school to the one sample problem may well be applied profitably to this problem.

32.3 Partially Parametric Alternatives to the t-Test

Assume we have a random sample $X_1, \ldots, X_n$ from a distribution initially thought to be $N(\mu, \sigma^2)$. We wish to test $H_0 : \mu = 0$ against $K : \mu \neq 0$ with σ a nuisance parameter. If normality has been rejected, testing could be based on nonparametric tests, but we now consider using the probability density function the goodness-of-fit testing has identified. We retain symmetry and simplicity by including only one h_i in (32.1), and that is h_4, giving, roughly, kurtosis differences from normality. This probability density function is denoted by $g(x; \theta_4, \mu, \sigma)$ in general and by $g_4(x)$ in Figure 32.1 that gives probability density functions of this form for varying θ_4. Note that we restrict consideration to $\theta_4 < 0$ to ensure the probability density function is proper.

We now conduct a simulation study to compare, for a random sample of $n = 50$ and a nominal size of 5%, the performances of three tests of H_0 against K:

- the t-test, which is known to be uniformly most powerful unbiased level α if the data are normal,
- the score test based on $g(x; \theta_4, \mu, \sigma)$ [derived in Carolan and Rayner (2000a)], and
- the Wilcoxon test.

The score test is quite complicated, and no details are given here: instead see Carolan and Rayner (2000a). Sizes and powers are based on 5,000 simulations.

Our simulations show that for all θ_4 the test sizes are comparable, that

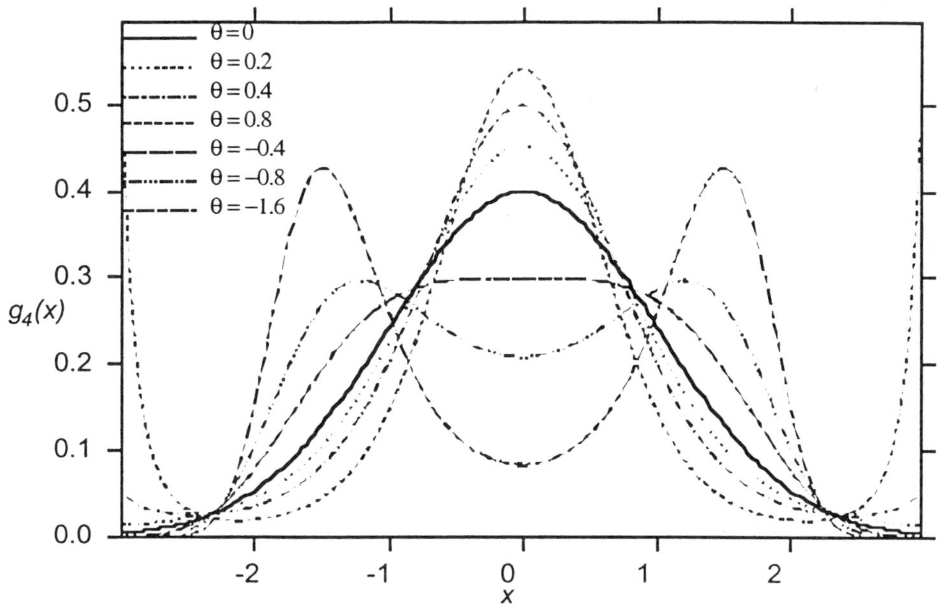

Figure 32.1: The probability distribution function $g(x; \theta_4, 0, 1)$ for varying values of θ_4

when $\theta_4 = 0$ there is no real difference between the power curves, and that as θ_4 increases the score test quickly becomes dominant. The Wilcoxon test is inferior to the t-test for $-1.2 < \theta_4 < 0$ and thereafter becomes the more powerful of the two. The results are most effectively shown graphically. See Figure 32.3.

It is interesting to note that one of the early criticisms of Pearson's test was that when it rejected a model, no alternative model was identified. The presentation of Pearson's test as a smooth test [see Rayner and Best (1989, Theorem 5.1.2)] overcomes this objection. The study reported above shows there may be considerable power gains available if testing is based on the model identified by goodness-of-fit testing.

A score test based on this model is not the only option. In fact Carolan and Rayner (2000b) found a Wald test to be preferable in a number of ways. In particular the Wald test was found to have better power properties for non-local alternatives. While the chi-squared asymptotic distribution is quite satisfactory for the score test with moderate to large sample sizes, resampling is recommended to obtain p-values for the Wald test. Both these tests require the numerical evaluation of maximum likelihood estimates and computational considerations limit the number of parameters that can be included in the model (particularly if resampling is to be used for p-values). In general it is impractical to include more than two extra parameters. Fortunately, Carolan and Rayner (2000b) showed that including just θ_4 can lead to large power gains

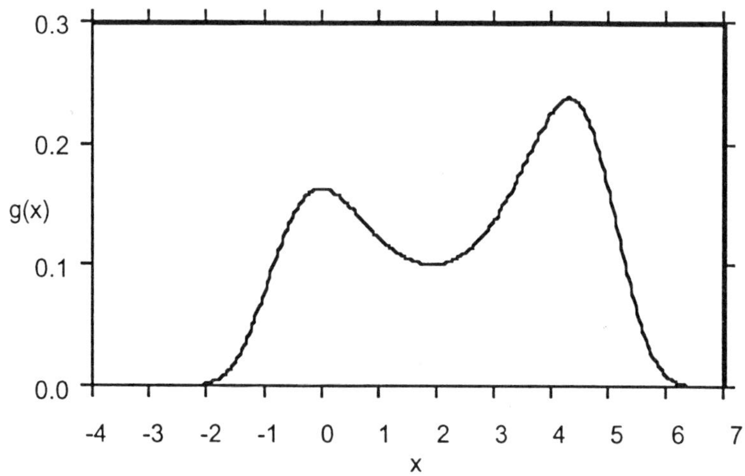

Figure 32.2: Probability density function of the bimodal distribution given by Equation (32.1) with modes at 0 and 4.32

and in general the gains to be made by including further parameters are small for data from all but the most extremely nonnormal distributions (including θ_6 may be worth while for distinctly trimodal data).

This work has been extended to cover the completely randomized, randomized blocks and balanced incomplete block designs. Papers on these designs are in preparation.

32.4 Tests for the Location of Modes

There is some literature on tests for the *number* of modes, but little on the *location* of modes. To construct such tests we consider probability density functions of the form (32.1) with $\sigma = 1$ and $q = 1$; including the θ_2 term models dispersion differences from $\sigma = 1$. In addition, for the probability density function to be proper, $q + k$ must be even, $\theta_2 < 0.5$ and $\theta_{q+k} < 0$. If also $h_r(x; \beta) = (x - \mu)^r$ then Carolan and Rayner (2001) show that μ is a mode of the distribution. There may be up to $k/2$ modes.

Score and various Wald tests of $H_0 : \mu = \mu_0$ against $K : \mu \neq \mu_0$ may be derived. One issue is when there are several modes, which maximum likelihood estimator is used? Carolan and Rayner (2001) suggest, depending upon the aims of the data analyst, either choosing the nearest mode of the fitted distribution or the mode of maximum height. Here we present some power curves using the nearest mode approach. Again due to the inadequacies of asymptotic approximations, resampling would be required to calculate p-values in practice.

As an example, consider testing $H_0 : \mu = 0$ against $K : \mu \neq 0$ given a sample of size n (taking the values 20, 50 and 100 below) from a bimodal distribution of the form (32.1) with modes at 0 and 4.32. To achieve this we put $q = 1\,(\beta = \mu), k = 3$ and $\theta = (0, 0.25, -0.03)^T$. See Figure 32.2.

The power functions have the interesting but sensible property of having power approximately equal to the size at both modes. It seems we are not so much testing "are the data consistent with a population mode of zero?" as "are the data consistent with a population mode?" See Figure 32.4.

This procedure addresses a problem previously given very little attention in the literature.

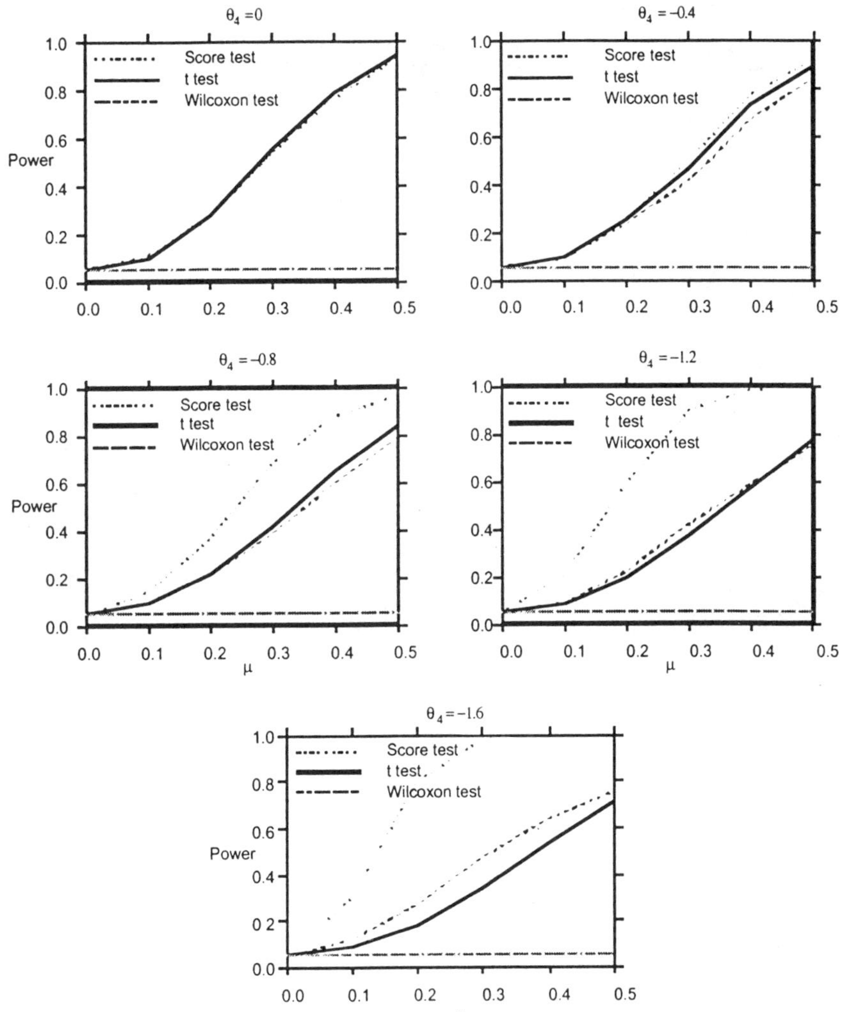

Figure 32.3: Comparison of t-test, Wilcoxon test and score test power curves for testing $H_0 : \mu = 0$ against $K : \mu \neq 0$ as the data becomes progressively more non-normal

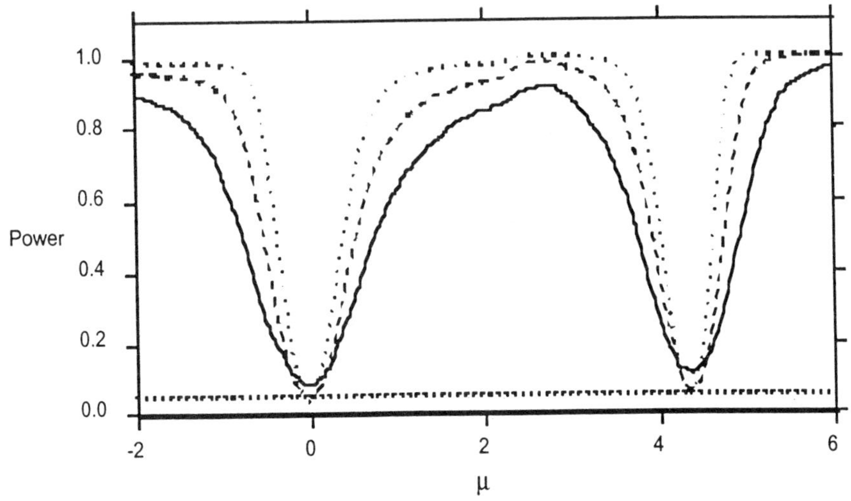

Figure 32.4: Comparison of power curves of the Wald test using the nearest mode technique for samples of size 20 (solid), 50 (dashes) and 100 (dots) from the bimodal distribution in Figure 32.2 above; 1000 simulations

References

1. Carolan, A. M. and Rayner, J. C. W. (2000a). One sample tests of location for nonnormal symmetric data, *Communications in Statistics—Theory and Methods*, **29**, 1569–1581.

2. Carolan, A. M. and Rayner, J. C. W. (2000b). Wald tests of location for symmetric nonnormal data, *Biometrical Journal*, **42**, 777–792.

3. Carolan, A. M. and Rayner, J. C. W. (2001). One sample score tests for the location of modes of nonnormal data, *Journal of Applied Mathematics and Decision Sciences*, **5**, 1–19.

4. Kopecky, K. J. and Pierce, D. A. (1979). Efficiency of smooth goodness-of-fit tests, *Journal of the American Statistical Association*, **74**, 393–397.

5. Neyman, J. (1937). Smooth test for goodness of fit, *Skandinavisk Aktuarietidskrift*, **20**, 150–199.

6. Rayner, J. C. W. and Best, D.J. (1989). *Smooth Tests of Goodness of Fit*, Oxford University Press: Oxford, England.

7. Rayner, J. C. W. and Best, D.J. (1996). Smooth extensions of Pearson's product moment correlation and Spearman's rho. *Statistics and Probability Letters*, **30**, 171-177.

8. Rayner, J. C. W. and Rayner, G. D. (1997). S-sample smooth goodness of fit tests, *Mathematical Scientist*, **22**, 106–116.

9. Rayner, J. C. W. and Rayner, G. D. (1998). S-sample smooth goodness of fit tests: Rederivation and Monte Carlo Assessment, *Biometrical Journal*, **40**, 651–663.

10. Thomas, D. R. and Pierce, D. A. (1979). Neyman's smooth goodness-of-fit test when the hypothesis is composite, *Journal of the American Statistical Association*, **74**, 441–445.

33

Exact Nonparametric Two-Sample Homogeneity Tests

Jean-Marie Dufour and Abdeljelil Farhat

CIRANO and C.R.D.E., Université de Montréal, Montréal, Québec, Canada

Abstract: In this paper, we study several tests for the equality of two unknown distributions. Two are based on empirical distribution functions, three on nonparametric probability density estimates, and the last ones on differences between sample moments. We suggest controlling the size of such tests (under nonparametric assumptions) by using permutational versions of the tests jointly with the method of Monte Carlo tests properly adjusted to deal with discrete distributions. In a simulation experiment, we show that this technique provides perfect control of test size, in contrast with usual asymptotic critical values.

Keywords and phrases: Monte Carlo tests, goodness-of-fit tests, nonparametric methods, two-sample problem

33.1 Introduction

A common problem in statistics consists in testing whether the distributions of two random variables are identical against the alternative that they differ in some way. More precisely, we consider two random samples $X_1, \dots, X_n$ and $Y_1, \dots, Y_m$ such that $F(x) = P[X_i \le x]$, $i = 1, \dots, n$, and $G(x) = P[Y_j \le x]$, $j = 1, \dots, m$. In this paper, we do not wish to impose additional restrictions on the form of the cumulative distribution functions (cdf) F and G, which may be continuous or discrete. We consider the problem of testing the null hypothesis $H_0 : F = G$ against the alternative $H_1 : F \neq G$.

H_0 is a nonparametric hypothesis, so testing H_0 requires a distribution-free procedure. Thus, many users who have to make such a confrontation resort to a goodness-of-fit test, usually the two-sample Kolmogorov-Smirnov (KS) test [Smirnov (1939)] or the Cramér-von Mises (CM) test [see Lehmann (1951),

Rosenblatt (1952) and Fisz (1960)]. Other procedures that have been suggested include permutation tests based on L_1 and L_2 distances between kernel-type estimators of the relevant probability density functions (pdf) [Allen (1997)] and tests based on the difference of the means of the two samples considered [see Pitman (1937), Dwass (1957), Efron and Tibshirani (1993)]. Except for the last procedure, which is designed to have power against samples that differ through their means, the exact and limiting distributions of the test statistics are not standard, and tables for the exact distributions are only available for a limited number of sample sizes. Thus these tests are usually performed with the help of tables based on asymptotic distributions. This leads to procedures that do not have the targeted size (which can easily be too small or too large) and may have low power.

In this paper, we aim at finding test procedures with two basic features. Namely, the latter should be: (1) truly distribution-free, irrespective of whether the underlying distribution F is discrete or continuous, and (2) exact in finite samples (i.e., they must achieve the desired size even for small samples). In this respect, it is important to note that the finite and large sample distributions of usual test statistics are not necessarily distribution-free under H_0. In particular, while the KS and CM statistics are distribution-free when the observations are $i.i.d$ with continuous distribution, this is not anymore the case when they follow a discrete distribution. For the statistics based on kernel-type density estimators, distribution-freeness does not obtain even for $i.i.d$ observations with a continuous distribution. This difficulty can be relaxed by considering a permutational version of these tests which uses the fact that all permutations of the pooled observations are equally likely when the observations are $i.i.d$ with a continuous distributions. The latter property, however, does not hold when the observations follow a discrete distribution. So none of the procedures proposed to date for testing H_0 satisfies the double requirement of yielding a test that is both distribution-free and exact.

Given recent progress in computing power, a way to solve this difficulty consists in using simulation-based methods, such as bootstrapping or Monte Carlo tests. The bootstrap technique however does not ensure that level will be fully controlled in finite samples. For this reason, we favor Monte Carlo (MC) test methods. MC tests were introduced by Dwass (1957) and Barnard (1963) and developed by Birnbaum (1974), Dufour (1995), Kiviet and Dufour (1996), Dufour *et al.* (1998), and Dufour and Kiviet (1998)].

In this paper, we *first* show how the size of all the two-sample homogeneity tests described above can be perfectly controlled for both *continuous* and *discrete* distributions on considering their permutational distribution and using the technique of MC tests properly adjusted to deal with discrete distributions. As a result, in order to implement these tests, it is not anymore necessary to establish the distributions of the test statistics, either in finite samples or asymptotically. *Second*, as a consequence of the great flexibility allowed by the

MC test technique in selecting test statistics, we suggest alternative procedures, in particular: (i) a statistic based on the L_∞ distances between kernel-type pdf estimators; (ii) extensions of the permutational test based on the difference of two-sample means to higher order moments, such as sample variances, asymmetry (based on third moments) and kurtosis sample coefficients. *Thirdly*, we present the results of an MC experiment showing clearly that usual large-sample critical values do not control size, while the MC versions of the tests achieve this aim perfectly. Further, we see that the new procedure introduced can lead to the true power, independently of the distributions nature, continuous or discrete.

Section 33.2 presents the test statistics studied. In Section 33.3, we explain how the technique of MC tests can be applied with all the statistics considered to control the size of the corresponding tests under nonparametric assumptions. Section 33.4 describes the results of our study, first for continuous distributions and then for discrete distributions. We conclude in Section 33.5.

33.2 Test Statistics

Let $X_1, \ldots , X_n$ be a sample of independent and identically (*i.i.d.*) distributed observations with common cdf $F(x) = P[X_i \leq x]$ and $Y_1, \ldots, Y_m$ a sample *i.i.d.* observations with cdf $G(x) = P[Y_i \leq x]$. The problem is to test the homogeneity hypothesis H_0 and, for that matter, our study will include the following test statistics. In all the tests presented below, H_0 is rejected when the test statistic is large.

The first two criteria are the KS and the CM statistics. The KS test was introduced by Smirnov (1939) and uses the statistic

$$KS = \sup_x |F_n(x) - G_m(x)| \tag{33.1}$$

where $F_n(x)$ and $G_m(x)$ are the usual empirical distribution functions (edf) associated with the X and Y samples respectively. It is well known that KS is distribution-free [see Conover (1971, page 313)] under H_0 when the common distribution function F is continuous, but its exact and limiting distributions are not standard. Birnbaum and Hall (1960) provide tables for its exact distribution in the case of equal sample sizes, whereas Massey (1952) does the same for unequal sample sizes. Further, it is important to note that KS is not distribution-free when F is a discrete distribution.

The two-sample CM statistic is defined as

$$CM = \frac{mn}{(m+n)^2}\left\{\sum_{i=1}^{n}[F_n(X_i) - G_m(X_i)]^2 + \sum_{j=1}^{m}[F_n(Y_j) - G_m(Y_j)]^2\right\}. \tag{33.2}$$

CM is also distribution-free under H_0 with F continuous and, again, the exact and limiting null distributions of CM are not standard. Anderson (1962) and Burr (1963, 1964) provide tables for the exact distribution in the case of small sample sizes ($n+m \leq 17$). Otherwise, a table of the asymptotic distribution is available from Anderson and Darling (1952).

The next three statistics are based on distances (L_1, L_2 and L_∞) between kernel-based pdf estimators. If f is the pdf associated with the cdf F, Allen (1997) considered the kernel-type density estimator

$$f_n(x) = \frac{C_X}{n} \sum_{i=1}^{n} K[C_X(x - X_i)] \tag{33.3}$$

where

$$C_X = n^{1/5}/(2S_X), \quad K(x) = \begin{cases} \frac{1}{2}, & \text{if } |x| \leq 1, \\ 0, & \text{if } |x| > 1, \end{cases}$$

and $S_X = \left[(n-1)^{-1} \sum_{i=1}^{n} \left(X_i - \bar{X}\right)^2\right]^{1/2}$ is the usual estimator of the population standard deviation. If g is the pdf associated with the cdf G, its estimator $g_m(x)$ is defined in a way analogous to (33.3). The L_1-distance test initially proposed by Allen (1997) is based on the statistic

$$\hat{L}_1 = \sum_{i=1}^{n} |f_n(X_i) - g_m(X_i)| + \sum_{j=1}^{m} |f_n(Y_j) - g_m(Y_j)| \, . \tag{33.4}$$

The L_2-distance and L_∞-distance tests are based on the statistics:

$$\hat{L}_2 = \left\{ \sum_{i=1}^{n} (f_n(X_i) - g_m(X_i))^2 + \sum_{j=1}^{m} (f_n(Y_j) - g_m(Y_j))^2 \right\}^{1/2} \tag{33.5}$$

$$\begin{aligned} \hat{L}_\infty &= \sup_x |f_n - g_m| \\ &= \max_{1 \leq i \leq n, 1 \leq j \leq m} \{|f_n(X_i) - g_m(X_i)|; \, |f_n(Y_j) - g_m(Y_j)|\} \, . \end{aligned} \tag{33.6}$$

For the case where both F and G are discrete, the pdf f and g are replaced by the probability mass functions $p(x) = P[X = x]$ and $q(x) = P[Y = x]$: each one is estimated with the help of formula (33.3) after which f_n and g_m are replaced by the these estimators in the $\hat{L}_1, \hat{L}_2$ and $\hat{L}_\infty$ statistics. In contrast with the KS and CM statistics, the finite sample distributions of the statistics $\hat{L}_1, \hat{L}_2$ and $\hat{L}_\infty$ are not distribution-free even when the distribution function F is continuous as well as (a fortiori) when F is discrete.

The next statistic to enter our study is the difference of the sample means $\hat{\theta}_1 = \bar{X} - \bar{Y}$. Permutation tests based on $\hat{\theta}_1$ were initially proposed by Fisher

(1935) and used by Dwass (1957) for testing the equality of means, but Efron and Tibshirani (1993, Chap. 15) suggested to extend their use, along with bootstrap tests, for testing the equality of two unknown distributions. Contrary to Allen (1997) who also considered bootstrap tests, the statistic based on the studentized difference of sample means

$$\hat{t} = \frac{(\bar{X} - \bar{Y})/\sqrt{\frac{1}{n} + \frac{1}{m}}}{\left\{\left[\sum_{i=1}^{n}(X_i - \bar{X})^2 + \sum_{j=1}^{m}(Y_j - \bar{Y})^2\right]/(n+m-2)\right\}^{1/2}} \tag{33.7}$$

will not be considered since our study is restricted to permutation tests and it is straightforward to see that such tests based on $\hat{\theta}_1$ and $\hat{t}$ are equivalent [see, for instance, Lehmann (1986)]. Further, we suggest here alternative test statistics based on comparing higher-order moments. Namely, the difference between unbiased estimators of sample variances,

$$\hat{\theta}_2 = \left| \frac{1}{n-1} \sum_{i=1}^{n} (X_i - \bar{X})^2 - \frac{1}{m-1} \sum_{i=1}^{m} (Y_i - \bar{Y})^2 \right|,$$

as well as statistics based on comparing sample skewness and kurtosis coefficients:

$$\hat{\theta}_3 = \left| \frac{1}{n} \sum_{i=1}^{n} \left(\frac{X_i - \bar{X}}{s_X} \right)^3 - \frac{1}{m} \sum_{i=1}^{m} \left(\frac{Y_i - \bar{Y}}{s_Y} \right)^3 \right|, \tag{33.8}$$

$$\hat{\theta}_4 = \left| \frac{1}{n} \sum_{i=1}^{n} \left(\frac{X_i - \bar{X}}{s_X} \right)^4 - \frac{1}{m} \sum_{i=1}^{m} \left(\frac{Y_i - \bar{Y}}{s_Y} \right)^4 \right| \tag{33.9}$$

where

$$s_X^2 = \frac{1}{n} \sum_{i=1}^{n} (X_i - \bar{X})^2 \text{ and } s_Y^2 = \frac{1}{m} \sum_{i=1}^{m} (Y_i - \bar{Y})^2. \tag{33.10}$$

Note that skewness and kurtosis coefficients play a central role in testing normality [see Jarque and Bera (1987) and Dufour *et al.* (1998)].

33.3 Exact Randomized Permutation Tests

Except for the Dwass (1957) procedure, all the tests described in the previous section involve imperfectly tabulated null distributions or are not distribution-free in finite samples. Consequently, the latter may lead to arbitrarily large size distortions. In view of obtaining distribution-free tests with known size in finite samples, we first note that truly distribution-free tests (for any given sample size) can be based on the statistics KS, CM, L_1, L_2, L_∞, $\hat{t}$, $\hat{\theta}_1$, $\hat{\theta}_2$, $\hat{\theta}_3$ and $\hat{\theta}_4$ by considering the distribution obtained on permuting in all possible ways (with equal probabilities) the $m+n$ grouped observations $X_1, \ldots, X_n, Y_1, \ldots, Y_m$. Since these permutations are equally probable under the null hypothesis H_0, irrespective of the unknown distribution F, any test which rejects H_0 by using an exact critical value obtained from its permutational distribution [i.e., its conditional distribution given the ordered statistics of the grouped observations] will have the same level conditionally (on the ordered statistics) as well as unconditionally.

If T designates a pivotal test statistic (i.e. its distribution does not depend on unknown parameters under the null hypothesis), we can proceed as follows to conduct an MC test. Denote by T_0 the test statistic computed from the observed sample. When the null hypothesis is rejected for large values of T_0, the associated critical region of size α may be expressed as $G(T_0) \leq \alpha$, where $G(x) = P[T \geq x | H_0]$ is the p-value function. Generate N independent samples $(X_{i1}, \ldots, X_{in}, Y_{i1}, \ldots, Y_{im})$, $1 \leq i \leq N$, drawn from the specified null distribution F_0. This leads to N independent realizations $T_i = T(X_{i1}, \ldots, X_{in}, Y_{i1}, \ldots, Y_{im})$, $1 \leq i \leq N$, from which we can compute an empirical p-value function:

$$\widehat{p}_N(x) = \frac{N\widehat{G}_N(x) + 1}{N+1} \tag{33.11}$$

where

$$\widehat{G}_N(x) = \frac{1}{N}\sum_{i=1}^{N} \mathbf{1}_{[0,\infty)}(T_i - x), \quad \mathbf{1}_A(x) = \begin{cases} 1, & x \in A \\ 0, & x \notin A \end{cases} .$$

The associated MC critical region is defined as

$$\widehat{p}_{N.}(T_0) \leq \alpha , \tag{33.12}$$

where $\widehat{p}_N(T_0)$ may be interpreted as an estimate of $G(T_0)$. When T has a continuous distribution, it can be shown that [see Dufour (1995)]:

$$P[\widehat{p}_N(T_0) \leq \alpha | H_0] = \frac{I[\alpha(N+1)]}{N+1} , \quad 0 \leq \alpha \leq 1 , \tag{33.13}$$

where $I[x]$ denotes the largest integer not exceeding x. Thus if N is chosen such that $\alpha(N+1)$ is an integer, the critical region (33.12) has the same size as the critical region $G(T_0) \leq \alpha$. The MC test so obtained is theoretically exact, irrespective of the number N of replications used.

The above procedure is closely related to the parametric bootstrap, with a fundamental difference however. Bootstrap tests are, in general, provably valid for $N \to \infty$. In contrast, we see from (33.13) that N is explicitly taken into consideration in establishing the validity of MC tests. Although the value of N has no incidence on size control, it may have an impact on power which typically increases with N.

Note that (33.13) holds for tests based on statistics with continuous distributions. In such a case, ties have non-zero probability. Nevertheless, the technique of MC tests can be adapted to discrete distributions by appeal to the following randomized tie-breaking procedure [see Dufour (1995, Section 2.2)]. Draw $N+1$ uniformly distributed variates $U_0, U_1, \ldots, U_N$, independently of the T_i's and arrange the pairs (T_i, U_i) following the lexicographic order:

$$(T_i, U_i) < (T_j, U_j) \Leftrightarrow [T_i < T_j \quad \textit{or} \quad (T_i = T_j \ \textit{and} \ U_i < U_j)] \, .$$

Then, proceed as in the continuous case and compute

$$\widetilde{p}_N(x) = \frac{N\widetilde{G}_N(x) + 1}{N+1} \, ,$$

where

$$\widetilde{G}_N(x) = 1 - \frac{1}{N}\sum_{i=1}^{N} \mathbf{1}_{[0,\infty)}(x - T_i) + \frac{1}{N}\sum_{i=1}^{N} \mathbf{1}_{[0]}(T_i - x)\, \mathbf{1}_{[0,\infty)}(U_i - U_0).$$

The resulting critical region $\widetilde{p}_N(T_0) \leq \alpha$ has the same level as the region $G(T_0) \leq \alpha$, provided again that $\alpha(N+1)$ is an integer. More precisely,

$$P[\hat{p}_N(T_0) \leq \alpha | H_0] \leq P[\widetilde{p}_N(T_0) \leq \alpha | H_0] = \frac{I[\alpha(N+1)]}{N+1}, \quad 0 \leq \alpha \leq 1.$$

If a null hypothesis ensures that the random sample is made up of exchangeable variables and if it should be rejected for large values of the test statistic, an MC test of that hypothesis is carried out in five steps: first, the test statistic is computed with the help of the observed sample which gives a value T_0, say; second, N permutations of the sample are chosen at random and without replacement from all possible permutations; third, the test statistic is recomputed for each of the permuted samples which gives the values $T_1, \ldots, T_N$, say; fourth, if R_0 designates the rank of T_0 among the set $\{T_0, T_1, \ldots, T_N\}$ [in the case of ties, one may resort to the randomization method suggested by Dufour (1995)], the p-value associated with the MC test of the null hypothesis is given by $1 - R_0/(N+1)$; lastly, a decision is reached according to the chosen level [see Dufour(1995)]. The fact that the procedure is randomized plays a central role in controlling the size of the test. In bootstrap-type procedures, one does as if the number of replications were infinite.

33.4 Simulation Study

In the simulation study presented here, all tests _ original as MC _ were performed at the 5% level using 10000 trials. This entails that the 95% confidence interval for the nominal level is [4.57%, 5.43%]. Furthermore, they were all conducted with equal sample sizes $m = n = 22$. As mentioned earlier, each MC test was carried out by picking at random $N = 99$ permutations of the original grouped sample and this was done by using the IMSL Program Library random number generator. In his simulation study, Allen (1997) used 2500 trials and each permutation or bootstrap test was carried out with 499 samples. A more extensive set of simulation results are available in Dufour and Farhat (2000).

For the first part of the study where F and G are both continuous, the following distributions were considered: normal $N(0, 1)$, exponential $Exp(0, 1.5)$, gamma $\Gamma(2, 1)$, beta $B(2, 3)$, logistic $Log(-1, 1)$, lognormal $\Lambda(4, 1.5)$ and uniform $U(0, 1)$. In this choice, care was taken to have at the same time simple parameters as well as appreciably different means and variances. Table 33.1 gives the list of those means and variances. From this preliminary list, four situations were considered: (i) the distributions were standardized, and thus had common null mean and unit variance; (ii) the distributions were centered to have common null mean but possibly different variances; (iii) the distributions were scaled to have common unit variance but different means; (iv) the distributions remained with different means and different variances. Whatever the situation, a null hypothesis is obtained each time F and G share the same distribution from the list and an alternative hypothesis is obtained each time F and G possess different distributions from that list.

Table 33.1: Continuous distributions with their means and variances.

Distribution	N	Exp	Γ	B	Log	Λ	U
Mean	0	1.5	2	.4	-1	168.17	.5
Variance	1	2.25	2	.04	0.55133^{-2}	240055	1/12

For the second part of the study where F and G are discrete, the five most commonly used distributions were retained: discrete uniform (DU), binomial (Bin), geometric (Geo), negative binomial ($Nbin$) and Poisson (P). Since it is a prohibitive task to find parameters that will simultaneously give rise to either common mean and common variance, the following three situations were considered: (i) the distributions were $DU(19)$, $Bin(20, 0.5)$, $Geo(0.1)$, $Nbin(8, 0.2)$, $P(10)$ and, thus had common mean 10 and variance 30, 5, 90, 2.5 and 10 respectively; (ii) the distributions were $DU(10)$, $Bin(33, 0.5)$, $Geo((\sqrt{34}-1)/16.5)$, $Nbin(3, (\sqrt{108}-3)/16.5)$, $P(8.25)$ and, thus had mean 5.5, 16.5, 3.42, 2.23 and 8.25 respectively but common variance 8.25; (iii) the distributions were $DU(10)$, $Bin(10, 0.1)$, $Geo(0.3)$, $Nbin(10, 0.2)$, $P(5)$ and, thus had mean 5.5, 1, 3.33, 50 and 5 respectively and variance 8.25, 0.9, 7.78, 200 and 5 respectively. $\hat{\theta}_1$ As a first check on the accuracy of our study, Tables 1 and 2 of Allen (1997) were done again adding, however, the CM, the $\hat{L}_\infty$ and MC tests based on higher-moments and by excluding the bootstrap tests. The results appear in Table 33.2 and they are quite similar to those of Allen(1997).

Table 33.2: Empirical level and power for tests of equality of two distributions: $m = 22$, $n = 22$ and $\alpha = 5\%$

	$F = N(0,1)$										
G	Original tests		MC tests								
	KS	CM	KS	CM	$\hat{\theta}_1$	$\hat{\theta}_2$	$\hat{\theta}_3$	$\hat{\theta}_4$	$\hat{L}_1$	$\hat{L}_2$	$\hat{L}_\infty$
$N(0,1)$	6.2	5.2	4.8	5.1	5.0	4.9	4.7	5.4	4.8	4.8	4.7
$N(0.2,1)$	7.7	6.7	6.2	6.3	6.4	4.7	5.0	5.1	5.3	5.2	5.1
$N(0.3,1)$	10.8	9.5	8.6	9.0	9.8	4.8	5.2	4.9	6.8	6.7	6.7
$N(0.4,1)$	16.1	15.8	13.5	15.0	16.2	4.7	5.6	5.5	10.2	10.1	9.2
$N(0.5,1)$	32.8	34.1	28.3	32.9	36.1	3.9	6.1	5.9	19.3	18.7	17.1
$N(0.7,1)$	54.3	57.5	48.9	55.9	60.3	3.1	6.1	6.5	34.8	33.9	30.8
$N(0,1.2^2)$	7.3	5.9	5.8	5.8	5.1	11.6	4.7	5.1	9.9	10.1	10.0
$N(0,1.4^2)$	9.1	6.8	7.2	6.6	5.0	28.5	4.2	5.2	22.7	23.4	23.6
$N(0,1.6^2)$	12.5	9.7	10.5	9.8	5.1	49.9	4.0	4.5	42.0	42.6	42.9
$N(0,1.8^2)$	15.4	11.5	13.1	11.6	5.3	66.2	3.2	3.9	58.9	59.7	59.8
$N(0,2.^2)$	20.4	16.1	17.3	15.6	5.1	80.2	2.6	3.4	74.6	75.3	74.5

Tables 33.3 and 33.4 contain the results of our study. The following conclusions can be drawn. As expected, the MC tests control size perfectly and are easily applicable. The original KS and CM tests, for which tables are available, show size distortions. Although in the case where both F and G are continuous, the CM test appears adequate and the KS test only exhibits light size distortions, the distortions become severe when both F and G are discrete.

The use of $\hat{\theta}_1$ to carry out equality tests of two distributions is erroneous. It is obvious that two distributions cannot be equal if they do not have the same mean but the converse is not true. Consequently, if the test based on $\hat{\theta}_1$ accepts the hypothesis H_0, it should not be interpreted as an acceptance of the fact that $F = G$ but rather that these distributions have equal means. The $\hat{L}_1$ and $\hat{L}_2$ tests behave almost identically and differ slightly from the $\hat{L}_\infty$ test.

In the same way, the power of the KS test is not very different from that of the CM test. In general, if we compare the powers of the tests based on edf's (KS and CM tests) with those based on pdf estimates ($\hat{L}_1, \hat{L}_2$ and $\hat{L}_\infty$ tests), we notice a great difference and we cannot conclude that a test stemming from one group is more powerful than all the tests in the other group. The edf tests are more powerful than those based on pdf estimates when two distributions have the same variance but different means. On the other hand, if the two distributions have the same mean but different variances, the tests based on pdf estimates are the most powerful.

As for the case where both F and G are discrete, the results for two from the three situations are presented successively in Table 33.4. As in the case where the distributions were continuous, the conclusions reached for the MC tests still apply. Moreover, the simulation confirms the result stated by Noether (1967) indicating that, if random variables are discrete, the KS test is still valid but becomes conservative. On the other hand, it reveals that the CM test is quite often tolerant although Conover (1971) indicates that it has a tendency to be conservative in the case of discrete distributions.

33.5 Conclusion

In this paper, we first showed that finite-sample distribution-free two-sample homogeneity tests, for both continuous and discrete distributions, can be easily obtained on combining two techniques: (1) by considering permutational versions of most proposed tests for that problem; (2) by implementing the permutation procedures as Monte Carlo tests with an appropriate tie-breaking technique to take account of the discreteness of the test null distributions. Second, due to the flexibility of the Monte Carlo test technique, we could easily introduce and implement several alternative procedures, such as permutation tests comparing higher-order moments. Other alternative procedures are described in Dufour and Farhat (2000). Thirdly, in a simulation study, it was shown that the procedures proposed work as expected from the viewpoint of size control, while the new suggestions made yielded power gains.

Note _ This paper is a summary of Dufour and Farhat (2000).

Table 33.3: Empirical level and power for MC tests of equality of two continuous distributions having same mean and same variance: $m = n = 22$ and $\alpha = 5\%$

G	Original tests		MC tests								
	KS	CM	KS	CM	$\hat{\theta}_1$	$\hat{\theta}_2$	$\hat{\theta}_3$	$\hat{\theta}_4$	$\hat{L}_1$	$\hat{L}_2$	$\hat{L}_\infty$
	$F = N$										
N	6.2	5.2	4.8	5.1	5.0	4.9	4.7	5.4	4.8	4.8	4.7
Exp	16.1	12.8	13.6	12.6	5.6	10.6	41.7	15.7	17.4	17.2	15.7
Γ	11.0	8.4	8.8	8.2	5.2	7.9	26.8	10.1	11.3	11.4	11.0
B	7.1	5.6	5.7	5.7	4.7	5.6	7.7	7.4	6.3	6.2	5.9
Log	6.4	5.2	5.1	5.0	4.7	5.5	5.4	6.3	5.2	5.4	5.4
Λ	77.3	69.1	71.6	65.0	6.0	59.0	70.3	63.1	68.8	67.9	65.6
U	8.2	5.9	6.7	5.9	5.2	6.1	6.7	17.1	6.7	6.8	7.2
	$F = Exp$										
Exp	6.1	5.0	4.9	5.0	4.9	5.2	5.4	5.2	4.8	4.9	4.8
Γ	7.0	5.7	5.7	5.5	5.0	6.3	7.6	6.2	6.2	6.1	6.4
B	13.6	9.8	11.6	9.9	5.1	12.8	36.6	20.9	15.6	16.0	16.1
Log	16.7	13.3	14.1	12.8	5.2	8.9	36.5	11.5	15.1	14.7	13.6
Λ	88.7	76.3	84.8	72.2	5.0	49.9	30.6	31.8	55.2	55.0	55.0
U	19.0	13.6	16.2	13.1	5.0	16.7	55.7	35.3	22.1	22.4	23.5
	$F = \Gamma$										
Γ	5.9	4.9	4.7	4.8	4.9	5.2	4.9	5.5	5.0	5.1	5.0
B	9.0	7.1	7.4	6.7	5.0	8.9	20.2	11.7	9.4	9.5	9.6
Log	11.3	8.7	9.2	8.7	4.8	6.5	22.1	6.5	9.1	9.1	8.6
Λ	84.3	72.5	80.1	67.8	5.2	53.6	43.0	43.4	60.7	60.1	59.7
U	12.8	9.4	10.6	9.0	5.2	12.6	35.9	25.6	15.0	15.2	15.5
	$F = B$										
B	6.5	5.6	5.2	5.5	5.3	5.2	5.1	4.9	5.3	5.2	5.4
Log	7.6	5.3	6.2	5.4	4.5	6.4	8.2	11.4	6.8	6.7	6.8
Λ	83.7	76.1	78.3	71.4	5.8	63.3	74.4	69.7	72.3	71.4	69.9
U	6.6	5.2	5.5	5.1	5.0	5.8	8.0	10.3	5.9	5.8	6.1
	$F = Log$										
Log	6.4	5.2	5.0	5.1	5.1	4.7	4.5	4.6	4.7	4.7	4.6
Λ	72.8	64.6	66.2	60.6	5.8	56.9	60.1	52.7	66.4	65.4	63.3
U	9.6	6.9	8.0	6.7	5.0	8.2	9.4	25.6	9.2	9.3	9.6
	$F = \Lambda$										
Λ	6.2	4.9	4.8	4.8	4.7	5.0	5.1	5.0	5.0	5.0	4.9
U	87.9	82.4	82.3	77.2	5.7	65.5	91.0	83.8	76.6	75.7	73.2

Table 33.4: Empirical level and power for tests of equality of two discrete distributions: $m = n = 22$ and $\alpha = 5\%$

G	Original tests		MC tests								
	KS	CM	KS	CM	$\hat{\theta}_1$	$\hat{\theta}_2$	$\hat{\theta}_3$	$\hat{\theta}_4$	$\hat{L}_1$	$\hat{L}_2$	$\hat{L}_\infty$
	Same mean but different variances										
	$F = UD$										
UD	3.8	5.4	4.7	5.1	5.1	5.1	4.8	5.5	5.2	5.2	4.9
Bin	38.3	50.0	51.5	45.1	5.5	99.2	7.0	13.3	96.8	96.9	97.1
Geo	14.1	19.4	16.6	17.9	6.5	35.6	34.6	8.7	22.3	23.2	26.0
$BinN$	69.2	79.3	82.6	71.3	5.5	100.	25.1	20.8	99.9	99.9	99.9
Poi	18.5	22.6	26.5	21.3	5.2	82.6	12.8	24.8	70.1	71.6	72.0
	$F = Bin$										
Bin	2.3	6.0	5.2	5.1	5.1	5.2	4.8	5.1	5.0	5.1	5.0
Geo	75.6	86.9	82.6	82.8	8.2	99.6	13.1	1.2	99.7	99.7	99.7
$BinN$	4.7	10.6	10.3	8.0	5.3	33.2	17.1	10.6	28.6	29.0	27.5
Poi	3.9	9.5	8.1	8.3	5.2	29.9	4.6	4.7	22.2	22.5	22.5
	$F = Geo$										
Geo	3.8	5.1	4.8	4.7	4.6	4.8	4.8	4.6	4.8	4.8	4.8
$BinN$	94.4	95.7	97.1	91.8	8.4	99.9	1.3	2.7	100.	100.	100.
Poi	53.4	61.3	61.6	57.8	7.8	94.3	10.9	2.4	93.3	93.9	93.8
	$F = BinN$										
$BinN$	1.8	9.7	5.3	5.4	5.2	5.0	4.7	5.2	5.1	5.1	5.1
Poi	11.5	27.5	24.2	21.8	5.1	78.7	10.2	9.8	63.5	64.7	66.7
	Different means but same variance										
	$F = UD$										
UD	2.9	5.6	4.9	4.9	4.9	5.2	4.8	5.2	4.9	4.9	4.8
Bin	100.	100.	100.	100.	100.	62.0	20.1	11.2	99.8	99.8	99.9
Geo	64.0	84.4	70.9	79.4	68.2	13.7	67.0	52.4	64.1	63.3	60.5
$BinN$	100.	100.	100.	100.	100.	0.2	27.1	33.1	100.	100.	100.
Poi	10.6	15.7	17.3	13.9	9.6	33.9	15.1	26.9	26.5	27.4	26.4
	$F = Bin$										
Bin	.9	23.3	5.1	5.1	5.1	4.9	4.6	4.9	4.9	4.9	4.8
Geo	84.5	95.3	94.9	84.2	99.6	15.1	3.5	0.9	65.8	66.8	71.3
$BinN$	100.	100.	100.	100.	100.	0.0	7.8	83.1	100.	100.	100.
Poi	100.	100.	100.	100.	100.	7.0	13.4	9.8	100.	100.	100.
	$F = Geo$										
Geo	2.1	12.3	4.8	4.7	4.7	4.9	4.8	4.9	4.7	4.6	4.7
$BinN$	100.	100.	100.	100.	100.	8.0	33.2	77.0	100.	100.	100.
Poi	68.3	79.2	76.1	74.5	57.1	12.8	40.7	23.6	55.0	55.6	56.1
	$F = BinN$										
$BinN$	1.9	7.8	5.2	5.1	5.2	4.8	4.8	5.1	5.0	4.9	5.0
Poi	100.	100.	100.	100.	100.	0.1	15.6	49.9	100.	100.	100.

References

1. Allen, D. L. (1997). Hypothesis testing using an L_1-distance bootstrap, *The American Statistician*, **51-2**, 145–150.

2. Anderson, T. W. (1962). On the distribution of the two-sample Cramér-von Mises criterion, *Annals of Mathematical Statistics*, **33**, 1148–1159.

3. Anderson, T. W. and Darling, D. A. (1952). Asymptotic theory of certain "goodness-of-fit" criteria based on processes, *Annals of Mathematical Statistics*, **23**, 193–212.

4. Barnard, G. A. (1963). Discussion on M. S. Bartlett, The spectral analysis of point processes, *Journal of the Royal Statistical Society, Series B*, **25**, 294.

5. Birnbaum, Z. W. (1974). *Reliability and Biometry*, pp. 441–445, Philadelphia: SIAM.

6. Birnbaum, Z. W. and Hall, R. A. (1960). Small sample distributions for multi-sample statistics of the Smirnov type, *Annals of Mathematical Statistics*, **31**, 710–720.

7. Burr, E. J. (1963). Distribution of the two-sample Cramér-von Mises criterion for small equal samples, *Annals of Mathematical Statistics*, **34**, 95–101.

8. Burr, E. J. (1964). Small samples distributions of the two-sample Cramér-von Mises' W^2 and Watson's U^2, *Annals of Mathematical Statistics*, **35**, 1091–1098.

9. Conover, W. J. (1971). *Practical Nonparametric Statistics*, New York: John Wiley & Sons.

10. Dufour, J.-M. (1995). Monte Carlo tests with nuisance parameters: A general approach to finite sample inference and nonstandard asymptotics in econometrics, Discussion paper, CRDE, Université de Montréal.

11. Dufour, J.-M. and Farhat, A. (2000). Exact nonparametric two-sample homogeneity tests for possibly discrete distributions, Discussion Paper, CRDE and CIRANO, Université de Montréal.

12. Dufour, J.-M. and Farhat, A. and Gardiol, L. and Khalaf, L. (1998). Simulation based finite sample normality tests in linear regressions, *The Econometrics Journal*, **1**, 154–173.

13. Dufour, J.-M. and Kiviet, J. F. (1998). Exact inference methods for first order autoregressive distributed lag models, *Econometrica*, **66**, 79–104.

14. Dwass, M. (1957). Modified randomization tests for nonparametric hypotheses, *Annals of Mathematical Statistics*, **28**, 181–187.

15. Efron, B. and Tibshirani, R. J. (1993). *An Introduction to the Bootstrap*, Vol. 57, New York: Chapman & Hall.

16. Fisher, R. A. (1935). *The Design of Experiments*, London: Oliver and Boyd.

17. Fisz, M. (1960). On a result by M. Rosenblatt concerning the Mises-Smirnov test, *Annals of Mathematical Statistics*, **31**, 427–429.

18. Jarque, C. M. and Bera, A. K. (1987). A test for normality of observations and regression residuals, *International Statistical Review*, **55**, 163–172.

19. Kiviet, J. F. and Dufour, J.-M. (1996). Exact tests in single equation autoregressive distributed lag models, *Journal of Econometrics*, **20-2**, 325–353.

20. Lehmann, E. L. (1951). Consistency and unbiasedness of certain nonparametric tests, *Annals of Mathematical Statistics*, **22**, 165–179.

21. Lehmann, E. L. (1986). *Testing Statistical Hypotheses, 2nd Edition*, New York: John Wiley & Sons.

22. Massey, F. J. (1952). Distribution table for the deviation between two sample cumulatives, *Annals of Mathematical Statistics*, **23**, 435–441.

23. Noether, G. E. (1967). *Elements of Nonparametric Statistics*, New York: John Wiley & Sons.

24. Pitman, E. J. G. (1937). Significance tests which may be applied to samples from any populations, *Journal of the Royal Statistical Society, Series A*, **4**, 119–130.

25. Rosenblatt, M. (1952). Limit theorems associated with variants of the von Mises statistic, *Annals of Mathematical Statistics*, **23**, 617–623.

26. Smirnov, N. V. (1939). Sur les écarts de la courbe de distribution empirique (Russian/French summary), *Rec. Math.*, **6**, 3–26.

34

Power Comparisons of Some Nonparametric Tests for Lattice Ordered Alternatives in Two-Factor Experiments

Thu Hoàng and Van L. Parsons

Université René Descartes, Paris, France
National Center for Health Statistics, Hyattsville, Maryland

Abstract: In biological or medical situations the expectations of response variables may be ordered by a rectangular grid partial ordering. For example, serum glucose as a function of body mass index and age would typically be assumed to be nondecreasing in each predictive variable. An order-restricted least squares approach to hypothesis testing may be implemented, but the practical implementation of estimation techniques and sampling theory tend to be complicated. However, advances in computer processing have now made computer intensive methods for such inference more practical.

In this paper we consider the problem of testing trend under the rectangular grid ordering whenever we have small sample sizes and nonparametric sampling assumptions. Our focus is upon the behavior of rank-based test statistics, in particular, the order-restricted version of the Kruskal-Wallis statistic. We compare the power of test statistics generated by order-restricted least squares to the power of more traditionally defined statistics.

Keywords and phrases: Isotonic regression, rectangular grid

34.1 Introduction

Data from two-factor completely random designs are typically analyzed using classical ANOVA methods. However, in certain two-factor experiments, the response is expected to increase as the levels of both factors increase. (Note, decreasing expectations and/or levels can be reparameterized to satisfy this framework.) This knowledge allows the researcher to implement order-restricted data analysis techniques on the experimental data. The usage of such techniques should result in testing procedures having more power than unrestricted F-tests. For example, consider a two-factor design for controlling high blood pressure by a hypertension treatment at increasing dosage and a regimen of physical exercise at increasing levels. Blood pressure should be expected to decrease if levels of both treatments increase. In such cases, we say that experimental responses are stochastically ordered with respect to the lattice order on the two factors. Discussion of such experiments appears in Higgins and Bain (1999), and we adopt some of their notation in presenting the analytical structures.

More precisely, consider a two-way experiment in which the increasing levels of the factors A and B are labeled by $i \in \{1, 2, \ldots I\}$ and $j \in \{1, 2, \ldots J\}$, respectively. We refer to the resulting $I \cdot J$ pairs of levels as the $I \times J$ cross-classification grid. A stochastic ordering of the experimental response random variable, X_{ij}, defined on the $I \times J$ grid can be imposed, and for this paper the case of distributional shift will be considered. We assume that the response has the form $X_{ij} = \mu_{ij} + e_{ij}$ where the e_{ij}'s are independent identically distributed random errors having a continuous distribution function. The trend function, μ_{ij}, on the $I \times J$ grid will be said to be isotone with respect to the lattice ordering $\preceq$ on the grid if $\mu_{ij} \leq \mu_{kl}$ whenever $(i, j) \preceq (k, l)$, i.e., $i \leq k$ and $j \leq l$.

In experimental situations involving small sample sizes and weak distributional assumptions about the samples, nonparametric procedures may be appropriate. Most of the previous work on nonparametric tests of ordered alternatives against homogeneity have been discussed for one-way layouts in completely random design or randomized complete block designs [Terpstra (1952), Jonckheere (1954), Büning and Kössler (1996), Chacko (1963) and Shorack (1967)]. Ordered two-factor completely random designs have been investigated to a lesser degree [Ager and Brent (1978) and Higgins and Bain (1999)]. Chacko (1963) and Shorack (1967) used an order-restricted least squares (ORLS) approach [see Robertson, Wright, and Dykstra (1988) for general discussion] to define test statistics. For two-factor designs we focus on a Kruskal-Wallis-type test statistic defined by ORLS procedures [see Robertson, Wright, and Dykstra (1988, Chapter 4.5)]. For this case the test statistic and its null distribution are difficult to compute, and thus, its usage has been deferred. Recent advances in computer processing have made ORLS methods more practical. With nonparametric tests, complicated null distributions can be simulated, and statistics involving heavy computation can be put to practical use. The authors are not aware of a power study of nonparametric ORLS tests such as the one herein.

34.2 Hypotheses and Test Statistics

Let X_{ijk}, $k = 1, 2, \ldots n_{ij}$, be the sample of size n_{ij} in cell (i, j) of the $I \times J$ cross-classification grid. The null hypothesis, $H_0 : \mu_{ij} = \mu$, is to be tested against the alternative hypothesis, H_1: The μ_{ij}'s are isotone with respect to the lattice order on the $I \times J$ grid, and strict inequality holds for at least one pair of cells.

Three types of "traditional" test statistics for trend will be assessed on the rectangular grid: Jonckheere-Kendall, lattice-rank, and linear rank type statistics. The two-factor Jonckheere-type statistic is of the form:

$$J = \sum_{\{\mathbf{a},\mathbf{b}\ :\ \mathbf{a} \preceq \mathbf{b}\}} U_{\mathbf{a},\mathbf{b}}$$

where $U_{\mathbf{a},\mathbf{b}}$ is the Mann-Whitney two-sample test statistic evaluated at distinct grid points $\mathbf{a} \preceq \mathbf{b}$.

Higgins and Bain (1999) considered a Kendall-type statistic on the $I \times J$ grid. For one-factor experiments, it has been shown [see Robertson, Wright, and Dykstra (1988, p. 202)] that the Kendall and Jonckheere statistics are equivalent, and the argument can be extended to the two-factor case. Higgins and Bain (1999) also discussed a Spearman-type lattice rank statistic

$$V^* = \sum_{i=1}^{I} \sum_{j=1}^{J} ij \sum_{k=1}^{n_{ij}} R_L(X_{ijk})$$

where $R_L(X_{ijk})$ is the lattice rank of X_{ijk}. The lattice rank for an observation in cell (i, j) is its rank among all observations in cells that are comparable to (i, j), i.e., all X_{ghk} such that $(g, h) \preceq (i, j)$ or $(i, j) \preceq (g, h)$.

Traditional linear rank statistics can also be defined for a lattice ordering, for example,

$$L = \sum_{i=1}^{I} \sum_{j=1}^{J} c_{ij} \sum_{k=1}^{n_{ij}} R_{ijk}$$

where c_{ij} are isotone coefficients with respect to the lattice order. We chose $c_{ij} = (\sqrt{i} + \sqrt{j})^2 = i + j + 2\sqrt{ij}$ to represent monotonic main effects and interaction structure. The case $c_{ij} = i + j$ turned out to be quite similar.

The statistics J, V^* and L all reject H_0 when large values are attained. These three statistics will be compared to a nonparametric version of the ORLS-motivated statistic, $\bar{\chi}^2_{01}$, for testing homogeneity against ordered alternatives [see Robertson, Wright, and Dykstra (1988, Chapter 2)]. A general ORLS version of $\bar{\chi}^2_{01}$ can be presented as follows:

If $\{(T_i, n_i/\sigma^2)\}_{i=1}^{k}$ are random variables and specified weights associated with the k elements of a finite partially ordered set, $(S, \preceq)$, then a test statistic for testing $H_0 : E(T_i)$ are constant versus $H_1 : E(T_i)$ satisfy the partial ordering (not all constant) is

$$\bar{\chi}^2_{01}(\mathbf{T}) = \sum_{i=1}^{k} \frac{n_i(T_i^* - \bar{T})^2}{\sigma^2}$$

where $(T_1^*, T_2^*, \dots T_k^*)$ is the isotonic regression of $\{(T_i, n_i/\sigma^2)\}_{i=1}^k$ [see Robertson, Wright, and Dykstra (1988, Definition 1.3.3)] and $\bar{T}$ is the weighted grand mean of the T_i's. One rejects H_0 for large values of $\bar{\chi}_{01}^2$.

For rank–transformed data with Wilcoxon scores Chacko (1963) and Shorack (1967) proposed analogues of the $\bar{\chi}_{01}^2$ test statistic, which can be thought of as ORLS versions of the Kruskal-Wallis test statistic. Shiraishi (1982) provided generalizations that cover scores other than Wilcoxon scores. For our study let R_{ijk} denote the Wilcoxon-scored rank of X_{ijk} based upon all $n = \sum_{i,j} n_{ij}$ observations on the $I \times J$ grid. We will also consider the median scores:

$$\phi(R_{ijk}) = \begin{cases} 0 & \text{if } 0 < R_{ijk} < (n+1)/2 \\ 1 & \text{if } (n+1)/2 \leq R_{ijk} \leq n \end{cases} .$$

The latter transformation is often used when sampling from heavy tailed distributions. For each cell (i, j) we compute the cell means for the scores

$$\bar{R}_{ij} = \sum_{k=1}^{n_{ij}} R_{ijk}/n_{ij}, \quad \bar{\phi}_{ij} = \sum_{k=1}^{n_{ij}} \phi(R_{ijk})/n_{ij}.$$

Two ORLS nonparametric statistics considered for testing H_0 versus H_1 are

$$\bar{\chi}_{01}^2(\mathbf{R}) = \frac{12}{n(n+1)} \sum_{i=1}^{I} \sum_{j=1}^{J} n_{ij} \left(\bar{R}_{ij}^* - \frac{n+1}{2} \right)^2 ,$$

$$\bar{\chi}_{01}^2(\phi_{\mathbf{R}}) = \frac{(n-1)}{np\,(1-p)} \sum_{i=1}^{I} \sum_{j=1}^{J} n_{ij} \left(\bar{\phi}_{ij}^* - p \right)^2 ,$$

$p = \frac{1}{2}$ for n even and $p = \frac{1}{2} + \frac{1}{2n}$ for n odd.

For additional motivation and null distribution properties for these nonparametric statistics the reader is referred to Robertson, Wright, and Dykstra (1988, pp. 204–206).

34.3 Test Statistic Power Evaluations

Any nonparametric comparison among the test statistics involves the dimensions of the $I \times J$ grid, the sample sizes per cell, and sampling distributions under an ordered alternative. We have run simulations under the following sampling distributional assumptions to give an indication of relative power.

1. We considered $I = 5$ and $J = 5$ to specify 25-cell grids with 1 and 4 observations per grid cell.

2. Shift alternative distributions, $X_{ijk} = \mu_{ij} + e_{ijk}$ where e_{ijk} were independent identically distributed random errors from the families:
   Standard Gaussian, double exponential, $H \overset{dist}{=} Z\exp(hZ^2/2)$, $h = 0.5, 1, 2$ [see Hoaglin, Mosteller, and Tukey (1985)] and Gamma (γ) with density function $\propto x^{\gamma-1}e^{-x}$, $\gamma = \frac{1}{2}, 1$ (exponential), 2, were considered.

   All the distributions for the error, e, were scaled to have unit variance except for the H distributions which were scaled so that $P(-1 < e < 1) = .6826$ to agree with a standard Gaussian. The Gamma distributions were selected to generate a skewed error.

3. The shape of the trend function, μ_{ij}, on the $I \times J$ grid was determined to be the most important factor in power comparisons, but having two factors now squares the order of magnitude for the number of cases needed for consideration over that of one dimension. For a two-factor design, the response as a function of the experimental levels is $\mu_{ij} = \theta + \alpha_i + \beta_i + \gamma_{ij}$, α and β are the main factors and γ the interaction. To represent such a broad range of increasing trends on the grid we considered simple discretized monotonic functions on the unit square and simple monotone step functions. The following list provides some basic trend shapes on the 25-cell grid which should provide some insight into relative merits of the proposed statistics.

   Tr1: One step: $\mu_{55} = 1$, $\mu_{ij} = 0$ elsewhere

   Tr2: Additive effect threshold: $\mu_{ij} = 1$, $i + j > 6$, $\mu_{ij} = 0$ elsewhere

   Tr3: Border one step: $\mu_{5j} = 1$, $j = 1, 2, ...5$, $\mu_{ij} = 0$ elsewhere

   Tr4: Angle corner one step: $\mu_{ij} = 0$, $i \leq 4$, $j \leq 4$, $\mu_{ij} = 1$ elsewhere

   Tr5: Angle three steps: $\mu_{ij} = 0$, $i, j \leq 3$, $\mu_{ij} = 1$, $i \leq 3$, $j \geq 4$, $\mu_{ij} = 2$, $j \leq 3$, $i \geq 4$, $\mu_{ij} = 3$, $i \geq 4$, $j \geq 4$

   Tr6: Uniform one main effect: $\mu_{ij} \propto (j-1)$

   Tr7: Uniform two main effects: $\mu_{ij} \propto (i-1) + (j-1)$

   Tr8: Early effects: $\mu_{ij} \propto (i-1)^{1/2} + (j-1)^{1/2}$

   Tr9: Late effects: $\mu_{ij} \propto (i-1)^2 + (j-1)^2$

   Tr10: Late effects + interaction: $\mu_{ij} \propto [(i-1) + (j-1)]^2$

   Tr11: Early+late effects: $\mu_{ij} \propto (i-1)^{1/2} + (j-1)^2$

   Our choice of the terms "early" and "late" for trends Tr8–Tr11 refer to the position of effects producing maximum change. Trends Tr1–Tr4 represent extreme cases of trend functions taking just two values.

4. We scaled the trend shapes of 3. above as in Robertson, Wright, and Dykstra (1988, Section 2.5) to represent a distance from the null hypothesis.

Denoting the basic shape trends as $\{\mu_{0,ij}\}_{I\times J}$, then for $\delta = 1, 2, 3, 4$ we defined *scaled* grids

$$\{\mu_{ij}\}_{I\times J} = \left\{\frac{\delta}{\Delta}\,\mu_{0,ij}\right\}_{I\times J} \quad \text{where } \Delta = \left(\sum_{I\times J} n_{ij}(\mu_{0,ij} - \bar{\mu}_0)^2\right)^{1/2}.$$

This transformation provided some degree of standardization among different trend functions and standardization for different sample sizes.

5. Based upon 100,000 simulations using a Fortran uniform random number generator and code [11], an approximate size $\alpha = 0.01$ critical region was established for each statistic. The $\bar{\chi}^2_{01}$ statistics were computed using the algorithm of Qian and Eddy (1996). The alternative distribution simulations were based upon 10,000 simulations. Publicly available Fortran algorithms were used in DATAPAC (1986). Furthermore, for a fixed error variable distribution, we used the same e_{ijk}'s for different μ_{ij}'s and test statistics. In general, this technique introduced a strong positive simulation correlation among the estimated powers for the different statistics. For example, the estimated correlations between estimated powers of the $\bar{\chi}^2_{01}(\mathbf{R})$ and J statistics were frequently 0.5 or greater for μ_{ij} close to H_0. This simulation feature helps to reduce the sampling error when making comparisons.

Comparing test statistics by their power levels is straightforward when the isotone function, μ_{ij}, error distribution and sample sizes are fixed, but comparing over different specifications becomes difficult. To facilitate global comparisons over different H_1 alternatives, a measure of relative efficiency will be used. John Tukey (personal communication) suggested the following measure for small sample nonparametric tests:

$$\text{eff} = \left(\frac{\Phi^{-1}(P_{H_1}(S' \in C')) - \Phi^{-1}(P_{H_0}(S' \in C'))}{\Phi^{-1}(P_{H_1}(S \in C)) - \Phi^{-1}(P_{H_0}(S \in C))}\right)^2$$

where $P_{H_0}(S \in C)$ is the size of the S test critical region, C, $P_{H_1}(S \in C)$ is the power of the S test for a distribution $F \in H_1$, and Φ^{-1} is the Gaussian quantile.

It should be noted that the traditional measure of efficiency is the ratio of sample sizes for two statistics needed to achieve the same power. The measure used above is not of that form; here both statistics are based upon the same sample sizes. However, some motivation for using eff for "large" sample sizes and shift alternatives can be informally stated as follows.

If $P_H(S \in C)) \approx \Phi(\,c(n,\alpha) - E(S_H)\,)/\sigma_0)$ then

$$\left[\Phi^{-1}(P_{H_1}(S \in C)) - \Phi^{-1}(P_{H_0}(S \in C))\right]^2 \approx \left[E(S_{H_1}) - E(S_{H_0})\right]^2/\sigma_0^2.$$

This quantity can be considered the distance of H_1 to H_0, and is often proportional to the sample size n. Under such conditions the magnitude of eff above would be somewhat consistent with a traditional measure of test statistic efficiency. For the small sample situation we consider herein, we believe that the magnitudes of eff and traditional efficiency would exhibit similar patterns. Another reason to use eff instead of power alone is that this measure adjusts for the possible different α-levels of the competitor statistics. For small sample sizes the discrete nature of the test statistics result in nominal α–levels being somewhat different from the achieved α–level.

Our main focus of this study is the comparison of the ORLS statistics to the more classically-defined competitors. The $\bar{\chi}^2_{01}(\mathbf{R})$ has been well-established and can be considered an omnibus test statistic for trend. In our work we let $S = \bar{\chi}^2_{01}(\mathbf{R})$ in the definition of eff and let S' be a competitor statistic.

34.4 Results and Conclusions

Our simulations are too numerous to present here, but Table 34.1 and Table 34.2 provide examples of the types of power and efficiencies obtained. Trend function shapes, firstly, and error variable distributions, secondly, are the most important factors in making global assessments about the different tests. Tables 34.1 and 34.2 present the two trend functions most likely to be of practical concern: a two-main-effects model and a one-main-effect model. Tables 34.3 and Table 34.4 provide the ranges of efficiencies over different shaped trend functions and error variables representing degrees of tail weight and skewness. The tabulated efficiencies are best used to discern general patterns of test statistic superiority. In Tables 34.3 and 34.4 we specify for each combination of trend function and distribution the superior statistic(s) in the column "Test choice". As a caveat we note that large efficiencies may be observed when the power of both $\bar{\chi}^2_{01}(\mathbf{R})$ and its competitor are both small, e.g. both powers ≤ 0.10, but superiority for such a case would be of little practical concern. Furthermore, for both powers small, the simulation error of the efficiency (see Appendix) could make a simulated 10% difference not significant. In our tables the results for the symmetric distributions with $\delta = 1$ tend to result in small powers for all the statistics and interpretation of the magnitudes of efficiencies must be treated with caution. Some conclusions are:

- Except for very heavy tailed distributions, the Jonckheere statistic performs relatively well when the trend function μ_{ij} is strictly increasing in both levels i and j. For example, in Tables 34.1 and 34.2 for the uniform two-main-effects model, Tr7, the Jonckheere statistic shows substantial increases in power over the $\bar{\chi}^2_{01}(\mathbf{R})$ statistic. While our targeted cell sample sizes were limited to $n = 1$ and $n = 4$, there appears to be a stronger relative superiority for the larger n. This superior behavior is also apparent for trends Tr7–Tr11 in Tables 34.3 and 34.4. While the L test and V^* test exhibit superiority in some cases, the Jonckheere statistic appears to perform well over a wider range of strictly increasing trend shapes.

- The one-step trend shapes Tr2, Tr3, Tr4 and Tr6 reverse the superiority just discussed. For these trend shapes the $\bar{\chi}^2_{01}(\mathbf{R})$ statistic shows superior power over the J, L and V^* statistics. The border one-step trend, Tr3, appears to be an extreme case. The one-main-effect trend, Tr6, is detailed in Tables 34.1 and 34.2.

- The angle three step trend, Tr5, represents a degree of trend somewhat between the one-step and strictly increasing trends. Here, the J and V^* statistics tend to have greater power than the $\bar{\chi}^2_{01}(\mathbf{R})$ statistic, but reversals are more frequent than for the stronger trends, especially for $n = 1$.

- Some caution must be used with the interpretation of the one-step, Tr1, trend. Here, cell parameter $\mu_{5,5}$ can be thought of as the only cell that deviates from H_0. The reduction of the data to ranks results in very low power for small sample sizes, even as $\mu_{5,5} \to \infty$. For trend Tr1 the maximum power was of the order 0.05 for $n = 1$ and about 0.20 for $n = 4$. When comparing the two different sample sizes for this case, the patterns of eff were less consistent between than for the other trend shapes.

- The V^* statistic is defined to detect strong trends that occur during late levels of the experimental treatments. Its best performance was for the "late effects + interaction trend", Tr10, and the one-step trend, Tr1. These observations are consistent with the discussion presented in Higgins and Bain (1999).

- As expected, for the very heavy tailed distribution, H, with $h = 2$, the $\bar{\chi}^2_{01}(\phi_{\mathbf{R}})$ statistic tends to superior power over the competitor statistics. This is most noticeable for the $n = 4$ sample size.

In conclusion the study seems to suggest that if extremely heavy tailed distributions are discounted, then the Jonckheere statistic or $\bar{\chi}^2_{01}(\mathbf{R})$ statistic would be a reasonable choice test statistic for testing a broad range of two factor ordered alternatives. The Jonckheere statistic might be favored if the researcher believed that an increase in any one factor level should result in a strictly increased response. For experiments where the effectiveness of one factor or many of levels are questionable, the $\bar{\chi}^2_{01}(\mathbf{R})$ statistic would be favored.

Table 34.1: Power and efficiency of test statistics compared to isotonized Kruskal Wallis statistic for $\alpha = 0.01$, 5×5 grids and one observation per cell

Trend	Distribution	Delta	Efficiency				Power				
			Jonc	Lin R	V*	Iso med	KW	Jonc	Lin R	V*	Iso med
							n = 1				
Uniform gradient	normal	1	1.07	1.07	1.03	0.67	0.07	0.08	0.08	0.08	0.05
		3	1.09	1.07	1.01	0.63	0.61	0.66	0.64	0.61	0.39
		4	1.08	1.08	1.01	0.61	0.86	0.90	0.89	0.87	0.63
	exponential	1	1.07	1.03	0.83	0.56	0.17	0.19	0.17	0.14	0.09
		3	1.15	1.10	1.04	0.74	0.79	0.86	0.84	0.81	0.65
		4	1.18	1.12	1.09	0.75	0.93	0.96	0.95	0.95	0.83
Two effects	gamma(.5)	1	1.11	1.01	0.85	0.68	0.33	0.37	0.33	0.28	0.22
		3	1.19	1.11	1.07	0.84	0.87	0.93	0.91	0.90	0.80
		4	1.24	1.13	1.10	0.85	0.95	0.98	0.97	0.97	0.91
	gamma(2)	1	1.05	1.01	0.89	0.60	0.11	0.12	0.11	0.10	0.07
		3	1.11	1.08	1.02	0.67	0.72	0.77	0.76	0.73	0.52
		4	1.13	1.10	1.08	0.68	0.91	0.94	0.94	0.93	0.75
	H(h=.5)	1	1.12	1.10	1.09	1.03	0.07	0.08	0.08	0.08	0.07
		3	1.15	1.12	1.05	0.90	0.48	0.56	0.53	0.50	0.43
		4	1.18	1.12	1.04	0.89	0.67	0.76	0.73	0.70	0.61
	H(h=1)	1	1.12	1.09	1.06	1.24	0.08	0.09	0.09	0.08	0.10
		3	1.21	1.13	1.06	1.08	0.44	0.54	0.50	0.47	0.47
		4	1.22	1.13	1.04	1.07	0.59	0.70	0.66	0.62	0.62
	H(h=2)	1	1.16	1.11	1.06	1.51	0.11	0.13	0.12	0.12	0.17
		3	1.30	1.17	1.06	1.31	0.42	0.54	0.49	0.44	0.54
		4	1.30	1.15	1.04	1.28	0.53	0.67	0.60	0.55	0.65
Uniform steps	normal	1	0.88	0.88	0.90	0.64	0.05	0.05	0.04	0.05	0.04
		3	0.78	0.73	0.74	0.63	0.39	0.31	0.28	0.29	0.24
		4	0.73	0.67	0.67	0.60	0.67	0.53	0.48	0.48	0.43
	exponential	1	0.88	0.79	0.70	0.57	0.10	0.09	0.08	0.07	0.06
		3	0.81	0.70	0.70	0.74	0.60	0.51	0.44	0.44	0.46
		4	0.82	0.68	0.71	0.75	0.80	0.72	0.62	0.64	0.66
One effect	gamma(.5)	1	0.92	0.76	0.68	0.70	0.20	0.19	0.15	0.13	0.14
		3	0.90	0.72	0.76	0.86	0.72	0.67	0.55	0.58	0.64
		4	0.89	0.71	0.77	0.87	0.86	0.82	0.71	0.75	0.80
	gamma(2)	1	0.91	0.86	0.78	0.61	0.06	0.06	0.06	0.05	0.04
		3	0.79	0.71	0.68	0.69	0.50	0.40	0.35	0.34	0.34
		4	0.77	0.67	0.67	0.69	0.74	0.63	0.54	0.55	0.56
	H(h=.5)	1	0.92	0.93	0.92	1.11	0.05	0.05	0.04	0.04	0.05
		3	0.86	0.77	0.79	0.88	0.31	0.27	0.23	0.24	0.27
		4	0.86	0.75	0.78	0.88	0.49	0.43	0.37	0.38	0.43
	H(h=1)	1	0.90	0.93	0.90	1.28	0.05	0.05	0.05	0.05	0.06
		3	0.92	0.79	0.83	1.06	0.29	0.27	0.22	0.24	0.31
		4	0.93	0.78	0.82	1.03	0.42	0.41	0.33	0.35	0.44
	H(h=2)	1	0.98	0.93	0.95	1.58	0.07	0.07	0.07	0.07	0.10
		3	1.00	0.80	0.87	1.26	0.28	0.29	0.23	0.25	0.36
		4	1.02	0.80	0.87	1.24	0.37	0.39	0.30	0.32	0.46

Power and effiency are computed for random variables of the form $cX + \mu\delta$ where μ is a trend function on the grid scaled by δ, the distance of μ to H_0, and X a random variable standardized by c. Distributions of X are symmetric: Gaussian Z, $H = Z\exp(hZ^2/2)$ for $h = .5, 1, 2$, or skewed: exponential, gamma(.5), gamma(2). In the case of uniform gradient trend $\mu_{ij} \propto (i-1) + (j-1)$ and for uniform steps $\mu_{ij} \propto (i-1)$

Table 34.2: Power and efficiency of test statistics compared to isotonized Kruskal Wallis statistic for $\alpha = 0.01$, 5×5 grids and four observations per cell

Trend	Distribution	Delta	Efficiency Jonc	Efficiency Lin R	Efficiency V*	Efficiency Iso med	Power KW	Power Jonc	Power Lin R	Power V*	Power Iso med
							n = 4				
Uniform gradient	normal	1	1.22	1.21	1.08	0.63	0.07	0.09	0.09	0.08	0.05
		3	1.18	1.16	1.09	0.67	0.63	0.71	0.70	0.67	0.44
		4	1.19	1.17	1.08	0.66	0.89	0.94	0.94	0.92	0.71
	exponential	1	1.21	1.15	1.04	0.41	0.19	0.23	0.22	0.19	0.08
		3	1.22	1.18	1.15	0.59	0.92	0.96	0.96	0.95	0.70
		4	1.26	1.21	1.21	0.67	0.99	1.00	1.00	1.00	0.93
Two effects	gamma(.5)	1	1.29	1.18	1.06	0.35	0.46	0.58	0.54	0.49	0.16
		3	1.26	1.16	1.17	0.81	0.99	1.00	1.00	1.00	0.96
		4	1.24	1.17	1.25	0.92	1.00	1.00	1.00	1.00	1.00
	gamma(2)	1	1.19	1.15	1.05	0.53	0.11	0.13	0.13	0.12	0.06
		3	1.19	1.17	1.12	0.60	0.80	0.88	0.87	0.85	0.55
		4	1.24	1.18	1.15	0.63	0.96	0.99	0.98	0.98	0.83
	H(h=.5)	1	1.25	1.22	1.12	1.00	0.07	0.09	0.09	0.08	0.07
		3	1.19	1.16	1.07	1.04	0.58	0.67	0.65	0.62	0.60
		4	1.21	1.15	1.06	1.02	0.83	0.90	0.88	0.85	0.83
	H(h=1)	1	1.25	1.21	1.09	1.35	0.08	0.10	0.10	0.09	0.11
		3	1.22	1.15	1.06	1.30	0.61	0.71	0.68	0.64	0.74
		4	1.24	1.15	1.06	1.26	0.82	0.90	0.88	0.85	0.91
	H(h=2)	1	1.22	1.20	1.11	1.87	0.13	0.16	0.16	0.15	0.25
		3	1.26	1.15	1.04	1.61	0.70	0.81	0.77	0.72	0.90
		4	1.29	1.14	1.03	1.54	0.86	0.94	0.90	0.87	0.97
Uniform steps	normal	1	1.02	0.98	0.92	0.62	0.05	0.05	0.05	0.05	0.04
		3	0.82	0.80	0.75	0.65	0.47	0.39	0.37	0.35	0.30
		4	0.76	0.73	0.69	0.63	0.77	0.64	0.62	0.59	0.55
	exponential	1	0.98	0.93	0.81	0.37	0.12	0.12	0.11	0.10	0.05
		3	0.82	0.75	0.74	0.54	0.82	0.73	0.69	0.68	0.52
		4	0.79	0.73	0.73	0.62	0.97	0.92	0.89	0.89	0.83
One effect	gamma(.5)	1	0.96	0.83	0.77	0.29	0.32	0.31	0.26	0.25	0.09
		3	0.85	0.74	0.78	0.72	0.96	0.93	0.88	0.90	0.88
		4	0.83	0.70	0.77	0.86	1.00	0.99	0.97	0.98	0.99
	gamma(2)	1	0.99	0.95	0.85	0.51	0.07	0.07	0.07	0.06	0.04
		3	0.81	0.77	0.73	0.57	0.66	0.55	0.53	0.50	0.39
		4	0.76	0.72	0.70	0.59	0.91	0.81	0.78	0.77	0.68
	H(h=.5)	1	1.02	0.99	0.94	1.12	0.05	0.05	0.05	0.05	0.05
		3	0.85	0.81	0.77	1.05	0.43	0.36	0.34	0.33	0.45
		4	0.80	0.76	0.72	1.05	0.69	0.58	0.56	0.53	0.72
	H(h=1)	1	1.01	1.00	0.92	1.47	0.06	0.06	0.06	0.05	0.08
		3	0.86	0.81	0.77	1.37	0.46	0.40	0.37	0.35	0.61
		4	0.83	0.76	0.74	1.34	0.70	0.61	0.56	0.55	0.83
	H(h=2)	1	1.05	1.02	0.93	2.19	0.08	0.09	0.08	0.08	0.18
		3	0.90	0.79	0.77	1.78	0.56	0.51	0.45	0.44	0.84
		4	0.89	0.76	0.76	1.72	0.75	0.69	0.61	0.61	0.95

Power and effiency are computed for random variables of the form $cX + \mu\delta$ where μ is a trend function on the grid scaled by δ, the distance of μ to H_0, and X a random variable standardized by a constant c. Distributions of X are symmetric: Gaussian Z, $H = Z\exp(hZ^2/2)$ for $h = .5, 1, 2$, or skewed: exponential, gamma(.5), gamma(2). In the case of uniform gradient trend $\mu_{ij} \propto (i-1) + (j-1)$ and for uniform steps $\mu_{ij} \propto (i-1)$ on cells (i,j)

Table 34.3: Comparing ranges of effiency of statistics and choosing a test for selected trend shapes and distributions and for $\alpha = 0.01$, 5×5 grids and one observation per cell

Trend	Distribution	Jonckheere test Min	Jonckheere test Max	Linear rank test Min	Linear rank test Max	V* test Min	V* test Max	Isotonic median test Min	Isotonic median test Max	Test choice
		n = 1								
One step	Gauss	1.21	1.34	1.14	1.24	2.15	2.26	0.18	0.36	V*
	Skewed	1.08	1.50	0.98	1.32	1.57	2.17	0.18	0.58	V*
	Heavy tail 1	1.13	1.45	1.11	1.40	1.87	2.34	0.20	0.50	V*
	Heavy tail 2	1.07	1.37	1.18	1.30	2.07	2.40	0.20	0.38	V*
Diagonal one step	Gauss	0.90	0.96	0.91	0.96	0.71	0.83	0.74	0.77	KW
	Skewed	0.90	0.98	0.89	0.94	0.66	0.73	0.53	0.81	KW
	Heavy tail 1	0.90	0.99	0.92	1.00	0.72	0.85	0.94	1.19	IM KW
	Heavy tail 2	0.94	1.00	0.94	1.01	0.77	0.84	1.30	1.76	IM
Border one step	Gauss	0.50	0.77	0.39	0.63	0.53	0.90	0.30	0.62	KW
	Skewed	0.49	0.90	0.39	0.77	0.51	0.85	0.33	1.47	KW
	Heavy tail 1	0.48	0.83	0.38	0.73	0.52	0.93	0.29	0.94	KW
	Heavy tail 2	0.56	0.85	0.47	0.69	0.62	0.83	0.42	1.11	KW
Angle corner one step	Gauss	0.66	0.91	0.56	0.84	0.64	0.99	0.61	0.69	KW
	Skewed	0.65	0.92	0.57	0.78	0.63	0.81	0.70	1.71	KW IM
	Heavy tail 1	0.64	0.95	0.55	0.82	0.64	0.92	0.71	1.03	KW
	Heavy tail 2	0.77	0.91	0.66	0.81	0.74	0.86	0.98	1.49	IM KW
Angle three steps	Gauss	0.94	1.08	0.91	1.05	0.96	1.15	0.62	0.71	KW V*
	Skewed	0.98	1.08	0.93	0.99	0.89	1.05	0.66	0.93	KW Jonc
	Heavy tail 1	0.96	1.09	0.92	1.03	0.98	1.12	0.74	1.06	V*
	Heavy tail 2	1.06	1.11	0.99	1.03	1.04	1.11	1.04	1.52	IM Jonc
Uniform steps (one effect)	Gauss	0.73	0.88	0.67	0.88	0.67	0.90	0.60	0.67	KW
	Skewed	0.77	0.92	0.67	0.86	0.67	0.78	0.57	0.87	KW
	Heavy tail 1	0.77	0.94	0.68	0.93	0.70	0.92	0.73	1.11	KW
	Heavy tail 2	0.90	1.02	0.78	0.93	0.82	0.95	1.03	1.58	IM KW
Uniform gradient (two uniformly increasing effects)	Gauss	1.07	1.09	1.06	1.08	1.01	1.03	0.61	0.67	Jonc
	Skewed	1.05	1.24	1.01	1.13	0.83	1.10	0.56	0.85	Jonc
	Heavy tail 1	1.11	1.18	1.09	1.12	1.00	1.09	0.73	1.03	Jonc
	Heavy tail 2	1.12	1.30	1.09	1.17	1.04	1.07	1.07	1.51	IM Jonc
Early effects	Gauss	1.03	1.08	1.08	1.09	0.89	0.97	0.61	0.66	L Jonc
	Skewed	1.00	1.14	0.98	1.10	0.74	0.99	0.51	0.86	Jonc
	Heavy tail 1	1.06	1.14	1.11	1.13	0.93	0.99	0.75	1.00	L Jonc
	Heavy tail 2	1.13	1.27	1.11	1.18	0.98	1.03	1.08	1.52	IM Jonc
Late effects	Gauss	1.04	1.06	0.98	1.00	1.05	1.07	0.60	0.65	V* Jonc
	Skewed	1.06	1.26	1.01	1.09	0.90	1.18	0.63	0.88	Jonc
	Heavy tail 1	1.08	1.14	1.00	1.07	1.06	1.11	0.72	0.97	Jonc
	Heavy tail 2	1.11	1.28	1.01	1.10	1.07	1.11	1.05	1.48	IM Jonc
Late effects + interaction	Gauss	1.04	1.06	1.06	1.09	1.15	1.18	0.57	0.65	V*
	Skewed	1.04	1.27	1.06	1.18	0.99	1.27	0.64	0.83	Jonc V*
	Heavy tail 1	1.09	1.16	1.09	1.15	1.17	1.24	0.71	0.96	V*
	Heavy tail 2	1.10	1.28	1.08	1.17	1.12	1.21	1.04	1.46	IM Jonc
Early + late effects	Gauss	1.04	1.05	1.04	1.06	0.97	1.01	0.59	0.68	L Jonc
	Skewed	1.04	1.20	0.97	1.10	0.82	1.08	0.56	0.81	Jonc
	Heavy tail 1	1.10	1.14	1.06	1.11	0.99	1.07	0.70	1.06	Jonc
	Heavy tail 2	1.13	1.27	1.09	1.13	1.02	1.05	1.01	1.48	IM Jonc

Distributions: Gauss; Skewed: exponential, gamma(.5), gamma(2); Heavy tail 1: double exponential, $H = Z\exp(hZ^2/2)$ for $h = .5$; Heavy tail 2: H for $h = 1, 2$

Table 34.4: Comparing effiency of statistics and choosing a test for selected trend shapes and distributions and for $\alpha = 0.01$, 5×5 grids and four observations per cell

Trend	Distribution	Jonckheere test Min	Jonckheere test Max	Linear rank test Min	Linear rank test Max	V^* test Min	V^* test Max	Isotonic median test Min	Isotonic median test Max	Test choice
					n = 4					
One step	Gauss	0.48	0.85	0.48	0.82	0.90	1.46	0.20	0.46	KW V^*
	Skewed	0.49	1.04	0.48	1.01	0.94	1.67	0.20	1.51	KW V^*
	Heavy tail 1	0.45	0.95	0.44	0.99	0.86	1.81	0.16	0.61	KW V^*
	Heavy tail 2	0.59	0.96	0.59	0.91	1.10	1.62	0.28	0.81	V^*
Diagonal one step	Gauss	0.99	1.21	0.95	1.15	0.76	0.90	0.65	0.68	Jonc KW
	Skewed	0.95	1.20	0.94	1.10	0.78	0.87	0.28	0.56	Jonc KW
	Heavy tail 1	0.98	1.24	0.95	1.19	0.76	0.93	0.96	1.23	Jonc KW
	Heavy tail 2	1.00	1.21	0.97	1.13	0.78	0.90	1.44	2.18	IM
Border one step	Gauss	0.44	0.99	0.39	0.86	0.49	0.98	0.56	0.73	KW
	Skewed	0.28	0.85	0.26	0.77	0.32	0.84	0.59	1.79	KW
	Heavy tail 1	0.41	0.95	0.36	0.84	0.46	0.98	0.69	1.24	KW
	Heavy tail 2	0.48	0.93	0.42	0.81	0.54	0.94	0.96	2.03	IM KW
Angle corner one step	Gauss	0.73	1.01	0.63	0.88	0.67	0.91	0.63	0.64	KW
	Skewed	0.64	0.99	0.56	0.89	0.63	0.92	0.41	0.91	KW
	Heavy tail 1	0.71	1.07	0.62	0.93	0.66	1.00	0.94	1.18	KW IM
	Heavy tail 2	0.75	1.06	0.66	0.90	0.70	0.94	1.40	2.20	IM KW
Angle three steps	Gauss	0.99	1.29	0.94	1.24	0.97	1.21	0.62	0.66	Jonc
	Skewed	0.95	1.17	0.89	1.08	0.98	1.09	0.36	1.06	Jonc
	Heavy tail 1	0.98	1.29	0.93	1.20	0.95	1.20	0.89	1.14	Jonc
	Heavy tail 2	1.03	1.27	0.94	1.20	0.95	1.21	1.34	2.11	IM Jonc
Uniform steps (one effect)	Gauss	0.76	1.02	0.73	0.98	0.69	0.92	0.62	0.65	KW
	Skewed	0.76	0.99	0.70	0.95	0.70	0.85	0.29	0.86	KW
	Heavy tail 1	0.75	1.02	0.72	0.99	0.69	0.94	0.90	1.12	KW
	Heavy tail 2	0.83	1.05	0.76	1.02	0.74	0.93	1.34	2.19	IM KW
Uniform gradient (two uniformly increasing effects)	Gauss	1.18	1.22	1.16	1.21	1.08	1.09	0.63	0.67	Jonc L
	Skewed	1.19	1.29	1.15	1.21	1.04	1.25	0.35	0.92	Jonc
	Heavy tail 1	1.19	1.28	1.15	1.25	1.06	1.19	0.87	1.11	Jonc
	Heavy tail 2	1.22	1.29	1.14	1.21	1.03	1.11	1.26	1.87	IM Jonc
Early effects	Gauss	1.11	1.21	1.14	1.21	0.94	1.01	0.63	0.65	L Jonc
	Skewed	1.12	1.21	1.11	1.20	0.92	1.10	0.30	0.77	L Jonc
	Heavy tail 1	1.10	1.27	1.11	1.28	0.95	1.07	0.85	1.12	L Jonc
	Heavy tail 2	1.16	1.24	1.13	1.25	0.96	1.03	1.25	1.84	IM L Jonc
Late effects	Gauss	1.12	1.21	1.06	1.13	1.10	1.15	0.60	0.65	Jonc
	Skewed	1.16	1.30	1.04	1.14	1.11	1.23	0.41	0.98	Jonc
	Heavy tail 1	1.12	1.28	1.05	1.18	1.08	1.24	0.87	1.12	Jonc
	Heavy tail 2	1.18	1.26	1.06	1.17	1.05	1.17	1.26	1.91	IM Jonc
Late effects + interaction	Gauss	1.12	1.24	1.14	1.24	1.21	1.26	0.63	0.65	V^*
	Skewed	1.15	1.31	1.12	1.23	1.22	1.30	0.47	1.08	V^* Jonc
	Heavy tail 1	1.14	1.26	1.13	1.27	1.18	1.34	0.85	1.10	V^* Jonc
	Heavy tail 2	1.19	1.25	1.15	1.23	1.13	1.28	1.22	1.85	IM V^* Jonc
One early + one late effects	Gauss	1.12	1.23	1.10	1.21	1.02	1.10	0.62	0.66	Jonc
	Skewed	1.14	1.25	1.10	1.14	1.02	1.15	0.35	0.97	Jonc
	Heavy tail 1	1.11	1.27	1.08	1.24	1.02	1.15	0.85	1.12	Jonc
	Heavy tail 2	1.16	1.24	1.10	1.18	1.00	1.09	1.22	1.85	IM Jonc

Distributions: Gauss; Skewed: exponential, gamma(.5), gamma(2); Heavy tail 1: double exponential, $H = Z\exp(hZ^2/2)$ for $h = .5$; Heavy tail 2: H for $h = 1, 2$

Appendix

The simulation standard error of eff can be approximated by applying a Taylor linearization on the functional form. For an estimator $\hat{p}$ of a proportion p we have $\mathrm{Var}(\Phi^{-1}(\hat{p})) \approx \left[\phi(\Phi^{-1}(p))\right]^{-2} p(1-p)/n$, where n is the number of simulations. For two estimated proportions on the same simulation we have the correlation coefficient $\rho\left(\Phi^{-1}(\hat{p}_1), \Phi^{-1}(\hat{p}_2)\right) \approx \rho(\hat{p}_1, \hat{p}_2)$. For a variable of the form : $T^2 = \left[\left(\Phi^{-1}(\hat{p}_2) - \Phi^{-1}(\alpha)\right) / \left(\Phi^{-1}(\hat{p}_1) - \Phi^{-1}(\alpha)\right)\right]^2$ with α treated as a constant, we have after additional Taylor linearizations an approximation for the standard error of T^2, $\mathrm{SE}(T^2)$, due to the simulation:
$\mathrm{SE}(T^2) \approx 2T^2 \left(CV_1^2 + CV_2^2 - 2\rho(\hat{p}_1, \hat{p}_2) CV_1 CV_2\right)^{1/2}$ where

$CV_i^2 = \left[\phi(\Phi^{-1}(p_i)) \cdot (\Phi^{-1}(\hat{p}_i) - \Phi^{-1}(\alpha))\right]^{-2} p_i(1-p_i)/n$. The value of ρ is estimated from the simulation. Using this approximation examples of estimates for the standard error of eff will be 0.0281 and 0.0127 when $p_1 = p_2 = 0.15$ and $p_1 = p_2 = 0.50$ respectively. (Here, we used $\rho = 0.30$).

References

1. Ager, J. W. and Brent, S. B. (1978). An index of agreement between a hypothesized partial order and an empirical rank order, *Journal of the American Statistical Association*, **73**, 827–830.

2. Büning, H. and Kössler, W. (1996). Robustness and efficiency of some tests for ordered alternatives in the c-sample location problem, *Journal of Statistical Computation and Simulation*, **55**, 337–352.

3. Chacko, V. J. (1963). Testing homogeneity against ordered alternatives, *The Annals of Mathematical Statistics*, **34**, 945–956.

4. DATAPAC (1986). A Fortran subroutine library for probability distributions, National Institute of Standards and Technology.

5. Higgins, J. J. and Bain, P. T. (1999). Non parametric tests for ordered alternatives in unreplicated two-factor experiment, *Journal of Nonparametric Statistics*, **11**, 307–318.

6. Hoaglin, D. C., Mosteller, F., and Tukey, J. W. (Eds.) (1985). *Exploring Data Tables, Trends, and Shapes*, New York: John Wiley & Sons.

7. Jonckheere, A. R. (1954). A distribution-free k-sample test against ordered alternatives, *Biometrica*, **41**, 133–145.

8. Marsaglia, G. and Zaman, A. (1987). Toward a universal random number generator, *Florida State University Report: FSU-SCRI-87–50.*

9. Qian, S. and Eddy, W. F. (1996). An algorithm for isotonic regression on ordered rectangular grids, *Journal of Computational and Graphical Statistics*, **5**, 225–235.

10. Robertson, T., Wright, F. T., and Dykstra, R. (1988). *Order Restricted Statistical Inference*, Chichester: John Wiley & Sons.

11. Shiraishi, T. (1982). Testing homogeneity against trend based on rank in one-way layout, *Communications in Statistics—Theory and Methods*, **11**, 1255–1268.

12. Shorack, G. R. (1967). Testing ordered alternatives in model I analysis of variance; Normal theory and nonparametric, *The Annals of Mathematical Statistics*, **38**, 1740–1753.

13. Terpstra, T. J. (1952). The asymptotic normality and consistency of Kendall's test against trend when ties are present in one ranking, *Proceedings of the Section of Science Koninklijke Nederlandse Akademie van Wetenschappen (A)*, **55**, *Indagationes Mathematicae*, **14**, 327–333.

35

Tests of Independence with Exponential Marginals

Paul Deheuvels

L.S.T.A., Université Paris VI, Bourg-la-Reine, France

Abstract: We present tests of independence for bivariate vectors with exponential marginals, in the setup of bivariate extreme value distributions. These rely on a new Karhunen-Loeve expansion due to Deheuvels and Martynov (2000).

Keywords and phrases: Test of independence, bivariate extreme values, Cramer-von Mises-type tests, Karhunen-Loeve expansions

35.1 Introduction

Let $\{(X_n, Y_n) : n \geq 1\}$ be independent and identically distributed [i.i.d] bivariate random vectors with exponential marginals. Assume, in addition, that the distribution of $(X, Y) = (X_1, Y_1)$, denoted by $E_A(\gamma, \nu)$, is such that, for constants $\gamma > 0$ and $\nu > 0$,

$$\mathbb{P}(X \geq \gamma x, Y \geq \nu y) = \exp\Big(- (x+y) A\Big(\frac{x}{x+y}\Big)\Big) \quad \text{for} \quad x > 0, y > 0, \quad (35.1)$$

where $\{A(u) : 0 \leq u \leq 1\}$ fulfills the assumptions

$(A.1)$ $\max\{u, 1-u\} \leq A(u) \leq 1$ for $0 \leq u \leq 1$;
$(A.2)$ A is convex on $[0, 1]$.

This model is discussed at length in Falk, Hüsler, and Reiss (1994, Section 4.2, pp. 111–118) [see also Deheuvels (1984) and Resnick (1987, Ch. 5)]. It is noteworthy [see, for example, Pickands (1981,1989) and Galambos (1987)] that the conditions $(A.1\text{–}2)$ are necessary and sufficient conditions for $\mathbb{P}(X \geq x, Y \geq y)$ to define, via (35.1) the *survival function* of a bivariate *extreme value* probability distribution for *minima*, with the following characteristic property. Whenever $(X, Y) \equiv E_A(\gamma, \nu)$ (this denoting the fact that (X, Y) follows the distribution $E_A(\gamma, \nu)$), then, for any constants $c > 0$ and $d > 0$, $\min(cX, dY)$ is exponential. This holds, in particular, for X (with $c = 1, d = \infty$) and Y (with $c = \infty, d = 1$), both of which are exponentially distributed with expectations given by $\mathbb{E}(X) = \gamma$ and $\mathbb{E}(Y) = \nu$.

We are concerned with testing the null hypothesis, denoted hereafter by $(H.0)$, that X and Y are *independent*, against the alternative. Under the $E_A(\gamma, \nu)$ model, the independence assumption $(H.0)$ may be reformulated into

$$(H.0) \qquad A(u) = 1 \quad \text{for} \quad 0 \le u \le 1.$$

There is a huge literature dealing either with models of bivariate distributions with exponential marginals [see, for example, Gumbel (1960a), Freund (1961), Marshall and Olkin (1967,1982), Downton (1970), Arnold (1975), Proschan and Sullo (1976), Johnson and Kotz (1977), Raftery (1984), Barnett (1985), Balakrishnan and Basu (1995)], or with bivariate extreme value models [see, for example, Geffroy (1958/59), Tiago de Oliveira (1958, 1984), Gumbel (1960b), de Haan and Resnick (1977), Pickands (1981, 1989), Leadbetter, Lindgren, and Rootzén (1983), Galambos (1987), Smith, Tawn, and Yuen (1990), Deheuvels (1991), Einmahl, de Haan, and Huang (1993)]. The latter correspond to limit laws obtained by taking coordinatewise minima (or maxima) for sequences of independent and identically distributed random vectors. In either case, it is of great practical usefulness to test independence of the marginals. The approach we will follow turns out to generate a series of theoretical results of interest.

Towards the aim of testing $(H.0)$ against the general alternative $(H.1)$ that $A(u) \ne 1$ for some $0 < u < 1$, we introduce the empirical process [see, for example, Deheuvels (1991)]

$$Z_n(u) = \frac{1}{\sqrt{n}} \sum_{i=1}^{n} \Big\{ \min\Big(\frac{X_i / \overline{X}_n}{u}, \frac{Y_i / \overline{Y}_n}{1-u} \Big) - 1 \Big\} \quad \text{for} \quad 0 \le u \le 1, \tag{35.2}$$

where we set

$$\overline{X}_n = \frac{1}{n} \sum_{i=1}^{n} X_i \quad \text{and} \quad \overline{Y}_n = \frac{1}{n} \sum_{i=1}^{n} Y_i. \tag{35.3}$$

We note that the statistic

$$\frac{1}{\widehat{A}_n(u)} = \frac{1}{n} \sum_{i=1}^{n} \min\Big(\frac{X_i / \overline{X}_n}{u}, \frac{Y_i / \overline{Y}_n}{1-u} \Big),$$

is a variant of an estimator of $1/A(u)$ due to Pickands (1981, 1989). The original Pickands estimator (see, e.g., Deheuvels (1991)) yields the maximal likelihood estimator of $1/A(u)$ under the $E(1,1)$ model. In (35.2) and elsewhere, we set $x/0 = \infty$ when $x > 0$. This convention entails that Z_n, as given in (35.2), defines a random variable with values in the Banach space $(C[0,1], \mathcal{U})$. The latter consists of the set $C[0,1]$ of continuous functions on $[0,1]$, endowed with the uniform topology $\mathcal{U}$ defined by the sup-norm $\|f\| = \sup_{0 \le u \le 1} |f(u)|$. It is noteworthy that the distribution of Z in (35.2) is independent of $\gamma > 0$ and $\nu > 0$. Denote by $\{Z(u) : 0 \leq u \leq 1\}$ a centered Gaussian process with covariance function

$$R(u,v) = R(v,u) = \frac{2v - u^2 - v^2}{(1-u)v} - 1 - (1-u)(1-v) - uv \quad \text{for} \quad 0 \leq u \leq v \leq 1. \tag{35.4}$$

Following the arguments of Deheuvels (1991), Deheuvels and Martynov (2000) have described the weak limiting behavior under $(H.0)$ of Z_n as $n \to \infty$ by proving the following theorem.

Theorem 35.1.1 *Under $(H.0)$, the empirical process $\{Z_n(u) : 0 \leq u \leq 1\}$ converges weakly in $(C[0,1], \mathcal{U})$ to $\{Z(u) : 0 \leq u \leq 1\}$ as $n \to \infty$.*

Given this result, it is natural to consider the general class of tests of independence of X and Y based upon statistics of the form $\Theta(Z_n)$, where Θ is an appropriate functional on $(C[0,1], \mathcal{U})$. Since, under suitable continuity conditions on Θ, the limiting distribution as $n \to \infty$ of $\Theta(Z_n)$ is that of $\Theta(Z)$, it is not too difficult to make use of tests of this type, as long as the distribution of $\Theta(Z)$ can be evaluated. In the following, we present some examples of the kind. For further details on this problem, we refer to Deheuvels and Martynov (1996,2000).

35.2 Karhunen-Loeve Expansions

We first recall the main properties of the Karhunen-Loeve [KL] expansion of a general centered Gaussian process $\{\mathcal{Z}(u) : 0 \leq u \leq 1\}$ with continuous covariance function

$$\mathcal{R}(u,v) = \mathbb{E}(\mathcal{Z}(u)\mathcal{Z}(v)) \quad \text{for} \quad 0 \leq u, v \leq 1,$$

and almost surely continuous sample paths. Under these assumptions [see, for example, Adler (1990, pp. 66–79), Kac and Siegert (1947), Shorack and Wellner (1986, pp. 206–218)] there exist constants $\lambda_1 \geq \lambda_2 \geq \ldots \geq 0$, together with continuous functions $e_1(t), e_2(t), \ldots$, on $[0,1]$ (the *eigenfunctions* of the *covariance kernel* $\mathcal{R}(u,v)$), such that the following properties $(K.1\text{–}2\text{–}3\text{–}4)$ are fulfilled.

$(K.1)$ The $\{e_k : k \geq 1\}$ are orthonormal in $L^2[0,1]$, i.e.,

$$\int_0^1 e_i(t)e_j(t)dt = \begin{cases} 1 & \text{if } i = j, \\ 0 & \text{if } i \neq j, \end{cases} \tag{35.5}$$

$(K.2)$ The $\{(\lambda_k, e_k) : k \geq 1\}$ form a complete set of solutions of the Fredholm-type equation in (λ, e),

$$\lambda e(u) = \int_0^1 \mathcal{R}(u,v)e(v)dv \quad \text{for} \quad 0 \leq u \leq 1; \tag{35.6}$$

$(K.3)$ We have

$$\mathcal{R}(u,v) = \sum_{k=1}^{\infty} \lambda_k e_k(u) e_k(v), \tag{35.7}$$

where the series on the right hand side of (35.7) is absolutely and uniformly convergent on $[0,1]^2$;

$(K.4)$ There exists a sequence $\{\omega_k : k \geq 1\}$ of independent $N(0,1)$ random variables such that the following *Karhunen-Loeve* [KL] expansion holds. For all $0 \leq u \leq 1$

$$\mathcal{Z}(u) = \sum_{k=1}^{\infty} \sqrt{\lambda_k}\, \omega_k e_k(u), \tag{35.8}$$

where the series is almost surely uniformly convergent on $[0,1]$.

The KL expansion induced by $(K.1–2–3–4)$ is of major interest for several reasons. Below, we discuss two of the most important applications. We limit ourselves to the non-trivial case where $\lambda_k > 0$ for all $k \geq 1$ and assume implicitly from now on that this condition is fulfilled.

First, the KL expansion (35.7)–(35.8) yields an explicit description of the *reproducing kernel Hilbert space* [RKHS] of $\mathcal{Z}$ [see, for example, Kuelbs (1976) and Adler (1990, Theorem 3.16)]. When $\mathcal{Z}$ is considered as a random variable with values in the Banach space $(C[0,1], \mathcal{U})$, the RKHS $\mathbb{H}$ of $\mathcal{Z}$ is the Hilbert subspace of $C[0,1]$ given by

$$\mathbb{H} = \Big\{ f : f(u) = \sum_{k=1}^{\infty} a_k \sqrt{\lambda_k}\, e_k(u), 0 \leq u \leq 1, \sum_{k=1}^{\infty} a_k^2 < \infty \Big\}, \tag{35.9}$$

with inner product

$$\langle f, g \rangle_{\mathbb{H}} = \sum_{k=1}^{\infty} a_k b_k = \sum_{k=1}^{\infty} \frac{1}{\lambda_k} \Big(\int_0^1 g(u) e_k(u) du \Big) \Big(\int_0^1 f(u) e_k(u) du \Big), \tag{35.10}$$

where

$$f(u) = \sum_{k=1}^{\infty} a_k \sqrt{\lambda_k}\, e_k(u) \quad \text{and} \quad g(u) = \sum_{k=1}^{\infty} b_k \sqrt{\lambda_k}\, e_k(u) \quad \text{for} \quad 0 \leq u \leq 1. \tag{35.11}$$

Via (35.10)–(35.11), the sequence $\{\sqrt{\lambda_k}\, e_k : k \geq 1\}$ yields a *convergent orthonormal sequence* [CONS] in $\mathbb{H}$, which is essential to analyze the structure of $\mathcal{Z}$ [refer to Ledoux and Talagrand (1991) and Lifshits (1995)].

Second, a statistically useful consequence of $(K.1\text{–}2\text{–}3\text{–}4)$, is the description of the distribution of

$$\mathcal{J}^2 = \int_0^1 \mathcal{Z}^2(u)du = \sum_{k=1}^{\infty} \lambda_k \omega_k^2, \tag{35.12}$$

given by its characteristic function

$$\mathbb{E}(\exp(iu\mathcal{J}^2)) = \prod_{k=1}^{\infty} \Big(1 - 2iu\lambda_k\Big)^{-1/2} \quad \text{for} \quad u \in \mathbb{R}. \tag{35.13}$$

The relations (35.12)–(35.13) have applications when $\{\mathcal{Z}(u) : 0 \le u \le 1\}$ is the weak limit (with respect to an appropriate topology) of a sequence $\{\mathcal{Z}_n(u) : 0 \le u \le 1\}$ of empirical processes. In this framework, the statistic

$$\mathcal{J}_n^2 = \int_0^1 \mathcal{Z}_n^2(u)du, \tag{35.14}$$

is typically of interest for tests of goodness-of-fit, since its distribution can be approximated for large values of n by that of $\mathcal{J}^2$. This, however, necessitates in practice a numerical evaluation of the quantiles of the latter distribution. It is not too difficult [see, for example, Johnson, Kotz, and Balakrishnan (1994, Section 18-8, pp. 444-450)] to invert numerically a finite product approximation of the right hand side of (35.13) leading to the desired values of $\mathbb{P}(\mathcal{J}^2 \le x)$. For discussions concerning the precision of this approach and related methods, refer to Imhof (1961) and Martynov (1975,1976). This, in turn, requires a prior *explicit* knowledge of the eigenvalues $\{\lambda_k : k \ge 1\}$, which are *implicit* in terms of $\mathcal{R}(u, v)$ via (35.6). In general, one knows only $\mathcal{R}(u, v)$ and the needed numerical evaluation of the λ_k's can only be made by tedious recursions, which do not allow us to achieve reasonable precision for higher order terms. Therefore, (35.12) is mostly useful for such applications when there exist simple closed-form expressions for the λ_k's. When it is not the case, one must use different techniques based on direct approximations of $\mathbb{P}(\mathcal{J} \le x)$ [see, for example, Martynov (1992) and Deheuvels and Martynov (1996)].

For most Gaussian processes of interest with respect to statistics, the values of the λ_k's are not known in explicit form [see, for example, Adler (1990, p. 76)]. Below, we give examples of Gaussian processes on $[0, 1]$ for which the constants $\{\lambda_k : k \ge 1\}$ in the KL expansion are known [refer to Csörgő (1979, 1981), Cotterill and Csörgő (1985), Deheuvels (1981), for examples of such processes indexed on $[0, 1]^d$ with $d \ge 2$].

– The (restriction to [0,1] of the) Wiener process $\{W(t) : t \ge 0\}$, with $\mathcal{Z} = W$ and $\lambda_k = 1/((k - \frac{1}{2})\pi)^2$ for $k \ge 1$ [see, e.g. Adler (1990, p. 77)];

– The Brownian bridge $\{B(t) : 0 \le t \le 1\}$, with $\mathcal{Z} = B$ and $\lambda_k = 1/(k\pi)^2$ for $k \ge 1$. When $\mathcal{Z}_n$ is the uniform empirical process on $[0, 1]$, $\mathcal{J}_n^2$ reduces to the

Cramér-von Mises statistic [see, for example, Shorack and Wellner (1986, Proposition 1 and Theorem 1, pp. 213–217), Durbin (1973, p. 32), Darling (1955, p. 15), Smirnov (1948), Anderson and Darling (1952), and Darling (1957)].

– The limiting process $\mathcal{Z}(t) = B(t)/\sqrt{t(1-t)}$ of the Anderson-Darling statistic, where $B(t)$ denotes a Brownian bridge [see, for example, Anderson and Darling (1954), Watson (1961, 1967), Shorack and Wellner (1986, pp. 148 and 224–227)]. In this case, $\lambda_k = 1/(k(k+1))$ for $k \geq 1$.

It turns out that in the framework of our study, the KL expansion of the centered Gaussian process Z with covariance function (35.4) has been obtained by Deheuvels and Martynov (2000). Their result is given in the next theorem. The following notation and facts from the theory of orthogonal polynomials will be needed.

The *Jacobi polynomials* [see, for example, Tricomi (1970, pp. 160–177) and Szegö (1967)] denoted by $P_n^{\alpha,\beta}(x)$ for $n \geq 0$ and $\alpha, \beta > -1$, with $x \in [-1, 1]$, yield the *modified Jacobi polynomials* [see, for example, Chihara (1978, (2.1), p. 143)] via the change of variable $u = (x+1)/2$. These are defined, for $n \geq 0$ and $0 \leq u \leq 1$, by

$$Q_n^{\alpha,\beta}(u) = P_n^{\alpha,\beta}(2u-1) = \frac{(-1)^n}{n!} \frac{1}{u^\beta(1-u)^\alpha} \frac{d^n}{du^n}\left\{u^{\beta+n}(1-u)^{\alpha+n}\right\}. \quad (35.15)$$

The modified Jacobi polynomials $\{Q_n^{\alpha,\beta} : n \geq 0\}$ fulfill the orthogonality relations [see, for example, Chihara (1978, (2.18), p. 148)], for $m, n \geq 0$,

$$\begin{aligned}\int_0^1 & Q_m^{\alpha,\beta}(u) Q_n^{\alpha,\beta}(u) u^\beta (1-u)^\alpha du \\ &= 0 \quad \text{when} \quad m \neq n, \qquad (35.16)\\ &= \frac{\Gamma(n+\alpha+1)\Gamma(n+\beta+1)}{(2n+\alpha+\beta+1)\Gamma(n+\alpha+\beta+1)\, n!} \quad \text{when} \quad m = n.\end{aligned}$$

For $\alpha = \beta = 2$, the polynomials, defined, for $n \geq 0$ and $0 \leq u \leq 1$, by

$$P_n(u) = Q_n^{2,2}(u) = \frac{(-1)^n}{n!} \frac{1}{u^2(1-u)^2} \frac{d^n}{du^n}\left\{u^{n+2}(1-u)^{n+2}\right\}, \quad (35.17)$$

fulfill the othogonality relations, for $m, n \geq 0$,

$$\begin{aligned}\int_0^1 u^2(1-u)^2 P_m(u) P_n(u) du &= 0 \quad \text{when} \quad m \neq n, \qquad (35.18)\\ = \Delta_n &= \frac{1}{2n+5} \times \frac{(n+1)(n+2)}{(n+3)(n+4)} \quad \text{when} \quad m = n.\end{aligned}$$

Theorem 35.2.1 *Let $\mathcal{Z} = Z$ and $\mathcal{R}(u, v) = R(u, v)$ be as in* (35.4). *Then, the properties* $(K.1\text{–}2\text{–}3\text{–}4)$ *hold with*

$$\lambda_k = \frac{6}{k(k+1)(k+2)(k+3)} \quad \textit{for} \quad k \geq 1, \quad (35.19)$$

and, for $k \geq 1$ *and* $0 \leq u \leq 1$,

$$\begin{aligned} e_k(u) &= \frac{u(1-u)}{\sqrt{\Delta_{k-1}}} P_{k-1}(u) = \Big\{(2k+3) \times \frac{(k+2)(k+3)}{k(k+1)}\Big\}^{1/2} \\ &\quad \times \frac{(-1)^{k-1}}{(k-1)!} \frac{1}{u(1-u)} \frac{d^{k-1}}{du^{k-1}} \Big\{u^{k+1}(1-u)^{k+1}\Big\}. \end{aligned} \tag{35.20}$$

Remark 35.2.1 For an explicit computation of the eigenfunctions $\{e_k : k \geq 1\}$ in (35.20), one may use a binomial expansion of $(u-1)^{j+1}$ in the formula (see, for example, Chihara (1978, (2.60), p. 144)]

$$\begin{aligned} e_k(u) &= -\Big\{(2k+3) \times \frac{(k+2)(k+3)}{k(k+1)}\Big\}^{1/2} \\ &\quad \times \sum_{j=0}^{k-1} \binom{k+1}{k-1-j} \binom{k+1}{j} u^{k-j}(u-1)^{j+1}. \end{aligned} \tag{35.21}$$

The first of these eigenfunctions are given by

$$\begin{aligned} e_1(u) &= \sqrt{30} \quad u(1-u) \\ e_2(u) &= -\sqrt{210} \quad u(1-u)(2u-1) \\ e_3(u) &= 3\sqrt{10} \quad u(1-u)(14u^2-14u+3) \\ e_4(u) &= -3\sqrt{2310} \quad u(1-u)(12u^3-18u^2+8u-1) \\ e_5(u) &= 2\sqrt{1365} \quad u(1-u)(33u^4-66u^3+45u^2-12u+1). \end{aligned} \tag{35.22}$$

Remark 35.2.2 1°) It follows from (35.20) that $e_k(1-u) = (-1)^{k+1}e_k(u)$ for $k \geq 1$. This, when combined with the version of (35.8) holding for Z_0, shows that

$$Z(1-u) = \sum_{k=1}^{\infty} \sqrt{\lambda_k}\,(-1)^{k+1}\omega_k e_k(u) =_d Z(u) = \sum_{k=1}^{\infty} \sqrt{\lambda_k}\,\omega_k e_k(u), \tag{35.23}$$

where "$=_d$" denotes equality in distribution. That

$$\{Z(1-u) : 0 \leq u \leq 1\} =_d \{Z(u) : 0 \leq u \leq 1\}$$

may be checked directly from the equality

$$R(u,v) = R(1-u, 1-v) \quad \text{for all} \quad 0 \leq u, v \leq 1.$$

2°) The equality (35.23) implies that the processes

$$Z^S(u) = \frac{1}{2}\Big\{Z(u) + Z(1-u)\Big\} = \sum_{\ell=1}^{\infty} \sqrt{\lambda_{2\ell+1}}\,\omega_{2\ell+1}e_{2\ell+1}(u), \tag{35.24}$$

and

$$Z^A(u) = \frac{1}{2}\Big\{Z(u) - Z(1-u)\Big\} = \sum_{\ell=1}^{\infty} \sqrt{\lambda_{2\ell}}\,\omega_{2\ell}e_{2\ell}(u), \tag{35.25}$$

are independent, with KL expansions given as above.

35.3 Applications to Tests of Independence

An immediate corollary of Theorems 35.1.1 and 35.2.1 is given below in terms of $\mathcal{J}_n^2$ and $\mathcal{J}^2$ defined respectively by

$$\mathcal{J}_n^2 = \int_0^1 Z_n^2(u)du \quad \text{and} \quad \mathcal{J}^2 = \int_0^1 Z^2(u)du. \tag{35.26}$$

Let $\{\lambda_k : k \geq 1\}$ be as in (35.19).

Corollary 35.3.1 *Under* $(H.0)$, *we have*

$$\lim_{n\to\infty} \mathbb{E}(\exp(iu\mathcal{J}_n^2)) = \mathbb{E}(\exp(iu\mathcal{J}^2)) = \prod_{k=1}^{\infty} \Big(1 - 2iu\lambda_k\Big)^{-1/2} \quad \textit{for} \quad u \in \mathbb{R}. \tag{35.27}$$

PROOF. Combine Theorems 35.1.1 and 35.2.1 with (35.12)–(35.13) and (35.26). ■

The statistic $\mathcal{J}_n^2$ in (35.26) allows us to test $(H.0)$ against $(H.1)$ by rejecting the null hypothesis when $\mathcal{J}_n^2$ exceeds a critical level $c_{n,\alpha}$, chosen in such a way that, for a specified $0 < \alpha < 1$, $\mathbb{P}(\mathcal{J}_n^2 \geq c_{n,\alpha}|(H.0)) = \alpha$. The evaluation of the exact values of $c_{n,\alpha}$ for the various possible choices of the risk level $\alpha \in (0,1)$, and the sample size $n \geq 1$, is beyond the scope of the present paper. Below, we limit ourselves to some selected values of the limiting constants c_α such that

$$\lim_{n\to\infty} \mathbb{P}(\mathcal{J}_n^2 \geq c_\alpha|(H.0)) = \mathbb{P}(\mathcal{J}^2 \geq c_\alpha) = \alpha. \tag{35.28}$$

In Deheuvels and Martynov (2000) c_α is tabulated with a precision of 10^{-3} for various values of α. For example $c_{10\%} = 0.770$ and $c_{5\%} = 1.053$.

Remark 35.3.1 We will not discuss here the efficiency of the test based upon $\mathcal{J}_n^2$, with respect to alternative methods [refer to Balakrishnan and Basu (1995)].

Among the many possible statistics based upon functionals of Z_n which (in addition to $\mathcal{J}_n$) may be used to test $(H.0)$ against $(H.1)$, one should mention the *principal component test statistics*

$$T_{n,k} = \sqrt{\lambda_k} \int_0^1 Z_n(u)e_k(u)du \quad \text{for} \quad k \geq 1. \tag{35.29}$$

A direct consequence of the above theorems is that, under $(H.0)$, for each $k \geq 1$,

$$T_{n,k} \to_d N(0,1), \tag{35.30}$$

and we may use this property to reject $(H.0)$ when $\pm T_{n,k}$ or $|T_{n,k}|$ exceeds the appropriate quantiles of the $N(0,1)$ law. The fact that the e_k have explicit expressions allows a simple use of this methodology. For example, making use of (35.22), we obtain readily, under $(H.0)$, that, as $n \to \infty$,

$$\begin{aligned} T_{n,1} &= 2\sqrt{30}\int_0^1 Z_n(u)u(1-u)du \\ &= \frac{\sqrt{30}}{\sqrt{n}}\sum_{i=1}^{n}\Big\{\frac{X_iY_i}{X_i\overline{Y}_n + Y_i\overline{X}_n} - \frac{1}{3}\Big\} \to_d N(0,1). \end{aligned} \tag{35.31}$$

This statistic has a particularly simple expression. Moreover, under $(H.1)$, it holds that

$$T_{n,1} = (1+o(1))\sqrt{n} \times 2\sqrt{30}\int_0^1 \Big\{\frac{1}{A(u)} - 1\Big\}u(1-u)du \to \infty \quad \text{a.s.},$$

so that the test of $(H.0)$ based upon $T_{n,1}$ is consistent. This property is not shared in general by $T_{n,k}$ for $k \geq 2$. For example for $k = 2$ and $A(u) = A(1-u)$ for $0 \leq u \leq 1$, we infer from (35.22) that

$$\int_0^1 \Big\{\frac{1}{A(u)} - 1\Big\}e_2(u)du = \int_0^1 \Big\{\frac{1}{A(u)} - 1\Big\}\sqrt{210}\,u(1-u)(1-2u)du = 0.$$

By (35.24)–(35.25), we infer from the above theorems the limiting distributions of the statistics

$$\mathcal{J}_n^S = \frac{1}{4}\int_0^1 \Big\{Z_n(u) + Z_n(1-u)\Big\}^2 du, \tag{35.32}$$

and

$$\mathcal{J}_n^A = \frac{1}{4}\int_0^1 \Big\{Z_n(u) - Z_n(1-u)\Big\}^2 du. \tag{35.33}$$

Under $(H.0)$, it holds that, for $u \in \mathbb{R}$,

$$\lim_{n\to\infty} \mathbb{E}(\exp(iu\mathcal{J}_n^S)) = \prod_{\ell=1}^{\infty}\Big(1 - 2iu\lambda_{2\ell+1}\Big)^{-1/2}, \tag{35.34}$$

and

$$\lim_{n\to\infty} \mathbb{E}(\exp(iu\mathcal{J}_n^A)) = \prod_{\ell=1}^{\infty}\Big(1 - 2iu\lambda_{2\ell}\Big)^{-1/2}. \tag{35.35}$$

Acknowledgement. I am grateful to the organizers of the conference for allowing me to present in this paper a series of results of a joint research program with G. Martynov.

References

1. Adler, R. J. (1990). *An introduction to continuity, extrema, and related topics for general Gaussian processes*, IMS Lecture Notes-Monograph Series **12**, Hayward, California: Institute of Mathematical Statistics.

2. Anderson, T. W. and Darling, D. A. (1952). Asymptotic theory of certain goodness of fit criteria based on stochastic processes, *Annals of Mathematical Statistics*, **23**. 193–212.

3. Anderson, T. W. and Darling, D. A. (1954). A test of goodness of fit, *Journal of the American Statistical Association*, **49**, 765–769.

4. Arnold, B. C. (1975). Multivariate exponential distributions based on hierarchical successive damage, *Journal of Applied Probability*, **12**, 142–147.

5. Balakrishnan, N. and Basu, A. P. (1995). *The Exponential Distribution, Theory, Methods and Applications*, Amsterdam: Gordon and Breach.

6. Barnett, V. (1985). The bivariate exponential distribution: A review and some new results, *Statistica Neerlandica.*, **39**, 343–356.

7. Basu, A. P. (1988). Multivariate exponential distributions and their applications in reliability, In *Handbook of Statistics* (Eds., B. K. Ghosh and P. K. Sen), pp. 581–592, New York: Marcel Dekker.

8. Chihara, T. S. (1978). *An Introduction to Orthogonal Polynomials*, New York: Gordon and Breach.

9. Cotterill, D. S. and Csörgő, M. (1985). On the limiting distribution of and critical values for the Hoeffding, Blum, Kiefer, Rosenblatt independence criterion, *Statistical Decisions*, **3**, 1–48.

10. Csörgő, M. (1979). Strong approximations of the Hoeffding, Blum, Kiefer, Rosenblatt multivariate empirical process, *Journal of Multivariate Analysis*, **9**, 84–100.

11. Csörgő, M. (1981). On the asymptotic distribution of the multivariate Cramér-von Mises and Hoeffding-Blum-Kiefer-Rosenblatt independence criteria, In *Statistical Distributions in Scientific Work*, Vol. 5 (Trieste, 1980), pp. 145–156, *ATO Adv. Study Inst. Ser. C: Math. Phys. Sci.*, **79**, Dordrecht: Reidel.

12. Darling, D. A. (1955). The Cramér-Smirnov test in the parametric case, *Annals of Mathematical Statistics*, **26**, 1–20.

13. Darling, D. A. (1957). The Kolmogorov-Smirnov, Cramér-von Mises tests, *Annals of Mathematical Statistics*, **28**, 823–838.

14. de Haan, L. and Resnick, S. I. (1977). Limit theory for multivariate sample extremes, *Z. Wahrscheinlichkeit. verw. Gebiete*, **40**, 317–337.

15. Deheuvels, P. (1981). An asymptotic decomposition for multivariate distribution-free tests of independence, *Journal of Multivariate Analysis*, **11**, 102-113.

16. Deheuvels, P. (1984). Probabilistic aspects of multivariate extremes, In *Statistical Extremes and Applications* (Ed., J. Tiago de Oliveira) pp. 117–130, Dordrecht: Reidel.

17. Deheuvels, P. (1991). On the limiting behavior of the Pickands estimator for bivariate extreme-value distributions, *Statistics & Probability Letters*, **12**, 429–439.

18. Deheuvels, P. and Martynov, G. V. (1996). Cramér–Von Mises–type tests with applications to tests of independence for multivariate extreme–value distributions, *Communications in Statistics—Theory and Methods*, **25**, 871–908.

19. Deheuvels, P. and Martynov, G. V. (2000).A Karhunen-Loeve decomposition of a Gaussian process generated by independent pairs of exponential random variables, submitted.

20. Downton, F. (1970). Bivariate exponential distributions in reliability theory, *Journal of the Royal Statistical Society, Series B*, **32**, 408–417.

21. Durbin, J. (1973). *Distribution theory for tests based upon the sample distribution function*, Regional Conference Series in Applied Mathematics, **9**, Philadelphia: S.I.A.M..

22. Einmahl, J. H. J., de Haan, L., and Huang, X. (1993). Estimating a multidimensional extreme-value distribution. *Journal of Multivariate Analysis*, **47**, 35–47.

23. Falk, M., Hüsler, J., and Reiss, R. D. (1994). *Laws of Small Numbers: Extremes and Rare Events*, Basel: Birkhäuser.

24. Freund, J. (1961). A bivariate extension of the exponential distribution, *Journal of the American Statistical Association*, **56**, 971–977.

25. Galambos, J. (1987). The Asymptotic Theory of Extreme Order Statistics, Second Edition, Malabar, Florida: Krieger.

26. Geffroy, J. (1958/59). Contribution à la théorie des valeurs extrêmes, *Pub. Inst. Statist. Univ. Paris*, **7/8**, 37–185.

27. Grad, A. and Solomon, H (1955) Distribution of quadratic forms and some applications, *Annals of Mathematical Statistics*, **26**, 464–477.

28. Gumbel. E. J. (1960a). Bivariate exponential distributions, *Journal of the American Statistical Association*, **55**, 698–707.

29. Gumbel, E. J. (1960b). Distribution des valeurs extrêmes en plusieurs dimensions, *Publ. Inst. Statist. Univ. Paris*, **9**, 171–173.

30. Imhof, J. P. (1961). Computing the distribution of quadratic forms in normal variables, *Biometrika*, **48**, 419–426.

31. Johnson, N. L. and Kotz, S. (1977). *Distributions in Statistics: Continuous Multivariate Distributions*, New York: John Wiley & Sons.

32. Johnson, N. L., Kotz, S., and Balakrishnan, N. (1994). *Continuous Univariate Distributions*, Volume 1, New York: John Wiley & Sons.

33. Kac, M. and Siegert, A. J. F. (1947). An explicit representation of a stationary Gaussian process, *Annals of Mathematical Statistics*, **26**, 189–211.

34. Kuelbs, J. (1976). A strong convergence theorem for Banach space valued random variables, *Annals of Probability*, **4**, 744–771.

35. Leadbetter, M. R., Lindgren, G., and Rootzén, H. (1983). *Extremes and Related Properties of Random Sequences and Processes*, New York: Springer-Verlag.

36. Ledoux, M. and Talagrand, M. (1991). *Probability in Banach Spaces*, New York: Springer-Verlag.

37. Lifshits, M. A. (1995). *Gaussian Random Functions*, Dordrecht: Kluwer.

38. Marshall, A. W. and Olkin, I. (1967). A multivariate exponential distribution, *Journal of the American Statistical Association*, **62**, 30–44.

39. Marshall, A. W. and Olkin, I. (1982). Domains of attraction of multivariate extreme value distributions, *Annals of Probability*, **10**, 168–177.

40. Martynov, G. V. (1975). Computation of distribution function of quadratic forms of normally distributed random variables, *Theory of Probability and its Applications*, **20**, 782–793.

41. Martynov, G. V. (1976). Computation of limit distributions of statistics for normality tests of type ω^2, *Theory of Probability and its Applications*, **21**, 1–13.

42. Martynov, G. V. (1992). Statistical tests based on empirical processes and related questions, *Journal of Soviet Mathematics*, **61**, 2195–2271.

43. Pickands, J. III (1981). Multivariate extreme value distributions, In *Bulletin of the International Statistical Institute, Proceedings of the 43rd. Session, Buenos Aires*, pp. 859–878.

44. Pickands, J. III (1989). Multivariate negative exponential and extreme value distributions, In *Extreme Value Theory* (Eds., J. Hüsler and R. D. Reiss), Lecture Notes in Statistics **51**, New York: Springer-Verlag.

45. Proschan, F. and Sullo, P. (1976). Estimating the parameters of a multivariate exponential distribution, *Journal of the American Statistical Association*, **71**, 465–472.

46. Raftery, A. E. (1984). A continuous multivariate exponential distribution, *Communications in Statistics—Theory and Methods*, **13**, 947–965.

47. Resnick, S. I. (1987). *Extreme Values, Regular Variation, and Point Processes*, New York: Springer-Verlag.

48. Shorack, G. R. and Wellner, J. A. (1986). *Empirical Processes with Applications to Statistics*, New York: John Wiley & Sons.

49. Sibuya, M. (1960). Bivariate extreme statistics. *Annals of the Institute of Statistical Mathematics*, **11**, 195–210.

50. Smirnov, N. V. (1936). Sur la distribution de ω^2, *C. R. Acad. Sci. Paris*, **202**, 449–452.

51. Smirnov, N. V. (1948). Table for estimating the goodness of fit of empirical distributions, *Annals of Mathematical Statistics*, **19**, 279–281.

52. Smith, R. L., Tawn, J. A., and Yuen, H. K. (1990). Statistics of multivariate extremes, *International Statistical Review*, **58**, 47–58.

53. Szegö, G. (1967). *Orthogonal Polynomials*, Third Edition, American Mathematic Society, Colloquium Publication, Vol. 23, New York: American Mathematical Society.

54. Tiago de Oliveira, J. (1958). Extremal distributions, *Revista da Fac. Ciencias Univ. Lisboa*, **7**, 215–227.

55. Tiago de Oliveira, J. (1984). Bivariate models for extremes; Statistical decision, In *Statistical Extremes and Applications* (Ed., J. Tiago de Oliveira), pp. 131–154, Dordrecht: Reidel.

56. Tricomi, F. G. (1970). *Vorlesungen über Orthogonalreihen*, New York: Springer-Verlag.

57. Watson, G. S. (1961). Goodness of fit test on a circle, *Biometrika*, **48**, 109–114.

58. Watson, G. S. (1967). Another test for the uniformity of a circular distribution, *Biometrika*, **54**, 675–676.

36

Testing Problem for Increasing Function in a Model with Infinite Dimensional Nuisance Parameter

M. Nikulin and V. Solev

University Bordeaux 2, Bordeaux, France
& Steklov Mathematical Institute, St. Petersburg, Russia
Steklov Mathematical Institute, St. Petersburg, Russia

Abstract: We consider the next statistical problem arising, for example, in accelerated life testing. Let X_1 be a random variable with density function $f(t)$, $\Psi(t)$ be an increasing absolutely continuous function, $\Phi(t) = \Psi^{-1}(t)$ be its inverse function, random variable X_2 be defined as $X_2 = \Phi(X_1)$. In order to test whether function $\Psi(t)$ belongs to a given parametric family when function f is completely unknown, we take two independent nonparametric estimators $\hat{f}_n$ of density function f and $\hat{g}_n$ of density function g of X_2 and compare the function $\hat{g}_n(t)$ with the function $\hat{f}_n(\Psi(\hat{\theta}_n; t))\psi(\hat{\theta}_n; t)$ for a minimum distance estimator $\hat{\theta}_n$. But at the begining we have to investigate the asymptotic behavior of the estimator $\hat{\theta}_n$. We consider a parametric minimum distance estimator for Ψ, when we observe (with a mechanism of independent censoring) two independent samples from the distributions of X_1 and X_2 respectively.

Keywords and phrases: Accelerated life testing, goodness-of-fit test, Kaplan-Meier estimator, kernel density estimator, minimum distance estimator, non-parametric estimation, nuisance parameter

36.1 Introduction

Let X_1 be a random variable with the distribution function $F(t)$ and density function $f(t)$, $\Psi(t)$ be an increasing absolutely continuous function, $\Phi(t)$ be the inverse function: $\Phi = \Psi^{-1}$. We put $X_2 = \Phi(X_1)$. The distribution function

$G(t)$ and the density function $g(t)$ of X_2 may be easily calculated

$$G(t) = F(\Psi(t)), \qquad g(t) = f(\Psi(t))\psi(t),$$
$$\text{where} \qquad \psi(t) = \tfrac{d}{dt}\,\Psi(t),\ \phi(t) = \tfrac{d}{dt}\,\Phi(t). \tag{36.1}$$

We consider the case when we have two independent samples

$$\mathbf{X}_1 = \{X_{1,1}, \cdots, X_{1,n_1}\} \quad \text{and} \quad \mathbf{X}_2 = \{X_{2,1}, \cdots, X_{2,n_2}\},$$

the first one with the distribution function $F(t)$ and the density function $f(t)$, and the second one with the distribution function $G(t) = F(\Psi(t))$ and the density function $g(t) = f(\Psi(t))\psi(t)$. Let

$$\mathbf{T}^1 = \{T_{1,1}, \cdots, T_{1,n_1}\} \quad \text{and} \quad \mathbf{T}^2 = \{T_{2,1}, \cdots, T_{2,n_2}\}$$

be another two independent samples of random variables with distribution functions $Q^1(t), Q^2(t)$ and density functions $q^1(t), q^2(t)$ respectively. We assume that $\mathbf{X_1}, \mathbf{X_2}, \mathbf{T^1}, \mathbf{T^2}$ are independent.

Now suppose we observe censored data: $\mathbf{\Gamma}_{1,1}, \cdots, \mathbf{\Gamma}_{1,n_1}$ and $\mathbf{\Gamma}_{2,1}, \cdots, \mathbf{\Gamma}_{2,n_2}$, where we denote

$$\begin{array}{ll} \mathbf{\Gamma}_{1,i} = (Y_{1,i}, K_{1,i}), & \mathbf{\Gamma}_{2,j} = (Y_{2,j}, K_{2,j}), \\ (i = 1, \cdots, n_1), & (j = 1, \cdots, n_2), \end{array} \tag{36.2}$$

where

$$Y_{k,j} = \begin{cases} X_{k,j}, & X_{k,j} \le T_{k,j} \\ T_{k,j}, & X_{k,j} > T_{k,j} \end{cases}, \qquad K_{k,j} = \begin{cases} 1, & X_{k,j} \le T_{k,j} \\ 0, & X_{k,j} > T_{k,j} \end{cases}, \tag{36.3}$$

$$j = 1, \cdots, n_k; \quad k = 1, 2.$$

We consider the estimation problem for the unknown function Φ in a situation, when the distribution function F is completely unknown and the increasing function Φ belongs to a given parametric family,

$$\Phi \in \mathcal{B} = \{\Phi : \Phi(t) = \Phi(\theta; t), \theta \in \Theta\},$$

where $\Theta \subset \mathbf{R}^d$. So further we shall write

$$\Phi(t) = \Phi(\theta; t), \Psi(t) = \Psi(\theta; t), \phi(t) = \phi(\theta; t), \psi(t) = \psi(\theta; t),$$

$$G(t) = G(\theta; t), g(t) = g(\theta; t),$$

but sometimes it will be convenient for us to omit the parameter θ. We suppose that

$$n_1 = n_1(n), \quad n_2 = n_2(n), \quad \text{where} \quad n = n_1 + n_2,$$

and

$$0 < \lim_{n\to\infty} \frac{n_1}{n} = p_1, \; 0 < \lim_{n\to\infty} \frac{n_2}{n} = p_2. \tag{36.4}$$

We note here that this problem is very natural in accelerated life testing, see for example, Bagdonavicius and Nikulin (1999, 2000), Gerville-Réache and Nikoulina (1999), Bagdonavicius, Gerville-Réache, Nikoulina and Nikulin (2000). It is convenient for us to admit that random variables $T_{1,j}$ or $T_{2,i}$ may take the value $+\infty$ with positive probability.

We shall use the Kaplan–Meier estimators $\mathcal{F}_{n_1}(t)$ and $\mathcal{G}_{n_2}(t)$ as the non-parametric estimators for $F(t)$ and $G(\theta;t)$,

$$1 - \mathcal{F}_{n_1}(t) = \prod_{j:K_{1,j}=1,\; Z_{1,j}\le t} \left(\frac{n-j}{n-i+1}\right),$$

$$1 - \mathcal{G}_{n_2}(t) = \prod_{j:K_{2,j}=1,\; Z_{2,j}\le t} \left(\frac{n-j}{n-i+1}\right),$$

(with the convention that $\prod_\emptyset = 1$), where $Z_{1,1} \le \ldots \le Z_{1,n_1}$ are the order statistics of the sample $\mathbf{Y}^1 = (Y_{1,1}, \cdots, Y_{1,n_1})$ and $Z_{2,1} \le \ldots \le Z_{2,n_2}$ are the order statistics of the sample $\mathbf{Y}^2 = (Y_{2,1}, \cdots, Y_{2,n_2})$.

We consider the kernel density estimator $\hat{f}_n(t)$ for estimating $f(t)$ at a fixed point $t \in \mathrm{R}$

$$\hat{f}_n(t) = \int \frac{1}{h_n}\mathcal{K}\left(\frac{t-u}{h_n}\right) d\mathcal{F}_{n_1}(u), \tag{36.5}$$

and the kernel density estimator $\hat{g}_n(t)$ for estimating $g(\theta;t)$ at a fixed point $t \in \mathrm{R}$

$$\hat{g}_n(t) = \int \frac{1}{h_n}\mathcal{K}\left(\frac{t-u}{h_n}\right) d\mathcal{G}_{n_2}(u), \tag{36.6}$$

where $\mathcal{K}$ is a suitably chosen nonnegative kernel function ($\mathcal{K} : \mathrm{R} \to \mathrm{R}$) such that $\int \mathcal{K}(x)\,dx = 1$, $h_n > 0$ is a bandwidth such that $h_n \to 0$ and $nh_n \to \infty$, as $n \to \infty$. More about the properties of $\hat{f}_n(t)$ and $\hat{g}_n(t)$ and their applications see, for example, Rosenblatt (1956), Parzen (1962), Bickel and Rosenblatt (1973), Bretagnolle and Huber (1979), Ghosh and Wei-Min (1991), Hall (1984), Horvath (1991), Czörgő, Gombay, and Horvath (1991), etc. Further (and here) for simplicity of notation all integrals $\int b(x)dx$ for any function $b : \mathrm{R} \to \mathrm{R}$, without set of integration are understood to be integrals $\int_{\mathrm{R}} b(x)dx$ on the Lebesgue measure. Let $[a, b]$ be an interval such that $F(a) = 0$,

$$I(\theta) = [a(\theta), b(\theta)], \quad \text{where} \quad a(\theta) = \Psi(\theta; a),\, b(\theta) = \Psi(\theta; b).$$

We suppose for simplicity that function $\Psi(\theta;\cdot)$ is defined on the real line and the image of $\Psi(\theta;\cdot)$ does not depend on θ. We define the function $f(\theta,\theta_1;t)$ by the relation

$$f(\theta,\theta_1;t)=g\left(\theta;\Phi(\theta_1;t)\right)\phi(\theta_1;t). \tag{36.7}$$

So if the operator A_θ is defined by

$$\left[A_\theta h\right](t)=h\left(\Psi(\theta;t)\right)\psi(\theta;t),$$

and therefore the operator A_θ^{-1} is defined by

$$\left[A_\theta^{-1}h\right](t)=h\left(\Phi(\theta;t)\right)\phi(\theta;t),$$

then

$$f(\theta,\theta_1;t)=\left[A_{\theta_1}^{-1}A_\theta f\right](t). \tag{36.8}$$

We put

$$\begin{aligned} g_n^*(\theta;t) &= \hat{f}_n\left(\Psi(\theta;t)\right)\psi(\theta;t),\\ f_n^*(\theta;t) &= \hat{g}_n\left(\Phi(\theta;t)\right)\phi(\theta;t). \end{aligned} \tag{36.9}$$

For a nonnegative function $r_0(t)$ we define the function $r_\theta(t)$ by the relations

$$r_0(t)\phi(\theta;t)=r_\theta(\Phi(\theta;t)) \quad \text{or} \quad r_\theta(t)\psi(\theta;t)=r_0(\Psi(\theta;t)).$$

It is clear that

$$\int\limits_{I(\theta)}|g_n^*(\theta;t)-\hat{g}_n(t)|^2 r_\theta(t)\,dt=\int\limits_I|\hat{f}_n(t)-f_n^*(\theta;t)|^2 r_0(t)\,dt.$$

We consider the minimum-distance estimator $\hat{\theta}_n$ of θ which is defined by the relation

$$\hat{\theta}_n=\arg\min_\theta\int\limits_I|\hat{f}_n(t)-f_n^*(\theta;t)|^2 r_0(t)\,dt. \tag{36.10}$$

It is evident that

$$\hat{\theta}_n=\arg\min_\theta\int\limits_{I(\theta)}|g_n^*(\theta;t)-\hat{g}_n(t)|^2 r_\theta(t)\,dt. \tag{36.11}$$

The estimator $\hat{\theta}_n$ may be considered as a candidate for $\sqrt{n}$–consistent estimator with the Gaussian limiting distribution.

In order to test whether the function $\Psi(t)$ belongs to a given parametric family

$$\Phi \in \mathcal{B} = \{\Phi : \Phi(t) = \Phi(\theta; t), \theta \in \Theta\},$$

where $\Theta \subset \mathbf{R}^d$, and the function f is completely unknown, we have to study the asymptotic properties of the statistics:

$$T_n = \int_I |\hat{f}_n(t) - f_n^*(\theta; t)|^2 r_0(t)\, dt$$

and we shall prove the asymptotic normality of T_n.

Now we introduce further notations and list all assumptions. We define the distribution functions $H(t), H(\theta; t), H_*(\theta; t)$ by the relation

$$\left.\begin{aligned} 1 - H(t) &= (1 - F(t))\,(1 - Q^1(t))\,, \\ 1 - H(\theta; t) &= (1 - F(\Psi(\theta; t)))\,(1 - Q^2(t)))\,, \\ 1 - H_*(\theta; t) &= (1 - F(t))\,(1 - Q^2(\Phi(\theta; t)))\,. \end{aligned}\right\}. \tag{36.12}$$

So we have

$$H_*(\theta; t) = H(\theta; \Phi(\theta; t)).$$

We assume, that

D1. Functions $H(t)$ and $H(\theta; t)$ satisfy the conditions:

$$b < t_* = \sup\{t : H(t) < 1\} \quad \text{and} \quad b < t_*(\theta) = \sup\{t : H_*(\theta; t) < 1\}.$$

It is clear that under **D1**

$$b(\theta) = \Psi(\theta; b) < \sup\{t : H(\theta; t) < 1\}. \tag{36.13}$$

We suppose also that kernel $\mathcal{K}$ satisfies the conditions

K1. Function $\mathcal{K}(t)$ is nonnegative, symmetric:

$$\mathcal{K}(t) = \mathcal{K}(-t) \quad \text{and} \quad \mathcal{K}(t) = 0, \quad \text{when} \quad |t| > 1.$$

K2. Function $\mathcal{K}(t)$ is twice differentiable.

K3. Integral $\int \mathcal{K}(t)\, dt = 1$.

K4. Integral $\int \left| \left[\frac{d^2}{dt^2}\mathcal{K}\right](t) \right| dt < \infty$. Notice that

$$\int t\mathcal{K}(t)\, dt = 0, \quad \text{and} \quad \int t^2\mathcal{K}(t)\, dt > 0.$$

We assume that the sequence of bandwidths satisfies the conditions

H1.

$$h_n \longrightarrow 0, \quad \frac{\log n}{h_n^2 n} \longrightarrow 0,$$

$$nh_n^4 \longrightarrow 0, \quad \text{when } n \longrightarrow \infty.$$

We suppose too that the density function $f(t)$, the weight $r_0(t)$ and the distribution functions $H(t), H(\theta; t)$ satisfy the following conditions on some appropriately chosen interval $[a, b]$:

F1. Function f is continuously differentiable on interval $[a, b+\varepsilon]$ for some $\varepsilon > 0$, function $\frac{d}{dt}f$ is absolutely continuous on $[a, b+\varepsilon]$ and

$$\sup_{t\in[a,b+\varepsilon]} \frac{\left[\frac{d}{dt}f\right]^2(t)}{f(t)} < \infty, \qquad \sup_{t\in[a,b+\varepsilon]} \left|\left[\frac{d^2}{dt^2}f\right](t)\right| < \infty.$$

F2. Functions f and g satisfy the condition

$$nh_{n_1}^{1/2} \int |f_n^{\mathcal{K}}(t) - \phi(\theta; t) g_n^{\mathcal{K}}(\theta; \Phi(\theta; t))|^2 \, dt \to 0, \text{ as } n \to \infty,$$

where

$$f_n^{\mathcal{K}}(t) = \int \frac{1}{h_n}\mathcal{K}\left(\frac{t-u}{h_n}\right) f(u)\, du,$$

and

$$g_n^{\mathcal{K}}(\theta; t) = \int \frac{1}{h_n}\mathcal{K}\left(\frac{t-u}{h_n}\right) g(\theta; u)\, du.$$

Q1. Functions $Q^1(t)$ and $Q^2(t)$ are absolutely continuous on $[a, b+\varepsilon]$ and

$$\sup_{t\in[a,b+\varepsilon]} |q^j(t)| < \infty, \quad \text{where} \quad q^j(t) = \frac{d}{dt}Q^j(t) \ (j = 1, 2);$$

r1. Function $r_0(t)$ is continuously differentiable on inCerval $[a, b]$ and supported on $[a, b]$.

r2. There exists such $C = C(\theta)$, that

$$r_\theta(t)C(\theta) \geq r_{\theta_1}(t) \quad \text{for all} \quad \theta_1 \in \Theta, t \in I(\theta) \cap I(\theta_1).$$

For a nonnegative function $r(t)$ we denote by L_r^2 the L^2-space generated by the measure with the density $r(t)$ and by $||\varphi(\cdot)||_r$ the norm of $\varphi(\cdot)$ in the space L_r^2. A function $\varphi(\theta; t)$ (as function on θ) is differentiable (in the space L_r^2) on θ in the open kernel of Θ_0 if for all θ_0, that belongs to Θ_0, there exists such vector-function $\boldsymbol{\nabla}\varphi$ that

$$\varphi(\theta; \cdot) - \varphi(\theta_0; \cdot) = \langle \boldsymbol{\nabla}\varphi(\theta_0; \cdot), \theta - \theta_0 \rangle + \rho_{\theta_0}(\theta; \cdot),$$

where

$$||\rho_{\theta_0}(\cdot, \theta)||_r \to 0, \quad \text{when} \quad ||\theta - \theta_0|| \to 0,$$

and $\langle \cdot, \cdot \rangle$, $||\cdot||$ are the inner product and the norm in $\mathbf{R}^d$. Similarly a function $\varphi(\theta; t)$ (as function on θ), that is differentiable (in the space L_r^2) on θ in the open kernel of Θ_0, is twice differentiable if for all $\theta_0 \in \Theta_0$ there exists such matrix-function $\mathbf{H}\varphi(\theta_0; t)$, that

$$\nabla\varphi(\theta; \cdot) - \nabla\varphi(\theta_0; \cdot) = \mathbf{H}\varphi(\theta_0; \cdot)(\theta - \theta_0) + \varrho_{\theta_0}(\theta; \cdot),$$

where

$$||\varrho_{\theta_0}(\cdot, \theta)||_2 \to 0 \quad \text{in the space } L_r^2, \text{ when } \quad ||\theta - \theta_0|| \to 0.$$

Here for matrix A we write $||A||_2^2 = \mathrm{tr}A^*A$, where A^* denotes the conjugate matrix and for

$$\text{matrix} \quad A = \begin{pmatrix} a_{11} & \cdots & a_{1d} \\ \cdots & \cdots & \cdots \\ a_{d1} & \cdots & a_{dd} \end{pmatrix}, \quad \text{and vector} \quad b = (b_1, \cdots, b_d)$$

we use the notation $Ab = (c_1, \cdots, c_d)$, where $c_j = \sum_{k=1}^d a_{jk}b_k$.

Further we will write

$$\nabla\varphi(\theta; \cdot) = \left(\frac{\partial}{\partial\theta_1}\varphi(\theta, \cdot), \cdots, \frac{\partial}{\partial\theta_d}\varphi(\theta; \cdot)\right),$$

and

$$\mathbf{H}\varphi(\theta; t) = \begin{pmatrix} \frac{\partial}{\partial\theta_1}\frac{\partial}{\partial\theta_1}\varphi(\theta; t) & \cdots & \frac{\partial}{\partial\theta_1}\frac{\partial}{\partial\theta_d}\varphi(\theta; t) \\ \cdots & \cdots & \cdots \\ \frac{\partial}{\partial\theta_d}\frac{\partial}{\partial\theta_1}\varphi(\theta; t) & \cdots & \frac{\partial}{\partial\theta_d}\frac{\partial}{\partial\theta_d}\varphi(\theta; t) \end{pmatrix}.$$

D4. Function $f(\theta, \theta_1; \cdot)$ is continuous on θ_1 in the space $L_{r_0}^2$ in some neighborhood of the true value θ and for any $\varepsilon > 0$

$$\delta(\varepsilon) = \inf_{\theta_*: ||\theta - \theta_*|| \geq \varepsilon} \int |f(t) - f(\theta, \theta_*; t)|^2 r_0(t) dt > 0.$$

D5. Functions $\Psi(\theta; t)$ and $\psi(\theta; t)$ are continuously differentiable on θ.

36.2 Consistency of the Estimator $\hat{\theta}_n$

For a nonnegative function r consider the L_r^2–distance between $\hat{f}_n(\cdot)$ and $f(\cdot)$. We put

$$\mathcal{J}_r(\hat{f}_n, f) = \int |\hat{f}_n(t) - f(t)|^2 r(t)\, dt. \tag{36.14}$$

Burke, Czörgő, and Horvath (1981) proved that under the above conditions the next proposal is true.

Proposition 36.2.1 *If function r is continuous then for some nonnegative constants*

$$\sigma^2(p) = \sigma^2\left(p, F, Q, r, \mathcal{K}\right) \quad \textit{and} \quad m(p) = m\left(p, F, Q, r, \mathcal{K}\right)$$

for $1 \leq p < \infty$ we have

$$h_n^{-1/2}\left\{(nh_n)^{p/2}\mathcal{I}_n(p) - m(p)\right\} \xrightarrow{\mathcal{D}} N\left(0, \sigma^2(p)\right), \tag{36.15}$$

where

$$\mathcal{I}_n(p) = \int |\hat{f}_n(t) - f(t)|^p r(t)\, dt.$$

Proposition 36.2.2 *Suppose that θ is the true value of unknown parameter. Under the conditions of Proposition 36.2.1 we suppose also, that there exists such $C = C(\theta)$, that*

$$r_\theta(t)C(\theta) \geq r(\theta_1; t) \quad \textit{for all} \quad \theta_1 \in \Theta, t \in I(\theta) \cap I(\theta_1).$$

Then

$$\mathcal{J}_{r_0}(f, f(\theta, \hat{\theta}_n; \cdot)) \xrightarrow{P} 0 \quad \textit{as} \quad n \longrightarrow \infty. \tag{36.16}$$

PROOF. Under the conditions of Proposition 36.2.1

$$\mathcal{J}_{r_0}(\hat{f}_n, f) \xrightarrow{P} 0 \quad \text{as} \quad n \longrightarrow \infty, \tag{36.17}$$

and

$$\mathcal{J}_{r_\theta}(\hat{g}_n, g) \xrightarrow{P} 0 \quad \text{as} \quad n \longrightarrow \infty. \tag{36.18}$$

Let $\hat{\theta}_n$ be the minimum distance estimator,

$$\hat{\theta}_n = \arg\min_\theta \int_I |\hat{f}_n(t) - f_n^*(\theta; t)|^2 r_0(t)\, dt,$$

for the true value of the parameter θ. We note that

$$\begin{aligned}
&\int_I |\hat{f}_n(t) - f_n^*(\hat{\theta}_n; t)|^2 r_0(t)\, dt \\
&\quad\leq \int_I |\hat{f}_n(t) - f_n^*(\theta; t)|^2 r_0(t)\, dt \\
&\quad\leq 2\left\{\int_I |f(t) - f_n^*(\theta; t)|^2 r_0(t)\, dt + \int_I |\hat{f}_n(t) - f(t)|^2 r_0(t)\, dt\right\} \\
&\quad= 2\left\{\int_I |g(\theta; t) - \hat{g}_n(t)|^2 r_\theta(t)\, dt + \int_I |\hat{f}_n(t) - f(t)|^2 r_0(t)\, dt\right\},
\end{aligned}$$

from which it follows that

$$\int_I |\hat{f}_n(t) - f_n^*(\hat{\theta}_n;t)|^2 r_0(t) \xrightarrow{P} 0 \quad \text{as} \quad n \longrightarrow \infty, \tag{36.19}$$

and therefore

$$\int_I |f(t) - f_n^*(\hat{\theta}_n;t)|^2 r_0(t) \xrightarrow{P} 0 \quad \text{as} \quad n \longrightarrow \infty. \tag{36.20}$$

By the same way it can be proved that

$$\int_I |f(\theta,\hat{\theta}_n;t) - f_n^*(\hat{\theta}_n;t)|^2 r_0(t) \xrightarrow{P} 0 \quad \text{as} \quad n \longrightarrow \infty. \tag{36.21}$$

So we have for the true value θ of the parameter

$$\int_I |f(t) - f(\theta,\hat{\theta}_n;t)|^2 r_0(t) \xrightarrow{P} 0 \quad \text{as} \quad n \longrightarrow \infty. \tag{36.22}$$

■

We suppose, that the function $f(\theta,\theta_1;\cdot)$ is continuous on θ_1 in the space $L^2_{r_0}$ in some neighborhood of the true value θ and for any $\varepsilon > 0$

$$\delta(\varepsilon) = \inf_{\theta_*:||\theta-\theta_*||\geq\varepsilon} \int |f(t) - f(\theta,\theta_*;t)|^2\, r_0(t)dt > 0.$$

Proposition 36.2.3 *We suppose, that the conditions of Proposition 36.2.2 are verified and for any $\varepsilon > 0$*

$$\delta(\varepsilon) > 0.$$

Then

$$P\left\{|\theta - \hat{\theta}_n| \geq \varepsilon\right\} \xrightarrow{P} 0 \quad \textit{as} \quad n \longrightarrow \infty. \tag{36.23}$$

PROOF. It is clear, that

$$P\left\{|\theta - \hat{\theta}_n| \geq \varepsilon\right\} \leq P\left\{\int |f(t) - f(\theta,\theta_*,t)|^2\, r_0(t)dt \geq \delta(\varepsilon)\right\}. \tag{36.24}$$

Since

$$\delta(\varepsilon) \longrightarrow 0, \quad \text{when } \varepsilon \to 0,$$

and, as it is follows from Proposition 36.2.2, if θ is the true value of parameter, then

$$P\left\{\int |f(t) - f(\theta,\hat{\theta}_n;t)|^2\, r_0(t)dt \geq \delta\right\} \longrightarrow 0, \quad \text{when } \delta \to 0,$$

and hence we obtain (36.23). ■

36.3 Asymptotic Behavior of Kernel Estimators of Densities

Let $\mathbf{X} = \{X_1, \cdots, X_n\}$ be a sample of i.i.d. random variables with the distribution function $F(t)$ and the density function $f(t)$ and $\mathbf{T} = \{T_1, \cdots, T_n\}$ be a sample of i.i.d. random variables with the distribution function $Q^1(t)$ and the density function $q^1(t)$. We assume that $\mathbf{X}$ and $\mathbf{T}$ are independent. Suppose we observe the random vectors

$$\mathbf{\Gamma}_i = (Y_i, K_i)\,,\ (i = 1, \cdots, n)\,,$$

which are defined by $\mathbf{X}, \mathbf{T}$ as it was done in (36.3). Following to (36.12) we put

$$H(t) = 1 - (1 - F(t))\left(1 - Q^1(t)\right). \tag{36.25}$$

Let

$$Y_{(1,n)} \leq Y_{(2,n)} \leq \cdots \leq Y_{(n,n)}$$

be the ordered statistics of the sample $\mathbf{Y} = \{Y_1, \cdots, Y_n\}$. We consider the Kaplan–Meier estimator $\hat{\mathcal{F}}_n(t)$ of the distribution function $F(t)$

$$1 - \hat{\mathcal{F}}_n(t) = \prod_{\substack{j\,:\,1 \leq j \leq n,\\ Y_{(j,n)} \leq t}} \left(\frac{n-j}{n-j+1}\right)^{K_j},$$

where we put $\prod\limits_{j \in S} c_j = 1$, if $S = \emptyset$. We consider also a kernel estimator of the density function $f(t)$,

$$\hat{f}_n(t) = \int \frac{1}{h_n}\mathcal{K}\left(\frac{t-u}{h_n}\right)\, d\mathcal{F}_n(t). \tag{36.26}$$

Notice that in the case, when $P\{X_j \leq T_j\} = 1$ the Kaplan–Meier estimator $\hat{\mathcal{F}}_n(t)$ coincides with the usual empirical distribution function $\hat{F}_n(t)$,

$$\hat{F}_n(t) = \sum_{i=1}^{n} \mathbb{1}_{(-\infty,t)}\,(X_i)\,.$$

We denote by $\mathcal{D}(t)$ the increasing function with density function $\rho(t)$,

$$\begin{aligned}\rho(t) &= \frac{f(t)}{(1-H(t))\,(1-F(t))} = \frac{f(t)}{(1-Q^1(t))\,(1-F(t))^2},\\ \mathcal{D}(a) &= 0.\end{aligned} \tag{36.27}$$

Burke, Czörgő, and Horvath (1981) proved that under the condition

$$b < t_* = \sup\{t : H(t) < 1\} \qquad \mathbf{D3}$$

there exists such a sequence of Wiener processes $\{W_n(t), t \geq a\}$, $W_n(a) = 0$, that

$$\sup_{t\in[a,b]} |\Delta_n(t) - W_n^*| = O\left(n^{-1/2}\log n\right), \quad \text{a.s.,} \tag{36.28}$$

where

$$W_n^*(t) = (1 - F(t))\, W_n(\mathcal{D}(t)), \quad \text{and} \quad \Delta_n(t) = \sqrt{n}\left(\hat{\mathcal{F}}_n(t) - F(t)\right).$$

Thus,

$$\begin{aligned} &\sqrt{n}\left(\hat{\mathcal{F}}_n(t) - F(t)\right) = W_n^*(t) + R_n(t), \\ &\text{where} \quad \sup_{t\in[a,b]} |R_n(t)| = O\left(n^{-1/2}\log n\right), \quad \text{a.s.} \end{aligned} \tag{36.29}$$

Let $W(t)$ be a Wiener process on interval $[a, \infty)$, $W(a) = 0$, and

$$W^*(t) = (1 - F(u))\, W(D(u)).$$

We consider the linear operator $\mathcal{A} : L_f^2 \longrightarrow L_f^2$, such, that for $y(\cdot) \in L_f^2$

$$\mathcal{A}[y](t) = y(t)(1 - F(t)) - Y(t), \quad \text{where} \quad Y(t) = \int_{[t,b]} f(t)y(t)\,dt.$$

It must be noted that the function $Y(t)$ is absolutely continuous,

$$\frac{d}{dt}Y(t) = -y(t)f(t)$$

and if

$$l(y) = \int_{[a,b]} f(t)y(t)\,dt = 0, \quad \text{then} \quad Y(a) = Y(b) = 0.$$

Lemma 36.3.1 *Suppose that $F(a) = 0$, $F(b) < 1$, then*

$$\int_{[a,b]} (\mathcal{A}[y](t))^2 \frac{f(t)dt}{(1 - F(t))^2} = \int_{[a,b]} y^2(t)f(t)\,dt - \left(\int_{[a,b]} y(t)f(t)\,dt\right)^2. \tag{36.30}$$

PROOF. Under the conditions $F(a)=0$, $F(b)<1$, integration by parts gives

$$2\int\limits_{[a,b]} y(t)Y(t)\frac{f(t)dt}{(1-F(t))} = Y^2(0) + \int\limits_{[a,b]} Y^2(t)\frac{f(t)dt}{(1-F(t))^2}. \tag{36.31}$$

Since

$$(\mathcal{A}\,[y]\,(t))^2\,\frac{f(t)}{(1-F(t))^2} = y^2(t)f(t) - 2y(t)Y(t)\frac{f(t)}{(1-F(t))} + Y^2(y)\frac{f(t)}{(1-F(t))^2},$$

from (36.31) we obtain (36.30). ■

Suppose that $Q_1(b)<1$,then

$$\rho(t) = \frac{f(t)}{(1-Q(t))\,(1-F(t))^2} \le C\frac{f(t)}{(1-F(t))^2} \quad \text{for} \quad C = \frac{1}{(1-Q(b))}.$$

Hence, under the conditions $H(b)<1, F(a)=0$, we have

$$\int\limits_{[a,b]} (\mathcal{A}\,[y]\,(t))^2\,\rho(t)dt \le C\int\limits_{[a,b]} y^2(t)f(t)\,dt,$$

Lemma 36.3.2 *Suppose that function $y(t)$ satisfies the condition*

$$\sigma^2(y) = \int\limits_{[a,b]} |\mathcal{A}\,[y]\,(t)|^2\rho(t)\,dt < \infty.$$

Then the relation

$$\boldsymbol{\Upsilon}(y) = \int\limits_{[a,b]} y(t)\,dW^*(t)$$

determines a Gaussian random variable with zero mean and variance $\sigma^2(y)$.

PROOF. It is clear that

$$dW^*(t) = -W\,(\mathcal{D}(t))\,f(t)\,dt + (1-F(t))\,dW\,(\mathcal{D}(t))\,.$$

Since for any bounded function $y(t)$ supported on such subinterval $[a_*,b_*]\subset[a,b]$, that the measure $d\mathcal{D}(t)$ is finite on $[a_*,b_*]$, the integral

$$\int\limits_{[a,b]} W\,(\mathcal{D}(t))\,f(t)y(t)\,dt$$

is well defined and

$$\int\limits_{[a,b]} W\,(\mathcal{D}(t))\,f(t)y(t)\,dt = \int\limits_{[a,b]} W\,(\mathcal{D}(t))\,dY(t),$$

where the function $Y(t)$ is defined on interval $[a, b]$ by the relation

$$Y(t) = \int_{[t,b]} f(s)y(s)\,ds.$$

Further we shall denote by S the set of all such functions $y(t)$. Integration by parts gives

$$\int_{[a,b]} W(\mathcal{D}(t))\, f(t)y(t)\,dt = \int_{[a,b]} Y(t)\,dW(\mathcal{D}(t))\,.$$

Hence for such function $y(t)$ integral

$$\int_{[a,b]} y(t)\,dW^*(t)$$

is well defined and

$$\begin{aligned}\int_{[a,b]} y(t)\,dW^*(t) &= \int_{[a,b]} [(1 - F(t)y(t) - Y(t))]\,dW(\mathcal{D}(t)) \\ &= \int_{[a,b]} [\mathcal{A}[y](t))]\,dW(\mathcal{D}(t))\,.\end{aligned}$$

Since the last integral is well defined for any $y(t)$ such that

$$||y(\cdot)||_*^2 = \int_{[a,b]} |\mathcal{A}(y(t))|^2 \rho(t)\,dt < \infty,$$

and the set S is dense in the space with semi-norm $||\cdot||_*$, we can define the integral on $dW^*(t)$ for any function φ, $||\varphi||_* < \infty$, by the relation

$$\boldsymbol{\Upsilon}(\varphi) = \int_{[a,b]} \varphi(t)\,dW^*(t) = \int_{[a,b]} [\mathcal{A}[\varphi](t)]\,dW(\mathcal{D}(t))\,.$$

■

Let $W(t)$ be a Wiener process and let

$$Z_n(t) = \int (1 - F(u))W(\mathcal{D}(u)) \left[\frac{\partial}{\partial u}\frac{1}{h_n}\mathcal{K}\left(\frac{t-u}{h_n}\right)\right] du.$$

Lemma 36.3.3 *Suppose that conditions* **D1**, **K1–K4**, **H1** *hold. Then, as* $n \to \infty$, *we have*

$$h_n^{-1/2}\left\{\int |h_n^{-1/2} Z_n(t)|^2 r(t)\,dt - m\right\} \to N(0, \sigma^2),$$

where

$$\sigma^2 = \sigma^2(f, H, \mathcal{K}) = \int\limits_{[a,b]} (1 - H(t))^{-2} f^2(u) r^2(u)\, du \left(\int K^2(u)\, du \right)^2,$$

$$m = m(f, H, \mathcal{K}) = \int\limits_{[a,b]} (1 - H(t))^{-1} f(u) r(u)\, du \int K^2(u)\, du.$$

This lemma was proved by Burke, Czörgő, and Horvath (1981).

Now we apply (36.29) also to the process $\hat{\mathcal{G}}_n(t)$:

$$\begin{aligned} &\sqrt{n}\left(\hat{\mathcal{G}}_n(t) - G(\theta; t)\right) = \tilde{W}_n(t) + \tilde{R}_n(t), \\ &\text{where} \quad \sup_{t \in [a,b]} |\tilde{R}_n(t)| = O\left(n^{-1/2} \log n\right), \quad \text{a.s.} \end{aligned} \tag{36.32}$$

Here

$$\tilde{W}_n(t) = (1 - G(\theta; t))\, W^n\left(\mathcal{D}_2(t)\right),$$

$\{W^n\}$ is a sequence of Wiener processes and we denote by $\mathcal{D}_2(t)$ the increasing function with density function $\rho_2(t)$,

$$\begin{aligned} \rho_2(t) &= \frac{g(\theta; t)}{(1 - Q^2(t))\,(1 - G(\theta; t))^2}, \\ \mathcal{D}_2(a(\theta)) &= 0. \end{aligned} \tag{36.33}$$

Thus, we have

$$\begin{aligned} \hat{f}_n(t) &= \int \frac{1}{h_n} \mathcal{K}\left(\frac{t-u}{h_n}\right) d\mathcal{F}_{n_1}(u) = \int \frac{1}{h_n} \mathcal{K}\left(\frac{t-u}{h_n}\right) dF(u) \\ &\quad + \frac{1}{\sqrt{n_1}} \int \frac{1}{h_n} \mathcal{K}\left(\frac{t-u}{h_n}\right) dW_n^*(u) + \frac{1}{\sqrt{n_1}} \int \frac{1}{h_n} \mathcal{K}\left(\frac{t-u}{h_n}\right) dR_n(u), \end{aligned}$$

and

$$\begin{aligned} f_n^*(\theta; t) &= \phi(\theta; t) \int \frac{1}{h_n} \mathcal{K}\left(\frac{\Phi(\theta; t) - u}{h_n}\right) d\mathcal{G}_{n_2}(u) \\ &= \phi(\theta; t) \int \frac{1}{h_n} \mathcal{K}\left(\frac{\Phi(\theta; t) - u}{h_n}\right) dG(\theta; u) \\ &\quad + \frac{1}{\sqrt{n_2}} \phi(\theta; t) \int \frac{1}{h_n} \mathcal{K}\left(\frac{\Phi(\theta; t) - u}{h_n}\right) d\tilde{W}_n(u) \\ &\quad + \frac{1}{\sqrt{n_2}} \phi(\theta; t) \int \frac{1}{h_n} \mathcal{K}\left(\frac{\Phi(\theta; t) - u}{h_n}\right) d\tilde{R}_n(u). \end{aligned}$$

Burke, Czörgő,and Horvath (1981) obtained the next result.

Lemma 36.3.4 *Under condition* **H1** *and conditions*

(1) function $K(t)$ is compactly supported and absolutely continuous,

(2) $\int |\frac{d}{dt} K(t)|\, dt < \infty$,

(3) and r_0 is a continuous nonnegative function such that,

$$\int_{[a,b]} r_0(t)\, dt < \infty,$$

(4) function $H(t)$ satisfies the condition **D3**,

we have

$$nh_n^{1/2} \int |\frac{1}{\sqrt{n}} \int \frac{1}{h_n} \mathcal{K}\left(\frac{t-u}{h_n}\right) dR_n(u)|^2 r_0(t) dt \xrightarrow{P} 0 \quad as \quad n \longrightarrow \infty.$$

By the same way we can prove the next lemma.

Lemma 36.3.5 *Under the conditions of Lemma 36.3.4 and the condition*

(5) $r(\theta; t)$ is a nonnegative function such that,

$$\int_{[a(\theta),b(\theta)]} r(\theta; t)\, dt < \infty,$$

we have

$$nh_n^{1/2} \int |\frac{1}{\sqrt{n_2}} \phi(\theta; t) \int \frac{1}{h_n} \mathcal{K}\left(\frac{\Phi(\theta; t) - u}{h_n}\right) d\tilde{R}_n(u)|^2 r_0(t) \xrightarrow{P} 0 \quad as \quad n \longrightarrow \infty.$$

From Lemma 36.3.4 and Lemma 36.3.5 follows the next lemma.

Lemma 36.3.6 *Under the conditions from Lemma 36.3.4 and Lemma 36.3.5 and the conditions* **F2** *we have*

$$nh_n \left\{ \int_I |\hat{f}_n(t) - f_n^*(\hat{\theta}_n; t)|^2 r_0(t)\, dt \right\}$$
$$= nh_n \left\{ \int_I |J_n^1(t) - J_n^2(\hat{\theta}_n; t)|^2 r_0(t)\, dt \right\} + r_n,$$

where

$$J_n^1(t) = \frac{1}{\sqrt{n_1}} \int \frac{1}{h_n} \mathcal{K}\left(\frac{t-u}{h_n}\right) dW_n^*(u),$$

$$J_n^2(\theta; t) = \frac{1}{\sqrt{n_2}} \phi(\theta; t) \int \frac{1}{h_n} \mathcal{K}\left(\frac{\Phi(\theta; t) - u}{h_n}\right) d\tilde{W}_n(u),$$

$$r_n \xrightarrow{P} 0 \quad as \quad n \longrightarrow \infty.$$

Here we suppose that the Wiener processes W_n and W^n are independent.

From this lemma and Lemma 36.3.3 we deduce that

$$h_n^{-1/2}\left[nh_n\left\{\int_I |\hat{f}_n(t) - f_n^*(\hat{\theta}_n;t)|^2 r_0(t)\,dt\right\} - m\right] \to N(0,\sigma^2),$$

with $m = \frac{1}{p_1}m_1 + \frac{1}{p_2}m_2$ and $\sigma^2 = \frac{1}{p_1^2}\sigma_1^2 + \frac{1}{p_2^2}\sigma_2^2$.

Acknowledgements. This research was supported by the Conseil Regional d'Aquitaine, Grant 20000204009 and by Russian Foundation for Basic Research, grant 99-01-00111, 00-015-019, RFBR–DFG grant 99-01-04027.

References

1. Bagdonavicius, V. and Nikulin, M. (1999). On semiparametric estimation of reliability from accelerated life testing, In *Statistical and Probabilistic Models in Reliability* (Eds., D. Ionescu and N. Limnios), pp. 75–89, Boston: Birkhäuser.

2. Bagdonavicius, V. and Nikulin, M. (2000). Semiparametric estimation in accelerated life testing. In *Recent Advances in Reliability Theory,* (Eds., N. Limnios and M. Nikulin), pp. 405–418, Boston: Birkhäuser.

3. Bagdonavicius, V., Gerville-Réache, L., Nikoulina, V., and Nikulin, M. (2000). Expériences accélérées: analyse statistique du modèle standard de vie accélérée, *Revue de Statistique Appliquée*, **XILYIII**, 5–38.

4. Bickel, P. and Rosenblatt, M. (1973). On some global measures of the deviations of density function estimators, *Annals of Mathematical Statistics,* **1**, 1071–1095.

5. Bretagnolle, J. and Huber, C. (1979). Estimation des densités: risque minimax, *Zeitschrift für Wahrscheinlichketstheorie und Verwandte Gebiete*, **47**, 119–137.

6. Burke, M., Czörgő, S., and Horvath, L. (1981). Strong approximation of some biometric estimates under random censorship, *Zeitschrift für Wahrscheinlichketstheorie und Verwandte Gebiete*, **56**, 86–112.

7. Czörgő, M., Gombay, E., and Horvath, L. (1991). Central limit theorems for L_p distance of kernel estimators of densities under random censorship, *Annals of Statistics*, **19**, 1813–1831.

8. Gerville-Réache, L. and Nikoulina, V. (1999). Analysis of reliability characteristics estimators in accelerated life testing, In *Statistical and Probabilistic Models in Reliability*, (Eds., D. Ionescu and N. Limnios), pp. 91–100, Boston: Birkhäuser.

9. Ghosh, B. K. and Wei-Min, H. (1991). The power and optimal kernel of the Bickel-Rozenblatt test for goodness of fit, *Annals of Mathematical Statistics*, **19**, 999–1009.

10. Hall, P. (1984). Central limit theorem for integrated square arrow of multivariate nonparametric density estimators, *Journal of Multivariate Analysis*, **14**, 1–16.

11. Horvath, L. (1991). On L_p−norms of multivariate density estimators, *Annals of Statistics*, **19**, 1933–1949. *Annals of Statistics*, **4**, 912–923.

12. Parzen, E. (1962). On estimation of a probability density function and mode, *Annals of Mathematical Statistics*, **33**, 1065–1076.

13. Rosenblatt, M. (1956). Remarks on some nonparametric estimates of a density function, *Annals of Mathematical Statistics*, **27**, 832–837.

37

The Concept of Generalized Asymptotic Deficiency and its Application to the Minimum Discrepancy Estimation

Masafumi Akahira

University of Tsukuba, Ibaraki, Japan

Abstract: The concept of (asymptotic) deficiency was defined as the (limit of) additional number of observations which is an integer. However, it would be convenient for the value of deficiency to be continualized in the higher order asymptotics. In order to do so, Hodges and Lehmann (1970) suggested using stochastic interpolation as one of the methods. In this paper, using the method, we consider the concept of generalized asymptotic deficiency and apply it to the minimum discrepancy estimation including the minimum chi-square one considered by Fisher (1928) in relation to the work of Pearson (1900).

Keywords and phrases: Asymptotic deficiency, risk stochastic interpolation, asymptotic variance, minimum chi-square estimator, maximum likelihood estimator

37.1 Introduction

In the paper, Hodges and Lehmann (1970) introduced the concept of (asymptotic) deficiency as follows. Let δ_n be a statistical procedure based on n observations, and a less effective procedure $\delta^*_{k_n}$ which requires a larger number k_n of observations to give equally good performance. For the additional number $k_n - n$ of observations needed by the procedure $\delta^*_{k_n}$, they called it deficiency. If $d := \lim_{n\to\infty}(k_n - n)$ exists, it is called the asymptotic deficiency. In the higher order asymptotics, the concept of asymptotic deficiency is very useful in discriminating asymptotically efficient estimators [see, for example, Akahira (1986, 1999a, 1999b)]. Then, it is desirable for the value of the asymptotic deficiency to be extended to be continualized in its relation to the higher order term of asymptotic variance of estimators. The method of stochastic interpolation on the continualization was remarked by Hodges and Lehmann (1970). In this chapter, using the method, we consider the concept of generalized asymptotic deficiency (by risk) of the statistical procedure and apply to the minimum discrepancy estimation of multinomial parameters which was studied by Fisher (1925), Rao (1961), Ponnapalli (1976), Tanaka and Akahira (1995, 1996) and others. The generalized asymptotic deficiency of the minimum chi-square estimator relative to the maximum likelihood estimator is also given.

37.2 The Concept of Generalized Asymptotic Deficiency

Let $\delta_{1,n} := \delta_{1,n}(\mathbf{X})$ and $\delta_{2,n} := \delta_{2,n}(\mathbf{X})$ be two statistical procedures based on the same sample $\mathbf{X} := (X_1, \ldots, X_n)$ of size n with risk $r(\delta_{1,n}) > 0$ and $r(\delta_{2,n}) > 0$ both converging to zero as $n \to \infty$. Let k be a positive number, and define π_k and K as

$$\pi_k := k - [k]$$

and

$$K := \begin{cases} [k] & \text{with probability (w.p.) } 1 - \pi_k, \\ [k] + 1 & \text{with probability } \pi_k, \end{cases}$$

where $[k]$ is the largest number not exceeding k. Then we have

$$E(K) = [k](1 - \pi_k) + ([k] + 1)\pi_k = k.$$

We also define

$$\delta_{2,k} := \begin{cases} \delta_{2,[k]} & \text{w.p. } 1 - \pi_k, \\ \delta_{2,[k]+1} & \text{w.p. } \pi_k, \end{cases} \tag{37.1}$$

$$r(\delta_{2,k}) := (1 - \pi_k) r(\delta_{2,[k]}) + \pi_k r(\delta_{2,[k]+1}). \tag{37.2}$$

Choose a sequence $\{k_n\}$ of positive numbers such that for some $\alpha > 0$

$$n^{\alpha} r(\delta_{2,k_n}) = n^{\alpha} r(\delta_{1,n}) + o\left(\frac{1}{n}\right). \tag{37.3}$$

If there exists

$$d := \lim_{n \to \infty} (k_n - n)$$

and it is independent of the particular sequence $\{k_n\}$ chosen, then it is called the generalized asymptotic (gen. as.) deficiency (by risk) of δ_{2,k_n} relative to $\delta_{1,n}$ [see also Hodges and Lehmann (1970)]. Note that the value of the gen. as. deficiency is not always integer.

Theorem 37.2.1 *If for some $a > 0$ and $b_j \in \mathbf{R}$*

$$n^\alpha r(\delta_{j,n}) = a + \frac{b_j}{n} + o\left(\frac{1}{n}\right), \qquad j = 1, 2, \tag{37.4}$$

then

$$\lim_{n\to\infty} (k_n - n) = \frac{b_2 - b_1}{\alpha a^{1/\alpha}}.$$

PROOF. Since the risks $r(\delta_{1,n})$ and $r(\delta_{2,n})$ are positive and converge to zero as $n \to \infty$, we have $k_n \to \infty$ as $n \to \infty$. Since, by (37.3)

$$r(\delta_{2,k_n}) = r(\delta_{1,n}) + o\left(\frac{1}{n^{\alpha+1}}\right),$$

it follows from (37.2) that

$$(1 - \pi_{k_n}) r(\delta_{2,[k_n]}) + \pi_{k_n} r(\delta_{2,[k_n]+1}) = r(\delta_{1,n}) + o\left(\frac{1}{n^{\alpha+1}}\right).$$

From (37.4) we have

$$\begin{aligned}&\left[(1 - \pi_{k_n})\left(\frac{k_n}{[k_n]}\right)^\alpha \left\{a + \frac{b_2}{[k_n]} + o\left(\frac{1}{[k_n]}\right)\right\}\right.\\&\qquad\left. + \pi_{k_n}\left(\frac{k_n}{[k_n]+1}\right)^\alpha \left\{a + \frac{b_2}{[k_n]+1} + o\left(\frac{1}{[k_n]+1}\right)\right\}\right]^{1/\alpha}\\&= \frac{k_n}{n}\left\{a + \frac{b_1}{n} + o\left(\frac{1}{n}\right)\right\}^{1/\alpha}.\end{aligned} \tag{37.5}$$

Since

$$\begin{aligned}\left(\frac{k_n}{[k_n]}\right)^\alpha &= \left(\frac{[k_n] + \pi_{k_n}}{[k_n]}\right)^\alpha = 1 + \frac{\alpha \pi_{k_n}}{[k_n]} + o\left(\frac{1}{[k_n]}\right),\\ \left(\frac{k_n}{[k_n]+1}\right)^\alpha &= \left(\frac{[k_n] + \pi_{k_n}}{[k_n]+1}\right)^\alpha = 1 - \frac{\alpha(1 - \pi_{k_n})}{[k_n]} + o\left(\frac{1}{[k_n]}\right),\end{aligned}$$

it follows from (37.5) that

$$\begin{aligned}&a^{\frac{1}{\alpha}} + a^{\frac{1}{\alpha}-1}\frac{b_2}{\alpha[k_n]} + o\left(\frac{1}{[k_n]}\right)\\&= \frac{k_n}{n}\left\{a^{\frac{1}{\alpha}} + a^{\frac{1}{\alpha}-1}\frac{b_1}{\alpha n} + o\left(\frac{1}{n}\right)\right\}.\end{aligned} \tag{37.6}$$

Since $\lim_{n\to\infty} k_n = \infty$, it is easily seen from (37.6) that $\lim_{n\to\infty} k_n/n = 1$. Subtracting $a^{1/\alpha}$ and multiplying n in (37.6), we have for large n

$$a^{1/\alpha}\frac{n}{[k_n]}\frac{b_2}{a\alpha} + o(1) = a^{1/\alpha}(k_n - n) + a^{1/\alpha}\frac{k_n}{n}\frac{b_1}{a\alpha} + o(1). \tag{37.7}$$

Since $\lim_{n\to\infty} n/[k_n] = 1$, it follows from (37.7) that

$$\frac{b_2}{a\alpha} = \lim_{n\to\infty}(k_n - n) + \frac{b_1}{a\alpha},$$

which implies

$$\lim_{n\to\infty}(k_n - n) = \frac{b_2 - b_1}{a\alpha}.$$

Thus we complete the proof. ∎

Assume that $X_1, X_2, \ldots, X_n, \ldots$ is a sequence of independent and identically distributed (i.i.d.) random variables according to a density $f(x,\theta)$ with respect to a σ-finite measure μ, where θ is a real-valued parameter. Let $\hat{\theta}_{1,n} := \hat{\theta}_{1,n}(\mathbf{X})$ and $\hat{\theta}_{2,n} := \hat{\theta}_{2,n}(\mathbf{X})$ be first order efficient estimators of θ, where $\mathbf{X} = (X_1, \ldots, X_n)$. Denote by $V_\theta(\hat{\theta}_{j,n})$ $(j = 1,2)$ the (as.) variances of the estimators $\theta_{j,n}$'s of θ. Then for any positive number k we define $\hat{\theta}_{2,k}$ and $V_\theta(\hat{\theta}_{2,k})$ like (37.1) and (37.2) by

$$\hat{\theta}_{2,k} := \begin{cases} \hat{\theta}_{2,[k]} & \text{w.p. } \pi_k := k - [k], \\ \hat{\theta}_{2,[k]+1} & \text{w.p. } 1 - \pi_k, \end{cases}$$

and

$$V_\theta(\hat{\theta}_{2,k}) := \pi_k V_\theta(\hat{\theta}_{2,[k]}) + (1 - \pi_k) V_\theta(\hat{\theta}_{2,[k]+1}),$$

and, in a similar way to (37.3), take k_n such that

$$nV_\theta(\hat{\theta}_{2,k_n}) = nV_\theta(\hat{\theta}_{1,n}) + o(1/n). \tag{37.8}$$

If, for each $j = 1, 2$, the (as.) variance $V_\theta(\hat{\theta}_{j,n})$ admits the expansion

$$V_\theta(\hat{\theta}_{j,n}) = \frac{1}{nI(\theta)} + \frac{\Delta_j(\theta)}{n^2} + o\left(\frac{1}{n^2}\right), \tag{37.9}$$

then, letting $\alpha = 1$, $a = 1/I(\theta)$ and $b_j = \Delta_j(\theta)$ $(j = 1,2)$ in Theorem 37.2.1 we have

$$\lim_{n\to\infty}(k_n - n) = I(\theta)\{\Delta_2(\theta) - \Delta_1(\theta)\} =: d(\hat{\theta}_{2,k_n}, \hat{\theta}_{1,n}), \tag{37.10}$$

which is the gen. (as.) deficiency (by as. variance) of $\hat{\theta}_{2,k_n}$ relative to $\hat{\theta}_{1,n}$, where $I(\theta)$ is the amount of the Fisher information of X_1, i.e.

$$I(\theta) := E_\theta\left[\{(\partial/\partial\theta)\log f(X_1,\theta)\}^2\right].$$

Example 37.2.1 Suppose that $X_1, X_2, \ldots, X_n, \ldots$ is a sequence of i.i.d. random variables according to the normal distribution with mean θ and variance 1, where $n \geq 2$. Then we consider two estimators

$$\begin{aligned} \hat{\theta}_{1,n} &:= \bar{X} = \frac{1}{n}\sum_{i=1}^{n} X_i, \\ \hat{\theta}_{2,n}^{(w)} &:= \frac{1}{n}\left\{X_1 + \cdots + X_{n-2} + wX_{n-1} + (2-w)X_n\right\}, \end{aligned}$$

as unbiased ones of θ, where $0 \leq w \leq 2$. Their variances are given by

$$V_\theta(\hat{\theta}_{1,n}) = \frac{1}{n}, \qquad V_\theta(\hat{\theta}_{2,n}^{(w)}) = \frac{1}{n} + \frac{2}{n^2}(w-1)^2. \tag{37.11}$$

Note that the Fisher information number $I(\theta)$ is equal to 1 and $\hat{\theta}_{1,n}$ is the UMVU estimator of θ. From (37.8) and (37.11) we have

$$k_n = \frac{1}{2}\left\{n + \sqrt{n^2 + 8n(w-1)^2}\right\} = n + 2(w-1)^2 + O\left(\frac{1}{n}\right),$$

hence

$$\lim_{n\to\infty}(k_n - n) = 2(w-1)^2, \tag{37.12}$$

which is the gen. as. deficiency (by variance) of $\hat{\theta}_{2,k_n}^{(w)}$ relative to $\hat{\theta}_{1,n}$. Here

$$\hat{\theta}_{2,k_n}^{(w)} = \begin{cases} \hat{\theta}_{2,[k_n]} & \text{w.p. } 1 - \pi_{k_n}, \\ \hat{\theta}_{2,[k_n]+1} & \text{w.p. } \pi_{k_n}, \end{cases}$$

and

$$V_\theta(\hat{\theta}_{2,k_n}^{(w)}) = (1 - \pi_{k_n})V_\theta(\hat{\theta}_{2,[k_n]}) + \pi_{k_n}V_\theta(\hat{\theta}_{2,[k_n]+1}).$$

On the other hand, from (37.9) and (37.11) we have $\Delta_1(\theta) \equiv 0$, $\Delta_2(\theta) \equiv 2(w-1)^2$, hence

$$d(\hat{\theta}_{2,k_n}, \hat{\theta}_{1,n}) = I(\theta)\{\Delta_2(\theta) - \Delta_1(\theta)\} = 2(w-1)^2. \tag{37.13}$$

It is easily seen from (37.12) and (37.13) that (37.10) holds. For example, if $w = 1/4$, it follows from (37.13) that $d(\hat{\theta}_{2,k_n}^{(1/4)}, \hat{\theta}_{1,n}) = 9/8$. We define

$$\hat{\theta}_{2,n+(9/8)}^{(1/4)} := \begin{cases} \hat{\theta}_{2,n+1} & \text{w.p. } 7/8, \\ \hat{\theta}_{2,n+2} & \text{w.p. } 1/8, \end{cases}$$

$$V_\theta(\hat{\theta}_{2,n+(9/8)}^{(1/4)}) = (7/8)V_\theta(\hat{\theta}_{2,n+1}) + (1/8)V_\theta(\hat{\theta}_{2,n+2}).$$

From Theorem 37.2.1 we have

$$\lim_{n\to\infty}(k_n - n) = 9/8.$$

This means that $\hat{\theta}_{2,k_n}^{(1/4)}$ asymptotically needs 9/8 more size (of sample) than $\hat{\theta}_{1,n}$ in the continualized one for $\hat{\theta}_{2,k_n}^{(1/4)}$ to be asymptotically equivalent to $\hat{\theta}_{1,n}$ in the variance.

37.3 An Application to the Minimum Discrepancy Estimation

Suppose that $\mathbf{Y} = (Y_1, \ldots, Y_k)$ is a random vector with the multinomial distribution $M(n; \pi_1(\theta), \ldots, \pi_k(\theta))$, where $\theta \in \Theta$ and $\pi_1, \ldots, \pi_k$ satisfy suitable regularity conditions. For each $j = 1, \ldots, n$, we put $\mathbf{X}_j = (X_{1j}, \ldots, X_{kj})$. Let $\mathbf{X}_1, \ldots, \mathbf{X}_n$ be i.i.d. random vectors with the multinomial distribution $M(1; \pi_1(\theta), \ldots, \pi_k(\theta))$. Then $\mathbf{S} := (\sum_{j=1}^n X_{1j}, \ldots, \sum_{j=1}^n X_{kj})$ is distributed as $M(n; \pi_1(\theta), \ldots, \pi_k(\theta))$, so we can identify $\mathbf{Y}$ with $\mathbf{S}$. We also denote the observed proportions by $p_i = \sum_{j=1}^n X_j/n$ $(i = 1, \ldots .k)$. Rao (1961) considered an estimator as a suitably chosen root of an equation $f(\theta; p) = f(\theta; p_1, \ldots, p_k) = 0$, where f satisfies certain conditions. Let M be a set of estimators $\hat{\theta}_f$ from solving the estimating equation for f. For l, $m = 0, 1, 2, \ldots$, let

$$\mu_{lm}(\theta) := \sum_{r=1}^{k} \left(\pi_r'(\theta)/\pi_r(\theta)\right)^l \left(\pi_r''(\theta)/\pi_r(\theta)\right)^m \pi_r(\theta),$$

particularly, $I(\theta) = \mu_{20}(\theta)$. Below we omit θ in $\mu_{lm}(\theta)$'s for simplicity. Now we consider the gen. as. deficiency (by as. variance), following Tanaka and Akahira (1996) [see also Ponnapalli (1976)]. Let the parameter space Θ be a finite open interval. A minimum discrepancy (m.d.) estimator $\hat{\theta}_g$ of θ is defined as one which minimizes the discrepancy function $D(\theta; \mathbf{p}) := \sum_{r=1}^k p_r g(\pi_r(\theta)/p_r)$ for a suitable function g. Most of usual estimators like maximum likelihood estimator (mle), minimum chi-square estimator (mcse), etc. are m.d. estimator [see Greenwood and Nikulin (1996) for the mcse]. Let L be a set of minimum discrepancy estimators $\hat{\theta}_g$ for g satisfying certain regularity conditions. The class L is regarded as a subset of M. Ponnapalli (1976) derived the asymptotic variance of $\hat{\theta}_g$ in L as

$$V_\theta(\hat{\theta}_g) = \frac{1}{nI} + \frac{B_g}{n^2} + o\left(\frac{1}{n^2}\right), \tag{37.14}$$

where

$$\begin{aligned}
B_g = \frac{1}{I^4} \Bigg[& \left\{ -I^3 + I\mu_{40} - I\mu_{21} - I\sum_{r=1}^{k} \left(\frac{\pi_r'}{\pi_r}\right)\left(\frac{\pi_r'''}{\pi_r}\right)\pi_r + \frac{7}{2}\mu_{11}^2 - \mu_{30}^3 \right\} \\
& + C_g \left\{ I^2 \sum_{r=1}^{k} \left(\frac{\pi_r'}{\pi_r}\right)^2 - I^2 \sum_{r=1}^{k} \frac{\pi_r''}{\pi_r} - 2I\mu_{40} + 3I\mu_{21} \right. \\
& \left. + 2I\mu_{11} \sum_{r=1}^{k} \frac{\pi_r'}{\pi_r} - I\mu_{30} \sum_{r=1}^{k} \frac{\pi_r'}{\pi_r} + 2\mu_{30}^2 - 4\mu_{11}\mu_{30} \right\} \\
& + C_g^2 \left\{ \frac{1}{2} I^2 \sum_{r=1}^{k} \left(\frac{\pi_r'}{\pi_r}\right)^2 - I\mu_{40} + \frac{1}{2}\mu_{30}^2 \right\} \Bigg] + o\left(\frac{1}{n^2}\right)
\end{aligned}$$

with $C_g := 2 + g'''(1)/g''(1)$. Here, the function g and the value of C_g corresponding to the various estimators are given by Table 37.1.

Table 37.1: Function g and value of C_g of various estimators

Estimator; $\hat{\theta}_g$	Function; g	Value of C_g
maximum likelihood; $\hat{\theta}_{ml}$	$-\log x$	0
minimum χ^2; $\hat{\theta}_{mcs}$	x^{-1}	-1
modified minimum χ^2; $\hat{\theta}_{mmcs}$	x^2	2
minimum Haldane discrepancy; $\hat{\theta}_{HmD_k}$	x^{k+1}	$k+1$
minimum Hellinger distance; $\hat{\theta}_{mHd}$	$-x^{1/2}$	$1/2$
minimum Kullback-Liebler separator; $\hat{\theta}_{KL}$	$x \log x$	1

Since any m.d. estimator $\hat{\theta}_g$ in L has a bias, in order to adjust the bias up to the order $o(1/n)$, let

$$\hat{\theta}_g^* := \hat{\theta}_g - \frac{1}{n}\xi_g(\hat{\theta}_g), \tag{37.15}$$

where

$$\xi_g(\theta) := \frac{1}{2I^2}\left\{C_g\left(\mu_{30} - I\sum_{r=1}^{k}\frac{\pi_r'}{\pi_r}\right) - \mu_{11}\right\}.$$

Then $E_\theta[\hat{\theta}_g^* - \theta] = o(1/n)$. Let L^* be a set of all the bias-adjusted estimators in L. Let $N_0 = \{(p_1, \ldots, p_k) | 0 \le p_r \le 1\ (r = 1, \ldots, k), \sum_{r=1}^k p_r = 1\}$ and $\hat{\Theta}_0 := \hat{\theta}_g(N_0)$. We assume that $\xi_g(\theta)$ and its derivative $\xi_g'(\theta)$ are continuous on $\bar{\hat{\Theta}}_0$ (closure of $\hat{\Theta}_0$). Then the following holds [see Tanaka and Akahira (1996)].

Theorem 37.3.1 *The gen. as. deficiency (by as. variance) of $\hat{\theta}_g^*$ relative to $\hat{\theta}_h^*$ is given by*

$$d(\hat{\theta}_g^*, \hat{\theta}_h^*) = \frac{Q_1}{2I^3}(C_g^2 - C_h^2),$$

where

$$Q_1 = \mu_{30}^2 - 2I\mu_{40} + I^2\sum_{r=1}^{k}\left(\frac{\pi_r'}{\pi_r}\right)^2 \ (\ge 0).$$

In particular, letting $\hat{\theta}_h^*$ be $\hat{\theta}_{ml}^*$, we have

$$d(\hat{\theta}_g^*, \hat{\theta}_{ml}^*) = \frac{Q_1}{2I^3}C_g^2.$$

Furthermore, $\hat{\theta}_{ml}^*$ has the minimum gen. as. deficiency (by as. variance) in L^*.

Example 37.3.1 We consider the case when $k = 3$ and $\pi_1(\theta) = \pi_2(\theta) = \theta$, $\pi_3(\theta) = 1 - 2\theta$, where $0 < \theta < 1/2$ [see also Tanaka and Akahira (1995)]. Then the mle and mcse are given by

$$\hat{\theta}_{ml} = \frac{1}{2}(p_1 + p_2), \quad \hat{\theta}_{mcs} = \left(2 + p_3\sqrt{\frac{2}{p_1^2 + p_2^2}}\right)^{-1},$$

respectively, and

$$I(\theta) = \frac{2}{\theta(1-2\theta)}, \quad Q_1(\theta) = \frac{4}{\theta^4(1-2\theta)^2}.$$

If $g(x) = x^{-1}$, then $\hat{\theta}_g = \hat{\theta}_{mcs}$ and $C_g = -1$ as is shown in Table 37.1. From Theorem 37.3.1 it follows that the gen. as. deficiency (by as. variance) of the bias-adjusted mcse $\hat{\theta}^*_{mcs}$ relative to the bias-adjusted mle $\hat{\theta}^*_{ml}$ is given by

$$d(\hat{\theta}^*_{mcs}, \hat{\theta}^*_{ml}) = \frac{1-2\theta}{4\theta} > 0,$$

where the bias-adjustment is due to (37.15). For example, if $\theta = 1/4$, $d(\hat{\theta}^*_{mcs}, \hat{\theta}^*_{ml}) = 1/2$. We newly denote $\hat{\theta}^*_{mcs}$ and $\hat{\theta}^*_{ml}$ by $\hat{\theta}^*_{mcs}(n)$ and $\hat{\theta}^*_{ml}(n)$ based on the sample $(X_1, \ldots, X_n)$ and define

$$\hat{\theta}^*_{mcs}\left(n + \frac{1}{2}\right) := \begin{cases} \hat{\theta}^*_{mcs}(n) & \text{w.p. } 1/2, \\ \hat{\theta}^*_{mcs}(n+1) & \text{w.p. } 1/2, \end{cases}$$

and

$$V_\theta\left(\hat{\theta}^*_{mcs}\left(n + \frac{1}{2}\right)\right) = \frac{1}{2}V_\theta(\hat{\theta}^*_{mcs}(n)) + \frac{1}{2}V_\theta(\hat{\theta}^*_{mcs}(n+1)).$$

From (37.10), we have

$$\lim_{n\to\infty}(k_n - n) = 1/2.$$

This means that $\hat{\theta}^*_{mcs}$ asymptotically needs 1/2 more size (of sample) than $\hat{\theta}^*_{ml}$ in the continualized one for $\hat{\theta}^*_{mcs}$ to be asymptotically equivalent to $\hat{\theta}^*_{ml}$ in the asymptotic variance up to the order $o(1/n^2)$.

References

1. Akahira, M. (1986). *The Structure of Asymptotic Deficiency of Estimators*, Queen's Papers in Pure and Applied Mathematics 75, Kingston, Canada: Queen's University Press.

2. Akahira, M. (1999a). The concept of normalized deficiency and its applications, *Statistics & Decisions*, **17**, 403–411.

3. Akahira, M. (1999b). On the normalized deficiency of estimators, *Metron*, **57**, 25–34.

4. Fisher, R. A. (1925). Theory of statistical estimation, *Proceedings of the Cambridge Philosophical Society*, **22**, 700–725.

5. Fisher, R. A. (1928). On a property connecting the χ^2 measure of discrepancy with the method of maximum likelihood, *Atti de Congresso Internazionale dei Mathematici*, Bologna, **6**, 94–100.

6. Greenwood, P. E. and Nikulin, M. S. (1996). *A Guide to Chi-Squared Testing*, New York: John Wiley & Sons.

7. Hodges, J. L. and Lehmann, E. L. (1970). Deficiency, *Annals of Mathematical Statistics*, **41**, 783–801.

8. Pearson, K. (1900). On the criterion that a given system of deviation is such that it can be reasonably supposed to have arisen from random sampling, *Philosophical Magazine*, **50**, 157–175.

9. Ponnapalli, R. (1976). Deficiency of minimum discrepancy estimators, *Canadian Journal of Statistics*, **4**, 33–50.

10. Rao, C. R. (1961). Asymptotic efficiency and limiting information, *Proceedings of the Fourth Berkeley Symposium on Mathematics, Statistics and Probability*, **1**, University of California Press, Berkeley, 531–546.

11. Tanaka, H. and Akahira, M. (1995). On the concept of deficiency and estimation of parameters of the multinomial distribution, (In Japanese), *Proc. Symp., Res. Inst. Math. Sci., Kyoto Univ.*, **916**, 52–74.

12. Tanaka, H. and Akahira, M. (1996). Deficiency of minimum discrepancy estimators of multinomial parameters, *Statistics & Decisions*, **14**, 241–251.

Index